AF678237

T. Aoki · H. Majima · Y. Takei · N. Tose (Eds.)

Algebraic Analysis of Differential Equations

from Microlocal Analysis to Exponential Asymptotics

Festschrift in Honor of Takahiro Kawai

Springer

Editors

Takashi Aoki
Department of Mathematics
Kinki University
Higashi-Osaka 577-8502, Japan
e-mail: aoki@math.kindai.ac.jp

Hideyuki Majima
Department of Mathematics
Ochanomizu University
Tokyo 112-8610, Japan
e-mail: majima@math.ocha.ac.jp

Yoshitsugu Takei
Research Institute for Mathematical Sciences
Kyoto University
Kyoto 606-8502, Japan
e-mail: takei@kurims.kyoto-u.ac.jp

Nobuyuki Tose
Faculty of Economics
Keio University
Yokohama 223-8521, Japan
e-mail: tose@econ.keio.ac.jp

Library of Congress Control Number: 2007939560

ISBN 978-4-431-73239-6
Springer Tokyo Berlin Heidelberg New York

Springer is a part of Springer Science+Business Media
springer.com
©Springer 2008

This work is subject to copyright. All rights are reserved, whether the whole or a part of the material is concerned, specifically the rights of translation, reprinting, reuse of illustrations, recitation, broadcasting, reproduction on microfilms or in other ways, and storage in data banks. The use of registered names, trademarks, etc. in this publication does not imply, even in the absence of a specific statement, that such names are exempt from the relevant protective laws and regulations and therefore free for general use.

Camera-ready copy prepared from the authors' LaTeXfiles.
Printed and bound by Shinano Co. Ltd., Japan.
SPIN: 12081080

Printed on acid-free paper

Dedicated to Professor Takahiro Kawai on the Occasion of
His Sixtieth Birthday

(⋆) Prof. Takahiro KAWAI
1. Takashi AOKI
2. Yasunori OKADA
3. Motoo UCHIDA
4. Yoshitsugu TAKEI
5. Takeshi MANDAI
6. Gen NAKAMURA
7. Yoshifumi ITO
8. Minoru NAKANO
9. Shinichi TAJIMA
10. Fumitsuna MARUYAMA
11. Shunsuke SASAKI
12. Sunao ŌUCHI
13. Susumu YAMAZAKI
14. Jiro SEKIGUCHI
15. Harris J. SILVERSTONE
16. Eric DELABAERE
17. Yutaka MATSUI
18. Setsuro FUJIIE
19. Carl M. BENDER
20. Frédéric PHAM
21. Naofumi HONDA
22. Tadashi MIYAMOTO
23. Masafumi YOSHINO
24. Hideshi YAMANE
25. Kiyoomi KATAOKA
26. Mikio SATO
27. Hideaki WAKAKO
28. Akira SHIRAI
29. Takuya WATANABE
30. Yoshiyasu YASUTOMI
31. Hajime NAGOYA
32. Hidetoshi TAHARA
33. Ryuichi ISHIMURA
34. André VOROS
35. Takashi TAKIGUCHI
36. Shun SHIMOMURA
37. Leon EHRENPREIS
38. Toshinori OAKU
39. Ovidiu COSTIN
40. Masaki KASHIWARA
41. Keiko FUJITA
42. Masaharu KOBAYASHI
43. Yoshishige HARAOKA
44. Hikosaburo KOMATSU
45. Yves LAURENT
46. Hidetaka SAKAI
47. Yousuke OHYAMA
48. Michio JIMBO
49. Louis BOUTET DE MONVEL
50. Toshiyuki TANISAKI
51. Takayuki ODA
52. Tetsuji TOKIHIRO
53. Gilles LEBEAU
54. Masakazu MURO
55. Akira SHUDO
56. Masaki HIBINO
57. Etsuro DATE
58. Jose Ernie C. LOPE
59. Johannes SJÖSTRAND
60. Nobuyuki TOSE
61. Hideyuki MAJIMA
62. Kyoji SAITO

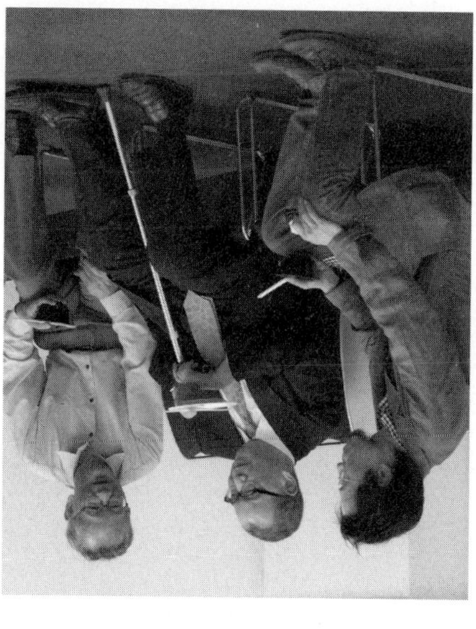

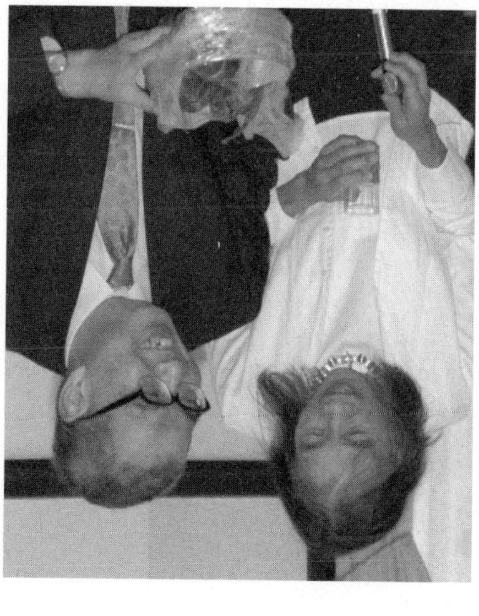

Preface

This is a collection of articles on algebraic analysis of differential equations and related topics. The contributors (T. Kawai excepted) and we, the editors, dedicate this volume to Professor Takahiro KAWAI, who is one of the creators of microlocal analysis, a central object of the subjects. To respect his many substantial contributions to the subjects, we have included some explanatory notes about the work of Professor T. Kawai in Part I of the volume.

Most of the contributed articles in Part II of this volume were read at the Conference "Algebraic Analysis of Differential Equations — from Microlocal Analysis to Exponential Asymptotics", which we organized from July 7 through July 14, 2005 at the Research Institute for Mathematical Sciences, Kyoto University to celebrate Professor T. Kawai's sixtieth birthday. Taking this opportunity, we would like to express once again our sincerest congratulations to Professor Kawai.

Kyoto in Japan,
June 27, 2007

Takashi Aoki
Hideyuki Majima
Yoshitsugu Takei
Nobuyuki Tose

Contents

List of Contributors

Takashi Aoki
Department of Mathematics
Kinki University
Higashi-Osaka 577-8502, Japan
aoki@math.kindai.ac.jp

Carl M. Bender
Physics Department
Washington University
St. Louis, MO 63130, USA
cmb@wustl.edu

Louis Boutet de Monvel
Université Pierre et Marie Curie
– Paris 6
France
boutet@math.jussieu.fr

Ovidiu Costin
Mathematics Department
The Ohio State University
Columbus, OH 43210, USA
costin@math.ohio-state.edu

Yuki Daido
Department of Mathematics
Hokkaido University
Sapporo 060-0810, Japan
daido@math.sci.hokudai.ac.jp

Eric Delabaere
Département de Mathématiques
UMR CNRS 6093,
Université d'Angers
2 Boulevard Lavoisier
49045 Angers Cedex 01, France
eric.delabaere@univ-angers.fr

Leon Ehrenpreis
Temple University
USA

Naofumi Honda
Department of Mathematics
Hokkaido University
Sapporo 060-0808, Japan
honda@math.sci.hokudai.ac.jp

Christopher J. Howls
University of Southampton
UK

Katsunori Iwasaki
Faculty of Mathematics
Kyushu University
6-10-1 Hakozaki, Higashi-ku
Fukuoka 812-8581, Japan
iwasaki@math.kyushu-u.ac.jp

Michio Jimbo
Graduate School of Mathematical Sciences
The University of Tokyo
Komaba, Tokyo 153-8914, Japan
jimbomic@ms.u-tokyo.ac.jp

Hyeonbae Kang
School of Mathematical Sciences
Seoul National University
Seoul 151-747, Korea
hkang@math.snu.ac.kr

Takahiro Kawai
Research Institute for Mathematical Sciences
Kyoto University
Kyoto 606-8502, Japan

Tatsuya Koike
Department of Mathematics
Graduate School of Science
Kyoto University
Kyoto 606-8502, Japan

Yves Laurent
Institut Fourier Mathématiques
UMR 5582 CNRS/UJF, BP 74
38402 St Martin d'Hères Cedex
France
Yves.Laurent@ujf-grenoble.fr

Gilles Lebeau
Département de Mathématiques
Université de Nice Sophia-Antipolis
Parc Valrose
06108 Nice Cedex 02, France
lebeau@math.unice.fr

Gen Nakamura
Department of Mathematics
Hokkaido University
Sapporo 060-0810, Japan
gnaka@math.sci.hokudai.ac.jp

Yukihiro Nishikawa
Hitachi Ltd.
Asagaya-kita 2-13-2, Suginami-ku
Tokyo 166-0001, Japan

Toshio Oshima
Graduate School of Mathematical Sciences
University of Tokyo
7-3-1, Komaba, Meguro-ku
Tokyo 153-8914, Japan
oshima@ms.u-tokyo.ac.jp

Sunao Ōuchi
Sophia University
Chiyoda-ku, Tokyo 102-8554, Japan
ouchi@mm.sophia.ac.jp

Shunsuke Sasaki
Mitsubishi UFJ Securities Co. Ltd.
Marunouchi 2-4-1, Chiyoda-ku
Tokyo 100-6317, Japan

Hujun Shen
Department of Chemistry
The Johns Hopkins University
3400 N. Charles St.
Baltimore, MD 21218, USA

Akira Shudo
Department of Physics
Tokyo Metropolitan University
Minami-Ohsawa, Hachioji
Tokyo 192-0397, Japan
shudo@phys.metro-u.ac.jp

Harris J. Silverstone
Department of Chemistry
The Johns Hopkins University
3400 N. Charles St.
Baltimore, MD 21218, USA
hjsilverstone@jhu.edu

Johannes Sjöstrand
CMLS
Ecole Polytechnique
FR 91120 Palaiseau, France
johannes@math.polytechnique.fr

Plamen Stefanov
Department of Mathematics
Purdue University
West Lafayette, IN 47907, USA
stefanov@math.purdue.edu

Hidetoshi Tahara
Department of Mathematics
Sophia University
Kioicho, Chiyoda-ku
Tokyo 102-8554, Japan
h-tahara@hoffman.cc.sophia.ac.jp

Yoshitsugu Takei
Research Institute for Mathematical Sciences
Kyoto University
Kyoto 606-8502, Japan
takei@kurims.kyoto-u.ac.jp

Nobuyuki Tose
Faculty of Economics
Keio University
Yokohama 223-8521, Japan
tose@econ.keio.ac.jp

Gunther Uhlmann
Department of Mathematics
University of Washington
Seattle, WA 98195, USA
gunther@math.washington.edu

André Voros
CEA
Service de Physique Théorique de Saclay (CNRS URA 2306)
F-91191 Gif-sur-Yvette CEDEX
France
voros@spht.saclay.cea.fr

Masafumi Yoshino
Department of Mathematics
Graduate School of Science
Hiroshima University
1-3-1 Kagamiyama
Higashi-hiroshima
Hiroshima 739-8526, Japan
yoshino@math.sci.hiroshima-u.ac.jp

Part I

The work of T. Kawai

Publications of Professor Takahiro Kawai

Papers

[1] On the theory of Fourier hyperfunctions and its applications to partial differential equations with constant coefficients, *J. Fac. Sci. Univ. Tokyo, Sect. IA*, **17** (1970), 467–517.

[2] Construction of elementary solutions for I-hyperbolic operators and solutions with small singularities, *Proc. Japan Acad.*, **46** (1970), 912–915.

[3] Pseudo-differential operators in the theory of hyperfunctions, *Proc. Japan Acad.*, **46** (1970), 1130–1134 (with M. Kashiwara).

[4] Construction of a local elementary solution for linear partial differential operators, I, *Proc. Japan Acad.*, **47** (1971), 19–23.

[5] Construction of a local elementary solution for linear partial differential operators, II, *Proc. Japan Acad.*, **47** (1971), 147–152.

[6] Boundary values of hyperfunction solutions of linear partial differential equations, *Publ. RIMS, Kyoto Univ.*, **7** (1971), 95–104 (with H. Komatsu).

[7] On the global existence of real analytic solutions of linear differential equations, I, *Proc. Japan Acad.*, **47** (1971), 537–540.

[8] On the global existence of real analytic solutions of linear differential equations, II, *Proc. Japan Acad.*, **47** (1971), 643–647.

[9] Construction of local elementary solutions for linear partial differential operators with real analytic coefficients (I) — The case with real principal symbols —, *Publ. RIMS, Kyoto Univ.*, **7** (1971), 363–397.

[10] Construction of local elementary solutions for linear partial differential operators with real analytic coefficients (II) — The case with complex principal symbols —, *Publ. RIMS, Kyoto Univ.*, **7** (1971), 399–426.

[11] Theorems on the finite-dimensionality of cohomology groups, I, *Proc. Japan Acad.*, **48** (1972), 70–72.

[12] Theorems on the finite-dimensionality of cohomology groups, II, *Proc. Japan Acad.*, **48** (1972), 287–289.

[13] On the global existence of real analytic solutions of linear differential equations (I), *J. Math. Soc. Japan*, **24** (1972), 481–517.

[14] On the structure of single linear pseudo-differential equations, *Proc. Japan Acad.*, **48** (1972), 643–646 (with M. Sato and M. Kashiwara).

[15] On the boundary value problem for elliptic system of linear differential equations, I, *Proc. Japan Acad.*, **48** (1972), 712–715 (with M. Kashiwara).

[16] On the propagation of analyticity of solutions of convolution equations, *J. Math. Kyoto Univ.*, **13** (1973), 67–72.

[17] Finite-dimensionality of cohomology groups attached to systems of linear differential equations, *J. Math. Kyoto Univ.*, **13** (1973), 73–95.

[18] On the global existence of real analytic solutions of linear differential equations, *Lecture Notes in Math.*, No. 287, Springer, 1973, pp. 99–121.

[19] Microfunctions and pseudo-differential equations, *Lecture Notes in Math.*, No. 287, Springer, 1973, pp. 265–529 (with M. Sato and M. Kashiwara).

[20] On the boundary value problem for elliptic system of linear differential equations, II, *Proc. Japan Acad.*, **49** (1973), 164–168 (with M. Kashiwara).

[21] On the boundary value problem for the elliptic system of linear differential equations, *Sém. Goulaouic-Schwartz, 1972–1973*, Exposé 19, 1973, (with M. Kashiwara).

[22] Theorems on the finite-dimensionality of cohomology groups, III, *Proc. Japan Acad.*, **49** (1973), 243–246.

[23] Some applications of micro-local analysis to the global study of linear differential equations, *Astérisque*, vol. 2 et 3, Soc. Math. France, 1973, pp. 229–243.

[24] On the global existence of real analytic solutions of linear differential equations (II), *J. Math. Soc. Japan*, **25** (1973), 644–647.

[25] Theorems on the finite-dimensionality of cohomology groups, IV, *Proc. Japan Acad.*, **49** (1973), 655–658.

[26] Theorems on the finite-dimensionality of cohomology groups, V, *Proc. Japan Acad.*, **49** (1973), 782–784.

[27] Structure of cohomology groups whose coefficients are microfunction solution sheaves of systems of pseudo-differential equations with multiple characteristics, I, *Proc. Japan Acad.*, **49** (1974), 420–425 (with M. Kashiwara and T. Oshima).

[28] Structure of cohomology groups whose coefficients are microfunction solution sheaves of systems of pseudo-differential equations with multiple characteristics, II, *Proc. Japan Acad.*, **50** (1974), 549–550 (with M. Kashiwara and T. Oshima).

[29] Pseudo-differential operators acting on the sheaf of microfunctions, *Lecture Notes in Math.*, No. 449, Springer, 1975, pp. 54–69.

[30] Micro-hyperbolic pseudo-differential operators, *Lecture Notes in Math.*, No. 449, Springer, 1975, pp. 70–82 (with M. Kashiwara).

[31] Theory of vector-valued hyperfunctions, *Publ. RIMS, Kyoto Univ.*, **11** (1975), 1–19 (with P. D. F. Ion).

[32] Removable singularities of solutions of systems of linear differential equations, *Bull. Amer. Math. Soc.*, **81** (1975), 461–463.

[33] Micro-local properties of $\prod_{j=1}^{n} f_{j+}^{s_j}$, *Proc. Japan Acad.*, **51** (1975), 270–272 (with M. Kashiwara).

[34] Micro-hyperbolic pseudo-differential operators, I, *J. Math. Soc. Japan*, **27** (1975), 359–404 (with M. Kashiwara).

[35] Microlocal study of S-matrix singularity structure, *Lecture Notes in Phys.*, No. 39, Springer, 1975, pp. 38–48 (with H. Stapp).

[36] On the propagation of analyticity of solutions of systems of linear differential equations with constant coefficients, *J. Fac. Sci. Univ. Tokyo, Sec. IA*, **22** (1975), 267–270.

[37] Vanishing of cohomology groups on completely k-convex sets, *Publ. RIMS, Kyoto Univ.*, **11** (1976), 775–784.

[38] Structure of a single pseudo-differential equation in a real domain, *J. Math. Soc. Japan*, **28** (1976), 80–85 (with M. Kashiwara and T. Oshima).

[39] On a conjecture of Regge and Sato on Feynman integrals, *Proc. Japan Acad.*, **52** (1976), 161–164 (with M. Kashiwara).

[40] Extension of solutions of systems of linear differential equations, *Publ. RIMS, Kyoto Univ.*, **12** (1976), 215–227.

[41] Extension of hyperfunction solutions of linear differential equations with constant coefficients, *Proc. Amer. Math. Soc.*, **59** (1976), 311–316.

[42] Finiteness theorem for holonomic systems of micro-differential equations, *Proc. Japan Acad.*, **52** (1976), 341–343 (with M. Kashiwara).

[43] Local extension of solutions of systems of linear differential equations with constant coefficients, *Comm. Pure Appl. Math.*, **30** (1977), 235–254 (with E. Bedford).

[44] Holonomic character and local monodromy structure of Feynman integrals, *Commun. Math. Phys.*, **54** (1977), 121–134 (with M. Kashiwara).

[45] Holonomic systems of linear differential equations and Feynman integrals, *Publ. RIMS, Kyoto Univ.*, **12** Suppl. (1977), 131–140 (with M. Kashiwara).

[46] Micro-analytic structure of the S-matrix and related functions, *Publ. RIMS, Kyoto Univ.*, **12** Suppl. (1977), 141–146 (with M. Kashiwara and H. Stapp).

[47] Discontinuity formula and Sato's conjecture, *Publ. RIMS, Kyoto Univ.*, **12** Suppl. (1977), 155–232 (with H. Stapp).

[48] A study of Feynman integrals by micro-differential equations, *Commun. Math. Phys.*, **60** (1978), 97–130 (with M. Kashiwara and T. Oshima).

[49] Invariance of cohomology groups under a deformation of an elliptic system of linear differential equations, *Proc. Japan Acad.*, **53** (1977), 144–145.

[50] Microlocal properties of local elementary solutions for Cauchy problems for a class of hyperbolic linear differential operators, *Publ. RIMS, Kyoto Univ.*, **14** (1978), 415–439 (with G. Nakamura).

[51] Holomorphic perturbation of a system of micro-differential equations, *Annali Scuola Normale Sup. Pisa, Series IV*, **5** (1978), 581–586.

[52] Micro-analyticity of the S-matrix and related functions, *Commun. Math. Phys.*, **66** (1979), 95–130 (with M. Kashiwara and H. Stapp).

[53] On a class of linear partial differential equations whose formal solutions always converge, *Ark. för Mat.*, **17** (1979), 83–91 (with M. Kashiwara and J. Sjöstrand).

[54] On holonomic systems for $\prod_{l=1}^{n}(f_l + \sqrt{-1}0)^{\lambda_l}$, *Publ. RIMS, Kyoto Univ.*, **15** (1979), 551–575 (with M. Kashiwara).

[55] On the characteristic variety of a holonomic system with regular singularities, *Adv. in Math.*, **34** (1979), 163–184 (with M. Kashiwara).

[56] Monodromy structure of solutions of holonomic systems of linear differential equations is invariant under the deformation of the system, *Publ. RIMS, Kyoto Univ.*, **15** (1979), 741–747 (with M. Kashiwara).

[57] Extended Landau variety and the singularity spectrum of position-space Feynman amplitudes, *Publ. RIMS, Kyoto Univ.*, **15** (1979), 767–772.

[58] The theory of holonomic systems with regular singularities and its relevance to physical problems, *Lecture Notes in Phys.*, No. 126, Springer, 1980, pp. 5–20 (with M. Kashiwara).

[59] Second-microlocalization and asymptotic expansions, *Lecture Notes in Phys.*, No. 126, Springer, 1980, pp. 21–76 (with M. Kashiwara).

[60] On holonomic systems of micro-differential equations III — systems with regular singularities —, *Publ. RIMS, Kyoto Univ.*, **17** (1981), 813–979 (with M. Kashiwara).

[61] On the holonomic character of Feynman integrals and the S-matrix, *Selected Studies (Einstein Volume)*, North-Holland, 1982, pp. 99–111.

[62] On the regular holonomic character of the S-matrix and microlocal analysis of unitarity-type integrals, *Commun. Math. Phys.*, **83** (1982), 213–242 (with H. Stapp).

[63] Poisson's summation formula and Hamburger's theorem, *Publ. RIMS, Kyoto Univ.*, **18** (1982), 833–846 (with L. Ehrenpreis).

[64] An example of a complex of linear differential operators of infinite order, *Proc. Japan Acad., Ser. A*, **59** (1983), 113–115.

[65] Linear differential equations of infinite order and theta functions, *Adv. in Math.*, **47** (1983), 300–325 (with M. Sato and M. Kashiwara).

[66] Microlocal analysis, *Publ. RIMS, Kyoto Univ.*, **19** (1983), 1003–1032 (with M. Kashiwara).

[67] Infra-red finiteness in quantum electro-dynamics, *Adv. Studies in Pure Math.*, Vol. 4, Kinokuniya, 1984, pp. 263–266 (with H. Stapp).

[68] Microlocal analysis of theta functions, *Adv. Studies in Pure Math.*, Vol. 4, Kinokuniya, 1984, pp. 267–289 (with M. Sato and M. Kashiwara).

[69] The Fabry-Ehrenpreis gap theorem for hyperfunctions, *Proc. Japan Acad., Ser. A*, **60** (1984), 276–278.

[70] The Poincaré lemma for a variation of polazired Hodge structure, *Proc. Japan Acad., Ser. A*, **61** (1985), 164–167 (with M. Kashiwara).

[71] Hodge structure and holonomic systems, *Proc. Japan Acad., Ser. A*, **62** (1986), 1–4 (with M. Kashiwara).

[72] On the global existence of real analytic solutions and hyperfunction solutions of linear differential equations, *Proc. Japan Acad., Ser. A*, **62** (1986), 77–79.

[73] A differential equation theoretic interpretation of a geometric result of Hartogs, *Proc. Amer. Math. Soc.*, **98** (1986), 222–224.

[74] On a class of linear differential operators of infinite order with finite index, *Adv. in Math.*, **62** (1986), 155–168 (with T. Aoki and M. Kashiwara).

[75] Systems of microdifferential equations of infinite order, *Proceedings of the Taniguchi Symposium, HERT (Kyoto)*, Kinokuniya, 1986, pp. 143–154.

[76] On a closed range property of a linear differential operator, *Proc. Japan Acad., Ser. A*, **62** (1986), 386–388 (with Y. Takei).

[77] The Fabry-Ehrenpreis gap theorem and systems of linear differential equations of infinite order, *Amer. J. Math.*, **109** (1987), 57–64.

[78] The Poincare lemma for variations of polarized Hodge structure, *Publ. RIMS, Kyoto Univ.*, **23** (1987), 345–407 (with M. Kashiwara).

[79] On the global existence of real analytic solutions and hyperfunction solutions of linear differential equations, *J. Fac. Sci. Univ. Tokyo, Sect. IA*, **34** (1987), 791–803.

[80] Microlocal analysis of infrared singularities, *Algebraic Analysis*, Vol. 1, Academic Press, 1988, pp. 309–330 (with H. Stapp).

[81] On the global existence of real analytic solutions of systems of linear differential equations, *Algebraic Analysis*, Vol. 1, Academic Press, 1988, pp. 331–344 (with Y. Takei).

[82] Systems of linear differential equations of infinite order — an aspect of infinite analysis —, *Theta Functions, Bowdoin 1987, Proc. Symp. in Pure Math.*, Vol. 49, Part 1, Amer. Math. Soc., 1989, pp. 3–17.

[83] A particular partition of unity — an auxiliary tool in Hodge theory —, *Theta Functions, Bowdoin 1987, Proc. Symp. in Pure Math.*, Vol. 49, Part 1, Amer. Math. Soc., 1989, pp. 19–26 (with M. Kashiwara).

[84] On the boundary value of a solution of the heat equation, *Publ. RIMS, Kyoto Univ.*, **25** (1989), 491–498 (with T. Matsuzawa).

[85] Bicharacteristical convexity and the semi-global existence of holomorphic solutions of linear differential equations with holomorphic coefficients, *Adv. in Math.*, **80** (1990), 110–133 (with Y. Takei).

[86] On the existence of holomorphic solutions of systems of linear differential equations of infinite order and with constant coefficients, *Int. J. of Math.*, **1** (1990), 63–82 (with D.C. Struppa).

[87] The Bender-Wu analysis and the Voros theory, *Special Functions*, Springer, 1991, pp. 1–29 (with T. Aoki and Y. Takei).

[88] The structure of cohomology groups associated with the theta-zerovalues, *Proc. of the Conference "Geometrical and Algebraical Aspects in Several Complex Variables"*, Editel (Italy), 1991, pp. 169–189 (with M. Kashiwara and Y. Takei).

[89] The complex-analytic geometry of bicharacteristics and the semi-global existence of holomorphic solutions of linear differential equations — A bridge between the theory of partial differential equations and the theory of holomorphic functions —, *Proc. of the Conference "Geometrical and Algebraical Aspects in Several Complex Variables"*, Editel (Italy), 1991, pp. 191–199 (with Y. Takei).

[90] Interpolation theorems in several complex variables and applications, *Lecture Notes in Math.*, No. 1540, Springer, 1993, pp. 1–9 (with C.A. Berenstein and D.C. Struppa).

[91] On the structure of Painlevé transcendents with a large parameter, *Proc. Japan Acad., Ser. A*, **69** (1993), 224–229 (with Y. Takei).

[92] Infrared catastrophe in a massless Feynman function, *Ann. Inst. Fourier*, **43** (1993), 1301–1310 (with H. Stapp).

[93] New turning points in the exact WKB analysis for higher-order ordinary differential equations, *Analyse algébrique des perturbations singulières*, tome I, Hermann, 1994, pp. 69–84 (with T. Aoki and Y. Takei).

[94] Secular equations through the exact WKB analysis, *Analyse algébrique des perturbations singulières*, tome I, Hermann, 1994, pp. 85–102 (with Y. Takei).

[95] Algebraic analysis of singular perturbations — On exact WKB analysis, *Sûgaku Expositions*, **8** (1995), 217–240 (with T. Aoki and Y. Takei). (This article originally appeared in Japanese in *Sugaku*, **45** (1993), 299–315.)

[96] Quantum electrodynamics at large distances, I. Extracting the correspondence-principle part, *Phys. Rev. D*, **52** (1995), 2484–2504 (with H. Stapp).

[97] Quantum electrodynamics at large distances, II. Nature of the dominant singularities, *Phys. Rev. D*, **52** (1995), 2505–2516 (with H. Stapp).

[98] Quantum electrodynamics at large distances, III. Verification of pole factorization and the correspondence principle, *Phys. Rev. D*, **52** (1995), 2517–2532 (with H. Stapp).

[99] WKB analysis of Painlevé transcendents with a large parameter, I, *Adv. in Math.*, **118** (1996), 1–33 (with Y. Takei).

[100] WKB analysis of Painleve transcendents with a large parameter, II. — Multiple-scale analysis of Painlevé transcendents, *Structure of Solutions of Differential Equations*, World Scientific, 1996, pp. 1–49 (with T. Aoki and Y. Takei).

[101] Exponential representation of a holomorphic solution of a system of differential equations associated with the theta-zerovalue, *Structure of Solutions of Differential Equations*, World Scientific, 1996, pp. 89–102 (with C.A. Berenstein, D.C. Struppa and Y. Takei).

[102] Interpolating varieties and the Fabry-Ehrenpreis-Kawai gap theorem, *Adv. in Math.*, **122** (1996), 280–310 (with C.A. Berenstein and D.C. Struppa).

[103] On the structure of Painlevé transcendents with a large parameter, II, *Proc. Japan Acad., Ser. A*, **72** (1996), 144–147 (with Y. Takei).

[104] On infrared singularities, *New Trends in Microlocal Analysis*, Springer, 1997, pp. 117–123 (with H. Stapp).

[105] WKB analysis of Painlevé transcendents with a large parameter, III. — Local reduction of 2-parameter Painlevé transcendents, *Adv. in Math.*, **134** (1998), 178–218 (with Y. Takei).

[106] On the exact WKB analysis for the third order ordinary differential equations with a large parameter, *Asian J. Math.*, **2** (1998), 625–640 (with T. Aoki and Y. Takei).

[107] On a complete description of the Stokes geometry for higher order ordinary differential equations with a large parameter via integral representations, *Toward the Exact WKB Analysis of Differential Equations, Linear or Non-Linear*, Kyoto Univ. Press, 2000, pp. 11–14 (with T. Aoki and Y. Takei).

[108] Can we find a new deformation of (SL_J) with respect to the parameters contained in (P_J) ? *Toward the Exact WKB Analysis of Differential Equations, Linear or Non-Linear*, Kyoto Univ. Press, 2000, pp. 205–208 (with T. Aoki and Y. Takei).

[109] Natural boundaries revisited through differential equations, infinite order or non-linear, *Toward the Exact WKB Analysis of Differential Equations, Linear or Non-Linear*, Kyoto Univ. Press, 2000, pp. 231–243.

[110] Overconvergence phenomena and grouping in exponential representation of solutions of linear differential equations of infinite order, *Adv. in Math.*, **161** (2001), 131–140 (with D.C. Struppa).

[111] On the exact steepest descent method — a new method for the description of Stokes curves, *J. Math. Phys.*, **42** (2001), 3691–3713 (with T. Aoki and Y. Takei).

[112] Exact WKB analysis of non-adiabatic transition probabilities for three levels, *J. Phys. A*, **35** (2002), 2401–2430 (with T. Aoki and Y. Takei).

[113] On infra-red singularities associated with QC photons, *Microlocal Analysis and Complex Fourier Analysis*, World Scientific, 2002, pp. 115–134 (with H. Stapp).

[114] On the exact WKB analysis of operators admitting infinitely many phases, *Adv. in Math.*, **181** (2004), 165–189 (with T. Aoki, T. Koike and Y. Takei).

[115] The exact steepest descent method — A new steepest descent method based on the exact WKB analysis, *Adv. Studies in Pure Math.*, Vol. 42, Math. Soc. Japan, 2004, pp. 45–61 (with T. Aoki and Y. Takei).

[116] On the exact WKB analysis of microdifferential operators of WKB type, *Ann. Inst. Fourier*, **54** (2004), 1393–1421 (with T. Aoki, T. Koike and Y. Takei).

[117] On the Stokes geometry of higher order Painlevé equations, *Astérisque*, Vol. 297, Soc. Math. France, 2004, pp. 117–166 (with T. Koike, Y. Nishikawa and Y. Takei).

[118] On WKB analysis of higher order Painlevé equations with a large parameter, *Proc. Japan Acad., Ser. A*, **80** (2004), 53–56 (with Y. Takei).

[119] On global aspects of exact WKB analysis of operators admitting infinitely many phases, *Contemporary Math.*, No. 373, Amer. Math. Soc., 2005, pp. 11–45 (with T. Aoki, T. Koike and Y. Takei).

[120] Virtual turning points and bifurcation of Stokes curves for higher order ordinary differential equations, *J. Phys. A*, **38** (2005), 3317–3336 (with T. Aoki, S. Sasaki, A. Shudo and Y. Takei).

[121] WKB analysis of higher order Painlevé equations with a large parameter — Local reduction of 0-parameter solutions for Painlevé hierarchies (P_J) $(J = \mathrm{I}, \mathrm{II}\text{-}1\mathrm{or}\mathrm{II}\text{-}2)$, *Adv. in Math.*, **203** (2006), 63-6-672 (with Y. Takei).

[122] Virtual turning points - A gift of microlocal analysis to the exact WKB analysis, in this volume (with T. Aoki, N. Honda, T. Koike, Y. Nishikawa, S. Sasaki, A. Shudo and Y. Takei).

Books

[B1] *Seminar on Micro-local Analysis*, Princeton Univ. Press, 1979 (with V. Guillemin and M. Kashiwara).

[B2] *Foundations of Algebraic Analysis*, Kinokuniya, 1980 (in Japanese) (with M. Kashiwara and T. Kimura). English translation: Princeton Univ. Press, 1986.

[B3] *Algebraic Analysis of Singular Perturbation Theory*, Iwanami, 1998 (in Japanese) (with Y. Takei). English translation: Amer. Math. Soc., 2005.

The work of T. Kawai on hyperfunction theory and microlocal analysis

Part 1 — Microlocal analysis and differential equations

Nobuyuki Tose
Center for Integrative Mathematical Sciences, Keio University, Japan

Takahiro Kawai talked about his first encounter with the theory of hyperfunctions recently on an occasion of 70th birthday of Hikosaburo Komatsu. We start, however, the story at the time before this encounter. After Mikio Sato invented the theory of hyperfunctions together with the notion of local cohomology groups in 1957, the theory did not prevail until Komatsu gave lectures in US and Europe. Kawai's experience with the theory began when Komatsu gave an invited talk on an annual meeting of the Mathematical Society of Japan around 1967. This invited talk by Komatsu was given just after his return to Japan from his long mission abroad and really fascinated Kawai so that he left all of his belongings in the conference room. At this point, Kawai's career as mathematician might have been oriented.

The name of Kawai remains in the note taken for the lectures given by Komatsu during the Japanese Academic year 1967/68 on Sato's hyperfunctions, in which we can now trace back the enthusiastic activities of Komatsu and his students. It is indeed in this period when Sato returned to the University of Tokyo. At that time the activities of Komatsu and his young followers called Sato back to the theory of hyperfunctions and Sato has established the notion of microlocal analysis, local analysis on the cotangent bundle. In fact, Sato was inspired by the work of Fritz John on the plane wave decomposition of the delta functions to derive the notion of microfunctions in 1969. What is more, in this period Maslov and Egorov discovered transformation theory of pseudo-differential operators which are compatible with contact transformations, which were reformulated later as quantized contact transformations and Fourier integral operators. In these splendid days, Kawai compiled, with Sato and Masaki Kashiwara, a historical volume called SKK, in which they proved the most important and fundamental theorem in microlocal analysis. They showed a general overdetermined system of pseudo-differential equations is microlocally a direct sum of partial de Rham systems, partial Cauchy-Riemann systems, and Lewy-Mizohata systems. Here the expression "microlocally" means that systems are transformed by quantized contact transformations. It is a remarkable fact that SKK constructed the theory including the construction of quantized contact transformations very algebraically in the ring of pseudo-differential operators using the division theorems of operators. Here at this point, this theory by SKK and a corresponding result in the C^∞ category by Lars Hörmander and J.J. Duistermaat have opened a door to a new era of analysis.

We go back to the time a little before SKK to explain Kawai's important contribution to the theory of hyperfunctions, in particular to partial differential equations. His master thesis is to construct the theory of Fourier hyperfunctions, with which he obtained many splendid results for partial differential equations with constant coefficients in the space of hyperfunctions. He proved many results on existence, ellipticity, hyperbolicity, propagation of regularities, and unique continuation properties. After these results, Kawai went further to the case where operators are variable coefficients. Actually, he succeeded in constructing fundamental solutions to the Cauchy problems for operators with simple characteristics. This result is based on Yusaku Hamada's result on the Cauchy problems in the complex domain with singular Cauchy data and made a dream of Jean Leray really happen. The dream of Jean Leray was to construct fundamental solutions to partial differential operators by integrating their holomorphic solutions. In fact, Kawai's construction utilized the plane wave expansion of delta functions to construct holomorphic solutions for plane wave as initial data.

One of the most general theorem for microfunctions solutions to pseudo-differential operators is due to Kawai and Kashiwara. They introduced microhyperbolic directions for pseudo-differential operators and obtained a very fundamental result on the solvability and the propagation of support for microfunction solutions. Their construction is based on the analytic continuation of holomorphic solutions across non-characteristic real hypersurfaces, which also realized a philosophy of Jean Leray. This method has been the most basic in studying microfunctions (or hyperfunctions) solutions to linear PDE with real analytic coefficients, and underlying geometric construction together with microhyperbolic direction entailed much later the microlocal study of sheaves due to Kashiwara and Pierre Schapira. Moreover this result by Kashiwara and Kawai was so decisive that it was reformulated in many ways. I have a good story to explain this. In fact, I was very surprised to hear the word "Kashiwara-Kawai" in a talk by Gilles Lebeau on boundary value problems with singular boundary. Lebeau mentioned these two names since he utilized an estimate obtained when microhyperbolic problems are reformulated in the framework of FBI transformation.

We can't overestimate Kawai's significant contribution to the theory of boundary value problems in the space of hyperfunctions. Kawai showed with Komatsu that any hyperfunction on a half space has its boundary value as a hyperfunction if it satisfies a non-characteristic differential equation. Furthermore he constructed with Kashiwara a beautiful theory of boundary value problems for elliptic systems of differential equations. It should be stressed that the boundary can be higher codimensional in this theory. I found, as a young student, the theory really revolutional since hyperfunction solutions near the boundary are related to microfunction solutions to tangential systems on the microlocal region corresponding to the boundary. This theory enjoyed applications in many ways such as extension and solvability of differential equations.

To conclude this part, I would like to devote two paragraphs to Kawai's contributions to holonomic systems. We can trace back to the idea of holonomic systems to a colloquium talk by Sato in the Department of Mathematics at the University of Tokyo. Sato proposed to control Riemann functions or, in other words, fundamental solutions by means of systems of differential equations. This dream of Sato came true in SKK and in the master thesis by Kashiwara. In SKK, the characteristic varieties of systems of (pseudo-)differential equations are shown to be involutive. This shows that maximally overdetermined systems of differential equations, later called holonomic systems are with Lagrangean characteristic varieties. In SKK, holonomic systems with simple characteristic are classified microlocally in a beautiful way and are utilized in a systematical way to control quantized contact transformations. On the other hand, in the master thesis by Kashiwara, the finite dimensionality of solution complexes are shown. From this result, holonomic systems has been considered to be an higher dimensional counterpart of ordinary differential equations. By the way, we should also make a stress on another Kawai's contribution to holonomic systems applied to the Feynman integrals. He considered, with Kashiwara, the Feynman integrals as solutions to holonomic systems to propose a very new insight both to mathematics and physics. I would like to mention that a continual collaboration of Kawai with Henry Stapp has given rise to important result in this direction.

Kawai also compiled with Kashiwara in 1981 a bible of holonomic systems with regular singularities from the viewpoint of microlocal analysis, in which they proved the most fundamental properties of regular holonomic systems. There is an astonishing result shown in this volume. All holonomic systems can be transformed into holonomic systems with regular singularities with the aid of differential operators of infinite order. This result was quite new in the sense that even the case of ordinary differential equations had not been discovered before Kashiwara and Kawai.

The work of T. Kawai on hyperfunction theory and microlocal analysis

Part 2 — Operators of infinite order and convolution equations

Takashi Aoki
Department of Mathematics, Kinki University, Japan

Takahiro Kawai has been interested in operators of infinite order and convolution equations from the beginning of his research career and he made significant contributions in these areas as well as in the theory of partial differential equations. He uses operators of infinite order most effectively in various fields and obtains a number of remarkable results.

His first paper [1] devoted to the theory of Fourier hyperfunctions and convolution equations was published in 1970. Here he introduces the sheaf of Fourier hyperfunctions and defines Fourier transformation for Fourier hyperfunctions. As applications, he discusses convolution equations and extends many results known in the Schwartz distribution theory to hyperfunctions. They are not only extensions, but have the most natural and beautiful forms. He also extends results known for partial differential equations of finite order to equations of infinite order. For example, microlocal ellipticity for linear partial differential operators with constant coefficients is extended to local convolution operators, namely, partial differential operators of infinite order with constant coefficients. He defines the characteristic set for a convolution operator $S*$ associated with a hyperfunction S supported by the origin and proves that the singularity spectrum (i.e., the support as a microfunction) of any hyperfunction solution u of the equation $S * u = f$, f being a given hyperfunction, is contained in the union of the characteristic set and the singularity spectrum of f. This is a natural extension of Sato's fundamental theorem to differential operators of infinite order with constant coefficients. That theorem of Sato declares the same statement for differential operators $P(x, D)$ of *finite order* with real analytic coefficients. Although the statements of these two theorems are similar, defining the characteristic set for the convolution operator $S*$ requires deep consideration of entire functions of infra-exponential type. This theorem of Kawai is extended to the analytic coefficient case in [74], which presents some conditions on a linear differential operator P of infinite order with holomorphic coefficients that guarantee the finiteness of the dimensions of the kernel and the cokernel of the map $P : \mathscr{O}_{\mathbb{C}^n,0} \to \mathscr{O}_{\mathbb{C}^n,0}$, where $\mathscr{O}_{\mathbb{C}^n,0}$ denotes the germ at the origin of the sheaf of holomorphic functions on $\mathbb{C}^n$.

In the epoch-making collaboration of M. Sato, T. Kawai and M. Kashiwara [19] in microlocal analysis, micro-differential ("pseudo-differential" is used there) operators of infinite order play an important role. One of the main theorems of [19] is that, in the complex domain, systems of partial differential equations with constant multiple characteristics can be transformed to a system with simple characteristics. One cannot obtain such a clear and beautiful theorem without using operators of infinite order. Similarly, in [60], the structure of general holonomic systems is clarified by using operators of infinite order. For any holonomic system $\mathscr{M}$, its regular part $\mathscr{M}_{\mathrm{reg}}$ is defined as a holonomic system with regular singularities and it is proved in [60] that

$$\mathscr{E}^{\infty} \underset{\mathscr{E}}{\otimes} \mathscr{M}_{\mathrm{reg}} = \mathscr{E}^{\infty} \underset{\mathscr{E}}{\otimes} \mathscr{M}.$$

Here $\mathscr{E}^{\infty}$ (resp. $\mathscr{E}$) denotes the sheaf of micro-differential operators of infinite order (resp. micro-differential operators of finite order). This isomorphism implies that irregular singularities can be reduced to regular singularities by using operators of infinite order.

The notion of holonomic systems of (micro-)differential equations of finite order was introduced by Sato and it has been one of fundamentals in algebraic analysis. Kawai extends the notion to systems of (micro-)differential operators of infinite order and obtains the notion of $\mathbb{R}$-holonomic systems. In [65] and [68], he establishes, with Sato and Kashiwara, some basic properties concerning finite dimensionality of solution spaces or solution complexes of systems of differential equations of infinite order for the Jacobi theta zerovalues and their generalizations which have $\mathbb{R}$-holonomicity. These works extend considerably the results of Sato given in a short paper [S]. Also, these works have a link with [63], a collaboration with L. Ehrenpreis, where some conditions are given that characterize a distribution which is a sum of distributions supported by the integral points and whose Fourier transform is again of the same form. As an application, another proof is given for the classical theorem of Hamburger on the characterization of the Riemann zeta function. After these works, Kawai (and his collaborators Kashiwara, Takei, Berenstein, Struppa and Aoki) published several papers concerning differential equations of infinite order ([64], [69], [74], [75], [77], [82], [88]). In [88], some parts of [64], [65] are extended to higher dimensional case, that is, the structure of some cohomology groups is determined which are associated with the theta-zerovalue

$$\vartheta(t) = \sum_{\nu \in \mathbb{Z}^n} \exp(\pi i \langle t\nu, \nu \rangle)$$

of the Riemann theta function.

In [69] and [77], Kawai gives a short and elegant proof, based on microlocal ellipticity of differential operators of infinite order, of the classical Fabry gap theorem and its generalization. In the simplest case, his theorem can be read as follows: Suppose that for a real positive sequence $\{a_n\}$, the series

$$\sum_{n=0}^{\infty} c_n \exp(ia_n x)$$

converges locally uniformly in the upper-half plane $\operatorname{Im} x > 0$ and defines a holomorphic function f there and that this series is "lacunary" in the sense that n/a_n converges to 0 when n tends to the infinity and $|a_n - a_m| \geq c|n-m|$ for some $c > 0$. If f is analytically continued to a neighborhood of the origin, then f has an analytic continuation to $\operatorname{Im} x > -\delta$ for some $\delta > 0$. His idea of proof is to consider f as a solution of an infinite-order differential equation $Pf = 0$ with

$$P = \prod_{n=0}^{\infty} \left(1 + \frac{1}{a_n^2} \frac{d^2}{dx^2}\right).$$

He observes that the characteristic set of this operator coincides with $i\mathbb{R}$. This means that P is microlocally elliptic outside $i\mathbb{R}$. Combining this fact and the theory of microhyperbolic systems due to Kashiwara and Schapira [KS], he obtains the analytic continuation of f. The idea is based on an observation by L. Ehrenpreis [E], which suggests that one can understand the Fabry gap theorem for one complex variable from the viewpoint of convolution operators, but Kawai is the first mathematician who has carried out this idea. A generalization of his theorem to higher dimensional problem is given in [101], [102], which are collaborations with C. A. Berenstein and D. Struppa. His interest in natural boundaries of holomorphic functions continues to [109], where he discusses several examples of differential equations, not only linear of infinite order, but also non-linear, and gives several observations on the relation between natural boundaries and differential equations. Also, his interest in differential equations of infinite order extends to [114], [116], which connect the exact WKB analysis with operators of infinite order.

References

[E] L. Ehrenpreis, Fourier Analysis of Several Complex Variables, Wiley-Interscience, 1970.

[KS] M. Kashiwara and P. Schapira, Microhyperbolic systems, Acta Math., **142** (1979), 1–55.

[S] M. Sato, Pseudo-differential equations and theta functions, Astérisque, **2** et **3** (1973), 286–291.

The work of T. Kawai on exact WKB analysis

— The pioneering work which combines two major fields "microlocal analysis" and "exponential asymptotics" —

Yoshitsugu Takei
Research Institute for Mathematical Sciences, Kyoto University, Japan

The year 1989 is a turning point in the research activities of T. Kawai.

In 1989, inspired by the report of Pham [P], M. Sato organized a seminar on "algebraic analysis of singular perturbation theory" and Kawai joined it as one of the chief participants. The subject of [P] is to understand Zinn-Justin's work [Z] from the viewpoint of exact WKB analysis initiated by Voros [V] or, equivalently, in the framework of Ecalle's theory of resurgent functions [E] which is expected to provide mathematical foundation of exact WKB analysis (cf. [DDP]). At that time both Sato and Kawai must have recognized that the following intimate relationship should exist between exact WKB analysis and microlocal analysis: In the cotangent bundle, a typical field of microlocal analysis, the role of the base manifold and that of the fiber are not fully interchangeable. For the fiber (i.e., cotangential component) only its direction is relevant. In other words, the dimension of the fiber is essentially smaller than that of the base manifold by one. (This asymmetry originates from the fact that "modulo analytic functions" is assumed at the beginning of discussions in microlocal analysis.) Exact WKB analysis can then be considered as an attempt to retrieve this lost one dimension of the fiber, because not only the behavior near infinity of a Borel transformed WKB solution but also its global structure as an analytic function are relevant in exact WKB analysis. As Kawai himself refers in an article written in Japanese, he started his research on exact WKB analysis with having this relationship between exact WKB analysis and microlocal analysis in mind and, in addition, dreaming that algebraic analysis of singular perturbation theory would help us to establish the structure theorem for non-linear differential equations in some future.

Thus Kawai has been engaged mainly in the research of exact WKB analysis or algebraic analysis of singular perturbation theory since 1989. He also continues his research on microlocal analysis and mathematical physics even after 1989: For example, the joint work with Struppa and Berenstein on holomorphic solutions and their exponential representations of linear differential equations of infinite order ([86], [90], [101], [102], [110]; here and in what follows [j] designates the reference [j] in *Publications of Professor Takahiro*

Kawai), the long-lasting joint work with Stapp on infra-red divergence problem in quantum electrodynamics from the viewpoint of microlocal analysis ([92], [96], [97], [98], [104], [113]), and so on. In what follows, however, I would like to recall and briefly explain the work of Kawai on exact WKB analysis only, as it is the principal theme in the latter part of his research activities.

Around 1989 mainly discussed in exact WKB analysis were the following one-dimensional Schrödinger equation with a large parameter η

$$\left(\frac{d^2}{dx^2} - \eta^2 Q(x)\right)\psi(x,\eta) = 0, \tag{1}$$

and its WKB solutions

$$\psi_\pm(x,\eta) = \exp\left(\pm\eta\int_{x_0}^{x}\sqrt{Q(x)}\,dx\right)\sum_{l=0}^{\infty}\psi_{\pm,l}(x)\eta^{-(l+1/2)}. \tag{2}$$

A key idea of exact WKB analysis is to consider the Borel transform of $\psi_\pm$ defined by

$$\psi_{\pm,B}(x,y) = \sum_{l=0}^{\infty}\frac{\psi_{\pm,l}(x)}{\Gamma(l+1/2)}\left(y+y_\pm(x)\right)^{l-1/2} \text{ with } y_\pm(x) = \pm\int_{x_0}^{x}\sqrt{Q(x)}\,dx \tag{3}$$

and to provide $\psi_\pm$ with an analytic meaning through its Borel sum

$$\Psi_\pm = \int_{-y_\pm(x)}^{\infty} e^{-\eta y}\psi_{\pm,B}(x,y)\,dy. \tag{4}$$

Although a WKB solution itself is divergent, its Borel transform does converge and defines an analytic function. By the use of the Borel transform $\psi_{\pm,B}$, WKB method becomes 'exact' to the effect that exponentially small terms appearing in the description of Stokes phenomena can be neatly handled. In fact, we can understand a Stokes phenomenon as an interaction between the integration path of the Borel sum and singular points of the Borel transform.

The relevance of Borel summation to WKB method was first observed by Bender-Wu [BW]. Together with Sato, T. Aoki and me, Kawai first tried to validate the conjecture of Bender-Wu on the secular equation for anharmonic oscillators by using exact WKB analysis à la Voros [V]. The details of this trial is reported in [94]. Through this trial Kawai also discovered the following intriguing relation between exact WKB analysis and microlocal analysis ([87]): A formal transformation which brings (1) to the Airy equation near a simple turning point (i.e., a simple zero of the potential $Q(x)$) can be endowed with an analytic meaning as a microdifferential operator (in the sense of [19]) if one considers its action on the Borel transform of a WKB solution. In particular, this entails that the singularity structure of a Borel transformed WKB solution of (1) near a simple turning point is the same as that of the Airy equation. Consequently it gives a new analytic proof of Voros' connection formula

$$\begin{cases} \Psi_+ \longrightarrow \Psi_+ \pm i\Psi_- \\ \Psi_- \longrightarrow \Psi_- \end{cases} \quad \text{or} \quad \begin{cases} \Psi_+ \longrightarrow \Psi_+ \\ \Psi_- \longrightarrow \Psi_- \pm i\Psi_+ \end{cases} \tag{5}$$

on a Stokes curve (i.e., an integral curve of the direction field $\operatorname{Im}\sqrt{Q(x)}dx = 0$) emanating from a simple turning point. (Here WKB solutions $\psi_\pm$ are normalized so that x_0 in (2) is a simple turning point in question.) This result of Kawai visualizes in a clear-cut way the existence of intimate relation between exact WKB analysis and microlocal analysis in the very fundamental part of the theory. Furthermore, pursuing Sato's suggestion that the argument of Voros should be regarded as a tool for understanding the general global structure of differential equations beyond the framework of eigenvalue problems, we next succeeded in finding a recipe for explicit computation of the monodromy group of a second order differential equation of Fuchsian type by applying exact WKB analysis ([95]); it claims that the monodromic structure of (1) can be described in terms of contour integrals of the logarithmic derivative of the WKB solution (to be more precise, of its odd part; see Chapters 2 and 3 of our monograph [B3] for details). As is exemplified by this impressive result, exact WKB analysis is really 'exact'.

After this success, Kawai, Aoki and I tried to extend exact WKB analysis in two different directions. First, suggested by M. Jimbo, we began to apply exact WKB analysis to the study of isomonodromic deformations of linear differential equations and Painlevé equations associated with them. To explain the results we have obtained, let us consider the simplest case, that is, the case of the following Painlevé I equation with a large parameter η here:

$$\frac{d^2\lambda}{dt^2} = \eta^2(6\lambda^2 + t). \tag{P_{I}}$$

This non-linear equation, or its Hamiltonian form

$$\frac{d\lambda}{dt} = \eta\frac{\partial K_{\mathrm{I}}}{\partial \nu}, \qquad \frac{d\nu}{dt} = -\eta\frac{\partial K_{\mathrm{I}}}{\partial \lambda} \tag{6}$$

with $K_{\mathrm{I}} = (\nu^2 - 4\lambda^3 - 2t\lambda)/2$, describes the compatibility condition of the following system of linear differential equations:

$$\left(\frac{\partial^2}{\partial x^2} - \eta^2 Q_{\mathrm{I}}\right)\psi = 0 \quad \text{with} \quad Q_{\mathrm{I}} = 4x^3 + 2tx + 2K_{\mathrm{I}} - \frac{\eta^{-1}\nu}{x-\lambda} + \frac{3\eta^{-2}}{4(x-\lambda)^2}, \tag{SL_{I}}$$

$$\frac{\partial\psi}{\partial t} = A_{\mathrm{I}}\frac{\partial\psi}{\partial x} - \frac{1}{2}\frac{\partial A_{\mathrm{I}}}{\partial x}\psi \quad \text{with} \quad A_{\mathrm{I}} = \frac{1}{2(x-\lambda)} \tag{D_{I}}$$

(cf. [JMU], [O]). Thanks to the existence of the large parameter η, (P_{I}) has a formal solution of the form

$$\lambda^{(0)}(t,\eta) = \lambda_0(t) + \eta^{-2}\lambda_2(t) + \eta^{-4}\lambda_4(t) + \cdots \tag{7}$$

called a “0-parameter solution”. Then, substituting $\lambda^{(0)}(t,\eta)$, or the corresponding solution $(\lambda^{(0)}(t,\eta), \nu^{(0)}(t,\eta))$ of (6), into the coefficients of the linear equations (SL_{I}) and (D_{I}), we find that the point $x = \lambda_0(t)$ is a double turning point (i.e., a double zero of the top order part of Q_{I}) of (SL_{I}). Furthermore, if we define a turning point (resp., Stokes curve) of (P_{I}) as a turning point (resp., Stokes curve) of the Fréchet derivative of (P_{I}) at a 0-parameter solution $\lambda^{(0)}(t,\eta)$, we can also confirm that the double turning point $x = \lambda_0(t)$ and a simple turning point $x = -2\lambda_0(t)$ of (SL_{I}) are connected by a Stokes curve γ when and only when t lies in a Stokes curve of (P_{I}). Such relations between the Stokes geometry (i.e., turning points and Stokes curves) of (SL_{I}) and that of (P_{I}) are universally observed for all Painlevé equations (P_J) and their underlying linear equations (SL_J) $(J = \mathrm{I}, \ldots, \mathrm{VI})$ ([91], [99]).

Making use of these geometric features, Kawai and I proved in [99] (see also its announcement paper [91]) that we can construct an invertible transformation which brings (SL_J) to (SL_{I}) on a neighborhood of the above mentioned Stokes curve γ and further that it describes an equivalence between a 0-parameter solution of (P_J) and that of (P_{I}) near a simple turning point of (P_J). Later in [103] and [105] we extended this result to a “2-parameter solution” (or “instanton-type solution”) of (P_J) of the form

$$\lambda_J(t,\eta) = \lambda_0(t) + \eta^{-1/2} \sum_{j \geq 0} \eta^{-j/2} \left(\sum_{k=0}^{j+1} a^{(j)}_{2k-j-1}(t) \exp((2k-j-1)\phi(t)\eta) \right), \tag{8}$$

which was constructed through multiple-scale analysis in [100]. (Here $\phi(t)$ and $a^{(j)}_l(t)$ are functions determined by the equation (P_J).) These results may be regarded as a non-linear counterpart of the fundamental fact that any linear differential equation of the form (1) can be brought to the Airy equation near a simple turning point by a suitable formal transformation. They also exemplify the efficiency of exact WKB analysis in clarifying the structure of non-linear differential equations.

The other direction we aimed after the success of exact WKB analysis for second order Fuchsian equations is to extend this analysis to higher order linear differential equations of the form

$$P(x, \partial_x, \eta)\psi = \left(\frac{d^m}{dx^m} + q_1(x)\eta \frac{d^{m-1}}{dx^{m-1}} + \cdots + q_m(x)\eta^m \right) \psi = 0 \tag{9}$$

where η again denotes a large parameter. In this direction we first succeeded in clarifying the local structure of WKB solutions of (9) near its simple turning points by using a local decomposition theorem for $P(x, \partial_x, \eta)$ together with the local transformation theorem to the Airy equation for second order differential equations ([93]). Note that this result is generalized several years later to “microdifferential operators of WKB type”, i.e., a class of operators including differential operators of infinite order and integral operators where a large parameter η is introduced in an appropriate manner ([114], [116]).

On the other hand, global structure of the Stokes geometry for higher order equations (9) is quite different from that for second order equations (1). As a matter of fact, using the following very simple third order equation

$$\left(\frac{d^3}{dx^3} + 3\eta^2\frac{d}{dx} + 2ix\eta^3\right)\psi = 0, \tag{10}$$

Berk et al. pointed out in [BNR] that, in addition to ordinary Stokes curves, we need to introduce "new Stokes curves" to describe where Stokes phenomena for Borel resummed WKB solutions occur. A natural question then arises: What is a new Stokes curve? Bringing in the microlocal analysis to the study of singularities of Borel transformed WKB solutions, Kawai gave the following intriguing answer to the above question in [93]: Let $P(x, \partial_x, \partial_y)$ be the Borel transform of the operator $P(x, \partial_x, \eta)$ in question and take a self-intersection point of the bicharacteristic curve of $P(x, \partial_x, \partial_y)$. If we call the x-component of such a self-intersection point (i.e., the projection of a self-intersection point to the x-space) a "new turning point", then a new Stokes curve is a Stokes curve (in the ordinary sense) that emanates from a new turning point —. A new turning point introduced in [93] is called a "virtual turning point" in recent literatures. Virtual turning points and new Stokes curves play a crucially important role in obtaining a complete description of the Stokes geometry for higher order equations. In [106], [107], [119] etc. we discussed how to describe the complete Stokes geometry for some concrete examples of higher order equations or microdifferential operators of WKB type. Note that a higher order differential equation generically has infinitely many virtual turning points, only finitely many of which are relevant to its Stokes geometry. It is a difficult task to determine which virtual turning points are relevant to the Stokes geometry for a given differential equation. However, if an integral representation of solutions is available as is the case with Laplace type equations, we can determine which virtual turning points are relevant by investigating the configuration of steepest descent paths passing through saddle points of the integral representation. In [111] (cf. [115] also) we extended this method to higher order differential equations with polynomial coefficients by considering what we call "exact steepest descent paths" for the inverse Laplace transform of a WKB solution of the Laplace transformed equation (with respect to the independent variable x) instead of ordinary steepest descent paths. (Roughly speaking, in addition to an ordinary steepest descent path, we consider additional steepest descent paths which are bifurcated from an ordinary one at its crossing points with Stokes curves of the Laplace transformed equation. An exact steepest descent path is by definition the collection of such bifurcated steepest descent paths.) Thus, from the practical point of view, the determination of relevant virtual turning points is now mostly possible with the aid of a computer. As an application of exact WKB analysis for higher order differential equations we also discussed the transition probability for level crossing problems of Landau-Zener type in [112].

Currently Kawai has been trying to generalize exact WKB analysis to higher order Painlevé equations together with his many collaborators such as Aoki, T. Koike, N. Honda, Y. Nishikawa, S. Sasaki, me, and others. So far we have obtained the following results:

(I) We confirmed that the Stokes geometry of a higher order Painlevé equation is related to that of its underlying linear equation in a similar manner to the case of traditional (i.e., second order) Painlevé equations when the underlying linear equation is of second order, or equivalently a 2×2 system ([117]).

(II) Using the above geometric features, we succeeded in proving an equivalence between a 0-parameter solution of a higher order Painlevé equation and that of the traditional Painlevé I equation (P_{I}) near a simple turning point of the first kind when the underlying linear equation is of second order ([118], [121]).

(III) Very recently the above equivalence result (II) near a simple turning point of the first kind is extended to an instanton-type solution containing sufficiently many free parameters. (The paper is now in preparation.)

(IV) As was first observed by Nishikawa, a new Stokes curve appears in general also for a higher order Painlevé equation similarly to the case of higher order linear equations. In [117] we explained the mechanism of appearance of new Stokes curves by using the Stokes geometry of the underlying linear equation when it is of second order.

In this way several known results for traditional Painlevé equations and higher order linear equations are successfully generalized to higher order Painlevé equations when the underlying linear equations are of second order. However, if the underlying linear equation is of higher order, the analysis becomes much more difficult. For example, in the case of Noumi-Yamada systems, a typical example of higher order Painlevé equations whose underlying linear equations are of higher order, virtual turning points of the underlying linear equation are also relevant to the description of Stokes geometry of a higher order Painlevé equation, as was observed in [120] and [122]. Thus there still remain several important problems to be discussed; completion of exact WKB analysis for Noumi-Yamada systems, local analysis near a turning point of the second kind which is a turning point peculiar to higher order Painlevé equations, global structure of the Stokes geometry (including new Stokes curves and virtual turning points) of higher order Painlevé equations, and so on. Kawai is now working intensively on these problems.

As we have observed so far, Kawai introduced the technique of microlocal analysis into exact WKB analysis and consequently combined microlocal analysis and exponential asymptotics. This is one of the greatest contributions of Kawai. In all these works one can find his keen insight and strong enthusiasm for mathematics.

On several occasions I have heard the following phrases from Kawai: "This is an appropriate problem for me to consider as an amusement for my old age!" I think he has now big stock of such problems. I sincerely wish he may enjoy mathematics and stimulate us as ever even after he is over sixty years old.

References

[BW] C.M. Bender and T.T. Wu: Anharmonic oscillator, *Phys. Rev.*, **184**(1969), 1231–1260.

[BNR] H.L. Berk, W.M. Nevins and K.V. Roberts: New Stokes' line in WKB theory, *J. Math. Phys.*, **23**(1982), 988–1002.

[DDP] E. Delabaere, H. Dillinger et F. Pham: Résurgence de Voros et périodes des courbes hyperelliptiques, *Ann. Inst. Fourier (Grenoble)*, **43**(1993), 163–199.

[E] J. Ecalle: Cinq applications des fonctions résurgentes, Prépublication d'Orsay 84T62, Univ. Paris-Sud, 1984.

[JMU] M. Jimbo, T. Miwa and K. Ueno: Monodromy preserving deformation of linear ordinary differential equations with rational coefficients. I, *Physica D*, **2**(1981), 306–352.

[O] K. Okamoto: Isomonodromic deformation and Painlevé equations, and the Garnier systems, *J. Fac. Sci. Univ. Tokyo*, Sect. IA, **33**(1986), 575–618.

[P] F. Pham: Resurgence, quantized canonical transformations, and multi-instanton expansions, *Algebraic Analysis*, Vol. II, Academic Press, 1988, pp. 699–726.

[V] A. Voros: The return of the quartic oscillator. The complex WKB method, *Ann. Inst. Henri Poincaré*, **39**(1983), 211–338.

[Z] J. Zinn-Justin: Instantons in quantum mechanics: Numerical evidence for a conjecture, *J. Math. Phys.*, **25**(1984), 549–555.

Part II

Contributed papers

Virtual turning points — A gift of microlocal analysis to the exact WKB analysis *

Takashi Aoki[1], Naofumi Honda[2], Takahiro Kawai[3], Tatsuya Koike[4], Yukihiro Nishikawa[5], Shunsuke Sasaki[6], Akira Shudo[7], and Yoshitsugu Takei[8]

[1] School of Science and Engineering, Kinki University, Higashi-Osaka 577-8502, Japan
[2] Department of Mathematics, Graduate School of Science, Hokkaido University, Sapporo 060-0810, Japan
[3] Research Institute for Mathematical Sciences, Kyoto University, Kyoto 606-8502, Japan
[4] Department of Mathematics, Graduate School of Science, Kyoto University, Kyoto 606-8502, Japan
[5] Hitachi Ltd., Asagaya-kita 2-13-2, Suginami-ku, Tokyo 166-0001, Japan
[6] Mitsubishi UFJ Securities Co. Ltd., Marunouchi 2-4-1, Chiyoda-ku, Tokyo 100-6317, Japan
[7] Department of Physics, Graduate School of Science, Tokyo Metropolitan University, Hachioji 192-0397, Japan
[8] Research Institute for Mathematical Sciences, Kyoto University, Kyoto 606-8502, Japan

Summary. Several aspects of the notion of virtual turning points are discussed; its background, its relevance to the bifurcation phenomena of a Stokes curve, its importance in the analysis of the Noumi-Yamada system (a particular higher order Painlevé equation) and a concrete recipe for locating them. Examples given here make it manifest that virtual turning points are indispensable in WKB analysis of higher order linear ordinary differential equations with a large parameter.

0 Introduction

Microlocal analysis and the exact WKB analysis are intimately related and they are often complementary. A typical example is the exact steepest descent method ([AKT3], [AKT4], [T]), where a global version of the quantized Legendre transformation is given in terms of exact steepest descent paths. Here in

Received 13 March, 2006. Accepted 30 April, 2006.
* This work has been partially supported by JSPS Grants-in-Aid No. 14077213, No. 14340042, No. 15540190, No. 15740088 and No. 16540148.

this report we discuss another important example of their interactions, namely the notion of virtual turning points. Since this notion does not find any precedents in traditional asymptotic analysis, we first explain why it is needed. As pointed out by Silverstone ([S]), notorious ambiguities in the connection problems in WKB analysis are resolved if we make use of the Borel resummation method; in a word we have to first specify the region (the so-called Stokes region) where the Borel sum of a WKB solution is well-defined. Parenthetically we note that the importance of the Borel resummation in WKB analysis is also shown by Bender-Wu ([BW]), Voros ([V]), Zinn-Justin ([Z]), Pham ([P]) and others from several points of view and that the exact WKB analysis means the WKB analysis based on the Borel resummation. (See [DDP], where the wording "exact semi-classical expansion" is also used.) In the description of Stokes regions for a second order linear ordinary differential operator with a large parameter η that is of the form $P = P(x, \eta^{-1}d/dx) = P(x, \eta^{-1}\xi)$, we need to consider only Stokes curves emanating from (ordinary, or traditional) turning points; it suffices to consider the union of integral curves of the direction field

$$\mathrm{Im}(\xi_j(x) - \xi_k(x))dx = 0 \tag{1}$$

that emanate from some traditional turning point a of type (j, k), i.e., a point a satisfying

$$\xi_j(a) = \xi_k(a), \tag{2}$$

where $\xi_j(x)$ and $\xi_k(x)$ are characteristic roots of the operator P. For higher order operators, however, the totality of Stokes curves emanating from turning points (i.e., points satisfying (2)) does not suffice to describe the Stokes region as Berk-Nevins-Roberts ([BNR]) first pointed out; we need a "new Stokes curve" that does not emanate from a traditional turning point if we want to find correct Stokes regions. As we discuss in Section 1, the needed "new Stokes curves" emanate from "new turning points", which were first detected in [AKT1] through microlocal analysis. The wording "a new turning point" has been superseded by "a virtual turning point" in recent literature, and here, and in what follows, we use this new wording. As the effect of a virtual turning point is inert in its immediate vicinity, we discuss in Section 2 its relevance to the bifurcation of a Stokes curve so that its effect may become impressively visible. In Section 3 we discuss how this relevance gives a neat interpretation of the strange and intriguing phenomenon that one of us (S. Sasaki; [Sa1], [Sa2]) has found in analyzing the Noumi-Yamada system ([NY]) with the help of a computer. In Appendix we briefly describe how to find a virtual turning point with the help of a computer.

As we discuss only some illuminating examples in this report, for the detailed argument we refer the reader to [AKT1], [AKT2], [AKT5], [AKSST], [Ho], [Sa1], and [Sa2]. We use the same notions and notations used in [KT].

1 The background of the introduction of a virtual turning point

Let P denote the following third order operator with a large parameter $\eta(\gg 1)$:

$$\eta^{-3}\frac{d^3}{dx^3} + 3\eta^{-1}\frac{d}{dx} + 2ix. \tag{3}$$

In what follows we call this operator the BNR operator after Berk-Nevins-Roberts ([BNR]), who first observed the importance of this operator in WKB analysis. The turning points of (3) are $x = \pm 1$, and the traditional Stokes geometry is given in Fig. 1.

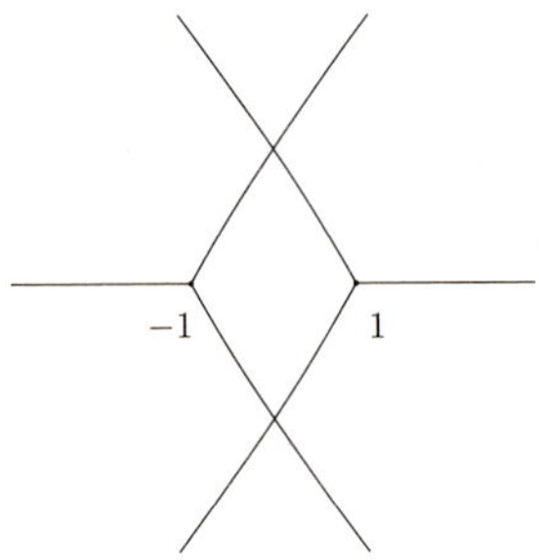

Fig. 1.

As Berk et al. ([BNR]) observed, we have to add two Stokes rays γ_1 and γ_2 to obtain the correct Stokes regions. See Fig. 2.

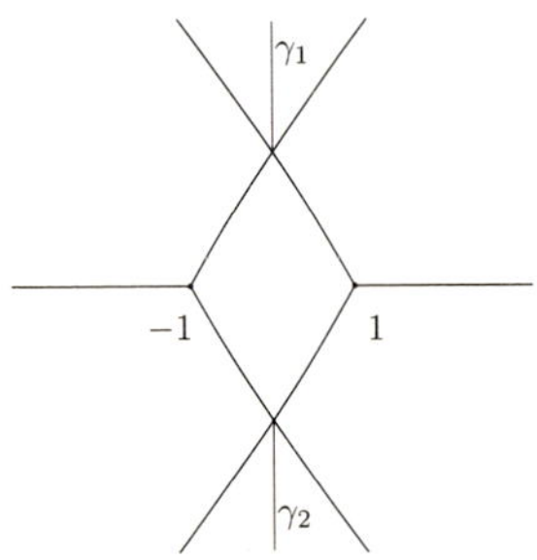

Fig. 2.

Then a natural question to be raised is: From what points do these rays emanate? Let us try to answer this question in terms of the singularity structure of the Borel transform $\psi_B(x, y)$ of a WKB solution $\psi(x, \eta)$ of the equation $P\psi = 0$. In view of the definition of the Borel sum, we know ([V]) that the Stokes phenomenon of ψ is due to the interplay of singularities of $\psi_B(x, y)$: the Stokes phenomenon is observed when the path $C(x)$ of integration used to define the Borel sum $\int_{C(x)} \exp(-y\eta)\psi_B(x, y)dy$ is hit by some "related" singular point of $\psi_B(x, y)$. Hence it is reasonable to surmise that a starting point of a Stokes curve should be the x-component of a point where two "related", or “cognate”, singularities of $\psi_B(x,y)$ coalesce. Actually a traditional turning point of a Schrödinger operator $\widetilde{P} = d^2/dx^2 - \eta^2 Q(x)$ is of this character: the Borel transform $\varphi_B(x,y)$ of a WKB solution φ of the equation $\widetilde{P}\varphi = 0$ has two singularities $s_\pm = \{(x,y); y = \pm \int_a^x \sqrt{Q}dx\}$ with $Q(a) = 0$, and they coalesce at $(x,y) = (a,0)$. Let us now raise the following question: In what sense are s_+ and s_- cognate? To answer this question, we have to understand the structure of singularities of $\varphi_B(x,y)$. Fortunately enough, a clear-cut answer to this question has been given by microlocal analysis: Assuming that the point a is a simple turning point, i.e., a is a simple zero of the potential $Q(x)$, there exists a non-singular bicharacteristic strip of the Borel transform $\widetilde{P}_B$ of $\widetilde{P}$ whose projection to the base manifold $\mathbb{C}^2_{(x,y)}$ is $s_+ \cup s_-$ near $(x,y) = (a,0)$. Here $\widetilde{P}_B$ is a partial differential operator given by $\partial^2/\partial x^2 - Q(x)\partial^2/\partial y^2$, and a bicharacteristic strip is, by definition, a solution curve of the Hamilton-Jacobi equation associated with $\widetilde{P}_B$. (See e.g. [CH, p.558].) Its projection to the base manifold is called a bicharacteristic curve. A fundamental result in microlocal analysis ([H], [SKK]) asserts that each bicharacteristic strip is the most “elementary” carrier of the singularities of solutions of a linear partial differential equation with simple characteristics. Hence a singular point in s_+ and that in s_- should be cognate, as they are both the projections of points in one and the same connected non-singular curve, a bicharacteristic strip. The next question is, then: Are there any other pairs of cognate singularities that coalesce? In a generic situation it is rather difficult to find such pairs. But the dimension of the base manifold is 2 in our case, and hence a bicharacteristic curve “generically” forms a self-intersection point.

Remark 1. We encounter self-intersection points of bicharcteristics “normally” even in higher dimensional case if we start with a subholonomic system with a large parameter instead of an ordinary differential equation with a large parameter. See [Sh] for such equations.

For example, in the case of the operator P given by (3), the associated bicharacteristic strip passing through $(x,y;\xi,\eta) = (1,0;-i,1)$ is given by

$$\begin{cases} x(t) = -4(t+1/2)(t^2+t-1/2) \\ y(t) = -6it^2(t+1)^2 \\ \xi(t) = -2it - i \\ \eta(t) = 1. \end{cases} \tag{4}$$

Hence its projection to the base space forms a (unique) self-intersection point at $(x,y) = (0,-3i/2)$. The situation is schematically illustrated in Fig. 3 with an appropriate labelling of solutions of the following characteristic equation of P:

$$\xi^3 + 3\eta^2\xi + 2ix\eta^3 = 0 \text{ with } \eta = 1. \tag{5}$$

The label j attached to a curve in Fig. 3 indicates that the curve is determined by the factor $\xi - \xi_j(x)\eta$ of the characteristic polynomial written in the form of $\prod_{l=1}^{3}(\xi - \xi_l(x)\eta)$. Note that the point A(resp., B) corresponds to the traditional

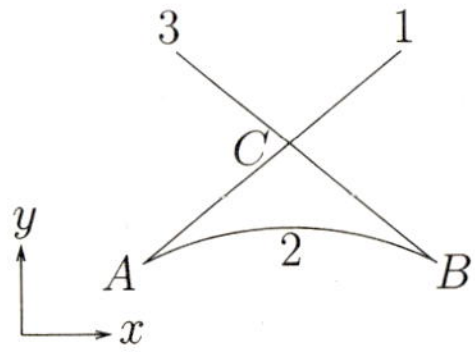

Fig. 3.

turning point $x = -1$(resp., 1). Thus the x-component of the point C, i.e., $x = 0$, is expected to play a role similar to a traditional turning point in the description of Stokes regions. Fortunately the actual situation is exactly as expected: the Stokes curve of type (1,3) that emanates from $x = 0$, that is,

$$\mathrm{Im} \int_0^x (\xi_1(x) - \xi_3(x))dx = 0 \tag{6}$$

contains Stokes rays γ_1 and γ_2 in Fig. 2. Furthermore, by using the reasoning of Voros ([V, p.244]) we find that the Stokes curve is inert near $x = 0$ (until it hits a crossing point of other Stokes curves) in the sense that no Stokes phenomena are observed there. We also note that the Voros argument ceases to work at the crossing points of Stokes curves as the singularity originating from the factor $\xi - \xi_2(x)\eta$ intervenes there. To emphasize the inert character of a portion of a Stokes curve, we usually use a dotted line to describe it. See Fig. 4.

Fig. 4.

Thus we have found the correct Stokes regions shown in Fig. 2 by making use of an ordinary Stokes curve that emanates from the hitherto undetected point $x = 0$, and we are now entitled to call the point $x = 0$

a virtual turning point (of the BNR operator),

a turning point which cannot be detected with the naked eye but whose effect may resurge when it hits a crossing point of Stokes curves, due to the interplay of three singularities that occurs there. Although we have so far discussed the particular operator (3), the reasoning goes as well in the general case.

Definition 1.1. ([AKT1],[AKT2],[AKKT]) *Let $P = P(x, \eta^{-1}d/dx, \eta^{-1})$ be a linear differential operator with a large parameter η that is of the following form:*

$$P_0(x, \eta^{-1}d/dx) + \eta^{-1}P_1(x, \eta^{-1}d/dx) + \eta^{-2}P_2(x, \eta^{-1}d/dx) + \cdots . \quad (7)$$

Assume that its Borel transform $P_B = P(x, \partial_y^{-1}\partial_x, \partial_y^{-1})$ is a well-defined microdifferntial operator and that its traditional turning points are all simple (in the sense of [AKKT]). Then a virtual turning point of P is, by definition, the x-component of a self-intersection point of a bicharacteristic curve associated with the operator P_B. If the crossing bicharacteristic curves are respectively associated with the factor $(\eta^{-1}\xi - \xi_j(x))$ and $(\eta^{-1}\xi - \xi_k(x))$ of $P_0(x, \zeta) = \prod_l (\eta^{-1}\xi - \xi_l(x))$, we say the virtual turning point is of type (j, k).

Definition 1.2. *Let τ be a virtual turning point of type (j, k) of the operator P in Definition 1.1. Then an integral curve of the direction field*

$$\mathrm{Im}(\xi_j(x) - \xi_k(x))dx = 0 \quad (8)$$

that emanates from τ is called a new Stokes curve of type (j, k), or just a Stokes curve of type (j, k).

Remark 2. A bicharacteristic strip is a curve in the complex cotangent bundle. Hence a virtual turning point is a complex-analytic notion; unlike Stokes curves or their crossing points, real structure is irrelevant. To avoid the possible confusion of the reader, we note that the assertion contrary to this remark in [HLO, p.2292, l.3] originates from their erroneous quotation of the wording 'a new turning point' ([HLO, p.2291]). We also note that they make a misleading claim in p.2291, $l.8 \sim l.10$; actually $f_0(a)$ and $f_2(a)$ coalesce at a "new turning point". (Logically speaking, their argument in p.2291, l.2 resulted in counting a virtual turning point as a (traditional) turning point, contrary to their intention. At the same time Fig. 4 of [HLO] indicates that they overlooked the relation $f_0(0) = f_2(0)$. They could have avoided losing their way in the logical labyrinth if they had noticed this relation.) Parenthetically we also note that we can actually characterize (either traditional or virtual) turning points by the comparison of phase functions evaluated at different saddle points if the differential equation in question admits some "nice" integral representation of its solutions. See [Sh] for the concrete examples related to the quantized Hénon map.

Remark 3. If some turning points of the operator P in question are double, some care is needed in the definition of virtual turning points. See [AKT5] for the details. We note that the care is needed due to the complexity of microlocal structure of solutions of the equation $P_B\psi_B = 0$; the operator P_B is an operator with multiple characteristics in this case.

2 The relevance of a virtual turning point to the bifurcation phenomenon of a Stokes curve

The Stokes geometry given in Fig. 4 of the BNR operator is described with η being real and positive. Let us now study what happens when we change $\arg\eta$. This amounts to considering the operator

$$\eta^{-3}\frac{d^3}{dx^3} + 3a^2\eta^{-1}\frac{d}{dx} + 2ia^3x \tag{9}$$

with a parameter a satisfying $|a| = 1$, keeping η to be positive. Note that Stokes curves do depend on $\arg\eta$ by their definition but that (virtual and traditional) turning points remain fixed. (See Remark 2.) The resulting Stokes geometry for (i) $\arg\eta = (\frac{1}{2} - \frac{1}{12})\pi$, (ii) $\arg\eta = \frac{1}{2}\pi$ and (iii) $\arg\eta = (\frac{1}{2} + \frac{1}{12})\pi$ are respectively given in Fig. 5 (i), (ii) and (iii).

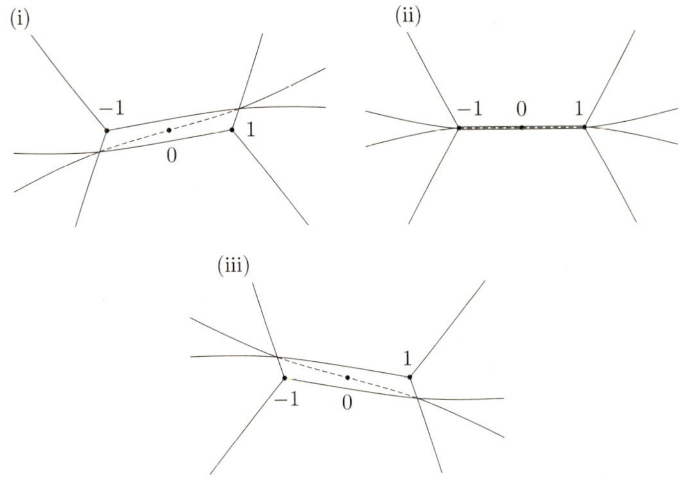

Fig. 5. The Stokes geometry of the BNR operator for (i) $\arg\eta = (\frac{1}{2} - \frac{1}{12})\pi$, (ii) $\arg\eta = \frac{1}{2}\pi$ and (iii) $\arg\eta = (\frac{1}{2} + \frac{1}{12})\pi$.

The bifurcation of a Stokes curve observed in Fig. 5 (ii) is due to the singularity that the direction field (1) acquires at a simple turning point. Impressively enough, the smooth transition between Fig. 5 (i) and Fig. 5 (iii) via Fig. 5 (ii) is attained with the addition of Stokes curves emanating from the virtual turning point $x = 0$. One should observe some clumsy transition if they were not added. A subtle and interesting fact is that Fig. 5 (ii) switches the relative location of a Stokes curve emanating from a traditional turning point and that from a virtual turning point. As we will see in Section 3, this fact plays an important role in understanding the intriguing fact which Sasaki ([Sa1]) has found in the study of the Noumi-Yamada system.

3 Deformation of the linear differential equations that underlie the Noumi-Yamada system

The Noumi-Yamada system ([NY]) is one of the Painlevé hierarchies, i.e., a family of higher order non-linear equations whose first member coincides with one of the second order Painlevé equations. The first member of the Noumi-Yamada hierarchy is the following $(NY)_2$, which is a symmetric form of the fourth Painlevé equation ($P_{\rm IV}$);

$$(NY)_2 : \eta^{-1}\frac{df_j}{dt} = f_j(f_{j+1} - f_{j+2}) + \alpha_j \quad (j = 0, 1, 2), \tag{10}$$

where $f_j = f_{j-3}$ $(j = 3, 4)$ and α_j $(j = 0, 1, 2)$ are constants that satisfy

$$\alpha_0 + \alpha_1 + \alpha_2 = \eta^{-1}. \tag{11}$$

Its underlying Lax pair, i.e., an overdetermined system of linear differential equations whose compatibility conditions are given by (10), is as follows:

$$-\eta^{-1}x\frac{\partial}{\partial x}\begin{pmatrix}\psi_0\\ \psi_1\\ \psi_2\end{pmatrix} = \begin{pmatrix}(2\alpha_1+\alpha_2)/3 & f_1 & 1\\ x & (-\alpha_1+\alpha_2)/3 & f_2\\ xf_0 & x & -(\alpha_1+2\alpha_2)/3\end{pmatrix}\begin{pmatrix}\psi_0\\ \psi_1\\ \psi_2\end{pmatrix}, \tag{12}$$

$$-\eta^{-1}\frac{\partial}{\partial t}\begin{pmatrix}\psi_0\\ \psi_1\\ \psi_2\end{pmatrix} = \begin{pmatrix}f_2 - t/2 & -1 & 0\\ 0 & f_0 - t/2 & -1\\ -x & 0 & f_1 - t/2\end{pmatrix}\begin{pmatrix}\psi_0\\ \psi_1\\ \psi_2\end{pmatrix}. \tag{13}$$

In what follows we regard (12) as the main equation containing a parameter t that is to be deformed obeying (13), and we study the Stokes geometry of (12) for each t. Our earlier study ([KT]) of the connection problems for the Painlevé transcendents indicates that, if t_0 lies on a Stokes curve γ properly defined for the Painlevé equation, the Stokes geometry of (12) should degenerate in the sense that some turning points are connected by a Stokes curve. A more accurate statement would be that the Stokes geometry topologically changes off the curve γ near $t = t_0$; actually the degeneration of the Stokes geometry is a symptom of such a change. In studying the Stokes geometry of (12), Sasaki ([Sa1]) found the following intriguing FACT 3.1:

FACT 3.1. *When t_0 moves along a Stoke curve γ of* (10)*, we observe the following phenomena in the Stokes geometry of* (12)*:*

(i) If t_0 is close to the starting point τ (i.e., a turning point of (10)*) of γ, a double turning point and a simple turning point are connected by a Stokes curve of* (12)*.*

(ii) If t_0 is far away from τ, no (traditional) turning points are connected by a Stokes curve of (12)*.*

The situation is illustrated in Fig. 6, where the Stokes geometry of (12) (with $\alpha_0 = 1 + 0.6i$ and $\alpha_1 = 0.2 - 0.1i$) is drawn at two points (i.e., at (i) $t_0 = -1.6104 - 0.2268i$ and (ii) $t_0 = -1.5783 - 0.4130i$) on one Stokes curve of (10) emanating from a turning point $\tau = -1.6276 - 0.0986i$. As can be readily seen, a double turning point d is connected with a simple turning point s_1 by a Stokes curve of (12) in Fig. 6 (i), while d is no longer connected with any simple turning point s_i in Fig. 6 (ii).

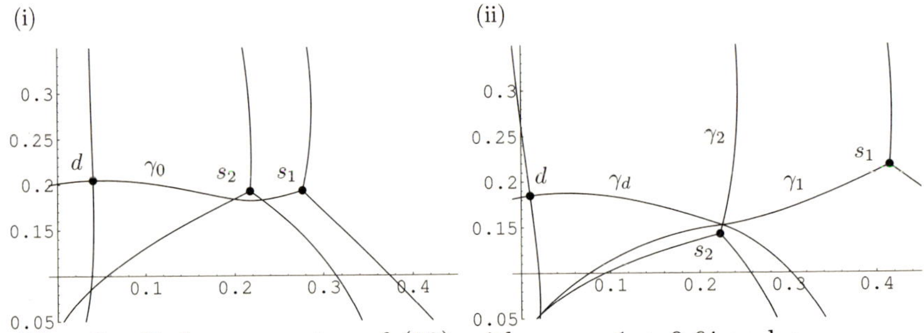

Fig. 6. Stokes geometry of (12) with $\alpha_0 = 1 + 0.6i$ and $\alpha_1 = 0.2 - 0.1i$ for (i) $t_0 = -1.6104 - 0.2268i$ and (ii) $t_0 = -1.5783 - 0.4130i$.

To understand FACT 3.1 properly, we next include relevant virtual turning points in Fig. 6. For this purpose we study the Stokes geometry of (12) for t_0 slightly away from γ: near $t = t_0$ that realizes Fig. 6 (i), the resulting configuration is either one of the following:

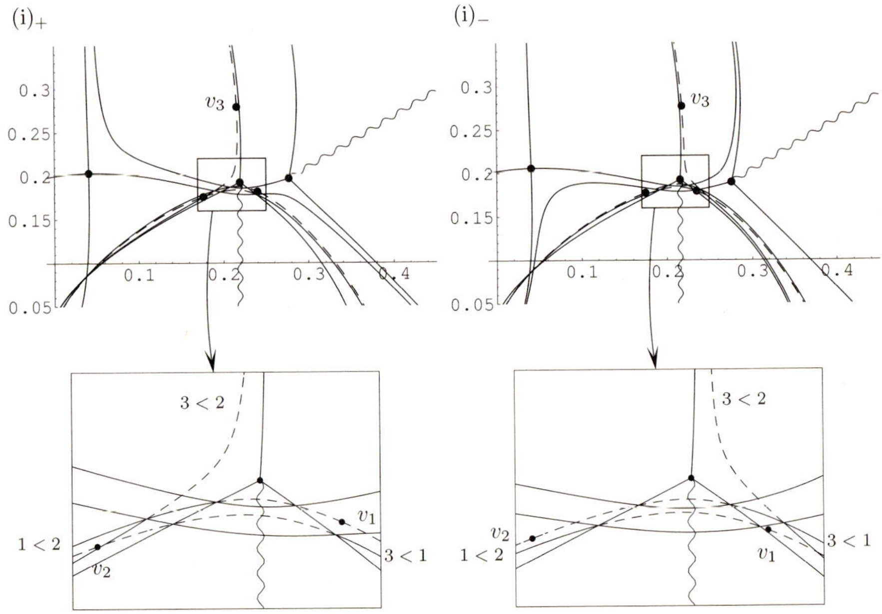

Fig. 7.

Similarly, near $t = t_0$ that realizes Fig. 6(ii), we find

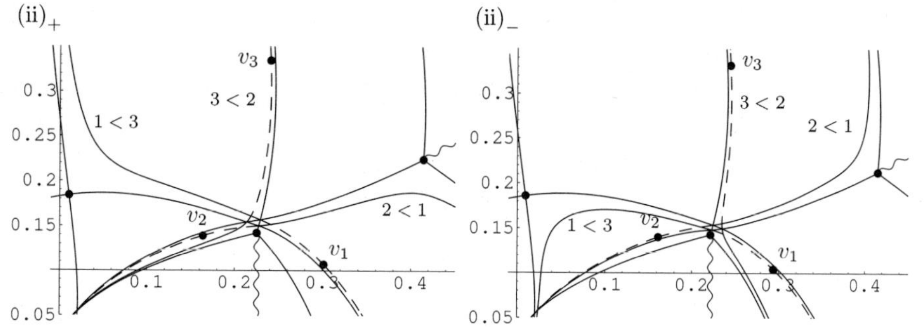

Fig. 8.

Here, and in what follows, a wiggly line designates a cut to fix the branch of a characteristic root, and the symbol $j > k$ attached to a Stokes curve indicates the dominance relation along the Stokes curve. (See, e.g., [AKT1], [AKSST] for the details.)

By letting the parameter t sit on the curve we then obtain the following Fig. 9 through the limiting procedure. (Cf. [AKSST, Fig. 2])

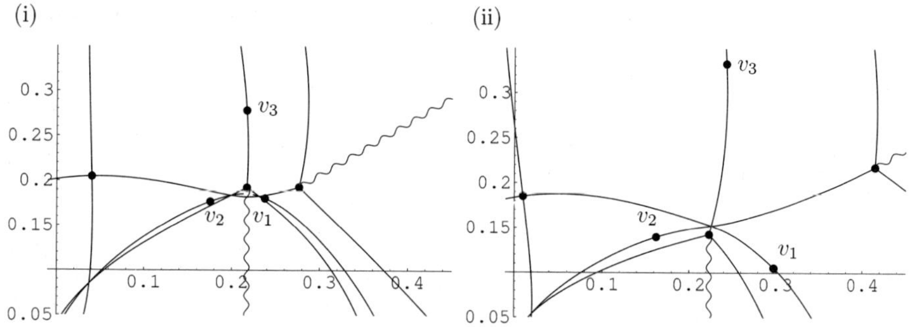

Fig. 9.

We now find the mechanism at the back of FACT 3.1: virtual turning points v_1, v_2 and v_3 should have been taken into account in Fig. 6 (ii). The degeneration of the Stokes geometry observed in Fig. 6 (i), that is, the existence of a pair of a double turning point d and a simple turning point s_1 which are connected by a Stokes curve, is superseded by another degeneration in Fig. 9 (ii), which is caused by the existence of a Stokes curve connecting the turning point d with a virtual turning point v_1 and that connecting s_1 with another virtual turning point v_2. We also note that Fig. 9 (i) and (ii) are switched by the following Fig. 10, which is observed when another simple turning point s_2 hits the Stokes curves in question and causes their bifurcation as we explained in Section 2.

Comparison of Fig. 7, 8 and 9 shows the following:

FACT 3.2. *Resolution of the degeneration in Fig. 9 (ii) by a tiny change of t induces the change of topological configurations of Stokes curves of* (12), *as is observed in Fig. 8* (ii)$_+$ *and* (ii)$_-$ *; it is exactly in the same manner as the result of the resolution of the degeneration observed in Fig. 9 (i).*

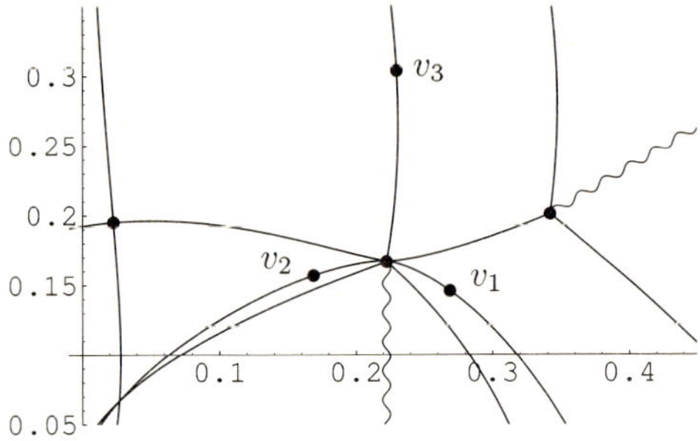

Fig. 10.

Thus we are forced to conclude that the role of virtual turning points is commensurate with that of traditional turning points in describing the Stokes geometry.

The same comparison manifests the smooth transition from Fig. 7 $(\mathrm{i})_+$ (resp. Fig. 7 $(\mathrm{i})_-$) to Fig. 8 $(\mathrm{ii})_+$ (resp. Fig. 8 $(\mathrm{ii})_-$) outside a small neighborhood of the simple turning point s_2; this is what the reasoning in Section 2 predicts, but such a smooth transition can never be observed without virtual turning points.

Remark 4. The Noumi-Yamada system $(NY)_l$ $(l \geq 4)$ is of higher order (higher than the second order), and the so-called Nishikawa phenomena ([KKNT]) are observed in its Stokes geometry. Its investigation requires some subtler study of Stokes geometry of the underlying linear differential equation (a counterpart of (12)). Such a study was initiated by Sasaki ([Sa2]) and its systematization is undertaken by Honda ([Ho]).

Appendix A practical recipe for locating a virtual turning point

If one wants to locate a virtual turning point following Definition 1.1, one needs to solve the Hamilton-Jacobi equation globally. In general it is a formidably difficult task. However, there is a practically satisfactory way to locate a virtual turning point with the help of a computer. For the reader's convenience we briefly describe the recipe. See also [AKKSST], [AKSST] and [Ho]; in particular [Ho] presents a systematic algorithm for describing a complete Stokes geometry for the underlying linear differential equation of $(NY)_4$. Probably the method of Honda ([Ho]) is applicable to general equations, beyond the framework of the Noumi-Yamada system.

To describe the recipe, let us first fix the situation to be considered: Suppose that a Stokes curve γ_1 of type (1,2) that emanates from a turning point τ_1 intersects at a point ι with another Stokes curve γ_2 of type (2,3) emanating from another turning point τ_2. In what follows τ_1 and τ_2 may be either virtual or traditional. (In case τ_1 or τ_2 is a simple turning point we need some care about the cut structure to fix the branches of solutions $\xi_j(x)$ of the characteristic equation.) Having the labeling of Fig. 3 in mind, a point x_* that satisfies

$$\int_{\tau_1}^{x_*} \xi_1 dx = \int_{\tau_1}^{\tau_2} \xi_2 dx + \int_{\tau_2}^{x_*} \xi_3 dx \tag{14}$$

is a virtual turning point. Next let us try to relate x_* with the point ι. Supposing that the cut structure is appropriately introduced if necessary, we use (14) to find the following:

$$\begin{aligned}\int_{\tau_1}^{x_*} (\xi_1 - \xi_2) dx &= \int_{\tau_1}^{\tau_2} \xi_2 dx + \int_{\tau_2}^{x_*} \xi_3 dx - \int_{\tau_1}^{x_*} \xi_2 dx \\ &= \int_{\tau_2}^{x_*} (\xi_3 - \xi_2) dx. \end{aligned} \tag{15}$$

(See [T, §3.3] for some diagrammatic interpretation of this relation.) By rewriting (15), we obtain

$$\begin{aligned} 0 = &\int_{\tau_1}^{\iota} (\xi_1 - \xi_2) dx + \int_{\iota}^{x_*} (\xi_1 - \xi_2) dx + \int_{\tau_2}^{\iota} (\xi_2 - \xi_3) dx \\ &+ \int_{\iota}^{x_*} (\xi_2 - \xi_3) dx \\ = &\int_{\iota}^{x_*} (\xi_1 - \xi_3) dx + \int_{\tau_1}^{\iota} (\xi_1 - \xi_2) dx + \int_{\tau_2}^{\iota} (\xi_2 - \xi_3) dx. \end{aligned} \tag{16}$$

Since ι is an intersection point of Stokes curves γ_1 and γ_2, (16) entails

$$\operatorname{Im} \int_{\iota}^{x_*} (\xi_1 - \xi_3) dx = 0. \tag{17}$$

Hence ι is most likely to lie in the Stokes curve of type (1,3) that emanates from x_*. Here we say "most likely" just because we have not confirmed that x_* and ι belong to the same connected component of the real one-dimensional curve defined by (17). This point is, however, almost automatically checked in the computer-assisted study. This reasoning can be converted to find out a virtual turning point relevant to ι: We first consider a curve γ defined by

$$\operatorname{Im} \int_{\iota}^{x} (\xi_1 - \xi_3) dx = 0. \tag{18}$$

Defining a function $\rho(x)$ by

$$\operatorname{Re} \int_{\iota}^{x} (\xi_1 - \xi_3) dx, \tag{19}$$

we seek for a point x_0 in γ at which the following relation holds:

$$\rho(x_0) = \int_{\tau_1}^{\iota} (\xi_2 - \xi_1)dx + \int_{\tau_2}^{\iota} (\xi_3 - \xi_2)dx. \tag{20}$$

Since $\rho(x)$ is monotonically decreasing or increasing on the real one-dimensional curve γ, we can normally (i.e., except for the case where ρ is bounded on γ) locate such a point x_0 in γ. (Note that the right-hand side of (20) is a real number, as ι is an intersection point of Stokes curves.) Then we have

$$\int_{\iota}^{x_0} (\xi_1 - \xi_3)dx = \int_{\tau_1}^{\iota} (\xi_2 - \xi_1)dx + \int_{\tau_2}^{\iota} (\xi_3 - \xi_2)dx. \tag{21}$$

Hence the comparison of (16) and (21) entails that x_0 is coincident with a virtual turning point x_*. Actually all the virtual turning points studied by Sasaki ([Sa1], [Sa2]) and Honda ([Ho]) have been detected by this method.

Furthermore virtual turning points thus detected play important roles in describing the phase function $\phi(t)$ used in the instanton expansion of a solution of the Noumi-Yamada system: for example the function $\phi(t)$ is given by

$$\int_{d(t)}^{v_1(t)} (\xi_1 - \xi_3)dx \tag{22}$$

in the situation of Fig. 9 (ii) ([Sa1], [AKSST]). Using the terminology of [Ho], we can go further to give another interpretation of the point v_1 in the following manner: We can associate a function $\phi_T(t)$ to each effective bi-directional binary tree T that corresponds to degeneration of the Stokes geometry, and we can locate the required virtual turning point $v(t)$ in an appropriate Stokes curve of type, say (j, k), which emanates from a turning point τ so that the following relation holds:

$$\phi_T(t) = \int_{\tau(t)}^{v(t)} (\xi_j - \xi_k)dx \quad \text{(up to sign)}. \tag{23}$$

Let us note that $\phi_T(t)$ is given by

$$\int_{d}^{C} (\xi_1 - \xi_3)dx + \int_{s_1}^{C} (\xi_2 - \xi_1)dx + \int_{s_2}^{C} (\xi_3 - \xi_2)dx \tag{24}$$

with an appropriate indexing ξ_j in the situation of Fig. 6 (ii), where C designates the crossing point of three Stokes curves observed there. We note that (21) and (24) immediately entail (23) in this case. This interpretation of a virtual turning point plays an important role in grasping the behavior of napping virtual turning points ([Sa2]) in the framework of Honda ([Ho]).

References

[AKKSST] T. Aoki, T. Kawai, T. Koike, S. Sasaki, S. Shudo and Y. Takei: A background story and some know-how of virtual turning points, RIMS Koukyuuroku, (ISSN 1880-2818), No.1424, 2005, pp.53-63.

[AKKT] T. Aoki, T. Kawai, T. Koike and Y. Takei: On global aspects of exact WKB analysis of operators admitting infinitely many phases. *Contemporary Math.*, No.373, 2005, pp. 11-47.

[AKSST] T. Aoki, T. Kawai, S. Sasaki, A. Shudo and Y. Takei: Virtual turning points and bifurcation of Stokes curves, *J. Phys. A: Math. Gen.*, **38** (2005), 3317-3336.

[AKT1] T.Aoki, T.Kawai and Y.Takei: New turning points in the exact WKB analysis for higher-order ordinary differential equations, Analyse Algébrique des Perturbations Singulières. I, Hermann, 1994, pp. 69-84.

[AKT2] ———: On the exact WKB analysis for the third order ordinary differential equations with a large parameter, *Asian J. Math.*, **2** (1998), 625-640.

[AKT3] ———: On the exact steepest descent method: A new method for the description of Stokes curves, *J. Math. Phys.*, **42** (2001), 3691-3713.

[AKT4] ———: The exact steepest descent method A new steepest descent method based on the exact WKB analysis, *Advanced Studies in Pure Math.*, No.42, 2004, pp. 45-61.

[AKT5] ———: Exact WKB analysis of non-adiabatic transition probabilities for three levels, *J. Phys. A: Math. Gen.*, **35** (2002), 2401-2430.

[BW] C. M. Bender and T. T. Wu: Anharmonic oscillator, *Phys. Rev.*, **184** (1969), 1231-1260.

[BNR] H. L. Berk, W. M. Nevins and K. V. Roberts: New Stokes' line in WKB theory, *J. Math. Phys.*, **23** (1982), 988-1002.

[CH] R. Courant and D. Hilbert: Methods of Mathematical Physics, II, Interscience, 1962.

[DDP] E. Delabaere, H. Dillinger and F. Pham: Exact semi-classical expansions for one dimensional quantum oscillators, *J. Math. Phys.*, **38** (1997), 6126-6184.

[H] L. Hörmander: Fourier integral operators I, *Acta Math.*, **127** (1971), 79-183.

[Ho] N. Honda: Toward the complete description of the Stokes geometry, in prep.

[HLO] C. J. Howls, P. J. Langman and A. B. Olde Daalhuis: On the higher-order Stokes phenomenon, *Proc. R. Soc. Lond. A*, **460** (2004), 2285-2303.

[KKNT] T. Kawai, T. Koike, Y. Nishikawa and Y. Takei: On the Stokes geometry of higher order Painlevé equations, *Astérisque*, No. 297, 2004, pp. 117-166.

[KT] T. Kawai and Y. Takei: Algebraic Analysis of Singular Perturbation Theory, Iwanami, Tokyo, 1998. (In Japanese; English translation, AMS, 2005)

[NY] M. Noumi and Y.Yamada: Higher order Painlevé Equations of type $A_l^{(1)}$, *Funkcial. Ekvac.*, **48** (1998), 483-503.

[P] F. Pham: Resurgence, quantized canonical transformations, and multi-instanton expansions, Algebraic Analysis, vol. II, Acad. Press, 1988, pp. 699-726.

[Sa1] S. Sasaki: On the role of virtual turning points in the deformation of higher order linear ordinary differential equations, RIMS Koukyuuroku, (ISSN 1880-2818), No. 1433, 2005, pp. 27-64. (In Japanese.)

[Sa2] ———: – II – On a new Stokes curve in Noumi-Yamada system, ibid., pp. 65-109. (In Japanese.)

[SKK] M. Sato, T. Kawai and M. Kashiwara: Microfunctions and pseudo-differential equations, *Lect. Notes in Math.*, No. 287, Springer, 1973, pp. 265-529.

[Sh] A. Shudo: A recipe for finding Stokes geometry in quantized Hénon map, RIMS Koukyuuroku, (ISSN 1880-2818), No. 1433, 2005, pp. 110-118.

[S] H. J. Silverstone: JWKB connection-formula problem revisited via Borel summation, *Phys. Rev. Lett.*, **55** (1985), 2523-2526.

[T] Y. Takei: Exact WKB analysis, and exact steepest descent method, – A sequel to "Algebraic analysis of singular perturbations", *Sûgaku*, **55** (2003), 350-367. (In Japanese. Its English translation will appear in *Sûgaku Expositions.*)

[V] A. Voros: The return of the quartic oscillator. The complex WKB method, *Ann. Inst. Henri Poincaré*, **39** (1983), 211-338.

[Z] J. Zinn-Justin: Instantons in quantum mechanics: Numerical evidence for a conjecture, *J. Math. Phys.*, **25** (1984), 549-555.

Regular sequences associated with the Noumi-Yamada equations with a large parameter*

Takashi Aoki[1] and Naofumi Honda[2]

[1] Department of Mathematics, Kinki University, Higashi-Osaka 577-8502, Japan
aoki@math.kindai.ac.jp

[2] Department of Mathematics, Hokkaido University, Sapporo 060-0808, Japan
honda@math.sci.hokudai.ac.jp

Summary. We consider the system of algebraic equations that defines the leading terms of formal solutions to the Noumi-Yamada equations of even order and prove that the polynomial sequence associated with the system is a regular sequence.

1 Introduction

The Noumi-Yamada equations are discovered by [NY1], [NY2] as the systems of non-linear ordinary differential equations that possess the affine Weyl group symmetry of type $A_l^{(1)}$. Takei introduced a large parameter in the Noumi-Yamada equations and slightly modified the equations to investigate them by using the exact WKB analysis [T]. In this article, we are concerned with the modified form by Takei. The appearance of the equations depends on the parity of l. For an even $l = 2m$ $(m = 1, 2, \dots)$, the system has the form

$$\eta^{-1}\frac{du_j}{dt} = u_j(u_{j+1} - u_{j+2} + \cdots - u_{j+2m}) + \alpha_j \qquad (NY)_{2m}$$

$(j = 0, 1, 2, \dots, 2m)$, where η is a large parameter, α_j are formal power series of η^{-1} with constant coefficients satisfying

$$\alpha_0 + \alpha_1 + \cdots + \alpha_{2m} = \eta^{-1} \qquad (1)$$

and the independent variable t and dependent variables u_j are normalized so that

---

Received 28 February, 2006. Revised 27 May, 2006, 31 May, 2006. Accepted 31 May, 2006.

* This work is supported in part by JSPS Grant-in-Aid No. 15540190, No. 17340042 and No. 18540197

$$u_0 + u_1 + \cdots + u_{2m} = t \tag{2}$$

holds. The indices j of u_j are considered to be elements of $\mathbb{Z}/(2m+1)\mathbb{Z}$, that is, $u_{j+2m+1} = u_j$.

On the other hand, for an odd $l = 2m+1$, the system is given as follows:

$$\eta^{-1}\frac{t}{2}\frac{du_j}{dt} = u_j \sum_{1\le r\le s\le m} (u_{j-1+2r}u_{j+2s} - u_{j+2r}u_{j+1+2s}) + \frac{\alpha_j}{2}t \quad (NY)_{2m+1}$$

$(j = 0, 1, 2, \ldots, 2m+1)$, where α_j are formal power series of η^{-1} with constant coefficients satisfying

$$\alpha_0 + \alpha_2 + \cdots + \alpha_{2m} = \alpha_1 + \alpha_3 + \cdots + \alpha_{2m+1} = \frac{\eta^{-1}}{2} \tag{3}$$

and u_j are normalized so that

$$u_0 + u_2 + \cdots + u_{2m} = u_1 + u_3 + \cdots + u_{2m+1} = \frac{t}{2} \tag{4}$$

holds. The indices j of u_j are considered to be elements of $\mathbb{Z}/(2m+2)\mathbb{Z}$, that is, $u_{j+2m+2} = u_j$.

The starting point of the exact WKB analysis for such systems is to construct formal solutions of the form

$$\hat{u}_j = \sum_{k=0}^{\infty} \eta^{-k} u_{j,k}(t). \tag{5}$$

Putting (5) into $(NY)_l$ and comparing the like powers of η, we find systems of algebraic equations which should be satisfied by $u_{j,k}(t)$'s. In particular, for $l = 2m$, the leading terms $u_{j,0}(t)$ should satisfy

$$\begin{cases} u_j(u_{j+1} - u_{j+2} + \cdots - u_{j+2m}) + \alpha_j = 0 \quad (0 \le j \le 2m-1), \\ u_0 + u_1 + \cdots + u_{2m} = t. \end{cases} \quad (NY)^0_{2m}$$

Here we abbreviated $u_{j,0}(t)$ to $u_j(t)$ and $\alpha_{j,0}$ (the leading term of α_j) to α_j. To construct formal solutions (5) to $(NY)_l$, we have to solve such systems of algebraic equations for the leading terms. In fact, Takei confirmed that the systems have finitely many solutions for $l = 2, 3, 4, 5$ [T]. In this article, we show that $(NY)^0_{2m}$ forms a regular sequence of algebraic equations and hence the number of solutions is finite for every fixed t and for general m.

2 Regular sequences and tame regular sequences

2.1 Local theory

Let X be an n-dimensional complex manifold and let x_0 be a point in X. We denote by $\mathcal{O}$ the sheaf of holomorphic functions on X. First we recall the definition of local regular sequences:

Definition 1. Let $f_0, f_1, \dots, f_l$ be elements in $\mathcal{O}_{x_0}$ vanishing at x_0. If f_k is not a zero divisor on $\mathcal{O}_{x_0}/(f_0, \dots, f_{k-1})$ for each $k = 0, \dots, l$, then the sequence $f_0, f_1, \dots, f_l$ is said to be a regular sequence at x_0. Here $(f_0, \dots, f_{k-1})$ denotes the ideal in $\mathcal{O}_{x_0}$ generated by $f_0, \dots, f_{k-1}$.

Let $V(x_0, f_0, \dots, f_k)$ denote the germ at x_0 of the analytic variety of common zeros of $f_0, f_1, \dots, f_k$. Since $\mathcal{O}_{x_0}$ is a Cohen-Macaulay ring, we have the following

Theorem 1. *Let l be a non-negative integer smaller than n and $f_0, f_1, \dots, f_l$ be elements in $\mathcal{O}_{x_0}$ vanishing at x_0. Then the following three conditions are equivalent:*

1. *The sequence $f_0, f_1, \dots, f_l$ is a regular sequence at x_0.*
2. *For each $k = 0, 1, \dots, l$, the dimension of $V(x_0, f_0, \dots, f_k)$ is equal to $n - k - 1$.*
3. *The dimension of $V(x_0, f_0, \dots, f_l)$ is $n - l - 1$.*

Thus, at least locally, the notion of regular sequences does not depend on the ordering of f_j's.

Definition 2. Let $f_0, f_1, \dots, f_l$ be elements in $\mathcal{O}_{x_0}$. The sequence $f_0, f_1, \dots, f_l$ is said to be a tame regular sequence at x_0 if for any integer k so that $0 \le k \le l$ and for any $(k+1)$ choice $f_{l_0}, f_{l_1}, \dots, f_{l_k}$ of elements in $\{f_0, f_1, \dots, f_l\}$, the element f_{l_k} is not a zero divisor on $\mathcal{O}_{x_0}/(f_{l_0}, \dots, f_{l_{k-1}})$.

Note that we do not assume x_0 to be a common zero of $f_0, f_1, \dots, f_l$. As in the case of regular sequences, we have

Theorem 2. *Let $f_0, f_1, \dots, f_l$ be elements in $\mathcal{O}_{x_0}$. Then the following two conditions are equivalent:*

1. *The sequence $f_0, f_1, \dots, f_l$ is a tame regular sequence at x_0.*
2. *For any $k = 0, \dots, l$ and for any $(k+1)$ choice $f_{l_0}, f_{l_1}, \dots, f_{l_k}$ of elements in $\{f_0, f_1, \dots, f_l\}$, the dimension of $V(x_0, f_{l_0}, \dots, f_{l_k})$ is equal to $n-k-1$ or $V(x_0, f_{l_0}, \dots, f_{l_k})$ is an empty set.*

Now let us study the relation between the notion of tame regular sequences and the Koszul complex. We denote by $\mathcal{L}^{\cdot} = K(f_0, f_1, \dots, f_l; \mathcal{O})$ the Koszul complex associated with the sequence $f_0, f_1, \dots, f_l$ with coefficients in $\mathcal{O}$.

Theorem 3 ([S], Appendix B.4). *If $f_0, f_1, \dots, f_l$ is a tame regular sequence at x_0, then we have*

$$H^k(\mathcal{L}^{\cdot}_{x_0}) = 0$$

for every $k \le l$.

If we look at the l-th order part of the Koszul complex, we have the following:

Theorem 4. *Suppose that $H^l(\mathcal{L}^{\cdot}_{x_0}) = 0$ holds. Then f_l is not a zero divisor on $\mathcal{O}_{x_0}/(f_0, \dots, f_{l-1})$.*

Combining Theorems 3 and 4 yields

Theorem 5. *Let f_0, f_1, ..., f_l be elements in $\mathcal{O}_{x_0}$. Then the following two conditions are equivalent:*

1. *The sequence f_0, f_1, ..., f_l is a tame regular sequence at x_0.*
2. *For any $k = 0, \ldots, l$ and for any $(k+1)$ choice $\{f_{l_0}, f_{l_1}, \ldots, f_{l_k}\}$ of elements in $\{f_0, f_1, \ldots, f_l\}$, $H^j(K(f_{l_0}, f_{l_1}, \ldots, f_{l_k}; \mathcal{O}_{x_0})) = 0$ holds for every $j \leq k$.*

2.2 Relation between global theory and local theory

Next we recall the notion of regular sequences in the theory of commutative algebras [M]. Let R be a commutative ring with unit. Let a_0, a_1, ..., a_l be elements of R.

Definition 3. The sequence a_0, a_1, ..., a_l is called a regular sequence if the following two conditions hold:

1. For any $k = 0, \ldots, l$, the element a_k is not a zero divisor on $R/(a_0, \ldots, a_{k-1})$.
2. $(a_0, \ldots, a_l) \neq R$.

Note that the notion of regular sequences depends on the ordering of a_0, a_1, ..., a_l.

Definition 4. The sequence a_0, a_1, ..., a_l is called a tame regular sequence if for any $k = 0, \ldots, l$ and for any $(k+1)$ choice $a_{l_0}, a_{l_1}, \ldots, a_{l_k}$ of elements in $\{a_0, a_1, \ldots, a_l\}$, the element a_{l_k} is not a zero divisor on $R/(a_{l_0}, \ldots, a_{l_{k-1}})$.

Note that the notion of tame regular sequences is independent of the ordering of a_j's and that the second condition in Definition 3 is not assumed. This notion can be stated in terms of the Koszul complex. Let us denote by $K(a_0, a_1, \ldots, a_l; R)$ the Koszul complex associated with the sequence a_0, a_1, ..., a_l with coefficients in R.

Theorem 6. *The following two conditions are equivalent:*

1. *The sequence a_0, a_1, ..., a_l is a tame regular sequence.*
2. *For any $k = 0, \ldots, l$ and for any $(k+1)$ choice $a_{l_0}, a_{l_1}, \ldots, a_{l_k}$ of elements in $\{a_0, a_1, \ldots, a_l\}$, $H^j(K(a_{l_0}, a_{l_1}, \ldots, a_{l_k}; R)) = 0$ holds for every $j \leq k$.*

Thus we can connect local theory with global theory. Hereafter let R denote the ring $\mathbb{C}[u_0, \ldots, u_{n-1}]$ of polynomials of u_0, u_1, ..., u_{n-1} with complex coefficients and let f_0, f_1, ..., f_l be elements of R.

Theorem 7. *The following two conditions are equivalent:*

1. *The sequence f_0, f_1, ..., f_l is a tame regular sequence in R.*
2. *For any point x_0 in $\mathbb{C}^n$, the sequence f_0, f_1, ..., f_l is a tame regular sequence at x_0.*

Proof. We denote respectively by $L^{\cdot}$ and by $\mathcal{L}^{\cdot}$ the Koszul complexes associated with the sequence $f_0, f_1, \ldots, f_l$ with coefficients in R and in $\mathcal{O}$. If the first condition holds, we have

$$H^k(L^{\cdot}) = 0$$

for every $k < l$. Since $\mathcal{O}$ is flat over R, taking tensor product with $\mathcal{O}$ over $L^{\cdot}$ yields the second.

Suppose that the second condition holds. We regard f_j's as meromorphic functions on the n-dimensional projective space $\mathbb{P}^n$. We set $H = \mathbb{P}^n - \mathbb{C}^n$ and denote by $\mathcal{O}[*H]$ the sheaf of meromorphic functions that have poles along H. Let $\mathcal{L}^{\cdot}$ denote the Koszul complex $K(f_0, f_1, \ldots, f_l; \mathcal{O}[*H])$. By the assumption, the support of $H^k(\mathcal{L}^{\cdot})$ is contained in H for $k \leq l$. We shall show that $H^k(\mathcal{L}^{\cdot})$ vanishes on $\mathbb{P}^n$ for $k \leq l$. Let p be a point in H. We take a defining function h of H. That is, we assume H is written in the form $\{h = 0\}$ near p. There exists a positive integer N so that $h^N f_i \in \mathcal{O}_{\mathbb{P}^n,p}$ for all i. Here $\mathcal{O}_{\mathbb{P}^n}$ denotes the sheaf of holomorphic functions on $\mathbb{P}^n$. We set $\tilde{f}_i = h^N f_i$ for $i = 0, 1, \ldots, l$ and consider the complex $\tilde{\mathcal{L}}^{\cdot} = K(\tilde{f}_0, \ldots, \tilde{f}_l; \mathcal{O}_{\mathbb{P}^n})$ near p. Since $\mathcal{L}^{\cdot}$ is quasi-isomorphic to $\mathcal{O}[*H] \underset{\mathcal{O}_{\mathbb{P}^n}}{\otimes} \tilde{\mathcal{L}}^{\cdot}$, and $\mathcal{O}[*H]$ is flat over $\mathcal{O}_{\mathbb{P}^n}$, we have

$$H^k(\mathcal{L}^{\cdot}) \simeq H^k(\mathcal{O}[*H] \underset{\mathcal{O}_{\mathbb{P}^n}}{\otimes} \tilde{\mathcal{L}}^{\cdot}) \simeq \mathcal{O}[*H] \underset{\mathcal{O}_{\mathbb{P}^n}}{\otimes} H^k(\tilde{\mathcal{L}}^{\cdot}). \tag{6}$$

The support of $H^k(\tilde{\mathcal{L}}^{\cdot})$ is contained in H for every $k \leq l$ and $H^k(\tilde{\mathcal{L}}^{\cdot})$ is a coherent $\mathcal{O}_{\mathbb{P}^n}$-module. Hence applying the Hilbert Nullstellensatz yields

$$\mathcal{O}[*H] \underset{\mathcal{O}_{\mathbb{P}^n}}{\otimes} H^k(\tilde{\mathcal{L}}^{\cdot}) = 0 \tag{7}$$

for $k \leq l$. On the other hand, by the vanishing theorem of cohomology with coefficients in the sheaf of holomorphic functions of tempered growth [Hö], we have

$$H^k(\mathbb{P}^n, \mathcal{O}[*H]) = \begin{cases} R & \text{for } k = 0, \\ 0 & \text{for } k \neq 0. \end{cases} \tag{8}$$

Thus we find

$$\Gamma(\mathbb{P}^n, \mathcal{L}^{\cdot}) = K(f_0, f_1, \ldots, f_l; R) \tag{9}$$

and combining this with (6), (7) and (8), we have $H^k(L^{\cdot}) = 0$ for $k \leq l$. This completes the proof of Theorem 7.

2.3 Principal symbols of polynomials

For a polynomial $f \in R = \mathbb{C}[u_0, \ldots, u_{n-1}]$ and for an integer i, we denote by $\sigma_i(f)$ the homogeneous part of degree i. If the degree of f is d, we abbreviate $\sigma_d(f)$ to $\sigma(f)$ and call it the principal symbol of f.

Theorem 8. *Let f_0, f_1, …, f_l be polynomials in R. If the sequence made of the principal symbols $\sigma(f_0)$, $\sigma(f_1)$, …, $\sigma(f_l)$ is a tame regular sequence, then the sequence f_0, f_1, …, f_l is a tame regular sequence of R.*

Proof. We shall prove that f_k is not a zero divisor on $R/(f_0, f_1, \dots, f_{k-1})$ for $k \leq l$. It suffices to prove the following lemma.

Lemma 1. *For any $R_j \in R$ $(j = 0, 1, \dots, k)$ with $\sum_{j=0}^{k} R_j f_j = 0$, there exist $g_{ij} \in R$ satisfying $g_{ij} = -g_{ji}$ and $R_i = \sum_j g_{ij} f_j$ for $i = 0, 1, \dots, k$.*

Proof. Let m_j denote the degree of f_j $(j = 0, 1, \dots, k)$ and set

$$\mu = \max_j \deg(R_j f_j).$$

We prove the lemma by induction on μ. The lemma trivially holds if $\mu < 0$. Suppose that the lemma is proved for $\mu - 1$. Taking σ_μ of the relation

$$\sum_{j=0}^{k} R_j f_j = 0, \tag{10}$$

we have

$$\sum_{j=0}^{k} \sigma_{\mu - m_j}(R_j)\sigma(f_j) = 0.$$

Since the sequence $\sigma(f_0)$, $\sigma(f_1)$, …, $\sigma(f_k)$ is a tame regular sequence, there exist $h_{ij} \in R$ satisfying

$$\deg(h_{ij}) = \mu - m_i - m_j,$$

$$h_{ij} = -h_{ji}$$

and

$$\sigma_{\mu - m_i}(R_i) = \sum_j h_{ij}\sigma(f_j).$$

Now we set

$$\tilde{R}_i = R_i - \sum_j h_{ij} f_j. \tag{11}$$

Eliminating R_i's by using (11) and (10), we have

$$\sum_i \sum_j h_{ij} f_j f_i + \sum_i \tilde{R}_i f_i = \sum_i \tilde{R}_i f_i = 0.$$

Since

$$\max_i \deg(\tilde{R}_i f_i) \leq \mu - 1,$$

we can use the assumption of induction to $\tilde{R}_i$'s and this proves the lemma for μ.

3 Regular sequences associated with $(NY)^0_{2m}$

Let us consider $(NY)^0_{2m}$ in the ring $\mathbb{C}[u_0, u_1, \ldots, u_{2m}, t]$ and the following system which consists of principal symbols of equations of $(NY)^0_{2m}$:

$$\begin{cases} u_j(u_{j+1} - u_{j+2} + \cdots - u_{j+2m}) = 0 \quad (0 \leq j \leq 2m-1), \\ u_0 + u_1 + \cdots + u_{2m} - t = 0. \end{cases} \tag{12}$$

We define polynomials f_j $(0 \leq j \leq 2m)$ as follows:

$$\begin{cases} f_j = u_j(u_{j+1} - u_{j+2} + \cdots - u_{j+2m}) \quad (0 \leq j \leq 2m-1), \\ f_{2m} = u_0 + u_1 + \cdots + u_{2m} - t. \end{cases} \tag{13}$$

Theorem 9. *The sequence f_0, ..., f_{2m} is a tame regular sequence at each point in $\mathbb{C}^{2m+2}$. Hence it is a tame regular sequence in $\mathbb{C}[u_0, u_1, \ldots, u_{2m}, t]$.*

Proof. We set $g_j = u_{j+1} - u_{j+2} + \cdots - u_{j+2m}$ for $0 \leq j \leq 2m-1$. Then the condition $f_j = 0$ is equivalent to the condition "$u_j = 0$ or $g_j = 0$". First we consider the case where $g_j = 0$ for all j. We write the coefficient matrix of the system of linear equations $(-1)^0 g_0 = 0, \ldots, (-1)^{2m-1} g_{2m-1} = 0, f_{2m} = 0$ with respect to unknowns $(-1)^{j+1}u_j$ and t:

$$\begin{pmatrix} 0 & 1 & 1 & \ldots & 1 & 1 & 1 & 0 \\ -1 & 0 & 1 & \ldots & 1 & 1 & 1 & 0 \\ -1 & -1 & 0 & \ldots & 1 & 1 & 1 & 0 \\ \vdots & & & \ldots & & & \vdots & \vdots \\ -1 & -1 & -1 & \ldots & 0 & 1 & 1 & 0 \\ -1 & -1 & -1 & \ldots & -1 & 0 & 1 & 0 \\ -1 & 1 & -1 & \ldots & -1 & 1 & -1 & -1 \end{pmatrix}. \tag{14}$$

Suppose that we choose $u_j = 0$ for some j's instead of $g_j = 0$. That is, let I be a subset of $\{0, 1, 2, \ldots, 2m-1\}$ and we consider the case where $u_j = 0$ for $j \in I$ and $g_j = 0$ for $j \notin I$. Then the coefficient matrix of the corresponding system of linear equations with respect to unknowns $(-1)^{j+1}u_j$ $(j \notin I)$ becomes

$$\begin{pmatrix} 0 & 1 & 1 & \ldots & 1 & 1 & 1 & 0 \\ -1 & 0 & 1 & \ldots & 1 & 1 & 1 & 0 \\ -1 & -1 & 0 & \ldots & 1 & 1 & 1 & 0 \\ \vdots & & & \ldots & & & \vdots & \vdots \\ -1 & -1 & -1 & \ldots & 0 & 1 & 1 & 0 \\ -1 & -1 & -1 & \ldots & -1 & 0 & 1 & 0 \\ * & * & * & \ldots & * & * & -1 & -1 \end{pmatrix}, \tag{15}$$

where $* = \pm 1$. Note that this matrix is obtained by removing i-th column and i-th row for each $i \in I$ from (14). If I has d elements, (15) has $2m+1-d$

rows and $2m+2-d$ columns. We remove $(2m+2-d)$-th column from (15) and have the following matrix:

$$\begin{pmatrix} 0 & 1 & 1 & \dots & 1 & 1 & 1 \\ -1 & 0 & 1 & \dots & 1 & 1 & 1 \\ -1 & -1 & 0 & \dots & 1 & 1 & 1 \\ \vdots & & & \dots & & \vdots & \\ -1 & -1 & -1 & \dots & 0 & 1 & 1 \\ -1 & -1 & -1 & \dots & -1 & 0 & 1 \\ * & * & * & \dots & * & * & -1 \end{pmatrix}. \tag{16}$$

Adding $(2m+1-d)$-th column to all of other columns yields the matrix

$$\begin{pmatrix} 1 & 2 & 2 & \dots & 2 & 2 & 1 \\ 0 & 1 & 2 & \dots & 2 & 2 & 1 \\ 0 & 0 & 1 & \dots & 2 & 2 & 1 \\ \vdots & & & \dots & & \vdots & \\ 0 & 0 & 0 & \dots & 1 & 2 & 1 \\ 0 & 0 & 0 & \dots & 0 & 1 & 1 \\ * & * & * & \dots & * & * & -1 \end{pmatrix}, \tag{17}$$

where $*$ denotes 0 or -2. The determinant of this matrix is clearly an odd integer. Hence it does not vanish and we see that the rank of (15) is $2m+1-d$. Thus for every l choice from the system of polynomials $f_0, \dots, f_{2m}$, the dimension of every irreducible component of the set of common zeros of these l polynomials is $2m+2-l$. Hence it follows from Theorem 2 that $f_0, \dots, f_{2m}$ is a tame regular sequence at every point in $\mathbb{C}^{2m+2}$.

Applying Theorems 7 and 8, we have

Theorem 10. *Let F_j $(0 \le j \le 2m)$ be polynomials defined by*

$$\begin{cases} F_j = u_j(u_{j+1} - u_{j+2} + \dots - u_{j+2m}) + \alpha_j \quad (0 \le j \le 2m-1), \\ F_{2m} = u_0 + u_1 + \dots + u_{2m} - t. \end{cases} \tag{18}$$

Then the sequence $F_0, \dots, F_{2m}$ is a tame regular sequence in $\mathbb{C}[u_0, \dots, u_{2m}, t]$.

Moreover, we can see that $(NY)^0_{2m}$ has solutions as follows. For a fixed t, $(NY)^0_{2m}$ has solutions in $\mathbb{P}^{2m+1}_u$ ([IIa]. Theorem 7.2). To see that there is no solution at the infinity, we replace u by u/λ in $(NY)^0_{2m}$. Then we have

$$\begin{cases} u_j(u_{j+1} - u_{j+2} + \dots - u_{j+2m}) + \alpha_j \lambda^2 = 0 \quad (0 \le j \le 2m-1), \\ u_0 + u_1 + \dots + u_{2m} = t\lambda. \end{cases} \tag{19}$$

Putting $\lambda = 0$ yields

$$\begin{cases} u_j(u_{j+1} - u_{j+2} + \cdots - u_{j+2m}) = 0 \quad (0 \le j \le 2m-1), \\ u_0 + u_1 + \cdots + u_{2m} = 0. \end{cases} \tag{20}$$

By the arguments in the proof of Theorem 9, we see that (20) has the unique solution $u_0 = u_1 = \cdots = u_{2m} = 0$ which does not lie in $\mathbb{P}_u^{2m+1}$. Hence we have

Theorem 11. *Under the same notation as in Theorem 10, the sequence* F_0, $\ldots$, F_{2m} *is a regular sequence in* $\mathbb{C}[u_0, \ldots, u_{2m}, t]$.

Remark 1. It is clear that the sequence F_0, $\ldots$, F_{2m} is a regular sequence in $\mathbb{C}[u_0, \ldots, u_{2m}]$ for every fixed t. Hence $(NY)^0_{2m}$ has a finite number of solutions for every fixed t. Under suitable generic condition, the number is 2^{2m} and the number of ramification points of solutions over t is $2m2^{2m}$. These will be proved in our forthcoming paper.

Remark 2. Of course we can consider $(NY)^0_{2m+1}$ for $(NY)_{2m+1}$. The sequence of polynomials corresponding to $(NY)^0_{2m+1}$ is, however, not a regular sequence in $\mathbb{C}[u_0, \ldots, u_{2m+1}, t]$ because the set of common zeros has an irreducible component contained in $\{t = 0\}$ with dimension greater than 1. But we can obtain similar finiteness theorems of solutions in this case.

References

[Ha] Hartshorne, R.: Algebraic Geometry, Graduate Texts in Mathematics 52, Springer-Verlag (1983)

[Hö] Hörmander, L.: An Introduction to Complex Analysis in Several Variables, North-Holland (1990)

[M] Matsumura, H.: Commutative Algebra, WA Benjamin (1970)

[NY1] Noumi, M., Yamada, Y.: Higher order Painlevé equations of type $A_l^{(1)}$. Funkcial Ekvac., **41**, 245–260 (1998)

[NY2] ______ : Symmetry in Painlevé equations. In: Howls, C. J., Kawai, T., Takei, Y. (eds) Toward the Exact WKB Analysis of Differential Equations, Linear or Non-Linear, Kyoto Univ. Press, 2000, pp. 245–260.

[S] Schapira, P.: Microdifferential Systems in the Complex Domain, Grundlehren Der Mathematischen Wissenschaften, Springer-Verlag (1985)

[T] Takei, Y.: Toward the exact WKB analysis for higher-order Painlevé equations – The case of Noumi-Yamada Systems –, Publ. RIMS, Kyoto Univ., **40** 709–730 (2004)

Ghost busting: Making sense of non-Hermitian Hamiltonians

Carl M. Bender

Physics Department, Washington University, St. Louis, MO 63130, USA
cmb@wustl.edu

Summary. The Lee model is an elementary quantum field theory in which mass, wave-function, and charge renormalization can be performed exactly. Early studies of this model in the 1950's found that there is a critical value of g^2, the square of the renormalized coupling constant, above which g_0^2, the square of the unrenormalized coupling constant, is *negative*. For g^2 larger than this critical value, the Hamiltonian of the Lee model becomes non-Hermitian. In this non-Hermitian regime a new state appears whose norm is negative. This state is called a *ghost*. It has always been thought that in this ghost regime the Lee model is an unacceptable quantum theory because unitarity appears to be violated. However, in this regime while the Hamiltonian is not Hermitian, it does possess $\mathcal{PT}$ symmetry. It has recently been discovered that a non-Hermitian Hamiltonian having $\mathcal{PT}$ symmetry may define a quantum theory that is unitary. The proof of unitarity requires the construction of a time-independent operator called $\mathcal{C}$. In terms of $\mathcal{C}$ one can define a new inner product with respect to which the norms of the states in the Hilbert space are positive. Furthermore, it has been shown that time evolution in such a theory is unitary. In this talk the $\mathcal{C}$ operator for the Lee model in the ghost regime is constructed in the $V/N\theta$ sector. It is then shown that the ghost state has a positive norm and that the Lee model is an acceptable unitary quantum field theory for all values of g^2.

1 Introduction

I showed in my talk in Kyoto in 2004 [Ben05] that a non-Hermitian Hamiltonian that possesses an unbroken $\mathcal{PT}$ symmetry describes a quantum theory that is unitary (probability-preserving). The key ingredient in the demonstration was the construction of a new operator called $\mathcal{C}$. In quantum mechanics the $\mathcal{C}$ operator is represented in position space as a sum over the appropriately normalized eigenfunctions of the Hamiltonian [BenBroJon02]: $\mathcal{C}(x, y) \equiv \sum_n \phi_n(x)\phi_n(y)$.

The problem is that using this definition to calculate $\mathcal{C}$ is difficult in quantum mechanics and impossible in quantum field theory. Thus, in re-

Received 30 January, 2006. Revised 25 March, 2006. Accepted 30 March, 2006.

cent papers [BenMeiWan03, BenJon04] an alternative method for calculating $\mathcal{C}$ directly in terms of the operator dynamical variables of the quantum theory was presented. This new method is general and applies to quantum-mechanical systems having several degrees of freedom [BenBroRefReu04]. More importantly, this method can be used to calculate $\mathcal{C}$ in quantum field theory [BenBroJon04-1, BenBroJon04-2, BenCavMilSha05]. We believe that the $\mathcal{C}$ operator should be regarded as fundamental because while the parity operator $\mathcal{P}$ in quantum field theory does not transform under the Lorentz group as a scalar [BenMeiWan05], the $\mathcal{C}$ operator *does* transform as a scalar [BenBraCheWan05-1].

We will show in this talk how to use the machinery that has been developed to study non-Hermitian $\mathcal{PT}$-symmetric quantum field theories to examine a famous model quantum field theory known as the Lee model. In 1954 the Lee model was proposed as a quantum field theory in which mass, wave function, and charge renormalization could be performed exactly and in closed form [Lee54, KälPau55, Sch61, Bar63]. Our reason for studying the Lee model here is that when the renormalized coupling constant is taken to be larger than a critical value, the Hamiltonian becomes non-Hermitian and a (negative-norm) ghost state appears. The appearance of the ghost state was assumed to be a fundamental defect of the Lee model. However, we show that the non-Hermitian Lee-model Hamiltonian is actually $\mathcal{PT}$ symmetric. When the states of this model are examined using the $\mathcal{C}$ operator, the ghost state is now found to be an ordinary physical state having positive norm.

Studying the Lee model as a non-Hermitian Hamiltonian was suggested by Kleefeld, who was the first to point out this transition to $\mathcal{PT}$ symmetry [Kle04]. His paper gives a beautiful and comprehensive history of non-Hermitian Hamiltonians.

2 Review of the Lee Model

In the Lee model the interaction of three spinless particles called V, N, and θ is described. The V and N particles are fermions and behave like nucleons, and the θ particle is a boson and behaves roughly like a pion. It is assumed that the V may emit a θ, but when it does so it becomes an N: $V \to N + \theta$. Also, an N may absorb a θ, but when it does so it becomes a V: $N + \theta \to V$. (This interaction models the phenomenon of a proton emitting a π^+ and becoming a neutron and a neutron absorbing a π^+ and becoming a proton.)

The Lee model is solvable because there is no crossing symmetry. That is, the N is forbidden to emit an anti-θ and become a V. Eliminating crossing symmetry makes the Lee model solvable because it introduces two conservation laws. First, *the number of N quanta plus the number of V quanta is fixed.* Second, *the number of N quanta minus the number of θ quanta is fixed.* These two highly constraining conservation laws decompose the Hilbert space of states into an infinite number of noninteracting sectors. The simplest sector is the vacuum sector. Because of the conservation laws there are no vacuum graphs and the bare vacuum is the physical vacuum. The next two sectors are the one-θ-particle and the one-N-particle sector. These two sectors are also trivial because the two conservation laws prevent any dynamical processes from occurring there. As a result, the masses of the N particle and of the θ particle are not renormalized; that is, the physical masses of these particles are the same as their bare masses.

The $V/N\theta$ sector is the lowest nontrivial sector. The physical states in this sector of the Hilbert space are linear combinations of the bare V and the bare $N\theta$ states, and these states consist of the one-physical-V-particle state and the physical N-θ-scattering states. To find these states one can look for the poles and cuts of the Green's functions. (The Feynman diagrams are chains of $N\theta$ bubbles connected by V lines.) The renormalization in this sector is easy to perform. Following the conventional renormalization procedure, one finds that the mass of the V particle is renormalized; that is, the mass of the physical V particle is different from its bare mass. In the Lee model one can calculate the unrenormalized coupling constant as a function of the renormalized coupling constant in closed form. There are many ways to define the renormalized coupling constant. For example, in an actual scattering experiment one could define the square of the renormalized coupling constant g^2 as the value of the $N\theta$ scattering amplitude at threshold.

The appearance of a ghost state in the $V/N\theta$ sector is the most intriguing aspect of the Lee model. To see how this state appears, one must perform coupling-constant renormalization. Expressing g_0^2, the square of the unrenormalized coupling constant, in terms of g^2, the square of the renormalized coupling constant, one obtains the graph in Fig. 1. In principle, the g is a physical parameter whose numerical value is determined by a laboratory experiment. If g^2 is measured to be near 0, then from Fig. 1 we can see that g_0^2 is also small. However, if the experiment gives a value of g^2 that is larger than this critical value, then the square of the unrenormalized coupling constant becomes negative. In this regime g_0 is imaginary and the Hamiltonian is non-Hermitian. Moreover, in this regime a new state appears in the $V/N\theta$ sector, and because its norm is negative, the state is called a *ghost*. There have been many attempts to make sense of the Lee model as a physical quantum theory in the ghost regime [KälPau55, Sch61, Bar63], but none of these attempts have been successful. Summarizing the situation, in Ref. [Bar63] Barton writes, "A non-Hermitean (*sic*) Hamiltonian is unacceptable partly because it may lead to complex energy eigenvalues, but chiefly because it implies a non-unitary S matrix, which fails to conserve probability and makes a hash of the physical interpretation."

Contrary to the view of Barton, we show that it is possible to give a physical interpretation for the $V/N\theta$ sector of the Lee model when g_0 becomes imaginary and H becomes non-Hermitian in the Dirac sense. (Dirac Hermitian conjugation, as indicated by the symbol $\dagger$, means combined transpose and complex conjugate.) The Lee model is a cubic interaction and there have been several studies of theories having a cubic interaction multiplied by an imaginary coupling constant and in all these studies it was found that the spectrum is real and positive. In two especially important cases it was noticed that the spectrum of Reggeon field theory is real and positive [AbaBroSugWhi75] and that the spectrum of Lee-Yang edge-singularity models is also positive [Fis78]. In this talk we use the methods developed in Ref. [BenBroJon04-1] to examine the Lee model and we summarize the procedure described in Ref. [BenBraCheWan05-2].

3 $\mathcal{PT}$-Analysis of the Lee Model

When the Hamiltonian for the Lee model is non-Hermitian, it is necessary to calculate the $\mathcal{C}$ operator for the Hamiltonian. We make use of the following three crucial properties of $\mathcal{C}$. First, the square of $\mathcal{C}$ operator is unity:

$$\mathcal{C}^2 = 1. \tag{1}$$

Second, $\mathcal{C}$ commutes with $\mathcal{PT}$,

$$[\mathcal{C}, \mathcal{PT}] = 0, \tag{2}$$

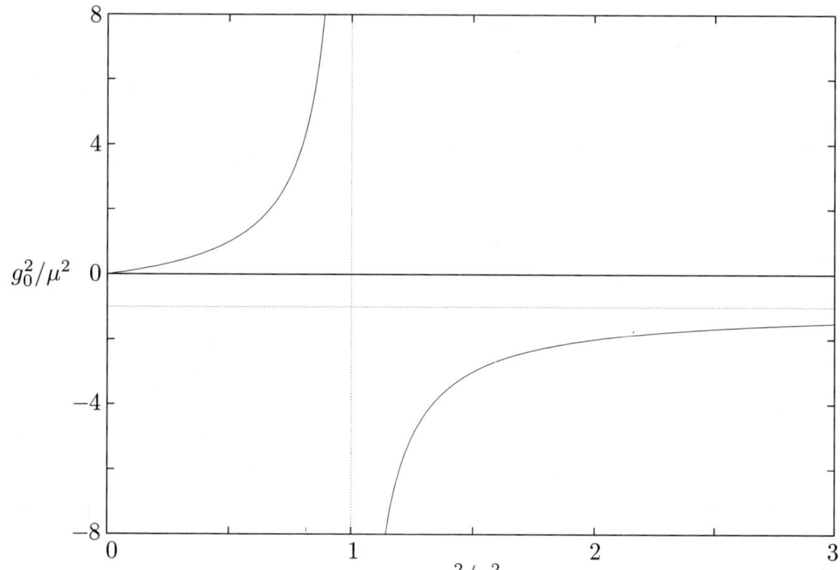

Fig. 1. Square of the unrenormalized coupling constant, g_0^2, plotted as a function of the square of the renormalized coupling constant g^2. When $g^2 = 0$ we have $g_0^2 = 0$, and as g^2 increases so does g_0^2. However, as g^2 increases past a critical value, g_0^2 abruptly becomes negative. In this regime g_0 is imaginary and the Hamiltonian becomes non-Hermitian.

and therefore $\mathcal{C}$ is $\mathcal{PT}$ symmetric. Third, $\mathcal{C}$ commutes with the Hamiltonian,

$$[\mathcal{C}, H] = 0, \tag{3}$$

and therefore the eigenstates of the Hamiltonian are also eigenstates of $\mathcal{C}$. In fact, states of H having a negative $\mathcal{PT}$ norm have eigenvalue -1 under $\mathcal{C}$, and eigenstates of H having a positive $\mathcal{PT}$ norm have eigenvalue $+1$ under $\mathcal{C}$. From these three properties one can use the $\mathcal{C}$ operator to construct the correct inner product for the Hilbert space of states of a Hamiltonian having an unbroken $\mathcal{PT}$ symmetry: $\langle A|B\rangle \equiv |A\rangle^{\mathcal{CPT}} \cdot |B\rangle$. This $\mathcal{CPT}$ inner product is associated with a *positive* norm. Furthermore, the usual time translation operator e^{iHt} preserves the inner product. Thus, with respect to this inner product, time evolution is unitary.

Here, we focus on the *quantum-mechanical* Lee model; the field-theoretic Lee model is discussed in Ref. [BenBraCheWan05-2]. In terms of creation and annihilation operators the Hamiltonian for the quantum-mechanical Lee model is

$$H = H_0 + g_0 H_1 = m_{V_0} V^\dagger V + m_N N^\dagger N + m_\theta a^\dagger a + \left(V^\dagger N a + a^\dagger N^\dagger V\right). \tag{4}$$

The bare states are the eigenstates of H_0 and the physical states are the eigenstates of the full Hamiltonian H. The mass parameters m_N and m_θ represent the *physical* masses of the one-N-particle and one-θ-particle states because these states do not undergo mass renormalization. However, m_{V_0} is the *bare* mass of the V particle.

The V, N, and θ particles are treated as pseudoscalars. To understand why, recall that in quantum mechanics the position operator $x = (a + a^\dagger)/\sqrt{2}$ and the momentum operator $p = i(a^\dagger - a)/\sqrt{2}$ both change sign under parity reflection $\mathcal{P}$:

$$\mathcal{P}x\mathcal{P} = -x, \quad \mathcal{P}p\mathcal{P} = -p. \tag{5}$$

Thus, $\mathcal{P}V\mathcal{P} = -V$, $\mathcal{P}N\mathcal{P} = -N$, $\mathcal{P}a\mathcal{P} = -a$, $\mathcal{P}V^\dagger\mathcal{P} = -V^\dagger$, $\mathcal{P}N^\dagger\mathcal{P} = -N^\dagger$, $\mathcal{P}a^\dagger\mathcal{P} = -a^\dagger$. Under time reversal $\mathcal{T}$, p and i change sign but x does not:

$$\mathcal{T}p\mathcal{T} = -p, \quad \mathcal{T}i\mathcal{T} = -i, \quad \mathcal{T}x\mathcal{T} = x. \tag{6}$$

Thus, $\mathcal{T}V\mathcal{T} = V$, $\mathcal{T}N\mathcal{T} = N$, $\mathcal{T}a\mathcal{T} = a$, $\mathcal{T}V^\dagger\mathcal{T} = V^\dagger$, $\mathcal{T}N^\dagger\mathcal{T} = N^\dagger$, $\mathcal{T}a^\dagger\mathcal{T} = a^\dagger$.

If the bare coupling constant g_0 is real, H in (4) is Hermitian: $H^\dagger = H$. However, when g_0 is imaginary, $g_0 = i\lambda_0 \quad (\lambda_0$ real), H is not Hermitian, but by virtue of the above transformation properties, H is $\mathcal{PT}$-symmetric: $H^{\mathcal{PT}} = H$.

Let us assume first that g_0 is real so that H is Hermitian and we examine the simplest nontrivial sector of the quantum-mechanical Lee model; namely, the $V/N\theta$ sector. We look for the eigenstates of the Hamiltonian H in the form of linear combinations of the bare one-V-particle and the bare one-N-one-θ-particle states. There are two eigenfunctions and two eigenvalues. We interpret the eigenfunction corresponding to the lower-energy eigenvalue as the physical one-V-particle state, and we interpret the eigenfunction corresponding with the higher-energy eigenvalue as the physical one-N-one-θ-particle state. (In the field-theoretic version of the Lee model this higher-energy state corresponds to the continuum of physical N-θ scattering states.) Thus, we make the *ansatz*

$$|V\rangle = c_{11}|1,0,0\rangle + c_{12}|0,1,1\rangle, \qquad |N\theta\rangle = c_{21}|1,0,0\rangle + c_{22}|0,1,1\rangle, \tag{7}$$

and demand that these states be eigenstates of H with eigenvalues m_V (the renormalized V-particle mass) and $E_{N\theta}$. The eigenvalue problem reduces to a pair of elementary algebraic equations:

$$c_{11}m_{V_0} + c_{12}g_0 = c_{11}m_V, \qquad c_{21}g_0 + c_{22}\left(m_N + m_\theta\right) = c_{22}E_{N\theta}. \tag{8}$$

The solutions to (8) are

$$\begin{aligned} m_V &= \frac{1}{2}\left(m_N + m_\theta + m_{V_0} - \sqrt{\mu_0^2 + 4g_0^2}\right), \\ E_{N\theta} &= \frac{1}{2}\left(m_N + m_\theta + m_{V_0} + \sqrt{\mu_0^2 + 4g_0^2}\right), \end{aligned} \tag{9}$$

where $\mu_0 \equiv m_N + m_\theta - m_{V_0}$. Notice that m_V, the mass of the physical V particle, is different from m_{V_0}, the mass of the bare V particle, because the V particle undergoes mass renormalization.

Next, we perform wave-function renormalization. Following Barton we define the wave-function renormalization constant Z_V by [Bar63] $\sqrt{Z_V} = \langle 0|V|V\rangle$. This gives

$$Z_V^{-1} = \frac{1}{2}g_0^{-2}\sqrt{\mu_0^2 + 4g_0^2}\left(\sqrt{\mu_0^2 + 4g_0^2} - \mu_0\right). \tag{10}$$

Finally, we perform coupling-constant renormalization. Again, following Barton we note that $\sqrt{Z_V}$ is the ratio between the renormalized coupling constant g and the bare coupling constant g_0 [Bar63]. Thus, $g^2/g_0^2 = Z_V$. After some elementary algebra we find that in terms of the renormalized mass and coupling constant, the bare coupling constant satisfies

$$g_0^2 = g^2/\left(1 - g^2/\mu^2\right), \tag{11}$$

where μ is defined as $\mu \equiv m_N + m_\theta - m_V$. We cannot freely choose g because the value of g is in principle taken from experimental data. Once g has been determined experimentally, we can use (11) to determine g_0. The relation in (11) is plotted in Fig. 1. Figure 1 reveals a surprising property of the Lee model: If g is larger than the critical value μ, then the square of g_0 is negative, and g_0 is imaginary.

As g approaches its critical value from below, the two energy eigenvalues in (9) vary accordingly. The energy eigenvalues are the two zeros of the secular determinant $f(E)$ obtained from applying Cramer's rule to (8) (see Fig. 2). As g (and g_0) increase, the energy of the physical $N\theta$ state increases (see Fig. 3). The energy of the $N\theta$ state becomes infinite as g reaches its critical value. As g increases past its critical value, the upper energy eigenvalue goes around the bend (see Fig. 4); it abruptly jumps from being large and positive to being large and negative. Then, as g continues to increase, this energy eigenvalue approaches the energy of the physical V particle from below (see Fig. 5).

When g increases past its critical value, the Hamiltonian H in (4) becomes non-Hermitian, but its eigenvalues in the $V/N\theta$ sector remain real. The eigenvalues remain real because H becomes $\mathcal{PT}$ symmetric, and cubic $\mathcal{PT}$-symmetric Hamiltonians that have been studied in the past have been shown to have real spectra [BenBroRefReu04]. However, in the $\mathcal{PT}$-symmetric regime it is no longer appropriate to interpret the lower eigenvalue as the energy of the physical $N\theta$ state. Rather, it is the energy of a new kind of state $|G\rangle$ called a *ghost*. As is shown in Refs. [KälPau55, Sch61, Bar63], the Hermitian norm of this state is *negative*. Until the writing of this paper, a satisfactory physical interpretation of the ghost state had not been found.

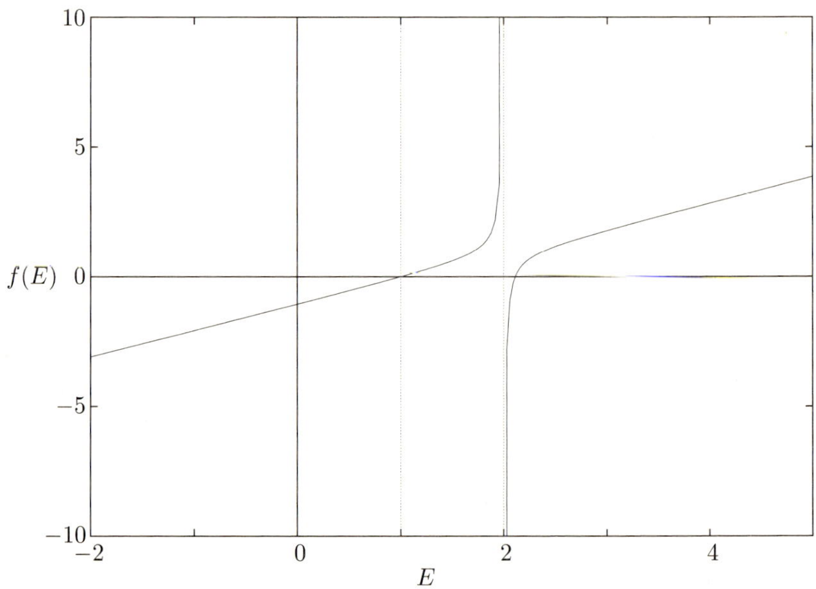

Fig. 2. Plot of the secular determinant $f(E)$ obtained by applying Cramer's rule to (8) for g_0 real and small. Values of E for which $f(E) = 0$ correspond to eigenvalues of the Hamiltonian (4). Observe that $f(E)$ has two zeros, the lower one corresponding to the energy of the physical V particle and the upper one corresponding to the energy of the physical $N\theta$ state.

We now show how a physical interpretation of the ghost state emerges easily when we use the methods developed in Ref. [BenBroJon02]. We begin by verifying that in the $\mathcal{PT}$-symmetric regime, where g_0 is imaginary, the states of the Hamiltonian are eigenstates of the $\mathcal{PT}$ operator, and we then choose the multiplicative phases of these states so that their $\mathcal{PT}$ eigenvalues are unity:

$$\mathcal{PT}|G\rangle = |G\rangle, \qquad \mathcal{PT}|V\rangle = |V\rangle.$$

It is then straightforward to verify that the $\mathcal{PT}$ norm of the V state is positive, while the $\mathcal{PT}$ norm of the ghost state is negative.

We then follow the procedures in Refs. [BenBroJon04-1, BenBroJon04-2] to calculate $\mathcal{C}$. In these papers it is shown that the $\mathcal{C}$ operator can be expressed as an exponential of a function Q multiplying the parity operator: $\mathcal{C} = \exp\left[Q(V^\dagger, V; N^\dagger, N; a^\dagger, a)\right]\mathcal{P}$. We then impose the operator equations in (1) - (3). The condition $\mathcal{C}^2 = 1$ gives

$$Q(V^\dagger, V; N^\dagger, N; a^\dagger, a) = -Q(-V^\dagger, -V; -N^\dagger, -N; -a^\dagger, -a). \qquad (12)$$

Thus, $Q(V^\dagger, V; N^\dagger, N; a^\dagger, a)$ is an odd function in total powers of $V^\dagger$, V, $N^\dagger$, N, $a^\dagger$, and a. Next, we impose the condition $[\mathcal{C}, \mathcal{PT}] = 0$ and obtain

$$Q(V^\dagger, V; N^\dagger, N; a^\dagger, a) = Q^*(-V^\dagger, -V; -N^\dagger, -N; -a^\dagger, -a), \qquad (13)$$

where $*$ denotes complex conjugation.

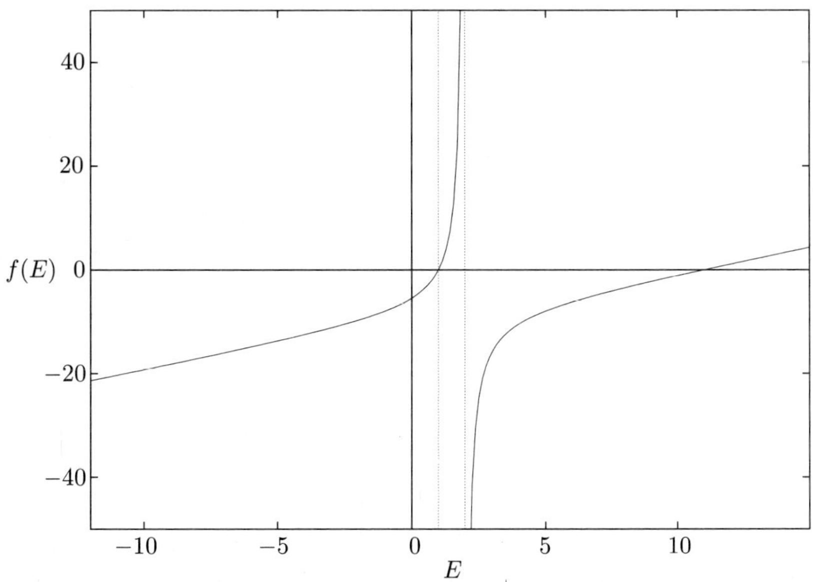

Fig. 3. Same as in Fig. 2 except that the value of g is larger. Note that the larger eigenvalue E (the larger value of E for which $f(E) = 0$), which corresponds to the physical $N\theta$ state, has moved up the real-E axis.

Finally, we impose the condition that the operator $\mathcal{C}$ commutes with H: $[\mathcal{C}, H] = 0$, which requires that

$$\left[e^Q, H_0\right] = g_0 \left[e^Q, H_1\right]_+ . \tag{14}$$

In Refs. [BenBroJon04-1, BenBroJon04-2] the $\mathcal{C}$ operator for an ix^3 quantum-mechanical model and for an $i\phi^3$ field-theory model were found approximately using low-order perturbation theory. However, for the Lee model it is possible to calculate the $\mathcal{C}$ operator exactly and in closed form. To do so, we seek a solution for Q as a formal Taylor series in powers of g_0:

$$Q = \sum_{n=0}^{\infty} g_0^{2n+1} Q_{2n+1} . \tag{15}$$

As in the case of the ix^3 and $i\phi^3$ models, only odd powers of g_0 appear in this series and Q_{2n+1} are all anti-Hermitian: $Q_{2n+1}^\dagger = -Q_{2n+1}$ [BenBroJon04-1, BenBroJon04-2]. From (14) we get

$$Q_{2n+1} = (-1)^n \frac{2^{2n+1}}{(2n+1)\mu_0^{2n+1}} \left(V^\dagger N a n_\theta^n - n_\theta^n a^\dagger N^\dagger V\right), \tag{16}$$

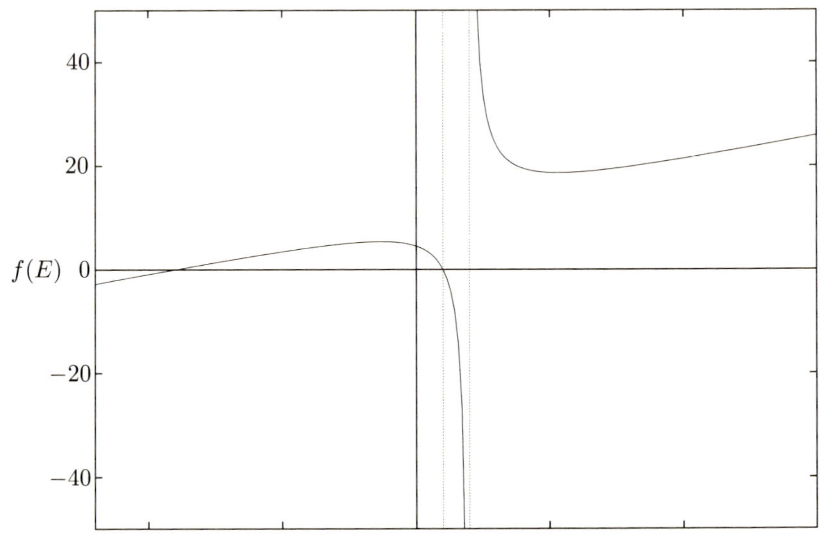

Fig. 4. Same as in Fig. 3 except that g is larger than its critical value and the unrenormalized coupling constant g_0 is imaginary. In this regime the Hamiltonian is non-Hermitian. Observe that the larger zero of $f(E)$ has moved out to infinity and is now moving up the negative real-E axis below the energy of the physical V particle. The $N\theta$ state has disappeared and has been replaced by a ghost state.

where $n_\theta = a^\dagger a$ is the number operator for θ-particle quanta.

We then sum over all Q_{2n+1} and obtain the *exact* result that

$$Q = V^\dagger N a \frac{1}{\sqrt{n_\theta}} \arctan\left(\frac{2g_0\sqrt{n_\theta}}{\mu_0}\right) \quad \frac{1}{\sqrt{n_\theta}} \arctan\left(\frac{2g_0\sqrt{n_\theta}}{\mu_0}\right) a^\dagger N^\dagger V. \quad (17)$$

We exponentiate this result to obtain the $\mathcal{C}$ operator. The exponential of Q simplifies considerably because we are treating the V and N particles as fermions and therefore we can use the identity $n_{V,N}^2 = n_{V,N}$. Our *exact* result for e^Q is

$$e^Q = \left[1 - n_V - n_N + n_V n_N + \frac{\mu_0 n_N (1 - n_V)}{\sqrt{\mu_0^2 + 4g_0^2 n_\theta}} + \frac{\mu_0 n_V (1 - n_N)}{\sqrt{\mu_0^2 + 4g_0^2 (n_\theta + 1)}} \right.$$
$$\left. + V^\dagger N a \frac{2g_0\sqrt{n_\theta}}{\sqrt{\mu_0^2 + 4g_0^2 n_\theta}} - \frac{2g_0\sqrt{n_\theta}}{\sqrt{\mu_0^2 + 4g_0^2 n_\theta}} a^\dagger N^\dagger V \right]. \quad (18)$$

We can also express the parity operator $\mathcal{P}$ in terms of number operators:

$$\mathcal{P} = e^{i\pi(n_V + n_N + n_\theta)} = (1 - 2n_V)(1 - 2n_N)\, e^{i\pi n_\theta}. \quad (19)$$

Combining e^Q and $\mathcal{P}$, we obtain the exact expression for $\mathcal{C}$:

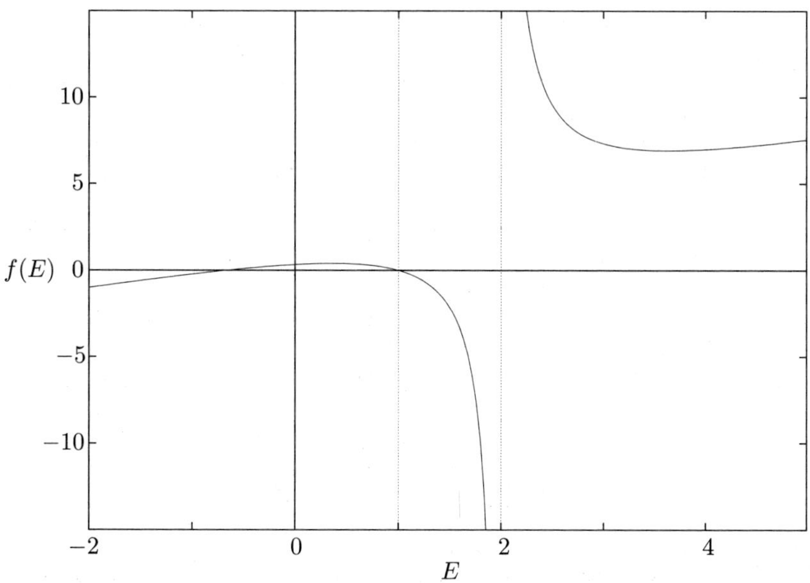

Fig. 5. Same as in Fig. 4 except that g has an even larger value. Note that as g continues to increase the ghost energy continues to move up towards the energy of the physical V particle.

$$\mathcal{C}=\left[1-n_V-n_N+n_Vn_N+\frac{\mu_0 n_N\left(1-n_V\right)}{\sqrt{\mu_0^2+4g_0^2n_\theta}}+\frac{\mu_0 n_V\left(1-n_N\right)}{\sqrt{\mu_0^2+4g_0^2\left(n_\theta+1\right)}}\right.$$
$$\left.+V^\dagger Na\frac{2g_0\sqrt{n_\theta}}{\sqrt{\mu_0^2+4g_0^2n_\theta}}-\frac{2g_0\sqrt{n_\theta}}{\sqrt{\mu_0^2+4g_0^2n_\theta}}a^\dagger N^\dagger V\right]\left(1-2n_V\right)\left(1-2n_N\right)e^{i\pi n_\theta}. \quad (20)$$

Using this $\mathcal{C}$ operator we calculate the $\mathcal{CPT}$ norm of the V state and of the ghost state and find that these norms are both positive. Furthermore, as is shown in Ref. [BenBroJon02], time evolution is unitary. Thus, we have verified that with the proper definition of the inner product, the quantum-mechanical Lee model is a physically acceptable and fully consistent quantum theory even in the ghost regime where the unrenormalized coupling constant is imaginary.

I am grateful to the U.S. Department of Energy for financial support.

References

[Ben05] Bender, C.M.: Quantum Mechanics Based on Non-Hermitian Hamiltonians. In: Takei, Y. (ed) Proceedings of the International Conference "Recent Trends in Exponential Asymptotics," Kyoto, July, 2004. RIMS Kokyuroku Symposium Series **1424**, 64-77 (2005). This paper gives a background of $\mathcal{PT}$-symmetric quantum mechanics and gives some references.

[BenBroJon02] Bender, C.M., Brody, D.C., Jones, H.F.: Complex Extension of Quantum Mechanics, Phys. Rev. Lett. **89**, 270401-4 (2002)

[BenMeiWan03] Bender, C.M., Meisinger, P.N., and Wang, Q.: Perturbative Calculation of the Hidden Symmetry Operator in $\mathcal{PT}$-Symmetric Quantum Mechanics, J. Phys. A: Math. Gen. **36**, 1973-83 (2003)

[BenJon04] Bender, C.M. and Jones, H.F.: Semiclassical Calculation of the $\mathcal{C}$ Operator in $\mathcal{PT}$-Symmetric Quantum Mechanics, Phys. Lett. A **328**, 102-9 (2004)

[BenBroRefReu04] Bender, C.M., Brod, J., Refig, A.T., Reuter, M.E.: The $\mathcal{C}$ Operator in $\mathcal{PT}$-Symmetric Quantum Theories, J. Phys. A: Math. Gen. **37**, 10139-10165 (2004)

[BenBroJon04-1] Bender, C.M., Brody, D.C., Jones, H.F.: Scalar Quantum Field Theory with Cubic Interaction, Phys. Rev. Lett. **93**, 251601-4 (2004)

[BenBroJon04-2] Bender, C.M., Brody, D.C., Jones, H.F.: Extension of $\mathcal{PT}$-Symmetric Quantum Mechanics to Quantum Field Theory with Cubic Interaction, Phys. Rev. D **70**, 025001 (2004) (19 pages)

[BenCavMilSha05] Bender, C.M., Cavero-Pelaez, I., Milton, K.A., Shajesh, K.V.: $\mathcal{PT}$-Symmetric Quantum Electrodynamics, Phys. Lett. B **613**, 97-104 (2005)

[BenMeiWan05] C. M. Bender, P. N. Meisinger, and Q. Wang: Wilson Polynomials and the Lorentz Transformation Properties of the Parity Operator, J. Math. Phys. **46**, 052302 (2005) (13 pages)

[BenBraCheWan05-1] Bender, C.M., Brandt, S.F., Chen, J., Wang, Q.: The $\mathcal{C}$ Operator in $\mathcal{PT}$-Symmetric Quantum Field Theory Transforms as a Lorentz Scalar, Phys. Rev. D **71**, 065010 (2005) (7 pages)

[Lee54] Lee, T.D.: Some Special Examples in Renormalizable Field Theory, Phys. Rev. **95**, 1329-34 (1954)

[KälPau55] Källén, G., Pauli, W.: On the Mathematical Structure of T. D. Lee's Model of a Renormalizable Field Theory, Mat.-Fys. Medd. **30**, No. 7 (1955)

[Sch61] Schweber, S.S.: An Introduction to Relativistic Quantum Field Theory. Row, Peterson and Co., Evanston (1961), Chap. 12

[Bar63] G. Barton: Introduction to Advanced Field Theory. John Wiley & Sons, New York (1963), Chap. 12

[Kle04] Kleefeld, K.: Non-Hermitian Quantum Theory and its Holomorphic Representations: Introduction and Applications, hep-th/0408028 and hep-th/0408097

[AbaBroSugWhi75] Abarbanel, H.D.I., Bronzan, J.D., Sugar, R.L., White, A.R.: Reggeon Field Theory: Formulation and Use, Phys. Rept. **21**, 119-182 (1975); Brower, R., Furman, M., Moshe, M.: Critical Exponents for the Reggeon Quantum Spin Model, Phys. Lett. B **76**, 213-9 (1978); Harms, B., Jones, S., Tan, C.-I.: Complex Energy Spectra in Reggeon Quantum Mechanics with Quartic Interactions, Nucl. Phys. B **171** 392-412 (1980) and New STructure in the Energy Spectrum of Reggeon Quantum Mechanics with Quartic Couplings, Phys. Lett. B **91B**, 291-5 (1980)

[Fis78] Fisher, M.E.: Yang-Lee Edge Singularity and ϕ^3 Field Theory, Phys. Rev. Lett. **40**, 1610-13 (1978); Cardy, J.L.: Conformal Invariance and the Yang-Lee Singularity in Two Dimensions, *ibid.* **54**, 1354-6 (1985); Cardy, J.L. and Mussardo, G.: S Matrix of the Yang-Lee Edge Singularity in Two Dimensions, Phys. Lett. B **225**, 275-8 (1989); Zamolodchikov, A.B.: Two-Point Correlation Function in Scaling Lee-Yang Model, Nucl. Phys. B **348**, 619-41 (1991)

[BenBraCheWan05-2] Bender, C.M., Brandt, S.F., Chen, J., Wang, Q.: Ghost Busting: $\mathcal{PT}$-Symmetric Interpretation of the Lee Model, Phys. Rev. D **71**, 025014 (2005) (11 pages)

Vanishing of the logarithmic trace of generalized Szegö projectors

Louis Boutet de Monvel

Université Pierre et Marie Curie - Paris 6, France
boutet@math.jussieu.fr

Summary. The logarithmic trace of Szegö projectors introduced by K. Hirachi [15] for CR structures and extended in [8] to contact structures vanishes identically.

Key words: CR manifolds, contact manifolds, Toeplitz operators, residual trace
Mathematics Subject Classification (2000): 32V05, 32A25, 53D10, 58J40

In [15] K. Hirachi showed that the logarithmic trace of the Szegö projector is an invariant of the CR structure. In [8] I showed that it is also defined for generalized Szegö projectors associated to a contact structure (definitions recalled below, sect.4), that it is a contact invariant, and that it vanishes if the base manifold is a 3-sphere, with arbitrary contact structure (not necessarily the canonical one). Here we show that it always vanishes. For this we use the fact that this logarithmic trace is the residual trace of the identity (definitions recalled below, sect.5), and show that it always vanishes, because the Toeplitz algebra associated to a contact structure can be embedded in the Toeplitz algebra of a sphere, where the identity maps of all 'good' Toeplitz modules have zero residual trace.

1 Notations

We first recall the notions that we will use. Most of the material below in §1-5 is not new; we have just recalled briefly the definitions and useful properties, and send back to the literature for further details (cf. [16, 17, 21, 18]).

If X is a smooth manifold we denote $T^\bullet X \subset T^*X$ the set of non-zero covectors. A complex subspace Z corresponds to an ideal $I_Z \subset C^\infty(T^\bullet X, \mathbb{C})$). Z is conic (homogeneous) if it is generated by homogeneous functions. It is

Received 27 February, 2006. Revised 8 November, 2006. Accepted 8 November, 2006.

smooth if I_Z is locally generated by k = codim Z functions with linearly independent derivatives. If Z is smooth, it is involutive if I_Z is stable by the Poisson bracket - in local coordinates $\{f,g\} = \sum(\frac{\partial f}{\partial \xi_j}\frac{\partial g}{\partial x_j} - \frac{\partial f}{\partial x_j}\frac{\partial g}{\partial \xi_j})$; it is $\gg 0$ if locally I_Z has generators u_i, v_j $(1 \leq i \leq p, 1 \leq j \leq q, p+q = \text{codim}\, Z)$ such that the v_j are real, u_i complex, and the matrix $\frac{1}{i}(\{u_k, \bar{u}_l)$ is hermitian $\gg 0$. The real part Z_R is then a smooth real submanifold of $T^\bullet X$, whose ideal is generated by the $\mathrm{Re}\, \mathrm{u_i}, \mathrm{Im}\, \mathrm{u_i}, \mathrm{v_j}$. If $\mathcal{I}$ is $\gg 0$, it is exactly determined by its formal germ (Taylor expansion) along the set of real points of Z_R.

A Fourier integral operator (FIO) from Y to X is a linear operator from functions or distributions on Y to same on X defined as a locally finite sum of oscillating integrals

$$f \mapsto Ff(x) = \int e^{i\phi(x,y,\theta)} a(x,y,\theta) f(y)\, d\theta dy \ ,$$

where ϕ is a phase function (homogeneous w.r. to θ), a a symbol function. Here we will only consider regular symbols, i.e. which are asymptotic sums $a \sim \sum_{k \geq 0} a_{m-k}$ where a_{m-k} is homogeneous of degree $m-k$, k an integer. m could be any complex number. There is also a notion of vector FIO, acting on sections of vector bundles, which we will not use here.

The canonical relation of F is the image of the critical set of ϕ $(d_\theta \phi = 0)$ by the differential map $(x,y,\theta) \mapsto (d_x\phi, -d_y\phi)$ - this is always assumed to be an immersion from the critical set onto a Lagrangian sub-manifold of $T^\bullet X \times T^\bullet Y^0$ (the sign ^{0} means that we have reversed the sign of the canonical symplectic form; likewise if $\mathcal{A}$ is a ring, $\mathcal{A}^0$ denotes the opposite ring). We will also use FIO with complex positive phase function: then the canonical relation is defined by its ideal (the set of complex functions $u(x,\xi,y,\eta)$ which lie in the ideal generated by the coefficients of $d_\theta\phi, \xi - d_x\phi, \eta + d_y\phi$ which do not depend on θ); it should not be confused with its set of real points.

Following Hörmander [16, 17], the degree of F is defined as $\deg F = \deg(ad\theta) - \frac{1}{4}(n_x + n_y + 2n_\theta)$, with n_x, n_y, n_θ the dimensions of the x, y, θ-spaces, $\deg(ad\theta)$ the degree of the differential form $ad\theta = \deg a + \sum \nu_k$ $(\nu_k = \deg \theta_k$, usually $\nu_k = 1$ but could be any real number - not all 0); this only depends on F and not on its representation by oscillating integrals). In what follows we will always require that the degree be an integer (which implies $m \in \mathbb{Z}/4$).

2 Adapted Fourier Integral Operators

The Toeplitz operators and Toeplitz algebras used here, associated to a CR or a contact structure, were introduced and studied in [4, 5], using the analysis of the singularity of the Szegö kernel (cf. [3, 19]), or in a weaker form, the "Hermite calculus" of [2, 11]. The terminology "adapted" is taken from [5]: lacking anything better I have kept it.

For $k = 1, 2$, let $\Sigma_k \subset T^\bullet X_k$ be smooth symplectic sub-cones of $T^\bullet X_k$ and $u : \Sigma_1 \to \Sigma_2$ an isomorphism.

Definition 1. *A Fourier integral operator A is adapted to u if its canonical relation C is complex $\gg 0$, with real part the graph of u. It is elliptic if its principal symbol does not vanish (on $\mathrm{Re}\, C$).*

As above, a conic complex Lagrangian sub-manifold Λ of $T^\bullet X$ is $\gg 0$ if its defining ideal $\mathcal{I}_\Lambda$ is locally generated by $n = \dim X$ homogeneous functions $u_1, \dots, u_n (n = \dim X)$ with independent derivatives, $u_1, \dots, u_k$ complex, $u_{k+1}, \dots, u_n$ real for some k $(1 \le k \le n)$, and the matrix $(\frac{1}{i}\{u_p, \bar{u}_q\})_{1 \le p,q \le k}$ is hermitian $\gg 0$; equivalently; the intersection $C \cap \bar{C}$ is clean, and on the tangent bundle the hermitian form $\frac{1}{i}\omega(U, \bar{V})$ is positive, with kernel the complexification of the tangent space of $\mathrm{Re}\, C$.

Pseudo-differential operators are a special case of adapted FIO ($X_1 = X_2 = X, u = Id_{T^\bullet X}$); so are Toeplitz operators on a contact manifold (see below).

Adapted FIO always exist (cf [5]), more precisely

Proposition 1. *For any symplectic isomorphism u as above, there exists an elliptic FIO adapted to u.*

In fact if Λ is a complex $\gg 0$ Lagrangian sub-manifold of $T^\bullet X$, in particular if it is real, it can always be defined by a global phase function with positive imaginary part ($\mathrm{Im}\, \phi \gtrsim \mathrm{dist}\, (., \Lambda_{\mathrm{R}})^2$) living on $T^\bullet X$: it is easy to see that such phase functions exist locally, and the positivity condition makes it possible to glue things together using a homogeneous partition of the unity. Once one has chosen a global phase function, it is obviously always possible to choose an elliptic symbol - of any prescribed degree (cf. also [5])[1]. Note that elliptic only means that the top symbol is invertible on the real part of the canonical relation, not that the operator is invertible mod. smoothing operators (for this the canonical relation must be real: X_1, X_2 have the same dimension, $\Sigma_k = T^\bullet X_k$ and C is the graph of an isomorphism).

3 Model Example

Here is a generic example of adapted FIO: let X_1, X_2, Z be three vector spaces

$$\Sigma_k = T_{X_k} X_k \times T^\bullet Z \subset T^\bullet(X_k \times Z) \quad (k = 1, 2,\ T_{X_k} X_k \text{ the zero section}),$$

$$u \text{ the identity map } \mathrm{Id}_{T^\bullet Z} : \ \Sigma_1 \to \Sigma_2 \,. \tag{1}$$

If $C \gg 0$ is a complex canonical relation with real part the graph of u, the complex formal germ along Σ of the restriction to C of the projection

[1] the intrinsic differential-geometric description of the symbol is elaborate: it is a section of a line bundle whose definition incorporates half densities and the Maslov index or an elaboration of this in the case of complex canonical relations. However on real manifolds this line bundle is always topologically trivial

$(x, \xi, z, \zeta, z', \zeta', y, \eta) \mapsto (x, z, \zeta', y)$ is an isomorphism (the dimensions are right, and it is an immersion: if $v = (0, \xi, 0, \zeta, z', 0, 0, \eta)$ is a complex vector with zero projection, it is orthogonal to $\bar{v}$ (because these vectors form a real Lagrangian space), so if it is tangent to $\mathcal{C} \gg 0$, it is tangent to the real part, i.e. the diagonal of $T^\bullet Z$, and this obviously implies $v = 0$).

So we can choose the phase function as

$$\phi =< z - z', \zeta' > +iq(x, z, \zeta', y)$$

where q is smooth complex function of x, z, ζ', y alone, homogeneous of degree 1 w.r. to ζ', vanishing of order 2 for $x = y = 0$, and $\mathrm{Re}\,\mathrm{q} \gtrsim (\mathrm{x}^2 + \mathrm{y}^2)\ |\zeta'|$ (it is easy to check that conversely any such phase function corresponds to a positive adapted canonical relation as above). The operator is

$$Ff(x, z) = \int e^{i<z-z',\zeta'>-q(x,z,\zeta',y)} a(x, z, \zeta', y) f(z', y) d\zeta' dz' dy \ , \qquad (2)$$

with a a symbol as above.

Since any symplectic sub-cone of a cotangent manifold is always locally equivalent to $T^\bullet Z \subset T^\bullet(X \times Z)$, the model above is universal i.e. any adapted FIO is micro-locally equivalent to $F_1 \circ A \circ F_2$ where F_1, F_2 are elliptic invertible FIO with real canonical relations, graphs of local symplectic isomorphisms, and F is as the model above.

For adapted FIO the Hörmander degree coincides with the degree in the scale of Sobolev spaces, i.e. if F is of degree s it is continuous $H^m(Y) \to H^{m-\mathrm{Re}\,s}(X)$; this is easily seen on the model example above (F is L^2 continuous if its degree is 0 i.e. a is of degree $-\frac{1}{4}(n_x + n_y)$). (This not true for general FIOs - in fact for a FIO with a real canonical relation $\mathcal{C}$, this is only true if $\mathcal{C}$ is locally the graph of a symplectic isomorphism.)

The following result also immediately follows from the positivity condition:

Proposition 2. *Let X_1, X_2, X_3 be three manifolds, $\Sigma_k \subset T^\bullet X_k$ symplectic sub-cones, u resp. v a homogeneous symplectic isomorphism $X_1 \to X_2$ resp. $X_2 \to X_3$, F, G FIO (with compact support) adapted to u, v. Then $G \circ F$ is adapted to $v \circ u$; its canonical relation is transversally defined and positive. It is elliptic if F and G are elliptic.*

This is mentioned in [5]; the crux of the matter is that if $Q(y)$ is a quadratic form with $\gg 0$ real part, the integral $\int e^{-|\xi|Q(y)} dy$ does not vanish: it is an elliptic symbol of degree $-\frac{1}{2}n_y$, equal to $\mathrm{disc}(\frac{Q}{\pi})^{-\frac{1}{2}} |\xi|^{-\frac{1}{2}\mathrm{n_y}}$.

4 Generalized Szegö projectors

These were called "Toeplitz projectors" in [7, 8]. C. Epstein suggested the present name, which is better. References: [3, 4, 5, 6].

Definition 2. *Let X be a manifold, $\Sigma \subset T^\bullet X$ a symplectic sub-cone. A generalized Szegö projector (associated to Σ) is an elliptic FIO S adapted to Id_Σ which is a projector ($S^2 = S$)*

(Note that "elliptic" (or "of degree 0") is part of the definition; otherwise there exist many non-elliptic projectors, of degree > 0 as FIOs). The case we are most interested in is the case where Σ is the half line bundle corresponding to a contact structure on X (i.e. the set of positive multiples of the contact form). But everything works as well in the slightly more general setting above.

We will not require here that S be an orthogonal projector; this makes sense anyway only once one has chosen a smooth density to define L^2-norms.

If S is a generalized Szegö projector, its canonical relation $\mathcal{C} \subset T^\bullet X \times T^\bullet X$ is idempotent, positive, and can be described as follows: the first projection is a complex positive involutive manifold Z_+ with real part Σ; the second projection is a complex negative manifold Z_- with real part Σ ($Z_- = \bar{Z}_+$ if S is selfadjoint). The characteristic foliations define fibrations $Z_\pm \to \Sigma$ (the fibers are the characteristic leaves; they have each only one real point so are "contractible" (they vanish immediately in imaginary domain), and there is no topological problem for them to build a fibration). Finally we have $\mathcal{C} = Z_+ \times_\Sigma Z_-$

Generalized Szegö projectors always exist, so as orthogonal ones (cf. [5, 6]). As mentioned in [8], generalized Szegö projectors mod. smoothing operators form a soft sheaf on Σ, i.e. any such projector defined near a closed conic subset of $T^\bullet X$ or Σ is the restriction of a globally defined such projector.

5 Residual trace and logarithmic trace

The residual trace was introduced by M. Wodzicki [23]. It was extended to Toeplitz operators and suitable Fourier integral operators by V. Guillemin [12] (cf. also [13, 24]). It is related to the first example of 'exotic' trace given by J. Dixmier [9].

Let $\mathcal{C}$ be a canonical relation in $T^\bullet X \times T^\bullet X$. A family $A_s (s \in \mathbb{C})$ of FIOs of degree s belonging to $\mathcal{C}$ is holomorphic if $s \mapsto (\Delta)^{-\frac{s}{2}} A_s$ is a holomorphic map from $\mathcal{C}$ to FIO of fixed degree (in the obvious sense). If A_s has compact support, the trace $\operatorname{tr} A_s$ is then well defined and depends holomorphically on s if $\operatorname{Re} s$ is small enough (A_s is then of trace class). Often, e.g. if the canonical relation is real analytic, this will extend as a meromorphic function of s on the whole complex plane, but this is not very easy to use because the poles are hard to locate and usually not simple poles.

Proposition 3. *If $\mathcal{C}$ is adapted to the identity Id_Σ, with $\Sigma \subset T^\bullet X$ a symplectic sub-cone, and A_s a holomorphic family, as above, then $tr\, A_s$ has at most simple poles at the points $s = -n - deg\, A_0 + k$, $k \geq 0$ an integer, $n = \frac{1}{2} dim\, \Sigma$ (the degree $deg\, A_0$ is defined as above)*

Proof: this is obviously true if A_s is of degree $-\infty$ (there is no pole at all). In general we can write A_s as a sum of FIO with small micro-support (mod smoothing operators), and a canonical transformation reduces us to the model case, where result is immediate.

Definition 3. *If A is a FIO adapted to $\mathcal{C}$, the residual trace $tr_{res}A$ is the residue at $s = 0$ of any holomorphic family A_s as above, with $A = A_0$.*

This does not depend on the choice of a family A_s: indeed if $A_0 = 0$, the family A_s is divisible by s i.e. $A_s = sB_s$ where B_s is another holomorphic family, and since $\operatorname{tr} B_s$ has only simple poles, $\operatorname{tr} A_s$ has no pole at all at $s = 0$.

Proposition 4. *The residual trace is a trace, i.e. if A and B are adapted Fourier integral operators, we have $tr_{res}AB = tr_{res}BA$.*

Indeed with the notations above $\operatorname{tr} AB_s$ and $\operatorname{tr} B_sA$ are well defined and equal for $\operatorname{Re} s$ small, so their meromorphic extensions and poles coincide.

Logarithmic trace (contact case)

Let $\Sigma \subset T^\bullet X$ be a symplectic half-line bundle, defining a contact structure on X. A complex canonical relation $\mathcal{C} \gg 0$ adapted to Id_Σ is always the conormal bundle of a complex hypersurface Y of $X \times X$, with real part the diagonal (rather the positive half) [2], so if A is a FIO adapted to Id_Σ, its Schwartz kernel can be defined by a one dimensional Fourier integral:

$$\tilde{A}(x, y) = \int_0^\infty e^{-T\phi(x,y)} a(x, y, T)\, dT\,,$$

with $\phi = 0$ an equation of the hypersurface Y, $\phi = 0$ on the diagonal, $\operatorname{Re}\phi \geq \mathrm{cst}\,\mathrm{dist}(.,\mathrm{diag})^2$, and a is a symbol: $a \sim \sum_{k\leq N} a_k(x, y)T^{k-1}$ $(N = \deg \mathrm{A})$.

Its singularity has a typically holonomic form:

$$f(x, y)(\phi + 0)^{-N} + g(x, y)\,\mathrm{Log}\,(\frac{1}{\phi + 0})\,, \tag{3}$$

with f, g smooth functions on $X \times X$, and in particular $g(x, x) = a_0(x, x)$.

Proposition 5. *With notations as above, the residual trace of A is the trace of the logarithmic coefficient:*

$$tr_{res}A = \int_X g(x, x)\,. \tag{4}$$

---

[2] indeed a complex vertical vector v is as before orthogonal to $\bar{v}$; if it is tangent to $\mathcal{C}$ at a point of $\mathcal{C}_R = \mathrm{diag}\,\Sigma$, it is tangent to the real part $\mathcal{C}_R = \mathrm{diag}\,\Sigma$ since $\mathcal{C} \gg 0$, but this implies that v is the radial vector, because the radial vector is the only vertical vector tangent to Σ. So the projection $\mathcal{C} \to X \times X$ is of maximal rank $2n - 1$ and the image is a hypersurface.

An obvious holomorphic family extending A (mod. a smoothing operator) is the family A_s with Schwartz kernel

$$\tilde{A}_s(x,y) = \int_1^\infty e^{-T\phi(x,y)} a(x,y,T)\, T^s dT\ .$$

Since $a_s(x,x,T) \sim \sum_{k \le N} T^{s+k-1} a_k(x,x)$ and $\phi(x,x) = 0$, we get

$$\tilde{A}_s(x,x) \sim \sum \frac{a_k(x,x)}{s+k}\ ,$$

with an obvious notation: the meromorphic extension of the trace has just simple poles at each integer $j \geq -N$, with residue $\int_X a_{-j}(x,x)$. In particular the residue for $s = 0$ is the logarithmic trace.

The residual trace is also equal to the logarithmic trace in the case of pseudo-differential operators, or in the model case. In general the residual trace is well defined, but I do not know if the logarithmic coefficient can be reasonably defined e.g. if the projection $\Sigma \to X$ is not of constant rank. For the equality with the residual trace, and for theorem 1 below, the sign is important: the logarithmic trace is the integral of the coefficient of Log $\frac{1}{\phi}$, not the opposite.

6 Trace on a Toeplitz algebra $\mathcal{A}$ and on End$_{\mathcal{A}}(M)$

If S is a generalized Szegö projector associated to $\Sigma \subset T^\bullet X$. The corresponding Toeplitz operators are the Fourier integral operators of the form $T_P = SPS$, P a pseudo-differential operator (equivalently, the set of Fourier integral operators A with the same canonical relation, such that $A = SAS$). They form an algebra $\mathcal{A}$ on which the residual trace is a trace: $\mathrm{tr}_{res} AB = \mathrm{tr}_{res} BA$. Mod. smoothing operators, this can be localized, and the Toeplitz algebra $\mathcal{A}_\Sigma$ is this quotient; it is a sheaf on Σ (or rather on the basis).

Proposition 6. *The Toeplitz algebra, so as the residual trace of S only depend on Σ and not on the embedding $\Sigma_k \subset T^\bullet X$*

Indeed if $\Sigma \to T^\bullet X'$ is another embedding, S' a corresponding Szegö projector, it follows from prop. 1 that there exist elliptic adapted FIO F, F' from X to X' resp. X' to X such that $F = S'FS, F' = SF'S', FF' \sim S, F'F \sim S$ so S, S' have the same residual trace, and $A \mapsto FAF'$ is an isomorphism of the two Toeplitz algebras. In the lemma wa could as well embed Σ in another symplectic cone endowed with a Toeplitz structure.

The definition of the residual trace extends in an obvious way to End$_{\mathcal{A}}(M)$ when M is a free $\mathcal{A}$-module (End$_{\mathcal{A}}(M)$ is isomorphic to a matrix algebra with coefficients in $\mathcal{A}^0$, where $\mathrm{tr}_{res}(a_{ij}) = \sum \mathrm{tr}_{res} a_{ii}$ is obviously a trace, independent of the choice of a basis of M). It extends further to the case where M is locally free (a direct summand of a free module), and to the case where M admits (locally or globally) a finite locally free resolution: if

$$0 \to L^N \xrightarrow{\mathrm{d}} \dots L^1 \xrightarrow{\mathrm{d}} L^0 \to 0$$

is such a resolution, i.e. a complex of locally free $\mathcal{A}$-modules L^j, exact in degree $\neq 0$, with a given isomorphism $\epsilon : L^0/dL^1 \to M$: Then $\mathrm{End}_{\mathcal{A}}(M)$ is isomorphic to $\mathrm{Rhom}^0(L, L)$, i.e. any $a \in \mathrm{End}_{\mathcal{A}}(M)$ extends as morphism $\tilde{a}$ of complexes of L ($\tilde{a} = (a_j), a_j \in \mathrm{End}_{\mathcal{A}}(L^j)$) (if a has compact support, $\tilde{a}$ can be chosen with compact support). Any two such extensions $\tilde{a}, \tilde{a}'$ differ by a super-commutator $[d, s] = ds + sd$ (s, of degree 1, can be chosen with compact support if $\tilde{a} - \tilde{a}'$ has compact support). The super-trace

$$\mathrm{supertr}_{res}(\tilde{a}) = \sum (-1)^j \mathrm{tr}_{res}(a_j)$$

is then well defined; it only depends on a, because the super-trace of a super-commutator $[s, d]$ vanishes: this defines the trace in $\mathrm{End}\,(M)$. Below we will use "good" modules, i.e. which have a global finite locally free resolution for which this is already seen on the principal symbols (i.e. M and the L_j are equipped with good filtrations for which $\mathrm{gr}\, d$ is a locally free resolution of $\mathrm{gr}\, M$ (in the analytic setting this always exists if M is coherent and has a global good filtration, and the base manifold is projective).

Alternative description of the residual trace

Let S be a generalized Szegö projector associated to a symplectic cone $\Sigma \subset T^\bullet X$. Then the left annihilator of S is a $\gg 0$ ideal $\mathcal{I}$ in the pseudo-differential algebra of X; its characteristic set is the first projection Σ_+ of the complex canonical relation of S; as mentioned earlier it is involutive $\gg 0$ with real part Σ. (there is a symmetric statement for the right annihilator).

Proposition 7. *Let M be the $\mathcal{E}_X$-module $M = \mathcal{E}/\mathcal{I}$. Then the Toeplitz algebra is canonically isomorphic to $\mathrm{End}_{\mathcal{E}}(M)$.*

Proof: let e_M be the image of $1 \in \mathcal{E}$ (it is a generator of M). It is elementary that $\mathrm{End}_{\mathcal{E}}(M)$ is identified with opposite algebra $([\mathcal{E} : \mathcal{I}]/\mathcal{I})^0$ where $[\mathcal{E} : \mathcal{I}]$ denotes the set of $P \in \mathcal{E}$ such that $\mathcal{I}P \subset \mathcal{I}$ (to P corresponds the endomorphism a_P such that $a_P(e) = Pe$). It is also immediate that the map u which to a_P assigns the Toeplitz operator $T_P = SPS = PS$ is an isomorphism (both algebras have a complete filtration by degrees, and the associated graded algebra in both cases is the algebra of symbols on Σ); clearly u is a homomorphism of algebras, of degree ≤ 0, and $\mathrm{gr}\, u = \mathrm{Id}$.

Now M is certainly "good" in the sense above: it is locally defined by transverse equations and has, locally, a resolution whose symbol is a Koszul complex. So the residual trace is well defined on $\mathrm{End}_{\mathcal{E}}(M)$. Since the trace on an algebra of Toeplitz type is unique up to a constant factor, there exists a constant C such that

$$\mathrm{tr}_{res} a_P = C\, \mathrm{tr}_{res} T_P \ . \tag{5}$$

Below we only only need $C \neq 0$; however with the conventions above we have:

Theorem 1. *The constant C above is equal to one ($C = 1$).*

Proof: to the resolution of M above corresponds a complex of pseudo-differential operators

$$0 \to C^\infty(x) \xrightarrow{\mathrm{D}} C^\infty(X, E_1) \to \ldots \xrightarrow{\mathrm{D}} C^\infty(X, E_N) \to 0 \ , \tag{6}$$

exact in degree > 0 and whose homology in degree 0 is the range of S (mod. smoothing operators), i.e. there exists a micro-local operator E on (E_k) such that $DE+ED \sim 1-S$ (cf. [2]; E is a pseudo-differential operator of type $\frac{1}{2}$, not a "classical" pseudo-differential operator, but it preserves micro-supports).

It is elementary that one can modify D, E, and if need be S, by smoothing operators so that (6) is exact (on global sections) in degree $\neq 0$, and $\ker D_0$ is the range of S. Then if $a_s = (a_k, s)$ is a holomorphic family of pseudo-differential homomorphisms of degree s, and T_{a_s} is the Toeplitz operator $T_{a_s} = a_{0,s}|_{\ker S}$, we have $\mathrm{tr}\, T_{a_s} = \sum(-1)^k \mathrm{tr}\, a_{k,s}$ for $\mathrm{Re}\, s \ll 0$ hence also equality for the meromorphic extensions and residues.

Here is an alternative proof (slightly more in the spirit of the paper because it really uses operators mod. C^∞ rather than true operators). Notice first that theorem 1 is (micro) local, and since locally all bundles are trivial, we can reason by induction on $\mathrm{codim}\, \Sigma$. Thus it is enough to check the formula for one example, where $\mathrm{codim}\, \Sigma = 2$. Note also that above we had embedded in the algebra of pseudo-differential operators on a manifold, but we could just as well embed in another Toeplitz algebra.

We choose Σ corresponding to the standard contact (CR) sphere S_{2n-1} of $\mathbb{C}^n$, embedded as the diameter $z_1 = 0$ in the sphere of $\mathbb{C}^{n+1}$

The Toeplitz space H_{n+1} is the space of holomorphic functions in the unit ball of $\mathbb{C}^{n+1}$ (more correctly: their restrictions to the sphere); we choose H_n the subspace of functions independent of z_1. There is an obvious resolution: $0 \to H_{n+1} \to H_{n+1} \to 0$ ($H_n = \ker \partial_1$). We choose on H_n the operator a, restriction of $\rho^- n$, with $\rho = \sum z_j \partial_{z_j}$ (this is the simplest operator with a nonzero residual - our convention is that ρ^σ kills constant functions for all σ).

Lemma 1. *On the sphere S_{2n-1} the residual trace of the Toeplitz operator ρ^{-n} is $tr_{res}\rho^{-n} = \frac{1}{(n-1)!}$.*

Proof: the standard Szegö kernel is $S(z,w) = \frac{1}{\mathrm{vol}\, \mathrm{S}_{2\mathrm{n}-1}}(1 - z.\bar{w})^{-n}$. Now we have the obvious identity

$$\rho(\rho+1)\ldots(\rho+n-1)\ \mathrm{Log}\, \frac{1}{1 - z.\bar{w}} = (n-1)!((1 - z\bar{w})^{-n} - 1) \ ,$$

so that the leading coefficient of the logarithmic part of ρ^{-n} is $\frac{1}{\mathrm{vol}\, \mathrm{S}_{2\mathrm{n}-1}(\mathrm{n}-1)!}$, whose integral over the sphere is $\frac{1}{(n-1)!}$.

On the sphere S_{2n+1} we have $\partial_1\rho = (\rho+1)\partial_1$ so we choose for $(\tilde{a})$ the pair $(a_0 = \rho^{-n}, a_1 = (\rho+1)^{-n})$. Since terms of degree $< -n-1$ do not contribute on S_{2n+1}, the super-trace is

$$\mathrm{supertr}_{res}(\tilde{a}) = \mathrm{tr}_{res}(\rho^{-n} - (\rho+1)^{-n}) = \mathrm{tr}_{res}(n\rho^{-n-1}) = \frac{n}{n!} = \mathrm{tr}_{res} a\ .$$

7 Embedding

If Σ_1, Σ_2 are two symplectic cones, with contact basis X_1, X_2, symplectic embeddings $\Sigma_1 \to \Sigma_2$ exactly correspond to contact embeddings $X_1 \to X_2$, i.e. an embedding $u : X_1 \to X_2$ such that the inverse image $u^*(\lambda_2)$ is a positive multiple of the contact form λ_1 (the corresponding symplectic map take the section $u^*\lambda_2$ of $T^\bullet X_1$ to the section λ_2 of $T^\bullet X_2$. With this in mind we have

Lemma 2. *If X is a compact oriented contact manifold, it can be embedded in the standard contact sphere.*

Proof: the standard contact $(2N-1)$-sphere of radius R has coordinates x_j, y_j $(1 \le j \le N, \sum x_j^2 + y_j^2 = R^2)$ and contact form $\lambda = \sum x_j dy_j - y_j dx_j$ (or a positive multiple of this). If X is a compact contact manifold, its contact form can always be written $2\sum_2^m x_j dy_j$ for some suitable choice of smooth functions x_j, y_j, or just as well $\sum_1^m x_j dy_j - y_j dx_j$, setting for instance $x_1 = 1, y_1 = \sum_2^m x_j y_j$ (m may be larger than the dimension). Adding suitably many other pairs (x_j, y_j) with $y_j = 0, x_N = (R^2 - \sum x_j^2 + y_j^2)^{\frac{1}{2}}$, for R large enough, we get an embedding in a contact sphere of radius R.

Theorem 2. *For any generalized Szegö projector Σ associated to a symplectic cone with compact basis, we have $tr_{res}S = 0$. In particular if Σ corresponds to a contact structure on a compact manifold, the logarithmic trace vanishes.*

By lemma 2 we can suppose that Σ is embedded in the symplectic cone of a standard odd contact sphere. Let $\mathcal{B}$ be the canonical Toeplitz algebra on the sphere: then by prop. 7, the Toeplitz algebra $\mathcal{A}$ of S is isomorphic to $\mathrm{End}_{\mathcal{B}}(M)$ where M is a suitable good $\mathcal{B}$ module. Now on the sphere any good locally free $\mathcal{B}$-module is stably free (any complex vector bundle on an odd sphere is stably trivial), and the Szegö projector has no logarithmic term, so and $\mathrm{tr}_{res} 1_M = 0$, for any free hence also for any good $\mathcal{B}$-module M.

This result is rather negative since it means that the logarithmic trace cannot define new invariants distinguishing CR or contact manifolds. Note however that that it is not completely trivial: it holds for the Toeplitz algebras associated to a CR or contact structure, as constructed in [5], but a contact manifold carries many other star algebras which are locally isomorphic to the Toeplitz algebra (I showed in [7] how Fedosov's classification of star products [10] can be adapted to classify these algebras). Any such algebra $\mathcal{A}$ carries a canonical trace, because the residual trace is invariant by all isomorphisms, so that local traces glue together. If the contact basis is compact, the trace $\mathrm{tr}_{res} 1_{\mathcal{A}}$ is well defined, but there are easy examples showing that it is not always zero.

References

1. Bayen F., Flato M., Fronsdal C., Lichnerowicz A., Sternheimer D. Deformation theory and quantization I, II, Ann. Phys 111 (1977), 61-151.
2. Boutet de Monvel, L. Hypoelliptic operators with double characteristics and related pseudo-differential operators Comm. Pure Appl. Math. 27 (1974), 585-639.
3. Boutet de Monvel, L.; Sjöstrand, J. Sur la singularité des noyaux de Bergman et de Szegö. Astérisque 34-35 (1976), 123-164.
4. Boutet de Monvel, L. On the index of Toeplitz operators of several complex variables. Inventions Math. 50 (1979), 249-272.
5. Boutet de Monvel, L.; Guillemin, V. The Spectral Theory of Toeplitz Operators Ann. of Math. Studies no. 99, Princeton University Press, 1981, 161 pp.
6. Boutet de Monvel, L. Symplectic cones and Toeplitz operators. Multidimensional complex analysis and partial differential equations (São Carlos, 1995), Contemp. Math., 205, 1997, 15-24.
7. Boutet de Monvel, Louis Related semi-classical and Toeplitz algebras. Deformation quantization (Strasbourg, 2001), IRMA Lect. Math. Theor. Phys., 1, de Gruyter, Berlin, 2002, 163–190.
8. Boutet de Monvel, L. Logarithmic trace of Toeplitz projectors Math Research Letters, 12, vol. 2-3, 401-412 (arXiv:math.CV/0412252v1).
9. Dixmier, J. Existence de traces non normales. C. R. Acad. Sci. Paris Sér. A-B 262 (1966) A1107-A1108.
10. Fedosov B.V. A simple geometrical construction of deformation quantization. J. Differential Geom. 40 (1994), no. 2, 213-238.
11. Guillemin, V. Symplectic spinors and partial differential equations. Géométrie symplectique et phys. mathématique , Éditions C.N.R.S. Paris (1975), 217-252.
12. Guillemin, V. Residue traces for certain algebras of Fourier integral operators. J. Funct. Anal. 115 (1993), no. 2, 391-417.
13. Guillemin, V. Wave-trace invariants. Duke Math. J. 83 (1996), no. 2, 287-352.
14. Hirachi, K. Construction of boundary invariants and the logarithmic singularity of the Bergman kernel. Ann. of Math. (2) 151 (2000), no. 1, 151-191.
15. Hirachi, K. Logarithmic singularity of the Szegö kernel and a global invariant of strictly pseudo-convex domains math.CV/0309176, to appear in Ann. Math.
16. Hörmander, L. Fourier integral operators I. Acta Math. 127 (1971), 79-183.
17. Hörmander, L. The analysis of linear partial differential operators I - II - III - IV. *Grundlehren der Math. Wiss. 256,257,274,275, Springer-Verlag (1985).*
18. Kashiwara, M.; Kawai, T.; Sato, M. Microfunctions and pseudodifferential equations, *Lecture Notes 287 (1973), 265-524, Springer-Verlag.*
19. Kashiwara, M. Analyse microlocale du noyau de Bergman. Séminaire Goulaouic-Schwartz 1976-77, exposé n8, École Polytechnique.

20. Kawai T., Kashiwara M. On holonomic systems of microdifferential equations III - systems with regular singularities, publ. RIMS, Kyoto University 17 (1981) 813-979.
21. Melin, A.; Sjöstrand, J. Fourier Integral operators with complex valued phase functions. Lecture Notes 459 (1974) 120-223.
22. Ponge R. Szegö projections and new invariants for CR and contact manifolds ArXiV math.DG/0601370.
23. Wodzicki, M. Noncommutative residue. I. Fundamentals. K-theory, arithmetic and geometry (Moscow, 1984–1986), 320–399, Lecture Notes in Math., 1289, Springer, Berlin, 1987.
24. Zelditch, S. Lectures on wave invariants. Spectral theory and geometry (Edinburgh, 1998), 284–328, London Math. Soc. Lecture Note Ser., 273, Cambridge Univ. Press, Cambridge, 1999.

Nonlinear Stokes phenomena in first or second order differential equations

Ovidiu Costin

Mathematics Department, The Ohio State University, Columbus OH 43210, USA
costin@math.ohio-state.edu

Summary. We study singularity formation in nonlinear differential equations of order $m \leqslant 2$, $y^{(m)} = A(x^{-1}, y)$. We assume A is analytic at $(0,0)$ and $\partial_y A(0,0) = \lambda \neq 0$ (say, $\lambda = (-1)^m$). If $m = 1$ we assume $A(0,\cdot)$ is meromorphic and nonlinear. If $m = 2$, we assume $A(0,\cdot)$ is analytic except for isolated singularities, and also that $\int_{s_0}^{\infty} |\Phi(s)|^{-1/2} d|s| < \infty$ along some path avoiding the zeros and singularities of Φ, where $\Phi(s) = \int_0^s A(0,\tau)d\tau$. Let $H_\alpha = \{z : |z| > a > 0, \arg(z) \in (-\alpha, \alpha)\}$.

If the Stokes constant S^+ associated to $\mathbb{R}^+$ is nonzero, we show that *all* y such that $\lim_{x\to+\infty} y(x) = 0$ are singular at $2\pi i$-quasiperiodic arrays of points near $i\mathbb{R}^+$. The array location determines and is determined by S^+. Such settings include the Painlevé equations P_I and P_{II}. If $S^+ = 0$, then there is exactly one solution y_0 without singularities in $H_{2\pi-\epsilon}$, and y_0 is entire iff $y_0 = A(z,0) \equiv 0$.

The singularities of $y(x)$ mirror the singularities of the Borel transform of its asymptotic expansion, $\mathcal{B}\tilde{y}$, a nonlinear analog of Stokes phenomena. If $m = 1$ and A is a nonlinear polynomial with $A(z,0) \not\equiv 0$ a similar conclusion holds even if $A(0,\cdot)$ is linear. This follows from the property that if f is superexponentially small along $\mathbb{R}^+$ and analytic in H_π, then f is superexponentially unbounded in H_π, a consequence of decay estimates of Laplace transforms.

Compared to [2] this analysis is restricted to first and second order equations but shows that singularities *always* occur, and their type is calculated in the polynomial case. Connection to integrability and the Painlevé property are discussed.

1 Introduction

Stokes transitions play an important role in the study of differential equations (see [1], [4], [10] and references therein). In linear differential equations, solutions that are small in some sector usually become exponentially large in complementary ones.

There is an interesting analog for analytic nonlinear equations which we study for first and second order ones.

Received 14 June, 2006. Revised 11 July, 2006. Accepted 11 July, 2006.

Assume that $A(z,y)$ is analytic at $(0,0)$, and that the following nondegeneracy condition holds: $\partial_y A(0,y) := \lambda \neq 0$. Let, for $m = 1, 2$,

$$y^{(m)} = A(1/x, y) \tag{1}$$

By a simple change of variable $y \mapsto y + a + bx^{-1} + cx^{-2}$ we can assume, without loss of generality, $A(z,0) = o(z^2)$. The conditions in [3] and [2] (which do not require $m \leqslant 2$) then apply. By a change of independent variable we can make $\lambda = (-1)^m$. Let $A_1'(0) = \alpha$.

There exists a one parameter family of solutions of (1) which decay algebraically as $x \to \infty$ [12], [3]. All these solutions can be written as generalized Borel summed transseries [3] (the notation here is similar to that in [2])

$$y(x) = \sum_{k\geq 0} C^k \mathrm{e}^{-kx} x^{\alpha k} y_k(x) = \sum_{k\geq 0} C^k \mathrm{e}^{-kx} x^{\alpha k} \mathcal{L}\mathcal{B}\tilde{y}_k(x) \equiv \mathcal{L}\mathcal{B}\tilde{y}(x) \tag{2}$$

where $\tilde{y}_k$ are formal series in integer powers of x^{-1}. The functions $y_k := \mathcal{L}\mathcal{B}\tilde{y}_k$, the generalized Borel sums of $\tilde{y}_k$, are analytic for $\Re(x) > \nu > 0$ uniformly in k. The function series in (2) is absolutely and uniformly convergent for $\Re(x) > \nu$ and defines an analytic function there. Any decaying solution can be represented uniquely in terms of Laplace transforms in the upper half plane, $y_k^+ = \mathcal{L}(\mathcal{B}\tilde{y})^+$ where $^+$ indicates the upper half plane continuation or in terms of the lower half plane continuations [3]

$$y(x) = \sum_{k\geq 0} C^{+k} e^{-kx} x^{\alpha k} y_k^+ = \sum_{k\geq 0} C^{-k} e^{-kx} x^{\alpha k} y_k^- \tag{3}$$

and we have ([3] Eq. (1.27))

$$C^+ - C^- = S^+ \tag{4}$$

where $S^+ = S_1$ in [3] is the Stokes constant associated to $\mathbb{R}^+$.

In this paper we prove, under some further assumptions, that all solutions, except at most one, that go to zero along some direction form arrays of singularities near $i\mathbb{R}$. After normalization of the equation, the array location is given by

$$x_n = 2n\pi i + \alpha \ln(2n\pi i) + \ln C^{\pm} - \ln \xi_s + o(1) \quad (\mathbb{N} \ni n \to \pm\infty) \tag{5}$$

for some $\xi_s \neq 0$. (If $C^- = 0$ ($C^+ = 0$), then y has no singularities in the lower (upper, respectively) half plane.) In [2], Eq. (28) we showed that such arrays occur provided a solution of an associated equation, (6) below, is singular. Here we show that the solution is singular indeed, under our assumptions.

There is a correspondence between singularities in inverse Laplace space and singularities in the original space, a nonlinear analog of Stokes phenomena: if we change variables to make $\partial_y A(0,0) = \lambda$, then n becomes n/λ. Singularities in Borel space are spaced by (exactly) $|\lambda|$, see [3], while in the x

plane their mirror singularities are, up to log corrections, as we see from (5), regularly spaced by $|\lambda|^{-1}$.

Part of the results rely on a general property that if an analytic function is superexponentially small along a ray, then it becomes superexponentially large along a complementary ray.

2 Main results

In [2] it is shown that there is a unique analytic solution F_0 of the associated equation

$$\sum_{k=1}^{m} \xi^k F_0^{(k)} = A(0, F_0) \tag{6}$$

with $\xi^{-1}F_0(\xi) \to 1$ as $\xi \to 0$. F_0 has the property that

$$y(x) = F_0(x^\alpha e^{-x}) + O(x^{-1}) \tag{7}$$

as $x \to \infty$ if $\xi = x^\alpha e^{-x}$ stays bounded, and if ξ_s is a singular point of F_0, then y is singular along the array (5). From (7) and the argument principle, if $F_0(\xi)$ has a pole or an algebraic or logarithmic branch point at ξ_s then the actual solution y of (1) has a quasiperiodic array of singularities, to leading order of the same type as that of F_0. The origin of the singularities of $y(x)$ is the fact that in (2) the exponentials e^{-kx} become $O(1)$ in the antistokes direction, and they can "pile up" to create blow-up of the solution. The singularities form $2\pi i$-quasiperiodic arrays since they are governed by the size of $x^\alpha e^{-x}$. If y^- is the solution analytic in the lower half plane, then $C^- = 0$ and S^+ can be calculated from (5) and (4).

Assumptions. We recall A is analytic at $(0,0)$. Let $H = A(0,\cdot)$ and $\Phi(s) = \int_0^s H(\tau)d\tau$. If $m = 1$ assume H is meromorphic and nonlinear. If $m = 2$ assume H is analytic except for isolated singularities and such that $\int_a^\infty |\Phi(s)|^{-1/2}ds < \infty$ along a path P avoiding the zeros and singularities of Φ. This includes the Painlevé equations P_I and P_{II}, where H is a nonlinear polynomial [2]. Consider the one-parameter family $\mathcal{F}$ of solutions y of (1) such that $\lim_{x\to+\infty} y(x) = 0$.

Theorem 1. *(i) Under the assumptions above, all functions in $\mathcal{F}$ except at most one are singular in the upper or lower half-plane. The singularities form arrays given by (5). If H is polynomial of degree $N > 1$, then the local behavior of y near a singular point x_s is to leading order $(x - x_s)^{m/(1-N)}$.*

(ii) If there exists an $y_0 \in \mathcal{F}$ analytic in a sector of opening more than π centered on $\mathbb{R}^+$, then there is no other such y_0, and $S^+ = 0$; y_0 is entire iff $y_0 \equiv A(x,0) \equiv 0$.

(iii) For $m = 1$, if $A(z,y)$ is polynomial of degree $N > 1$ in y and $A(z,0) \not\equiv 0$, all solutions in $\mathcal{F}$ except at most one have infinitely many singularities in $H_{\pi+\epsilon}$, even if $A(0,\cdot)$ is linear. The singularity type is as in (i).

2.1 General decay estimates

We state separately some useful and relatively basic estimates, helpful in proving Theorem 1 (iii)[1].

Lemma 2. *If the function $f \not\equiv 0$ is analytic in a half plane $\{z : \Re(z) > \beta\}$ and for all $a > 0$ we have $f(t) = O(e^{-at})$ for $t \in \mathbb{R}^+$, $t \to +\infty$, then for any $\epsilon \geqslant 0$ small enough $f(z)e^{-bz^{1-\epsilon}}$ is unbounded in the closed right half plane for all $b > 0$.*

Lemma 3. *Assume $F \in L^1(\mathbb{R}^+)$ and for some $\epsilon > 0$ we have*

$$\mathcal{L}F(x) = O(e^{-\epsilon x}) \quad \text{as} \quad x \to +\infty \tag{8}$$

Then $F = 0$ a.e. on $[0, \epsilon]$. (The result is sharp as discussed after the proof.)

Corollary 4. *Assume $F \in L^1$ and $\mathcal{L}F = O(e^{-ax})$ as $x \to +\infty$ for all $a > 0$. Then $F = 0$ a.e. on $\mathbb{R}^+$.*

Corollary 5. *Assume $\alpha > \frac{1}{2}$, $f \not\equiv 0$ is analytic in $S = \{z : |z| > R, 2|\arg z| \leqslant \pi/\alpha\}$ and for any $a > 0$ we have $f(t) = o(e^{-at^{\alpha}})$ as $t \to +\infty$. Then for any $b, p > 0$ and $0 < \alpha' < \alpha$, the function $f(z)e^{-bz^{\alpha'}}$ is unbounded in S.*

3 Proofs

3.1 Proof of Theorem 1

(i) Let first $m = 1$. We show by contradiction that F_0 in (6) is not entire. Assume $g(t) := F_0(e^t)$ was entire. We have $g' = H(g)$. First, H must have roots or else $G = 1/H$ would be entire and then the function Q given by

$$Q(g) = \int G(g)dg + C = t \tag{9}$$

would be entire, and injective since Q has an (entire) inverse. Thus Q is linear.

Assume, without loss of generality, that $H(0) = 0$. Next we show that if $H(\tau) = 0$ then $g(t_0) \neq \tau$ for all $t_0 \in \mathbb{C}$. Indeed, if not, $g'(t_0) = 0$ and we would have

$$\frac{(g(t) - \tau)'}{g(t) - \tau} = \frac{H(g(t)) - H(\tau)}{g(t) - \tau} \to H'(\tau) \text{ as } t \to t_0 \tag{10}$$

But this is a contradiction, since t_0 is a zero of finite order of $g(t) - \tau$ and the left side of (10) tends to infinity. Since $g(t) \neq 0$ for all $t \in \mathbb{C}$, we have $g = e^h$

[1] Lemma 2 resembles Carlson's Lemma [11] pp. 185, but does not appear to imply it or to follow from it. Also, the referee points out that similar estimating methods appear in [5].

with h entire, and since g is nontrivial it cannot avoid any other value, thus 0 is the only root of H. Then $h' = e^{-h}H(e^h)$ where now the right hand side is meromorphic and everywhere nonzero, a case we have already analyzed.

Let now $m = 2$, $g(t) = F_0(e^{-t})$. We have $g'' = H(g)$. Then, multiplying by g' and integrating once, we get $g'^2 = \Phi(g) + C$. Using the condition $g \sim e^{-t}$ for large t and the analyticity of H for small g we get $C = 0$. Choosing a determination of the log and a t_0 such that $g(t_0)$ is not a zero or a singularity of Φ we get

$$J(g(t_0), g(t)) := \int_{g(t_0)}^{g(t)} \Phi(s)^{-1/2} ds = t - t_0 \tag{11}$$

We take the path P towards infinity. We have that $J = \lim_{g\to\infty} J(g(t_0), g)$, $g \in P$, is finite and $t_s := t_0 + J$ is clearly a singular point of g.

(ii) It follows from (4) that unless $S^+ = 0$ every solution develops singularities in the upper or lower half plane (in both, except when $C_- = 0$ or $C^+ = 0$). Conversely, if $S^+ = 0$, it follows from [3] that $\mathcal{B}\tilde{y}$ is analytic in the $\mathbb{C} \setminus \mathbb{R}^-$ (Theorem 1) and exponentially bounded in distributions in $H_{2\pi-\epsilon}$ (Theorem 2). Thus $\mathcal{L}\mathcal{B}\tilde{y}$ is a solution of the original differential equation with the asymptotic expansion $\tilde{y}$ in any sector of opening less than $2\pi + \pi/2$ centered on $\mathbb{R}^+$. In particular, this solution cannot be entire unless it is identically zero, and then $A_0 \equiv 0$. The type of singularity follows from (11) and the discussion at the beginning of §2.

For part (iii) we need the following result.

Lemma 6. *Assume y solves*

$$y' = \sum_{k=0}^{N} B_k(x) y^k \tag{12}$$

$N > 1$, in a sector S where the analytic coefficients B_k are $O(x^{q_k})$ and $B_N \not\equiv 0$. Assume furthermore that for a sequence $x_j \to \infty$ in S, $y(x_j)$ grows faster than polynomially in x_j. Then for all j large enough, within a distance $O(x_j^{-q_N} y(x_j)^{1-N})$ of x_j there is a singularity of y.

Note. Heuristically, near a large x_j, the dominant balance is of the form $y' \sim C y^N x^{q_N}$, $C \neq 0$, which forms a singularity as described.

Proof. Let $x = x_j + \sigma$ and $\eta = 1/y$. The equation for σ is

$$\frac{d\sigma}{d\eta} = -\frac{\eta^{N-2}}{\sum_{k=0}^{N} \eta^k B_{N-k}(x_j + \sigma)} = G(\sigma, \eta) \tag{13}$$

with the initial condition $\sigma(1/y(x_j)) = 0$. It is straightforward to check that in our assumptions, for large enough x_j, $G(\sigma, \eta)$ is analytic in the polydisk $\{(\sigma, \eta) : |\sigma| < 1, |\eta| < 2/|y(x_j)|\}$ where furthermore $|G(\sigma, \eta)| = O(\eta^{N-2} x_j^{-q_N})$. Then, in the disk $\{\eta : |\eta| < 2/|y(x_j)|\}$ there exists a unique analytic solution σ and $\sigma = O(\eta^{N-1} x_j^{-q_N})$. It is clear that $x_j + \sigma(0)$ is a zero of η and thus a singularity of y. □

Proof of Theorem 1 (iii) Assume that there is a solution y_0 of such that $y_0 \to 0$ as $x \to +\infty$ with finitely many singularities in a sector S of opening $\pi + 2\epsilon$ centered on $\mathbb{R}^+$. (There may be an exceptional solution, for instance when the Stokes constant is zero and the asymptotic series converges.) Let y_2 be another solution of $y' = A(x^{-1}, y)$ such that $y_2 \to 0$ as $x \to \infty$ and let $\delta = y_2 - y_1$. The equation for δ is of the form (12) with $B_0 = 0$. From the general theory of differential equations [12] or from [3] it follows that $\delta = O(e^{-x})$ as $x \to \infty$. By Lemma 2 there is a sequence of $x_j \to \infty$ in S so that $\delta \geq const. \exp(|x_j|^{1-\epsilon})$. Near every x_j with j large enough there is, by Lemma 6, a singularity of δ and thus of y_2.

3.2 Connection with integrability

It is seen that if H is a polynomial of degree $N > 2$ if $m = 1$ or $N > 3$ if $m = 2$ a one parameter family of solutions forms arrays of branch-point singularities in $\mathbb{C}$ and the Painlevé test fails. For other values of p, the corrections $F_j, j > 1$ [2] can be calculated to determine the exact type of singularity. More can be done however. Assuming we are in the case where singularities are branch points, in view of the exact description of type and location of singularities it is possible to calculate the monodromy of solutions along a curve in $\mathbb{C}$ winding among sufficiently many singularities to show dense branching, a concept proposed by Kruskal, which can be used to show absence of continuous first integrals as in [7], [8]. This will be the subject of a different paper.

3.3 Proof of Lemma 3

We write

$$\int_0^\infty e^{-px} F(p)dp = \int_0^\epsilon e^{-px} F(p)dp + \int_\epsilon^\infty e^{-px} F(p)dp \tag{14}$$

we note that

$$\left| \int_\epsilon^\infty e^{-px} F(p)dp \right| \leqslant e^{-\epsilon x} \int_\epsilon^\infty |F(p)|dp \leqslant e^{-p\epsilon} \|F\|_1 = O(e^{-\epsilon x}) \tag{15}$$

Therefore

$$g(x) = \int_0^\epsilon e^{-px} F(p)dp = O(e^{-\epsilon x}) \quad \text{as} \quad x \to +\infty \tag{16}$$

The function g is manifestly entire. Let $h(x) = e^{\epsilon x} g(x)$. Then by assumption h is entire and uniformly bounded for $x \in \mathbb{R}$ (since by assumption, for some x_0 and all $x > x_0$ we have $|h| \leqslant C$ and by continuity $\max|h| < \infty$ on $[0, x_0]$). The function is bounded by $\|F\|_1$ for $x \in i\mathbb{R}$. By Phragmén-Lindelöf's theorem (first applied in the first quadrant and then in the fourth quadrant, with $\beta = 1, \alpha = 2$) h is bounded in the closed right half plane. Now, for $x = -s < 0$ we have

$$e^{-s\epsilon} \int_0^\epsilon e^{sp} F(p)dp \leqslant \int_0^\epsilon |F(p)| \leqslant \|F\|_1 \tag{17}$$

Again by Phragmén-Lindelöf (and again applied twice) h is bounded in the closed left half plane thus bounded in C, and it is therefore a constant. But, by the Riemann-Lebesgue lemma, $h \to 0$ for $x = is$ when $s \to +\infty$. Thus $h \equiv 0$. Therefore, with χ_A the characteristic function of A,

$$\int_0^\epsilon F(p)e^{-isp}dp = \hat{\mathcal{F}}(\chi_{[0,\epsilon]}F) = 0 \tag{18}$$

for all $s \in \mathbb{R}$ entailing the conclusion.
Note. In the opposite direction, by Laplace's method it is easy to check that for any small $\epsilon > 0$ we have $\mathcal{L}e^{-p^{-\frac{1-\epsilon}{\epsilon}}} = o\left(e^{-x^{1-\epsilon}}\right)$ and for any n $\mathcal{L}\left(e^{-E_{n+1}(1/p)}\right) = o(e^{-x/L_n(x)})$ where E_n is the n-th composition of the exponential with itself and L_n is the n-th composition of the log with itself. □

3.4 Proof of Proposition 2

By a change of variable we may assume that $\beta = -1$. Assume that for some $b > 0$ $e^{-bz^{1-\epsilon}} f(z)$ was bounded in the closed right half plane. Then $\psi(z) = (1+z)^{-2}e^{-bz^{1-\epsilon}} f(z)$ satisfies the assumptions of Lemma 8. But then $\psi(z) = \mathcal{L}\mathcal{L}^{-1}\psi(z)$ satisfies the assumptions of Corollary 4 and $\psi \equiv 0$.

Note 7. *There is indeed loss of exponential rate: the entire function* $\Gamma(x)^{-1} = O(e^{-n \ln n})$ *on* $\mathbb{R}^+$ *but is bounded by* $O(e^{|x|\pi/2})$ *in the closed right half plane.*

4 Appendix: overview of Laplace transform properties

For convenience we provide some standard results on Laplace transforms.

Remark 1 (Uniqueness). *Assume* $F \in L^1(\mathbb{R}^+)$ *and* $\mathcal{L}F = 0$ *for a set of* x *with an accumulation point. Then* $F = 0$ *a.e.*

Proof. By analyticity, $\mathcal{L}F = 0$ in the open right half plane and by continuity, for $s \in \mathbb{R}$, $\mathcal{L}F(is) = 0 = \hat{\mathcal{F}}F$ where $\hat{\mathcal{F}}F$ is the Fourier transform of F (extended by zero for negative values of p). Since $F \in L^1$ and $0 = \hat{\mathcal{F}}F \in L^1$, by the known Fourier inversion formula [6], $F = 0$. □

Lemma 8. *Assume that* $c \geqslant 0$ *and* $f(z)$ *is analytic in* $H_c := \{z : \Re z \geqslant c\}$. *Assume further that* $g(t) := \sup_{c' \geqslant c} |f(c' + it)| \in L^1(\mathbb{R}, dt)$. *Let*

$$F(p) = \frac{1}{2\pi i} \int_{c-i\infty}^{c+i\infty} e^{px} f(x) dx =: (\mathcal{L}^{-1} f)(p) \tag{19}$$

Then for any $x \in \{z : \Re z > c\}$ *we have* $\mathcal{L}F = \int_0^\infty e^{-px} F(p) dp = f(x)$

Note that for any $x' = x_1' + iy_1' \in \{z : \Re z > c\}$

$$\int_0^\infty dp \int_{c-i\infty}^{c+i\infty} \left| e^{p(s-x')} f(s) \right| d|s| \leqslant \int_0^\infty dp e^{p(c-x_1')} \|g\|_1 \leqslant \frac{\|g\|_1}{x_1' - c} \tag{20}$$

and thus, by Fubini we can interchange the orders of integration:

$$\begin{aligned} U(x') &= \int_0^\infty e^{-px'} \frac{1}{2\pi i} \int_{c-i\infty}^{c+i\infty} e^{px} f(x) dx \\ &= \frac{1}{2\pi i} \int_{c-i\infty}^{c+i\infty} dx f(x) \int_0^\infty dp e^{-px'+px} = \frac{1}{2\pi i} \int_{c-i\infty}^{c+i\infty} \frac{f(x)}{x' - x} dx \end{aligned} \tag{21}$$

Since $g \in L^1$ there exist subsequences $\{\tau_n\}, \{-\tau_n'\}$ tending to infinity such that $|g(\tau_n)| \to 0$. Let $x' > \Re x = x_1$ and consider the box $B_n = \{z : \Re z \in [x_1, x'], \Im z \in [-\tau_n', \tau_n]\}$ with positive orientation.

$$\int_{B_n} \frac{f(s)}{x' - s} ds = -f(x') \tag{22}$$

while, by construction,

$$\lim_{n\to\infty} \int_{B_n} \frac{f(s)}{x' - s} ds = \int_{x'-i\infty}^{x'+i\infty} \frac{f(s)}{x' - s} ds - \int_{c-i\infty}^{c+i\infty} \frac{f(s)}{x' - s} dx \tag{23}$$

On the other hand, by dominated convergence, we have

$$\int_{x'-i\infty}^{x'+i\infty} \frac{f(s)}{x' - s} ds \to 0 \text{ as } x' \to \infty \tag{24}$$

Acknowledgments. The work was partially supported by NSF grant 0406193.

References

[1] T. Aoki, T. Kawai, S. Sasaki, A. Shudo and Y. Takei, *Virtual turning points and bifurcation of Stokes curves for higher order ordinary differential equations.* J. Phys. A 38 (2005), no. 15, 3317–3336.

[2] O. Costin and R. D. Costin, *On the formation of singularities of solutions of nonlinear differential systems in antistokes directions.* Inventiones Mathematicae 145 (2001), no. 3, 425–485.

[3] O. Costin, *On Borel summation and Stokes phenomena for rank one nonlinear systems of ODE's.* Duke Math. J. 93 (1998), no. 2, 289–344.

[4] A. Its, A. R. Fokas and A. S. Kapaev, *On the asymptotic analysis of the Painlevé equations via the isomonodromy method.* Nonlinearity 7 (1994), no. 5, 1291–1325.

[5] R. Paley and N. Wiener, *Fourier transforms in the complex domain.* New York, American Mathematical Society (1934).

[6] W. Rudin, *Real and Complex Analysis.* Third Edition, Mc-Graw-Hill (1987).

[7] R. D. Costin, *Integrability Properties of Nonlinearly Perturbed Euler Equations.* Nonlinearity 10 (1997), no. 4, 905–924.

[8] R. D. Costin, *Integrability Properties of a Generalized Lame Equation; Applications to the Henon-Heiles System.* Meth. Appl. An. 4 (1997), no. 2, 113–123.

[9] G. G. Stokes, *On the discontinuity of arbitrary constants which appear in divergent developments.* Trans. Camb. Phil. Soc. X (1864), 106–128.

[10] Y. Sibuya, *Global theory of a second order linear ordinary differential equation with a polynomial coefficient.* Amsterdam, North-Holland Pub. Co.; New York, American Elsevier Pub. Co. (1975).

[11] E. C. Titchmarsh, *The theory of functions.* Oxford University Press (1964).

[12] W. Wasow, *Asymptotic expansions for ordinary differential equations.* Interscience Publishers (1968).

Reconstruction of inclusions for the inverse boundary value problem for non-stationary heat equation

Yuki Daido[1], Hyeonbae Kang[2], and Gen Nakamura[3]

[1] Department of Mathematics, Hokkaido University, Sapporo 060-0810, Japan
daido@math.sci.hokudai.ac.jp
[2] School of Mathematical Sciences, Seoul National University, Seoul 151-747, Korea
hkang@math.snu.ac.kr
[3] Department of Mathematics, Hokkaido University, Sapporo 060-0810, Japan
gnaka@math.sci.hokudai.ac.jp

Summary. An inverse problem for identifying an inclusion inside an isotropic, homogeneous heat conductive medium is considered. The shape of inclusion can change time dependently. For the one space dimensional case, we developed an analogue of the probe method known for inverse boundary value problems for elliptic equations and gave a reconstruction scheme for identifying the inclusion from the Neumann to Dirichlet map.

1 Introduction

We consider an inverse boundary value problem identifying an unknown inclusion inside heat conductive media Ω by boundary measurements. The input and output of the measurements are the heat flux and temperature on the boundary $\partial\Omega$ of Ω, respectively. This kind of problem arises from thermography.

For the one space dimensional case, we will give a reconstruction scheme for identifying the unknown inclusion which can depend on time by infinitely many measurements. It is an analogue of the probe method [3] which is a well known reconstruction scheme to identify an unknown inclusion for a stationary heat equation. If our reconstruction scheme can be numerically realized, it gives an approximate reconstruction of the unknown inclusion by finitely many measurements. Hence, it may have a practical application not only for thermography but also for inverse problem identifying an unknown phase in a phase transition problem such as Stefan problem, because we allow the unknown inclusion to depend on time.

Received 20 February, 2006. Revised 7 November, 2006, 31 March, 2007. Accepted 31 March, 2007.

Next we formulate the problem in detail for any space dimension $n = 1, 2, 3$ and state our result. Let $\Omega \subset \mathbb{R}^n$ $(n = 1, 2, 3)$ be a bounded domain considered as an isotropic heat conductive medium. We denote its boundary by $\partial\Omega$ and assume that it is of class C^2 if $n = 2, 3$. It has an inclusion which can depend on time $t \in [0, T]$ with $T > 0$. More precisely, for each $t \in [0, T]$, we are given an inclusion $D(t)$ which is a bounded domain such that $\overline{D(t)} \subset \Omega$, $\Omega \setminus \overline{D(t)}$ is connected and the boundary $\partial D(t)$ of $D(t)$ is of class C^2 if $n = 2, 3$. Moreover, $\partial D(t)$ is C^1 with respect to $t \in [0, T]$. That is $\cup_{0 \le t \le T} \partial D(t)$ is a piecewise C^1 manifold with boundary $\partial D(0)$, $\partial D(T)$ in $\Omega \times [0, T]$. The two dimensional figures of Ω and $D := \bigcup_{0 \le t \le T} D(t) \times \{t\}$ are given below.

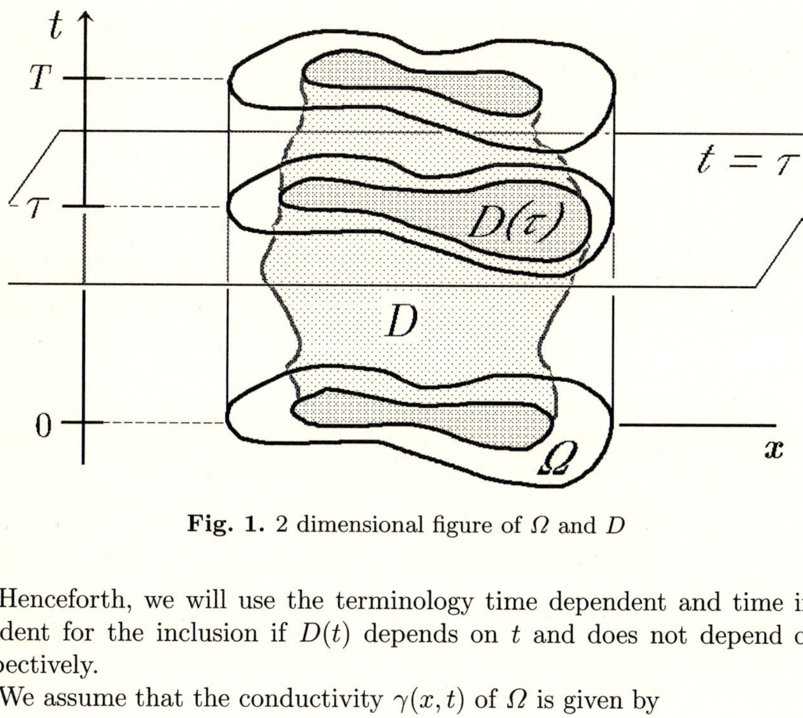

Fig. 1. 2 dimensional figure of Ω and D

Henceforth, we will use the terminology time dependent and time independent for the inclusion if $D(t)$ depends on t and does not depend on t, respectively.

We assume that the conductivity $\gamma(x, t)$ of Ω is given by

$$\gamma(x, t) = 1 + (k - 1)\chi_{D(t)} \tag{1.1}$$

for each $0 \le t \le T$ with a constant $k > 0$ $(k \ne 1)$, where $\chi_{D(t)}$ is the characteristic function of $D(t)$. This means that the conductivity is k inside D and 1 outside D.

For simplicity, we will use the following notations. That is, for any $E \subset \mathbb{R}^n$ and $T_0, T_1 \in \mathbb{R}$ $(T_0 < T_1), T > 0$, we denote $E_{(T_0, T_1)} := E \times (T_0, T_1)$ and $E_T := E \times (0, T)$. Also, we use the following notations for Sobolev spaces. That is, for $p, q \in \mathbb{Z}_+ := \mathbb{N} \cup \{0\}$ or $p = \frac{1}{2}$, $H^p(\Omega), H^p(\partial\Omega)$ and $H^{p,q}(\Omega_T)$ denote the usual Sobolev spaces where p and q in $H^{p,q}(\Omega_T)$ denote the regularity with respect to x and t, respectively (cf.[7]). The dual spaces of $H^p(\Omega)$ and $H^p(\partial\Omega)$ are denoted by $H^p(\Omega)^*$ and $H^p(\partial\Omega)^*$, respectively. Moreover, for an open set $U \subset \mathbb{R}^{n+1}$ with Lipschitz boundary and $p,q \in \mathbb{Z}_+$, $H^{p,q}(U)$ is defined likewise $H^{p,q}(\Omega_T)$. That is $g \in H^{p,q}(U)$ if and only if there exists $\boldsymbol{g} \in H^{p,q}(\mathbb{R}^{n+1}) := \{\boldsymbol{g} \in \mathcal{D}'(\mathbb{R}^{n+1}); \|\boldsymbol{g}\|_{H^{p,q}(\mathbb{R}^{n+1})} := \|\{(1+|\xi|^2)^{\frac{p}{2}} + (1+\tau^2)^{\frac{q}{2}}\}\hat{\boldsymbol{g}}(\xi,\tau)\|_{L^2(\mathbb{R}^{n+1})} < \infty\}$ such that $\boldsymbol{g}\big|_U = g$. Here the norm $\|g\|_{H^{p,q}(U)} := \inf_{\boldsymbol{g}|_U = g,\, \boldsymbol{g} \in H^{p,q}(\mathbb{R}^{n+1})} \|\boldsymbol{g}\|_{H^{p,q}(\mathbb{R}^{n+1})}$.

Now, as for the forward problem of our inverse problem, we consider the boundary value problem

$$\begin{cases} (P_D u)(x,t) := \partial_t u(x,t) - \operatorname{div}_x\big(\gamma(x,t)\nabla_x u(x,t)\big) = 0 \quad \text{in} \quad \Omega_T \\ \partial_\nu u(x,t) = f(x,t) \quad \text{on} \quad \partial\Omega_T, \quad u(x,0) = 0 \end{cases} \tag{1.2}$$

for given $f \in L^2((0,T);(H^{\frac{1}{2}}(\partial\Omega))^*)$. The physical meaning of u and f are the temperature and heat flux, respectively.

We look for a weak solution $u \in H^{1,0}(\Omega_T)$ to (1.2). The definition of the weak solution is as follow.

Definition 1.1 (Weak solution).

If $u \in H^{1,0}(\Omega_T)$ satisfies

$$\int_{\Omega_T} (-u\partial_t\varphi + \gamma(x,t)\nabla_x u \cdot \nabla_x\varphi)\, dx\, dl = \int_{\partial\Omega_T} f\varphi\big|_{\partial\Omega_T}\, d\sigma\, dt \tag{1.3}$$

for all $\varphi \in W(\Omega_T) := \{u \in H^{1,0}(\Omega_T); \partial_t u \in L^2((0,T);(H^1(\Omega))^*)\}$ with $\varphi = 0$ at $t = T$, we call u a weak solution to (1.2).

Then, we have the well known unique solvability of (1.2).

Theorem 1.2 (Unique solvability [8]).

For given $f \in L^2((0,T);(H^{\frac{1}{2}}(\partial\Omega))^*)$, there exists a unique solution $u = u(f) \in W(\Omega_T)$ to (1.2).

Based on Theorem 1.2, we define the Neumann to Dirichlet map Λ_D which we take as our measured data.

Definition 1.3 (Neumann-to-Dirichlet map).

Let $u(f) \in W(\Omega_T)$ be the solution to (1.2). We define the Neumann-to-Dirichlet map $\Lambda_D : L^2((0,T);(H^{\frac{1}{2}}(\partial\Omega))^*) \to L^2((0,T);H^{\frac{1}{2}}(\partial\Omega))$ by

$$\Lambda_D(f) := u(f) \quad on \quad \partial\Omega_T. \tag{1.4}$$

The physical meaning of the measurement Λ_D is to measure the temperature induced from inputting current or heat flux infinitely many times.

Now, we consider the inverse problem:

(IP) **Suppose that k, D are unknown. Reconstruct D from Λ_D.**

In this paper we only consider reconstructing D and ignore reconstructing k. Of course it is important to reconstruct k. It will be our next project.

Our main theorem is the following.

Theorem 1.4. *If $n = 1$, there is a reconstruction scheme for the inverse problem* (IP) *if D satisfies the geometric assumption given in Section 4. The details of the reconstruction scheme will be given later.*

There are several results already known for the case taking the Dirichlet-to-Neumann map Π_D as measured data. The Dirichlet-to-Neumann map Π_D is defined by changing the role of Neumann data and Dirichlet data in the definition of Neumann-to-Dirichlet map. Mathematically, taking Λ_D or Π_D as a measurement does not make any difference. We remark that if the inclusion is time independent, we can reduce the problem to an inverse boundary value problem for stationary heat equation by analytically continuing the solution of (1.2) to the time interval $[0, \infty)$. But it is better to avoid any analytic continuation, because it is an ill-posed procedure. Hence, we will not touch any result based on such kind of argument. Bellout proved the local uniqueness and stability for time independent inclusions ([1]). For time dependent inclusions, Elayyan and Isakov proved the global uniqueness taking the so called localized Dirichlet-to-Neumann map Π'_D as measured data ([2]). Here, the localized Dirichlet-to-Neumann map Π'_D is defined by $\Pi'_D(f) := (\Pi_D(f))\big|_{\Gamma_T}$, where Γ is a nonempty open subset of $\partial\Omega$. Clearly, for one space dimension, $\Pi'_D = \Pi_D$. These results all concerned with the uniqueness or stability, and there was not any result for the reconstruction (i.e. our inverse problem (IP)). Hence, our result is the first result for the reconstruction.

Next possible problems are (i) a numerical realization of our reconstruction scheme and (ii) the generalization of our result to higher space dimension. However, we have not achieved any result yet. The difficulties of the both problems are as follows. For (i), the numerical realization of the Runge's approximation given below for the heat equation becomes more difficult than that for the Laplace equation, and for (ii), analyzing the behavior of the reflected solution given below becomes more complicated.

In this paper we give a reconstruction scheme and an outline of the proof of its validity. The details of the proof will be given elsewhere.

The rest of this paper is organized as follows. In Section 2, we define the pre-indicator function based on Runge's approximation and give its representation in terms of the reflected solution. This can be done for any space dimension. In Section 3, restricting to one space dimensional case, we first analyze the behavior of reflected solution. Next, based on this, we define two mathematical testing machineries called indicator functions from pre-indicator function for detecting the time independent unknown inclusion and time dependent inclusion, respectively. Finally, by using these indicator functions, we give reconstruction schemes for identifying the time independent inclusion and time dependent unknown inclusion, respectively.

2 Pre-indicator Function

In this section we define the pre-indicator function for any space dimension. We expressed everything for general space dimension to show what extent we can do for general space dimension.

For $(y,s),(y',s') \in \Omega_T \setminus \overline{D}$ such that $(y,s) \neq (y',s')$, let $G_{(y,s)}(x,t)$ and $G^*_{(y',s')}(x,t)$ be

$$G_{(y,s)}(x,t) = \begin{cases} \dfrac{1}{[4\pi(t-s)]^{n/2}} \exp\left[-\dfrac{|x-y|^2}{4(t-s)}\right] & (t>s) \\ 0 \qquad (t \le s), \end{cases} \tag{2.1}$$

$$G^*_{(y',s')}(x,t) = \begin{cases} 0 \qquad (s' \le t) \\ \dfrac{1}{[4\pi(s'-t)]^{n/2}} \exp\left[-\dfrac{|x-y'|^2}{4(s'-t)}\right] & (t<s'). \end{cases} \tag{2.2}$$

In order to define the pre-indicator function we use the following Runge's approximation theorems to approximate $G_{(y,s)}$ and $G^*_{(y',s')}$ by the homogeneous solutions of $P_\emptyset$ and $P^*_\emptyset$ in $\Omega_{(-\varepsilon,T+\varepsilon)}$, respectively. Here, $P_\emptyset$ is P_D in (1.2) with $D=\emptyset$ and $P^*_\emptyset$ is the adjoint operator of $P_\emptyset$.

Theorem 2.1 (Runge's approximation theorem 1).
For $T_0' < T_0 < T_1 < T_1'$, let U be an open subset of $\Omega_{(T_0',T_1')}$ such that

$$\begin{cases} \partial U \text{ is Lipschitz,} \\ \overline{U} \subset \Omega_{(T_0',T_1')} \text{ and } \Omega_{(T_0',T_1')} \setminus \overline{U} \text{ is connected.} \end{cases} \tag{2.3}$$

Then, for any open subset V of $\Omega_{(T_0',T_1')}$ such that $\overline{U} \subset V \subset \overline{V} \subset \Omega_{(T_0',T_1')}$ and any $v \in H^{2,1}(V)$ satisfying

$$P_\phi v := \partial_t v - \Delta_x v = 0 \quad in \quad V, \quad v\big|_{(T_0',T_0]} = 0, \tag{2.4}$$

there exists a sequence $\{v^j\} \subset H^{2,1}(\Omega_{(T_0',T_1')})$ such that

$$P_\phi v^j = 0 \quad in \quad \Omega_{(T_0',T_1')}, \quad v^j\big|_{(T_0',T_0]} = 0 \tag{2.5}$$

and

$$v^j \to v \ (j \to \infty) \quad in \quad L^2(U). \tag{2.6}$$

We also have the same kind of theorem for the adjoint problem.

Theorem 2.2 (Runge's approximation theorem 2).
For $T_0' < T_0 < T_1 < T_1'$, let Ω and U be as above. Then, for any open subset V of $\Omega_{(T_0',T_1')}$ such that $\overline{U} \subset V \subset \overline{V} \subset \Omega_{(T_0',T_1')}$ and any $\varphi \in H^{2,1}(V)$ satisfying

$$P_\phi^* \varphi := -\partial_t \varphi - \Delta_x \varphi = 0 \quad in \quad V, \; \varphi\big|_{[T_1, T_1')} = 0, \tag{2.7}$$

there exists a sequence $\{\varphi^j\} \subset H^{2,1}(\Omega_{(T_0', T_1')})$ *such that*

$$P_\phi^* \varphi^j = 0 \quad in \quad \Omega_{(T_0', T_1')}, \; \varphi^j\big|_{[T_1, T_1')} = 0 \tag{2.8}$$

and

$$\varphi^j \to \varphi \; (j \to \infty) \quad in \quad L^2(U). \tag{2.9}$$

These theorems can be proven by adapting the proof of Runge's approximation theorem given in [5].

By Runge's approximation theorem (Theorem 2.1, 2.2) and the interior regularity theorem for $P_\emptyset, P_\emptyset^*$ ([6]), we can select sequences $\{v^j_{(y,s)}\}$, $\{\varphi^j_{(y',s')}\} \subset H^{2,1}(\Omega_{(-\varepsilon, T+\varepsilon)})$ such that

$$\begin{cases} P_\phi v^j_{(y,s)} = 0 \ \text{ in } \Omega_{(-\varepsilon,T+\varepsilon)}, \; v^j_{(y,s)}\big|_{(-\varepsilon,0]} = 0, \\ v^j_{(y,s)} \to G_{(y,s)} \; (j \to \infty) \ \text{ in } H^{2,1}(U) \end{cases} \tag{2.10}$$

and

$$\begin{cases} P_\phi^* \varphi^j_{(y',s')} = 0 \ \text{ in } \Omega_{(-\varepsilon,T+\varepsilon)}, \; \varphi^j_{(y',s')}\big|_{[T,T+\varepsilon)} = 0, \\ \varphi^j_{(y',s')} \to G^*_{(y',s')} \; (j \to \infty) \ \text{ in } H^{2,1}(U) \end{cases} \tag{2.11}$$

for each open set $U \subset \Omega_{(-\varepsilon,T+\varepsilon)}$ with Lipschitz boundary such that $\Omega_{(-\varepsilon,T+\varepsilon)} \setminus \overline{U}$ is connected and $(y,s), (y',s') \notin U$. We call this U and these $\{v^j_{(y,s)}\}, \{\varphi^j_{(y',s')}\}$ an approximation domain and Runge's approximation functions, respectively. Henceforth, for example, $v^j_{(y,s)}\big|_{(-\varepsilon,0]}$ and $U\big|_{[0,T]}$ denote the restriction of $v^j_{(y,s)}$ and U to $\mathbb{R}^n \times (-\varepsilon, 0]$ and $\mathbb{R}^n \times [0,T]$, respectively.

Remark 2.3. *Later giving a reconstruction scheme for the one space dimensional case, we will move (y,s) along a straight line C_s parallel to the x axis given by $C_s := \{(y(\lambda), s); 0 \le \lambda \le 1\}$ for fixed $s \in (0,T)$ with end points $y(0),\ y(1) \in \partial\Omega$. Since the approximation domain U has to satisfy $(y,s) \notin U$ and $\overline{D} \subset U$, we want to have $\Omega_{(-\varepsilon,T+\varepsilon)} \setminus \overline{U}$ small and narrow as much as possible. Hence, for fixed $j \in \mathbb{N}$ and $(y,s) = (y(\lambda), s) \in C_s \setminus \overline{D}$ with $0 < \lambda < 1$, we take $U = U_j$ and $\{v^j_{(y,s)}\}$ in (2.10) as follows.*

(i) Each U_j satisfies (2.3).

(ii) $\{U_j\}$ satisfies $\overline{U_j} \subset U_{j+1}$ $(j \in \mathbb{N})$ and $\bigcup_{j=1}^{\infty} U_j = \Omega_{(-\varepsilon,T+\varepsilon)} \setminus C_s(\lambda)$, where $C_s(\lambda) := \{(y(\lambda'), s); 0 \le \lambda' \le \lambda\}$.

(iii) Each $v^j_{(y,s)}$ satisfies

$$\|v^j_{(y,s)} - G_{(y,s)}\|_{L^2(U_j)} < \frac{1}{j} \quad (j = 1, 2, \cdots). \tag{2.12}$$

In the same way, we take $U = U_j$ and $\{\varphi^j_{(y',s')}\}$ in (2.11) by replacing (y,s) by (y',s').

Using these Runge's approximation functions $\{v^j_{(y,s)}\}, \{\varphi^j_{(y',s')}\}$, we define the pre-indicator function as follow.

Definition 2.4 (Pre-indicator function).
Let $(y,s),(y',s') \in \Omega_T \setminus \overline{D}$ such that $(y,s) \neq (y',s')$, and $\{v^j_{(y,s)}\}$, $\{\varphi^j_{(y',s')}\} \subset H^{2,1}(\Omega_{(-\varepsilon,T+\varepsilon)})$ be Runge's approximation functions as above. We define the pre-indicator function $I(y',s';y,s)$ by

$$I(y',s';y,s) := \lim_{j\to\infty} \int_{\partial\Omega_T} \{\partial_\nu v^j_{(y,s)}\big|_{\partial\Omega_T} \varphi^j_{(y',s')}\big|_{\partial\Omega_T} - \Lambda_D(\partial_\nu v^j_{(y,s)}\big|_{\partial\Omega_T})\partial_\nu \varphi^j_{(y',s')}\big|_{\partial\Omega_T}\}\, d\sigma\, dt. \tag{2.13}$$

Next we analyze the behavior of the pre-indicator function. To start, we have to represent the pre-indicator function in terms of the reflected solution which is given in the following lemma.

Lemma 2.5 (Reflected solution).
Let $\{v^j_{(y,s)}\} \subset H^{2,1}(\Omega_{(-\varepsilon,T+\varepsilon)})$ be Runge's approximation functions and $u^j_{(y,s)} := u(\partial_\nu v^j_{(y,s)}\big|_{\partial\Omega_T})$, $w^j_{(y,s)} := u^j_{(y,s)} - v^j_{(y,s)}$. Then $w^j_{(y,s)}$ has a limit $w_{(y,s)} \in W(\Omega_T)$ which satifies

$$\begin{cases} P_D w_{(y,s)} = (k-1)\,\mathrm{div}_x(\chi_{D(t)} \nabla_x G_{(y,s)}) \text{ in } \Omega_T \\ \partial_\nu w_{(y,s)} = 0 \text{ on } \partial\Omega_T,\ w_{(y,s)}(x,0) = 0\,. \end{cases} \tag{2.14}$$

We call $w_{(y,s)}$ the reflected solution.

Theorem 2.6 (Representation formula).
For $(y,s),(y',s') \in \Omega_T \setminus \overline{D}$ such that $(y,s) \neq (y',s')$,

$$I(y',s';y,s) = -w_{(y,s)}(y',s') - \int_{\partial\Omega_T} w_{(y,s)} \partial_\nu G^*_{(y',s')}\, d\sigma\, dt. \tag{2.15}$$

This can be proven by using the convergence of $\{w^j_{(y,s)}\} \subset H^{1,0}(\Omega_T)$ for $(y,s) \in \Omega_T \setminus \overline{D}$, the integration by parts and

$$\lim_{\varepsilon\to 0} \int_\Omega w_{(y,s)}(x,s')G_{(y',0)}(x,\varepsilon)\, dx = w_{(y,s)}(y',s'). \tag{2.16}$$

3 One Dimensional and Time-independent Case

Let $\Omega = (a_0,a_1), D(t) = (d_0,d_1)$ $(a_0 < d_0 < d_1 < a_1)$. For simplicity, we set $w_{+,1} := w_{(y,s)}\big|_{(d_1,a_1)_T}, w_- := w_{(y,s)}\big|_{(d_0,d_1)_T}$ and $w_{+,0} := w_{(y,s)}\big|_{(a_0,d_0)_T}$.

Let us focus on identifying d_1. We first take the Laplace transform $w_{(y,s)}$ with respect to t and then solve a transmission boundary value problem. Next,

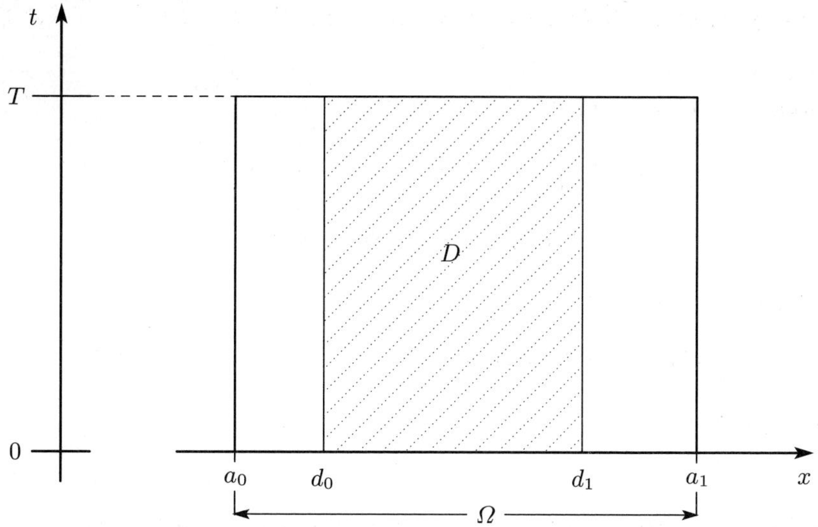

Fig. 2. 1 dimensional and time-independent case

we apply the inverse Laplace transform to the solution of the transmission boundary value problem. Then, we can extract the dominant part $w_{+,1;p}$ of $w_{+,1}$ (cf.[4]). It is given by

$$w_{+,1;p}(x,t) := \frac{1-\sqrt{k}}{1+\sqrt{k}} G_{(y,s)}(-x+2d_1,t), \tag{3.1}$$

$$R_{(y,s)}(x,t) := w_{+,1}(x,t) - w_{+,1;p}(x,t) \tag{3.2}$$

such that

$$|R_{(y,s)}(x,t)| \text{ is uniformly bounded in } [d_1,a_1]. \tag{3.3}$$

Next, we define an indicator function $\mathcal{I}_{\mathrm{ind}}(y,s;\varepsilon)$ for the time-independent case as follow.

Definition 3.1 (Indicator function for the time-independent case). *For $(y,s) \in \Omega_T \setminus \overline{D}$, $\varepsilon > 0$, we define an indicator function $\mathcal{I}_{\mathrm{ind}}(y,s;\varepsilon)$ for the time-independent case by*

$$\mathcal{I}_{\mathrm{ind}}(y,s;\varepsilon) := |I(y+\varepsilon, s+\varepsilon^2; y,s)|. \tag{3.4}$$

Now we observe that the integrand of the second term of (2.15) has no singularity on $\partial\Omega_T$. Hence the second term of (2.15) is bounded as $(y',s') \to (y,s)$. We also observe that

$$w_{+,1;p}(y+\varepsilon, s+\varepsilon^2) = \frac{1-\sqrt{k}}{1+\sqrt{k}} G_{(y,s)}(-(y+\varepsilon)+2d_1, s+\varepsilon^2). \tag{3.5}$$

Then, by recalling (2.15) and (3.4), we have the following result for identifying d_1.

Theorem 3.2. *For any* $0 < s < T$,

$$\lim_{\varepsilon \to 0} \mathcal{I}_{\mathrm{ind}}(y, s; \varepsilon) = \infty \text{ when } y = d_1, \tag{3.6}$$

$$\lim_{\varepsilon \to 0} \mathcal{I}_{\mathrm{ind}}(y, s; \varepsilon) < \infty \text{ when } y > d_1. \tag{3.7}$$

Therefore d_1 *is given by*

$$d_1 = \inf\{y < a_1; \lim_{\varepsilon \to 0} \mathcal{I}_{\mathrm{ind}}(y', s; \varepsilon) < \infty \text{ for any } y' \in (y, a_1)\}. \tag{3.8}$$

Remark 3.3. *We can also identify* d_0 *in a similar way.*

4 One Dimensional and Time-dependent Case

Let $\Omega = (a_0, a_1)$ and $D(t)$ be given by $D(t) = (d_0(t), d_1(t))$ $(a_0 < d_0(t) < d_1(t) < a_1)$ for each $0 \le t \le T$ with $d_0, d_1 \in C^1([0, T])$.

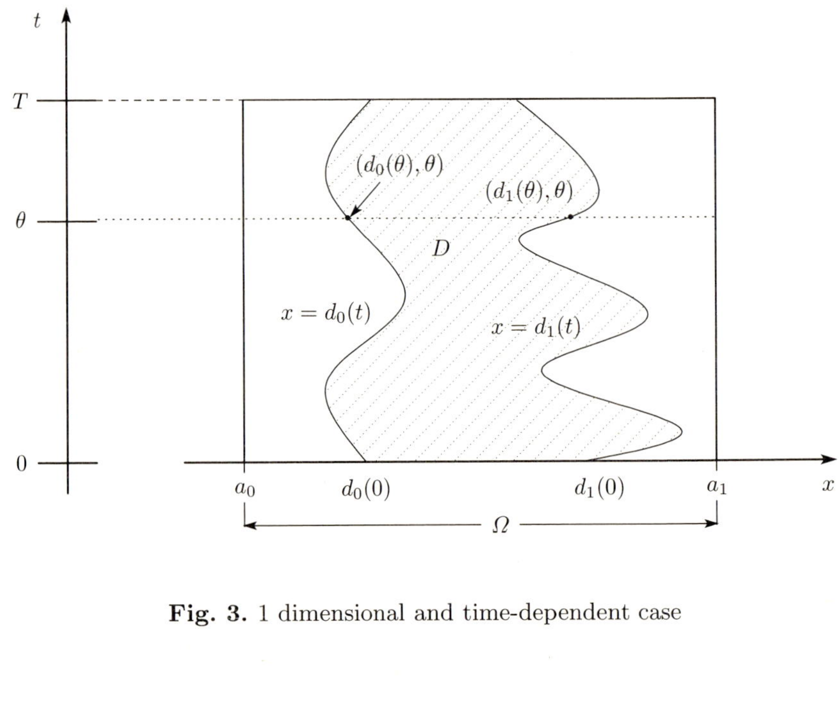

Fig. 3. 1 dimensional and time-dependent case

Let us focus on identifying $d_1(t)$. To start, for fixed $t = \theta$, let $w^{\theta}_{(y,s)}$ be the reflected solution satisfying (2.14) in Section 2 with $D = D(\theta)_T$ and $w^\theta_{+,1} := w^\theta_{(y,s)}\big|_{(d_1(\theta),a_1)_T}$. Also, we define $w_{+,1;p}(x,t)$ by

$$w_{+,1;p}(x,t) := w^t_{+,1;p}(x,t) = \frac{1-\sqrt{k}}{1+\sqrt{k}} G_{(y,s)}(-x+2d_1(t),t), \tag{4.1}$$

where $w^\theta_{+,1;p}$ is the dominant part of $w^\theta_{+,1}$.

Then, the following lemma shows that $w_{+,1;p}(x,t)$ is the dominant part of $w_{(y,s)}$.

Lemma 4.1. $w_{(y,s)} - w_{+,1;p}$ *is bounded in* $H^{1,0}(E_1)$ *as* (y,s) *tends to* $x = d_1(t)$*, where* $E_1 := \bigcup_{0<t<T} (d_1(t),a_1) \times \{t\}$.

Now we define $\mathcal{S}_{r;\alpha}(y,s)$ as an open sector with a vertex (y,s), a radius r, an angle 2α and a center axis parallel to the x axis whose figure is given below.

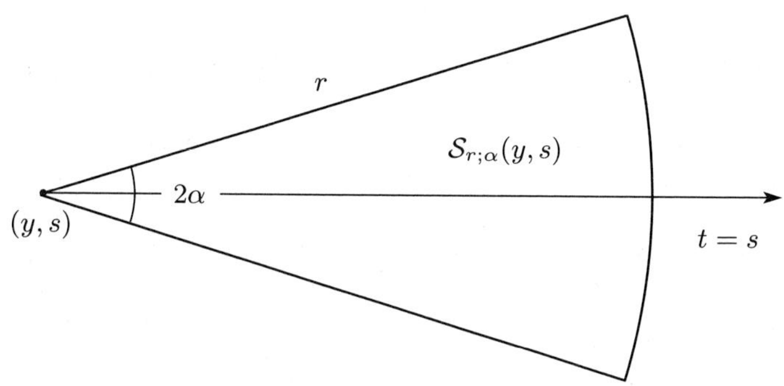

Fig. 4. The sector $\mathcal{S}_{r;\alpha}(y,s)$

Geometric assumption for Theorem 1.4

We can take small $r > 0, \alpha > 0$ such that for each $0 < s < T$, $\mathcal{S}_{r;\alpha}(y,s)$ touches $\partial_x D$ at only one point $\{(d_1(t),t); 0 < t < T\}$ as $y \downarrow d_1(s)$ along the line $t = s$.

Definition 4.2 (Indicator function for the time-dependent case). *For* $(y,s) \in \Omega_T \setminus \overline{D}$*, we define an indicator function* $\mathcal{I}_{\mathrm{dep}}(y,s)$ *for the time-dependent case by*

$$\mathcal{I}_{\mathrm{dep}}(y,s) := \|\nabla_{y'} I(\cdot,\cdot;y,s)\|_{L^2(\mathcal{S}_{r;\alpha}(y,s))}. \tag{4.2}$$

Theorem 4.3. *For each* $0 < s < T$,

$$\mathcal{I}_{\mathrm{dep}}(y,s) \to \infty \text{ as } y \downarrow d_1(s), \tag{4.3}$$

$$\mathcal{I}_{\mathrm{dep}}(y,s) < \infty \text{ when } y > d_1(s). \tag{4.4}$$

Therefore $d_1(s)$ *is given by*

$$d_1(s) = \inf\{y' < a_1; \mathcal{I}_{\mathrm{dep}}(y,s) < \infty \text{ for any } y' \in (y,a_1)\}. \tag{4.5}$$

This theorem follows from (2.15) in Section 2, (4.1) in Section 4, Lemma 4.1 and the following behavior on $\|\nabla_x w_{+,1;p}\|_{L^2(\mathcal{S}_{r;\alpha}(y,s))}$:

$$\|\nabla_x w_{+,1;p}\|_{L^2(\mathcal{S}_{r;\alpha}(y,s))} \to \infty \text{ as } y \downarrow d_1(s). \tag{4.6}$$

for each $0 < s < T$.

Remark 4.4. *We can also identify* $d_0(t)$ *in a similar way.*

References

1. H. Bellout, Stability result for the inverse transmissivity problem, J. Math. Anal. Appli. No.168 13-27 (1992).
2. A. Elayyan and V. Isakov, On uniqueness of recovery of the discontinuous conductivity coefficient of a parabolic equation, SIAM. J. Math. Anal. 28 No.1 49-59 (1997).
3. M. Ikehata, Reconstruction of inclusion from boundary measurements, J.Inverse and Ill-Posed Problems, 10 (2002) 37-65.
4. V. D. Kupradze , Three-dimensional problems of the mathematical theory of elasticity and thermoelasticity, Appl. Math. Mech. 25 (1979).
5. P. D. Lax, A stability theorem for solutions of abstract differetial equations, and its application to the study of the local behavior of solutions of elliptic equations, Comm. Pure and Appl. Math. Vol. IX 747-766 (1956).
6. G. M. Lieberman, Second order parabolic differential equations, World Scientific (1996).
7. J. L. Lions and E. Magenes, Non-homogeneous boundary value problems and applications II, Springer-Verlag (1972).
8. J. Wloka, Partial differentisl equations, Cambridge Univ. Press (1987).

Exact WKB analysis near a simple turning point

Eric Delabaere

Département de Mathématiques, UMR CNRS 6093, Université d'Angers, 2 Boulevard Lavoisier, 49045 Angers Cedex 01, France
eric.delabaere@univ-angers.fr

Summary. We extend and propose a new proof for a reduction theorem near a simple turning point due to Aoki *et al*, in the framework of the exact WKB analysis. Our scheme of proof is based on a Laplace-integral representation derived from an existence theorem of holomorphic solutions for a singular linear partial differential equation.

Key words: Resurgence theory, Exact WKB methods, Simple turning point, Singular linear PDE
Mathematics Subject Classification (2000): 34M60, 34M37, 35A07

1 Introduction

This article is devoted to a new proof of a reduction theorem near a simple turning point in the framework of the exact WKB analysis. Our main motivation stems from the following problem: consider the Schrödinger equation

$$\frac{d^2Y}{dq^2} = \eta^2 V(q) Y \tag{1}$$

where V is an analytic function near the origin. We assume that the origin is a simple zero for V, that is $V(q) \sim q$ near 0 (by rescaling η if necessary). Thinking of η as a large parameter, this assumption on V means that $q = 0$ is a *simple turning point* for the formal WKB solutions. These WKB solutions are linear combinations of elementary formal WKB solutions of (1) defined (locally in q) as:

$$Y_{wkb}(q,\eta) = \left(\sum_{k=0}^{\infty} Y_k(q)\eta^{-k} \right) \exp\left(-\eta \int^q \sqrt{V(t)}dt \right), \tag{2}$$

Received 14 November, 2005. Revised 23 February, 2006. Accepted 23 February, 2006.

and they are uniquely defined up to normalization. In the framework of the WKB analysis, equation (1) can be reduced into the Airy equation:

Theorem 1. *There exists a sequence* $\left(s_k(q)\right)_{k\geq 0}$ *of holomorphic functions near the origin such that, under the transformation*

$$s(q,\eta) = \sum_{k\geq 0} s_k(q)\eta^{-k}, \quad Y(q,\eta) = \left(\frac{\partial s}{\partial q}\right)^{-\frac{1}{2}} y\big(s(q,\eta),\eta\big), \tag{3}$$

the equation (1) is transformed into the Airy equation

$$\frac{d^2y}{ds^2} = \eta^2 sy, \tag{4}$$

for q (resp. s) *near the origin. Under the transformation (3), a formal WKB solution of (4) is transformed into a formal WKB solution of (1).*

At a formal level, theorem 1 goes back to [Silverstone 85]. As a rule, the series expansion $s(q,\eta)$ in (3) is divergent, more precisely:

- In [Aoki et al. 91], T. Aoki, T. Kawai and Y. Takei show that the series expansion $s(q,\eta)$ is in fact a *local resurgence constant*, that is its minor $\sum_{k\geq 1} s_k(q)\frac{\xi^{k-1}}{\Gamma(k)}$ defines a germ of holomorphic function at $(0,0)$. We mention that this result has been extended in [Aoki et al. 93] to the case where in equation (1) the potential function V is a local resurgence constant.
- In [Pham 00], F. Pham shows that the formal expansion $s(q,\eta)$ is *resurgent in* η *with regular dependence in* q. This result is based on the assumption that $V(q)$ is an entire function "sufficiently well behaved near infinity" and makes use of a theorem of Ecalle [Ecalle 84].

In this article our main goal will be to recover the result of Aoki *et al* [Aoki et al. 91] in a "pedestrian" way which avoids the Sato's microdifferential calculus. In doing so, we mention that our scheme of proof can be extended up to a "semi-global" analysis, in the direction of the result of Pham [Pham 00], but this will be published elsewhere (see [Rasoamanana 06]).

Following [Aoki et al. 91] and [Pham 00], the first natural step in the direction of theorem 1 is to straighten out the local geometry near the origin by defining a change of variable $q \leftrightarrow z$ so as to transform the differential form $\sqrt{V(q)}dq$ into $\sqrt{z}dz$ (the associated cotangent map transforms the equation of the Lagrangian submanifold $P^2 - V(q) = 0$ into $p^2 - z = 0$). Thus introducing the transformation:

$$z(q) = \left(\frac{3}{2}\int_0^q V(t)^{\frac{1}{2}}dt\right)^{\frac{2}{3}}, \quad Y(q,\eta) = \left(\frac{dz}{dq}\right)^{-\frac{1}{2}} \Phi(z,\eta), \tag{5}$$

equation (1) becomes the "master" equation

$$\frac{d^2\Phi}{dz^2} - \eta^2 z\Phi = F(z)\Phi, \tag{6}$$

with $F(z) = \frac{z}{2V(q)}\{z,q\}_{|q = q(z)}$, where $\{z,q\}$ is the Schwarzian derivative of z with respect to q. Concerning $F(z)$, one has the following property:

Lemma 1. *If $V(q)$ is holomorphic near 0 with a simple zero at $q = 0$, then $F(z)$ is holomorphic near the origin.*

By the transformation (5), the exact WKB analysis for equation (1) thus translates into its analogous analysis for the associated equation (6). This is what we shall do in the next sections.

The paper is organized as follows. In section 2, we introduce a family of formal WKB solutions for (6). The particular Airy case is detailed and analyze in term of Borel-resummation, so as to introduce the necessary definitions. Section 3 is the main part of the paper. We show how such a formal WKB solution can be considered as the local decomposition of a *confluent function* in a germ of Stokes sector. The construction of these confluent functions is based on an existence theorem (theorem 2) of holomorphic solutions for a singular linear partial differential equation. Section 4 explains how one can recover and extend the result of Aoki *et al* [Aoki et al. 91] from the results of section 3.

Note: in the exact WKB analysis, all main objects (like Borel (pre)summation, Stokes sectors, etc ...) are related to a given direction α, which can be thought of as an argument. In what follows, otherwise mentioned, it will be assumed that $\alpha = 0$, so that $\Re(\eta) > 0$ (and $|\eta|$ large enough).

2 WKB analysis: formal aspects

One easily sees that the following expansion,

$$\Phi_{wkb}(z,\eta) = \frac{e^{-\frac{2}{3}\eta z^{3/2}}}{z^{\frac{1}{4}}}\left(1 + g_1(z)\eta^{-1} + g_2(z)\eta^{-2} + \cdots\right), \tag{7}$$

is a formal WKB solution of equation (6) as soon as the g_n's satisfy some "transport" differential equations which we do not write here. The expansion (7) is multivalued in z and depends on the choice of a determination for $z^{3/2}$ (as well as for $z^{\frac{1}{4}}$). Since equation (6) is invariant under the mapping $\eta \mapsto -\eta$, we deduce that

$$\Phi_{wkb}(z,-\eta) \tag{8}$$

is another formal WKB solution, and moreover $\{\Phi_{wkb}(z,\eta), \Phi_{wkb}(z,-\eta)\}$ defines a basis of formal WKB solutions for equation (6).

To analyze the g_n's, and somehow to normalize them, one can use another representation for these WKB expansions. Writing $\Phi_{wkb}(z,\eta)$ under the form $\Phi_{wkb}(z,\eta) = \exp\left(-\eta \int^z P(t,\eta)dt\right)$, equation (6) translates into:

$$\eta \frac{dP}{dz} + \eta^2 \left(z - P^2\right) + F(z) = 0. \tag{9}$$

Writing $P(z,\eta) = \sum_{n\geq 0} p_n(z)\eta^{-n}$, one easily shows by induction that:

$$p_0(z) = z^{\frac{1}{2}}, \quad p_1(z) = \frac{1}{4z}, \quad p_n(z) \in z^{-\frac{3n-1}{2}}\mathbb{C}\{z\},\ n \geq 2. \tag{10}$$

Introducing the decomposition $P = P_{even} + P_{odd}$, $\begin{cases} P_{even} = \sum_{k\geq 0} p_{2k}\eta^{-2k} \\ P_{odd} = \sum_{k\geq 0} p_{2k+1}\eta^{-2k-1} \end{cases}$, we infer from (9) that $P_{odd} = \dfrac{\eta^{-1}}{2}\dfrac{P'_{even}}{P_{even}}$ where $P'_{even} = \dfrac{dP_{even}}{dz}$. Therefore:

$$\Phi_{wkb}(z,\eta) = \frac{C(\eta)}{\sqrt{P_{even}(t,\eta)}} \exp\left(-\eta \int^z P_{even}(t,\eta)dt\right), \quad \text{with } C(\eta) \in \mathbb{C}[[\eta^{-1}]]. \tag{11}$$

Comparing the two representations (7) and (11), one gets:

Proposition 1. *The formal WKB solutions (7) of (6) can be normalized in such a way that for every $n \geq 0$, $g_n(z) \in z^{-\frac{3n}{2}}\mathbb{C}\{z\}$.*

Definition 1. *The formal WKB solutions described in proposition 1 will be called the* elementary WKB solutions for the equation (6).

When $F(z) = 0$, equation (6) is nothing but the Airy equation. Among the elementary WKB solutions, it will be convenient to specify the following one:

Definition 2. *The following elementary WKB solution,*

$$A_{wkb}(z,\eta) = \frac{e^{-\frac{2}{3}\eta z^{3/2}}}{z^{\frac{1}{4}}} \left(1 + \sum_{n=1}^{\infty} \left(-\frac{3}{4}\right)^n \frac{\Gamma(n+\frac{1}{6})\Gamma(n+\frac{5}{6})}{2\pi\Gamma(n+1)} z^{-\frac{3n}{2}}\eta^{-n}\right) \tag{12}$$

will be called the Airy WKB symbol.

Proposition 2. *The Airy WKB symbol is resurgent Borel-resummable in η, with regular dependence on $z \neq 0$.*

Let us we give some explanations (see [Jidoumou 94, Delabaere et al. 93, Delabaere et al. 97, Delabaere et al. 99] for more details). Since it is assumed that the direction of Borel-summation is $\alpha = 0$, this prescribes the Stokes lines and the Stokes sectors, as drawn on figure 1.a.
As long as z stays in one of the Stokes sectors, the Airy WKB symbol is Borel-resummable. For instance, let us fix the following convention:

Convention: drawing a cut as on Fig. 1.a, we fix the determination of $z^{3/2}$ (*resp.* $z^{1/4}$) such that $z^{3/2}$ (*resp.* $z^{1/4}$) is real positive along L_0. We note $A^+_{wkb}(z,\eta)$ the determination of $A_{wkb}(z,\eta)$ thus defined, and $A^-_{wkb}(z,\eta) := A^+_{wkb}(z,-\eta)$.

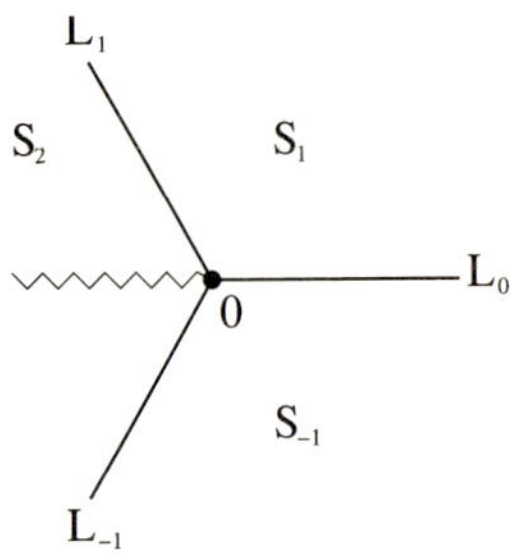

Fig. 1.a

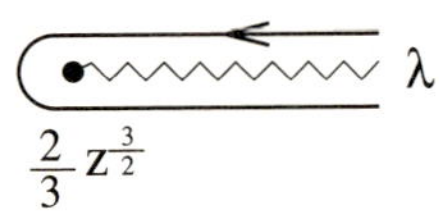

Fig. 1.b

Fig. 1. Fig. 1.a : L_0, L_1 and $L{-1}$ are the Stokes lines (in the z-plane) related to the direction $\alpha = 0$. The 3 Stokes sectors are the open connected sectors bounded by the Stokes lines (forget the wavy line). Fig. 1.b : the countour of integration in the ξ-plane. The wavy lines are cuts.

For z in the Stokes sector S_1 (*resp.* S_{-1}) drawn on figure 1.a, the Airy WKB symbol A^+_{wkb} is Borel-resummable. This Borel-sum, denoted by $\mathrm{s}_0\left(A^+_{wkb}\right)(z,\eta)$, is holomorphic in (z,η), $\Re(\eta) > 0$ and $z \in S_1$ (*resp.* S_{-1}). Note that the function $\mathcal{A}$,

$$\mathcal{A}(z,\eta) = \mathrm{s}_0\left(A^+_{wkb}\right)(z,\eta) \tag{13}$$

extends analytically as an entire function in z. As a matter of fact, $\mathcal{A}(z,\eta) = 2\sqrt{\pi}\eta^{1/6}Airy(z\eta^{2/3})$, where *Airy* is the Airy function.
Conversely, since the Borel-resummability induces a one-to-one correspondence between the formal expansion and its Borel-sum, one can associate to $\mathcal{A}$ its *decomposition* A^+_{wkb} for $z \in S_1$ (*resp.* S_{-1}):

$$\mathcal{A}(z,\eta) \overset{\sigma_{S_1}}{\longrightarrow} A^+_{wkb}(z,\eta), \quad \left(resp.\ \mathcal{A}(z,\eta) \overset{\sigma_{S_{-1}}}{\longrightarrow} A^+_{wkb}(z,\eta)\right). \tag{14}$$

Here, the fact that the decomposition of $\mathcal{A}(z,\eta)$ in S_1 and S_{-1} is given by the same formal expansion shows that the Stokes line is "inactive". This is no more true when, coming from S_1 (*resp.* S_{-1}) one crosses the Stokes lines L_1 (*resp.* L_{-1}): for z on these lines, a Stokes phenomenon occurs, which is completely described by the action of the following alien derivative:

$$\dot{\Delta}_{-\frac{4}{3}z^{3/2}} A^{+}_{wkb}(z,\eta) = \ell A^{+}_{wkb}(z,\eta) = -iA^{-}_{wkb}(z,\eta) \tag{15}$$

where ℓ is the analytic continuation in z around 0 anticlockwise. This means that the decomposition of $\mathcal{A}$ for $z \in S_2$ (say) becomes:

$$\mathcal{A}(z,\eta) \overset{\sigma_{S_2}}{\longrightarrow} A^{+}_{wkb}(z,\eta) - \ell A^{+}_{wkb}(z,\eta) = A^{+}_{wkb}(z,\eta) + iA^{-}_{wkb}(z,\eta) \tag{16}$$

Similarly, for $z \in L_0$, one has:

$$\dot{\Delta}_{+\frac{4}{3}z^{3/2}} A^{-}_{wkb}(z,\eta) = \ell A^{-}_{wkb}(z,\eta) = -iA^{+}_{wkb}(z,\eta). \tag{17}$$

The Stokes phenomenon can be thought in term of the singular locus of a major as follows: the Borel-sum of A_{wkb} for $z \in S_1$ (say) can be defined as an integral,

$$\mathrm{S}_0\left(A_{wkb}\right)(z,\eta) = \int_{\lambda} e^{-\eta\xi} \overset{\vee}{A}_{wkb}(z,\xi)\, d\xi. \tag{18}$$

where $\overset{\vee}{A}_{wkb}(z,\xi)$ is a major associated to the Airy WKB symbol. This major is holomorphic on the universal covering of $\mathbb{C}^2 \backslash \Gamma$, where the singular support Γ is the algebraic curve $\Gamma = \{(z,\xi),\ 9\xi^2 = 4z^3\}$. The contour of integration λ is drawn on figure 1.b for $z \in S_1$, and its deformation for $z \in S_2$ after the crossing of the Stokes line L_1 is drawn on figure 2.

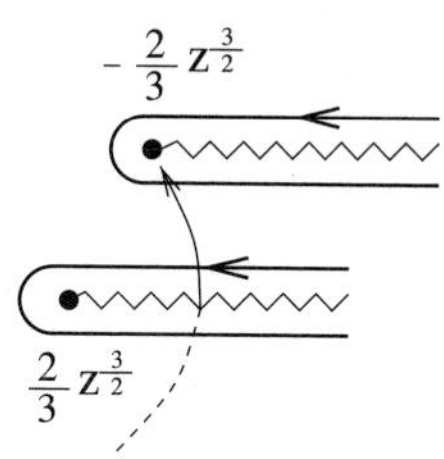

Fig. 2. Effect of the Stokes phenomenon described by (15) in term of the deformation of the integration contour for the Borel sum (18).

Note that the above integral representation (18) can be deduced from the usual representation for the Airy function, namely (up to a normalization factor):

$$\int e^{-\eta S(z,\widehat{z})}\, d\widehat{z} \quad \text{where} \quad S(z,\widehat{z}) = z\widehat{z} - \frac{1}{3}\widehat{z}^3. \tag{19}$$

Our analysis in section 3 will be based on an extension of this integral representation.

3 Local exact WKB analysis

We now go back to the elementary WKB solutions described in proposition 1. We would like to "realize" the Borel-Ritt theorem, that is to construct analytic functions whose asymptotics are govern by (at least a family of) these elementary WKB symbols.
Since the principal symbol $p^2 - z$ of the operator defining equation (6) is just the Airy operator, one could start from the above representation (19), thinking of $S(z,\widehat{z}) = z\widehat{z} - \frac{1}{3}\widehat{z}^3$ as the generating function of a canonical transformation $(p,z) \leftrightarrow (\widehat{p},\widehat{z})$ in the cotangent space whose effect is to straighten out the Lagrangian submanifold $\widehat{p} = p^2 - z = 0$. Pursuing in that direction leads to consider, as in [Pham 88], solutions by means of a quantization of the canonical transformation, i.e., to look for solutions in the form $\Phi(z,\eta) = \int e^{-\eta S(z,\widehat{z})}\widehat{\varphi}(\widehat{z},\eta)\,d\widehat{z}$. However, we rather look here for solutions of (6) defined as Borel sums, $\Phi(z,\eta) = \int e^{-\eta\xi}\,\overset{\vee}{\Phi}(z,\xi)\,d\xi$, where $\overset{\vee}{\Phi}(z,\xi)$ should be a major of a convenient confluent microfunction (in the sense of [Delabaere et al. 99, Jidoumou 94]). What we mean by "convenient" is the following: on the previous integral representation, derivating under the integration symbol and formally integrating by parts, we translate the fact that Φ is a solution of (6) to the demand that $\overset{\vee}{\Phi}$ should satisfy the equation:

$$\frac{\partial^2 \overset{\vee}{\Phi}}{\partial z^2} - z\frac{\partial^2 \overset{\vee}{\Phi}}{\partial \xi^2} = F(z)\,\overset{\vee}{\Phi}\,. \tag{20}$$

Rather than solving the previous PDE (20), we now combine the two previous ideas; making the change of variable $\xi \leftrightarrow \widehat{z}$ defined by $\xi = S(z,\widehat{z})$, we get the integral representation:

$$\Phi(z,\eta) = \int_\gamma e^{-\eta S(z,\widehat{z})}\Psi(z,\widehat{z})\,d\widehat{z}, \qquad \overset{\vee}{\Phi}(z,\xi)|_{\xi} = S(z,\widehat{z}) = \frac{\Psi(z,\widehat{z})}{z-\widehat{z}^2}, \tag{21}$$

where the path of integration γ can be thought of, for the moment, as an endless path, ending at infinity in the valleys where $\Re(\eta S(z,\widehat{z})) \to +\infty$.
Introducing $\widetilde{\Psi}(z,\widehat{z}) = \overset{\vee}{\Phi}(z,\xi)$, one easily deduces from (20) that $\widetilde{\Psi}$ should be solution of the following linear PDE:

$$\frac{\partial^2\widetilde{\Psi}}{\partial z^2} - \frac{2\widehat{z}}{z-\widehat{z}^2}\frac{\partial^2\widetilde{\Psi}}{\partial z\partial\widehat{z}} - \frac{1}{z-\widehat{z}^2}\frac{\partial^2\widetilde{\Psi}}{\partial\widehat{z}^2} = F(z)\widetilde{\Psi}. \tag{22}$$

3.1 Construction of the major

Having in mind the steepest-descent method, we expect the function $\Psi(z,\widehat{z})$ in (21) to be holomorphic near the locus $\frac{\partial S(z,\widehat{z})}{\partial\widehat{z}} = 0$ defining the saddle points. We thus introduce the transformation

$$(z,\widehat{z}) \leftrightarrow (x = z - \widehat{z}^2, \widehat{z}), \qquad \psi(x,\widehat{z}) := \Psi(z,\widehat{z}) = \widetilde{\Psi}(z,\widehat{z})(z-\widehat{z}^2). \tag{23}$$

Equation (22) then translates into the following one for ψ:

$$x^2\frac{\partial^2\psi}{\partial x^2} + 2x\widehat{z}\frac{\partial^2\psi}{\partial x\partial\widehat{z}} - x\frac{\partial^2\psi}{\partial\widehat{z}^2} - 2\widehat{z}\frac{\partial\psi}{\partial\widehat{z}} - x^2F(x+\widehat{z}^2)\psi = 0. \tag{24}$$

We now look for an holomorphic solution of equation (24) for x near zero (and $\widehat{z}$ near 0 as well). Since in equation (24) $x=0$ is a singular point, this cannot be simply derived from the Cauchy-Kovalevska theorem.
We thus start looking for a formal solution of (24) under the form

$$\psi(x,\widehat{z}) = \sum_{n\geq 0} a_n(\widehat{z})x^n. \tag{25}$$

We introduce the Taylor expansion $F(x+\widehat{z}^2) = \sum_{n\geq 0} f_n(\widehat{z}^2)x^n$ which converges for $|x|$ and $|\widehat{z}|$ small enough, by the holomorphy of F near 0. It will be convenient to introduce also the product

$$F(x+\widehat{z}^2)\psi(x,\widehat{z}) = \sum_{n\geq 0} b_n(\widehat{z})x^n, \qquad b_n(\widehat{z}) = \sum_{k=0}^{n} a_k(\widehat{z})f_{n-k}(\widehat{z}^2). \tag{26}$$

Plugging (25) and (26) into (24) and identifying the powers in x, one gets:

$$\frac{\partial a_0}{\partial\widehat{z}} = 0, \qquad 2\widehat{z}\frac{\partial a_n}{\partial\widehat{z}} + na_n = \frac{1}{n-1}\left(\frac{\partial^2 a_{n-1}}{\partial\widehat{z}^2} + b_{n-2}\right), \quad \text{for } n\geq 2. \tag{27}$$

In what follows, we shall fix the normalization of ψ so that $a_0(\widehat{z}) = 1$. One easily gets the following lemma:

Lemma 2. *Let $h(\widehat{z})$ be an holomorphic function near the origin. Then there exists a unique formal series expansion $\psi(x,\widehat{z}) = \sum_{n\geq 0} a_n(\widehat{z})x^n$ solution of (24) such that the $a_n(\widehat{z})$ are holomorphic functions near $\widehat{z}=0$, with $a_0(\widehat{z}) = 1$ and $a_1(\widehat{z}) = h(\widehat{z})$. In this case, one has*

$$a_n(\widehat{z}) = \frac{1}{n-1}\int_0^1 u^{n-1}\left(a''_{n-1}(u^2\widehat{z}) + b_{n-2}(u^2\widehat{z})\right)du, \quad \text{for } n\geq 2, \tag{28}$$

where the b_n's are defined by (26). Furthermore, if $h(z)$ is even, then every $a_n(z)$ is even.

As far as we know, the known theories (see, e.g., [Gérard et al. 96] and [Kashiwara et al. 79]) to analyze the convergence of the formal solutions of the singular PDE (24) described in Lemma 2 do not apply in our case. To show the convergence, we shall use the following result (the proof is left to the reader):

Lemma 3. *The formal series expansion given by Lemma 2 represents the Taylor series of an holomorphic function $\psi(x,\widehat{z})$ near $(x,\widehat{z})=(0,0)$ if and only if $\varphi(x,\widehat{z}) := \dfrac{\psi(x,\widehat{z})}{x} - \dfrac{1}{x}$ satisfies the following integral equation:*

$$\begin{aligned}\varphi(x,\widehat{z}) = h(\widehat{z}) &+ \int_0^1 du \int_0^{ux} dt F(t+u^4\widehat{z}^2)\\ &+ \int_0^1 du \int_0^{ux} dt\Big(\partial_2^2\varphi(t,u^2\widehat{z}) + tF(t+u^4\widehat{z}^2)\varphi(t,u^2\widehat{z})\Big).\end{aligned} \tag{29}$$

Theorem 2. *Let $h(\widehat{z})$ be an holomorphic function near the origin. Then there exists a unique holomorphic function $\psi(x,\widehat{z})$ near $(0,0)$, solution of equation (24), and satisfying the following initial data:*

$$\psi(0,\widehat{z}) = 1, \qquad \frac{\partial\psi}{\partial x}(0,\widehat{z}) = h(\widehat{z}). \tag{30}$$

Proof. The proof will be derived from more or less standard techniques (see, e.g., [Trèves 75], §17). We first remind the following well-known result. If W is a bounded open subset of $\mathbb{C}^n$, $n \geq 1$, and E a Banach space, we denote by $H(\overline{W},E)$ the space of functions $f : Z \mapsto f(Z) \in E$ which are continuous in $Z \in \overline{W}$ and holomorphic in W. We introduce on the space $H(\overline{W},E)$ the maximum norm: $\|f\|_W = \sup_{Z\in W} |f(Z)|$. Then $(H(\overline{W},E), \|.\|_W)$ is a Banach vector space.

In what follows, $D(0,l) \subset \mathbb{C}$ denotes the open disc centered on 0 with radius $l>0$. It is assumed that: $F \in H(\overline{V}),\mathbb{C})$ with $V = D(0,\tau)$. Also $R>0$ and $r_1 := r_0 + d_0 > 0$ are such that $R + r_1^2 \leq \tau$. This implies that

$$\forall u \in [0,1],\, \forall (x,\widehat{z}) \in D(0,R)\times D(0,r_1),\, x + u^4\widehat{z}^2 \in D(0,\tau). \tag{31}$$

For $0 \leq s \leq 1$ we note $U_s := D(0,r_s)$ with $r_s := r_0 + sd_0$. For $u \in [0,1]$ and $x \in \overline{D(0,R)}$ we introduce the function $L : (u,x) \mapsto L(u,x)$ defined by

$$L(u,x) : \psi(\widehat{z}) \mapsto \partial^2\psi(u^2\widehat{z}) + xF(x+u^4\widehat{z}^2)\psi(u^2\widehat{z}). \tag{32}$$

We consider $L(u,x)$ as a linear operator acting on the Banach space $H(\overline{U_s},\mathbb{C})$ with values in $H(\overline{U_{s'}},\mathbb{C})$, $L(u,x) : H(\overline{U_s},\mathbb{C}) \to H(\overline{U_{s'}},\mathbb{C})$, where $0 \leq s' < s \leq 1$. By the Cauchy's formula, since $r_s - u^2 r_{s'} = (1-u^2)r_0 + (s-u^2s')d_0$, one gets:

$$\|\partial^2\psi_{(u)}\|_{U_{s'}} \leq \frac{2}{\big((1-u^2)r_0 + (s-u^2s')d_0\big)^2}\|\psi\|_{U_s}$$

where $\partial^2\psi_{(u)} : \widehat{z} \mapsto \partial^2\psi(u^2\widehat{z})$. This implies that, for every $(u,x) \in [0,1]\times\overline{D(0,R)}$,

$$\|L(u,x)\psi\|_{U_{s'}} \leq \left(\frac{2}{\big((1-u^2)r_0 + (s-u^2s')d_0\big)^2} + |x|\|F\|_V\right)\|\psi\|_{U_s},$$

then

$$\|L(u,x)\psi\|_{U_{s'}} \leq \frac{C}{\big((1-u^2)r_0+(s-u^2s')d_0\big)^2}\|\psi\|_{U_s}, \tag{33}$$

with $C = 2 + r_1^2 R\|F\|_V$. We now assume that $h \in H(\overline{D(0,r_1)}, \mathbb{C})$ and introduce:

$$\begin{cases} \theta_0(x) := \widehat{z} \mapsto h(\widehat{z}) + \displaystyle\int_0^1 du \int_0^{ux} dt\, F(t+u^4\widehat{z}^2) \\ \theta_{k+1}(x) = \theta_0(x) + \displaystyle\int_0^1 du \int_0^{ux} dt\, L(u,t)\theta_k(t), \quad k \geq 0. \end{cases} \tag{34}$$

Obviously (34) defines a sequence $(\theta_k)_k$ of holomorphic functions in $x \in D(0,R)$, continuous in $x \in \overline{D(0,R)}$, valued in $H(\overline{U_s}, \mathbb{C})$, for every $0 \leq s < 1$:

$$\forall k \geq 0,\ \theta_k \in H\left(\overline{D(0,R)}, H(\overline{U_s}, \mathbb{C})\right). \tag{35}$$

We also set:

$$\begin{cases} \delta_0(x) := \theta_0(x) \\ \delta_{k+1}(x) := \theta_{k+1}(x) - \theta_k(x) = \displaystyle\int_0^1 du \int_0^{ux} dt\, L(u,t)\delta_k(t), \quad k \geq 0. \end{cases} \tag{36}$$

We first observe that, for any $0 \leq s < 1$, and every $x \in \overline{D(0,R)}$,

$$\|\delta_0(x)\|_{U_s} \leq M, \qquad M = \|h\|_{D(0,r_1)} + \frac{R}{2}\|F\|_V. \tag{37}$$

Next, we show the following lemma:

Lemma 4. *For every $0 \leq s < 1$, for every $k \in \mathbb{N}$ and every $x \in \overline{D(0,R)}$,*

$$\|\delta_k(x)\|_{U_s} \leq M\left(\frac{Ce|x|}{2r_0d_0(1-s)}\right)^k. \tag{38}$$

Proof. We show the lemma by induction on k. The case $k=0$ is given by (37). Let assume that (38) is satisfied for a given $k \in \mathbb{N}$ and every $0 \leq s < 1$. For every $0 \leq s' < s < 1$, we deduce from (36) and (33) that

$$\|\delta_{k+1}(x)\|_{U_{s'}} \leq \int_0^1 du \int_0^{ux} |dt| \frac{C}{\big((1-u^2)r_0+(s-u^2s')d_0\big)^2}\|\delta_k(t)\|_{U_s}. \tag{39}$$

Using our hypothesis on $\delta_k(x)$, we deduce that

$$\|\delta_{k+1}(x)\|_{U_{s'}} \leq \int_0^1 du \int_0^{u|x|} dt \frac{C}{\big((1-u^2)r_0+(s-u^2s')d_0\big)^2} M\left(\frac{Cet}{2r_0d_0(1-s)}\right)^k \tag{40}$$

Integrating, and since $\displaystyle\int_0^1 \frac{u}{\left((1-u^2)r_0 + (s-u^2s')d_0\right)^2}\,du \leq \frac{1}{2r_0d_0(s-s')}$, we deduce that

$$\|\delta_{k+1}(x)\|_{U_{s'}} \leq M\left(\frac{Ce}{2r_0d_0(1-s)}\right)^k \frac{|x|^{k+1}}{k+1}\frac{C}{2r_0d_0(s-s')}. \tag{41}$$

Choosing $s = s' + \dfrac{1-s'}{k+1}$, so that $1-s = \dfrac{k}{k+1}(1-s')$, one obtains:

$$\|\delta_{k+1}(x)\|_{U_{s'}} \leq M\left(\frac{Ce}{2r_0d_0(1-s')}\right)^k \left(1+\frac{1}{k}\right)^k |x|^{k+1}\frac{C}{2r_0d_0(1-s')}. \tag{42}$$

We use the remark that $\left(1+\dfrac{1}{k}\right)^k \leq e$ to conclude that

$$\|\delta_{k+1}(x)\|_{U_{s'}} \leq M\left(\frac{Ce|x|}{2r_0d_0(1-s')}\right)^{k+1}. \tag{43}$$

This proves the lemma. □

It follows from lemma 4 that the series $\sum_{k\geq 0}\delta_k(x)$ is absolutly convergent in $H(\overline{U_s},\mathbb{C})$ (for every $0 \leq s < 1$) and uniformly in $x \in K$, where K is any compact set of the open disc $|x| < \min\left(\dfrac{2r_0d_0(1-s)}{Ce}, R\right)$. By construction, its sum $\theta(x)$ satisfies the identity

$$\theta(x) = \theta_0(x) + \int_0^1 du \int_0^{ux} dt\, L(u,t)\theta(t), \tag{44}$$

so that $\varphi(x,\widehat{z}) := \theta(x)(\widehat{z})$ is an holomorphic solution of equation (29). □

3.2 Associated confluent function

We now go back to the function $\Psi(z,\widehat{z})$ related to $\psi(x,\widehat{z})$ by (23). From lemma 2 and theorem 2 one deduces the following result:

Proposition 3. *We assume that F and h are holomophic functions near the origin. Then there exists a unique holomorphic function $\Psi(z,\widehat{z})$ near $(0,0)$ satisfying the condition $\Psi(z,\widehat{z})\big|_{z=\widehat{z}^2} = 1$ and $\dfrac{\partial\Psi}{\partial z}(z,\widehat{z})\big|_{z=\widehat{z}^2} = h(\widehat{z})$, and such that $\widetilde{\Psi}(z,\widehat{z}) := \dfrac{\Psi(z,\widehat{z})}{z-\widehat{z}^2}$ is solution of the linear PDE (22).*

This easily translates into the following proposition (cf. [Delabaere et al. 99, Jidoumou 94]):

Proposition 4. *We assume that F and h are holomophic functions near the origin. Then the function $\overset{\vee}{\Phi}(z,\xi)$ defined by*

$$\overset{\vee}{\Phi}(z,\xi)_{|\xi = S(z,\widehat{z})} := \frac{\Psi(z,\widehat{z})}{z-\widehat{z}^2} \tag{45}$$

with Ψ described by proposition 3 is solution of (20) and is a major of a confluent microfunction at $(0,0)$ with singular support in the algebraic curve $\Gamma = \{(z,\xi),\ 9\xi^2 = 4z^3\}$.

Following the proposition 3, $\Psi(z,\widehat{z})$ is holomorphic in a neighbourhood of the origin in $\mathbb{C}^2$, say for $(z,\widehat{z}) \in D(0,\frac{r^2}{4}) \times D(0,r)$ with $r > 0$ small enough. Therefore, the integral (21) is well defined provided that we truncate the path of integration γ. The truncated path will be denoted by $\overline{\gamma}$, as on figure 3.

In the integral representation

$$\Phi(z,\eta) = \int_{\overline{\gamma}} e^{-\eta S(z,\widehat{z})}\Psi(z,\widehat{z})\,d\widehat{z}, \tag{46}$$

making the change of variable $\xi = S(z,\widehat{z})$, we get the corresponding integral, where the path $\overline{\lambda}$ is drawn on figure 3.

$$\Phi(z,\eta) = \int_{\overline{\lambda}} e^{-\eta\xi}\,\overset{\vee}{\Phi}(z,\xi)\,d\xi. \tag{47}$$

A consequence of proposition 4 is that the Laplace integral (47) represents a *confluent function* $\mathbf{\Phi}(z,\eta)$ *with singular support in* Γ, in the sense of [Delabaere et al. 99], §0.4 and 0.6, [Jidoumou 94], Appendice 3.

The local decomposition of the confluent function $\mathbf{\Phi}(z,\eta)$ can be deduced from the integral representation (46) as a consequence of the steepest-descent method. We describe what we get for the germs of Stokes sectors as drawn on figure 1.a.

Lemma 5. *For z in the germ of Stokes sector S_1, the local decomposition of the confluent function $\mathbf{\Phi}(z,\eta)$ yields a single formal WKB expansion, namely:*

$$\mathbf{\Phi}(z,\eta) \xrightarrow{\sigma_{S_1}} i\frac{\sqrt{\pi}}{\sqrt{\eta}}\Phi^{+}_{wkb}(z,\eta), \tag{48}$$

If h is even, then $\Phi^{+}_{wkb}(z,\eta)$ is the determination in S_1 of an elementary WKB solution $\Phi_{wkb}(z,\eta)$ for the equation (6), in the sense of the Definition 1.

Proof. For the integral representation (46), z being in the germ of Stokes sector S_1 corresponds to the situation described on Fig. 3.1a. Deforming the path of integration $\overline{\gamma}$ under the flow $\nabla\,(\Re(\eta S(z,\widehat{z}))$ (the end points of $\overline{\gamma}$ remaining fixed), one sees that only the saddle point $\widehat{z} = \sqrt{z}$ contributes to the decomposition. This gives the formula (48). Note that the formal WKB expansion

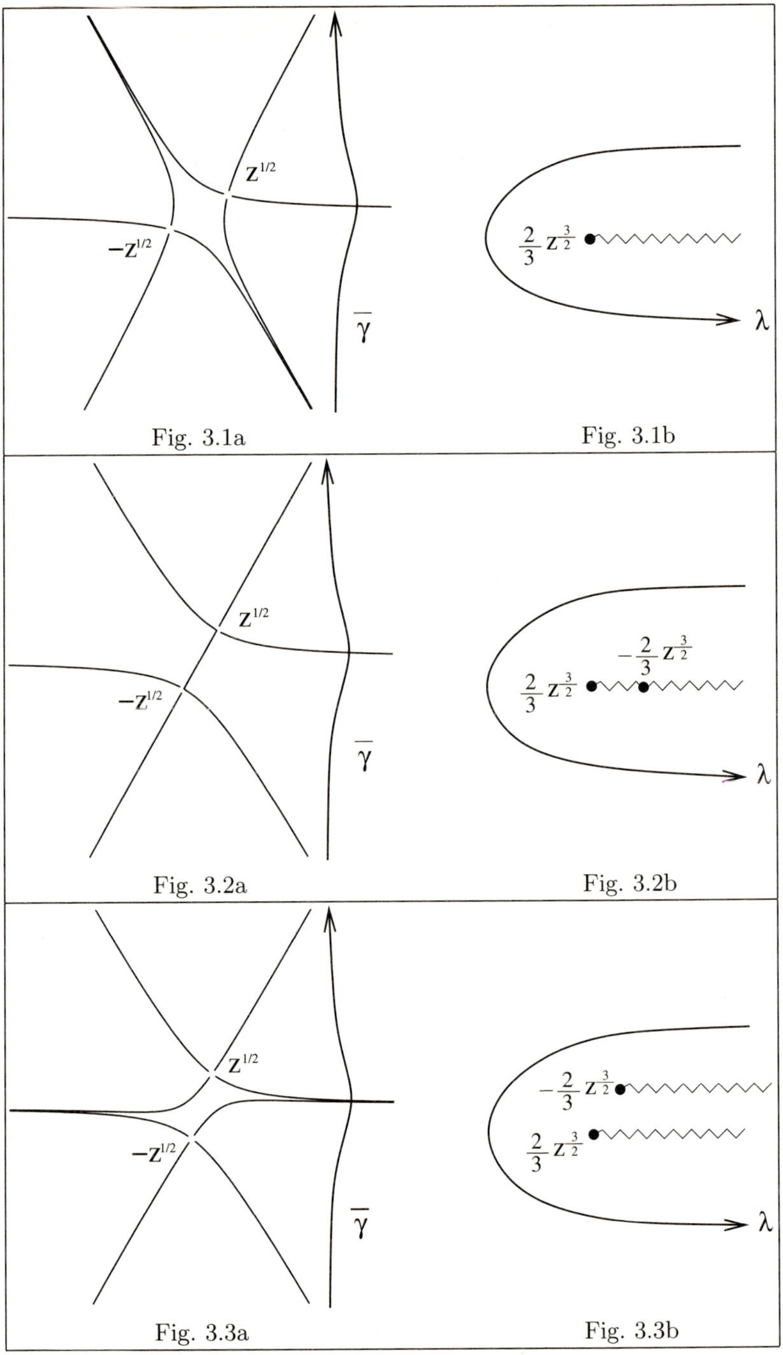

Fig. 3. On the left pictures the steepest-descent pattern and the path $\overline{\gamma}$, and on the right pictures its image λ by the transformation $\widehat{z} \mapsto \xi = S(z, \widehat{z})$ (the wavy lines are cuts). Fig. 3.1 for $z \in S_1$, Fig. 3.2 for z on L_1, Fig. 3.3 for $z \in S_2$ (see figure 1).

thus obtained is a formal solution of equation (6) since the major $\overset{\vee}{\Phi}(z,\xi)$ is a solution of (20) (cf. Prop. 4). The fact that, when h is even, this WKB solution is an an elementary WKB solution $\Phi_{wkb}(z,\eta)$ for the equation (6) is derived from an explicit calculation left to the reader. □

In the same way, one can show that for z on the Stokes line L_1 a Stokes phenomenon occurs (see Fig. 3.2) so that, for z in the germ of Stokes sector S_2 (see Fig. 3.3a), the local decomposition of the confluent function $\mathbf{\Phi}(z,\eta)$ now involves a sum of two formal WKB expansions:

$$\mathbf{\Phi}(z,\eta) \xrightarrow{\sigma_{S_2}} i\frac{\sqrt{\pi}}{\sqrt{\eta}}\Phi^{+}_{wkb}(z,\eta) - \frac{\sqrt{\pi}}{\sqrt{\eta}}\Phi^{-}_{wkb}(z,\eta), \tag{49}$$

where $\Phi^{-}_{wkb}(z,\eta) = \Phi^{+}_{wkb}(z,-\eta)$.

Conversely, let us consider the elementary WKB solution $\Phi_{wkb}(z,\eta)$ for the equation (6). By (48), and up to the factor $i\frac{\sqrt{\pi}}{\sqrt{\eta}}$, a determination of this elementary WKB solution appears to be the local decomposition in a germ of Stokes sector of a confluent function. But more than this, every determination of $\Phi_{wkb}(z,\eta)$ in every germ of Stokes sector can be thought of as the local decomposition in this germ of Stokes sector of a confluent function (up to a factor $c\frac{\sqrt{\pi}}{\sqrt{\eta}}$, $c \in \{\pm 1, \pm i\}$): on the integral representation (46), this just amounts in choosing a convenient truncated path of integration $\overline{\gamma}$. Moreover, since all we have done so far can be applied by choosing another direction α than 0, what we get is the following result.

Theorem 3. *When F is holomorphic near the origin, there exists a family of* local Airy type *elementary WKB solutions $\Phi_{wkb}(z,\eta)$ of equation (6) (in the sense of Definition 1).*

We refer the reader to [Jidoumou 94] (§2.3 and Appendice 3) for a precise definition of an elementary WKB symbol of local Airy-type. Here we concentrate on the consequence of the above theorem 3, just applying a theorem of Jidoumou [Jidoumou 94] (§2.3):

Theorem 4. *We note $\Phi_{wkb}(z,\eta)$ an elementary WKB solution given by theorem 3. Then for $z \neq 0$ in a neighbourhood of 0, one has the following unique decomposition,*

$$\Phi_{wkb}(z,\eta) = a(z,\eta)A_{wkb}(z,\eta) + b(z,\eta)\frac{\partial A_{wkb}}{\partial z}(z,\eta),$$

where $A_{wkb}(z,\eta)$ is the Airy WKB symbol (12), whereas a and b (depending on Φ_{wkb}) are local resurgence constants.

Here a "local resurgence constant" should be understood in the following way:

Definition 3. *In this paper, a* local resurgence constant $c(Z,\eta)$ *is a formal WKB expansion* $c(Z,\eta) = \sum_{n\geq 0} c_n(Z)\eta^{-n}$ *such that its minor* $\sum_{n\geq 1} c_n(Z)\frac{\xi^{n-1}}{\Gamma(n)}$ *defines a germ of holomorphic function at* $(Z,\xi) = (0,0) \in \mathbb{C}^m \times \mathbb{C}$, $m \geq 1$.

4 Conclusion

Just copying the reasoning of [Pham 00] §2.4, one easily deduces the following result from the theorems 4:

Theorem 5. *We assume that* $F(z)$ *in (6) is holomorphic near the origin. Then there exists a local resurgence constant* $s(z,\eta)$ *such that, under the transformation*

$$s(z,\eta) = \sum_{k\geq 0} s_k(z)\eta^{-k}, \quad s_0(z) = z, \quad \Phi(z,\eta) = \left(\frac{\partial s}{\partial z}\right)^{-\frac{1}{2}} y(s(z,\eta),\eta) \tag{50}$$

the equation (6) is transformed into the equation (4), for z (resp. s) *near the origin. Furthermore, under the transformation (50), an elementary WKB solution of (4) is transformed into an elementary WKB solution of (6).*

In particular, this theorem gives theorem 1 through the transformation (5). Therefore:

Corollary 1. *When* $V(q)$ *is holomorphic near the origin, with a simple zero at the origin, then, in theorem 1, the mapping* $(q,\eta) \mapsto s(q,\eta)$ *is a local resurgence constant.*

Note that it is straightforward to extend our results to the equation

$$\frac{d^2\Phi}{dz^2} - \eta^2 z\Phi = F(z,\beta)\Phi, \tag{51}$$

where F depends holomorphically on $(z,\beta) \in \mathbb{C}^2$ near the origin. In this case, the theorems 3 and 4 become:

Theorem 6. *There exists a family of elementary WKB solutions* $\Phi_{wkb}(z,\beta,\eta)$ *of equation (51) which are of Local Airy type, with regular dependence on* β *near the origin. For such an elementary WKB solution* $\Phi_{wkb}(z,\beta,\eta)$, *and for* $z \neq 0$ *in a neighbourhood of* 0 *and* β *near the origin, one has the following unique decomposition,*

$$\Phi_{wkb}(z,\beta,\eta) = a(z,\beta,\eta)A_{wkb}(z,\eta) + b(z,\beta,\eta)\frac{\partial A_{wkb}}{\partial z}(z,\eta),$$

where $A_{wkb}(z,\eta)$ *is the Airy WKB symbol (12), whereas* a *and* b *are local resurgence constants.*

The reader will easily translates theorem 5 in a similar way. This has the following consequence: since substituting to β a small Gevrey-1 expansion $\beta(\eta) = \sum_{k\geq 1} \beta_k \eta^{-k}$ in a local resurgence constant yields a local resurgence constant [Delabaere et al. 99], one recovers the result of [Aoki et al. 93].

Theorem 7. *We consider the differential equation*

$$\frac{d^2\Phi}{dz^2} - \eta^2 z\Phi = F(z,\eta)\Phi, \tag{52}$$

where $F(z,\eta)$ is a local resurgence constant. Then there exists a local resurgence constant $s(z,\eta)$ such that, under the transformation

$$s(z,\eta) = \sum_{k\geq 0} s_k(z)\eta^{-k}, \quad s_0(z) = z, \qquad \Phi(z,\eta) = \left(\frac{\partial s}{\partial z}\right)^{-\frac{1}{2}} y\big(s(z,\eta),\eta\big) \tag{53}$$

*the equation (52) is transformed into the equation (4), for z (*resp. *s) near the origin. Furthermore, under the transformation (53), an elementary WKB solution of (4) is transformed into an elementary WKB solution of (52).*

References

[Aoki et al. 91] T. Aoki, T. Kawai, Y. Takei: The Bender-Wu analysis and the Voros theory. Special functions (Okayama, 1990), 1–29, ICM-90 Satell. Conf. Proc., Springer, Tokyo (1991).

[Aoki et al. 93] T. Aoki, J. Yoshida: Microlocal reduction of ordinary differential operators with a large parameter. Publ. Res. Inst. Math. Sci. **29** (1993), no. 6, 959–975.

[Candelpergher et al. 91] B. Candelpergher, C. Nosmas, F. Pham: Approche de la résurgence. Actualités mathématiques, Hermann, Paris (1993).

[Delabaere et al. 93] E. Delabaere, H. Dillinger, F. Pham: Résurgence de Voros et périodes des courbes hyperelliptiques. Annales de l'Institut Fourier **43** (1993), no. 1, 163-199.

[Delabaere et al. 97] E. Delabaere, H. Dillinger, F. Pham: Exact semi-classical expansions for one dimensional quantum oscillators. Journal Math. Phys. **38** (1997), 12, 6126-6184.

[Delabaere et al. 99] E. Delabaere, F. Pham: Resurgent methods in semi-classical asymptotics. Ann. Inst. Henri Poincaré, Sect. A **71** (1999), no 1, 1-94.

[Ecalle 84] J. Ecalle: Cinq applications des fonctions résurgentes. preprint 84T 62, Orsay, (1984).

[Gérard et al. 96] R. Gérard, H. Tahara: Singular nonlinear partial differential equations. Aspects of Mathematics. Friedr. Vieweg & Sohn, Braunschweig (1996).

[Jidoumou 94] A. O. Jidoumou: Modèles de résurgence paramétrique: fonctions d'Airy et cylindro-paraboliques. J. Math. Pures Appl. (9) **73** (1994), no. 2, 111–190.

[Kashiwara et al. 79] M. Kashiwara, T. Kawai, J. Sjöstrand: On a class of linear partial differential equations whose formal solutions always converge. Ark. Mat. **17** (1979), no. 1, 83–91.

[Pham 88] F. Pham: Resurgence, Quantized canonical transformations and multi-instanton expansions. Algebraic Analysis II, vol. in honor of M. Sato, R.I.M.S. Kyoto, Kashiwara Kawai ed., Acad. Press, 699-726 (1988).

[Pham 00] F. Pham: Multiple turning points in exact WKB analysis (variations on a theme of Stokes). Toward the exact WKB analysis of differential equations linear or non-linear (C. Howls, T. Kawai, Y. Takei ed.), Kyoto University Press (2000),71-85.

[Rasoamanana 06] J.-M. Rasoamanana: Étude résurgente d'une classe d'équations différentielles de type Schrödinger. PhD thesis, Université d'Angers, France, to appear.

[Silverstone 85] H. Silverstone: JWKB connection-formula problem revisited via Borel summation. Phys. Rev. Lett. **55** (1985), no. 23, 2523–2526.

[Trèves 75] F. Trèves: Basic linear partial differential equations. Pure and Applied Mathematics, Vol. 62. Academic Press, New York-London (1975).

The Borel transform

Dedicated to Takahiro Kawai whose power and insights have been an inspiration for me.

Leon Ehrenpreis

Temple University, USA

Summary. We formulate the Borel transform in terms of functional analysis, in particular in terms of Analytically Uniform (AU) spaces. This enables us to extend the ideas of Borel to functions of several complex variables. Combining these ideas with nonlinear Fourier analysis leads to a generalization of the convexity that appears in Borel's work.

1 Introduction

Let

$$f(z) = \sum a_m z^m \tag{1}$$

be holomorphic near the origin. Then

$$\phi(t) = \sum \frac{a_m}{m!} t^m \tag{2}$$

is an entire function of exponential type. Borel's idea is that the Laplace transform (for z fixed)

$$F(z) = \int_0^\infty e^{-t}\phi(zt)dt \tag{3}$$

$$F(z) = \frac{1}{z}\int_0^\infty e^{-t/z}\phi(t)dt$$

provides an analytic (holomorphic) continuation of f across any arc of the circle on which f is holomorphic.

Of course, we have to assume that the integral in (3) converges in a suitable sense, which means that ϕ does not grow fast along the ray $\{tz\}$ as $t \to \infty$. This is the main point in the theory. For ϕ is defined by power series and it

Received 27 February, 2006. Revised 28 March, 2006. Accepted 28 March, 2006.

is extremely difficult to determine, via power series, the growth at infinity of $\phi(tz)$ (z fixed, $t \to \infty$) in any way which distinguishes directions.

For the simplest example of this difficulty, imagine trying to prove that

$$e^{-t} = \sum (-1)^n \frac{t^n}{n!} \tag{4}$$

is bounded as $t \to \infty$. This is related via (2) to the analytic continuation of

$$f(z) = \sum (-z)^n \tag{5}$$

We can recognize the fact that a function is rational from the recursion relation for its power series coefficients. This can be related to growth properties of its Borel transform as in the case of e^{-t}.

In general it is easier to work with the function F directly. Thus, although one doesn't know how to use the power series to prove e^{-t} is bounded as $t \to +\infty$, one can use the "product structure"

$$e^{-t} = \lim \left(1 - \frac{t}{N}\right)^N . \tag{6}$$

The formula

$$e^{it} = \lim \left(1 + \frac{it}{N}\right)^N \tag{7}$$

can also be used to prove e^{it} is bounded on the real axis.

Problems of this sort were popular in the early 1960's. For example it is difficult, but possible, to prove

$$\cos t = \prod \left(1 - \frac{t^2}{[(n+1/2)\pi]^2}\right) \tag{8}$$

is bounded from the product form, (as far as I remember no one was able to obtain the bound 1 !!)

Thus it seems that a product method should have more success. The Wiener-Feynmann integral gives a product method for the heat and Schrodinger equations. In [3], this was extended to general linear partial differential equations. This "probabilistic method" is used to prove existence theorems for various equations by showing suitable boundedness of the related products.

The WKB method (see [5]) which is the centerpiece of this conference, involves an application of the Borel transform to the Schrodinger equation

$$\left[\frac{d^2}{dx^2} - \eta^2 Q(x)\right] \psi(x, \eta) = 0. \tag{9}$$

We take the Borel transform in η^{-1} of (a slight modification) of ψ. In the WKB method the power series is completed by use of the logarithmic derivative S of ψ which satisfies a Riccati equation

$$S^2 + S' = \eta^2 Q. \tag{10}$$

(It is interesting that a Ricatti equation related to the Riemann ζ function is the main tool in Selberg's elementary proof of the prime number theorem. This is studied in detail in [4].)

It is essential that we study ψ near $\eta = \infty$ since $\eta = 1/h$ where h is Planck's constant and we are interested in what happens near the classical limit. To apply Fourier analysis to (9) it makes sense to replace η by h^{-1} and then replace h by $\partial/\partial u$, thus obtaining a $4th$ order partial differential equation which is difficult to handle; it is better to use Mellin transform since the change $\eta \to \eta^{-1}$ is readily understood in the framework of the Mellin transform. Thus we write

$$\psi(x,\eta) = \int \eta^{-is} \underline{\psi}(x,s)ds \tag{11}$$

so that, formally

$$\eta^2 \psi(x,\eta) = \int \eta^{-is+2} \underline{\psi}(x,s)ds \tag{12}$$

$$\begin{aligned} \eta^2 \psi(x,\eta) &= \int \eta^{-i(s+2i)} \underline{\psi}(x,s)ds \\ &= \int \eta^{-is} \underline{\psi}(x,s-2i)ds. \end{aligned}$$

Equation (9) becomes the differential-difference equation

$$\left[\frac{d^2}{dx^2} - \tau_{2i} Q(x) \right] \underline{\psi}(x,s) = 0 \tag{13}$$

where τ is the translation operator in s.

We shall discuss the relation of (13) to the Borel transform in another work. The essential point is that the passage between Fourier and Mellin transform is governed by the Γ function.

2 Functional Analysis and the Borel Transform

We can formulate the Borel transform in the language of AU spaces (see [1], which will be referred to as FA).

Let us return to the notation of (1), (2), ... The function $g(z) = z^{-1} f(z^{-1})$ is holomorphic near $z = \infty$. As such it defines an element of H' which is the dual of the space H of entire functions. More precisely, we inject functions $g(z)$ which are holomorphic at ∞ into H' by

$$g.h = \int_\Gamma ghdz \tag{14}$$

for any $h \in H$. Here Γ is a simple contour which surrounds the origin and lies in the region of analyticity of g.

This injection of g into H' is related to the classical theorem which asserts (imprecisely) that there is a duality between holomorphic functions on a domain Ω and on its complement $c(\Omega)$. We say "imprecisely" because certain conditions have to be put on Ω and in the behavior of the functions in question at the boundary of Ω. We shall not attempt to make these concepts precise as there are many possibilities and the reader can make his choice.

To avoid confusion we shall denote the injection of $g = f(z^{-1})$ into H' by $\jmath(f)$. Suppose f is holomorphic on a connected set Ω containing the origin in its interior. Then (14) defines $\jmath(f)$ as an element of $H'(c\Omega^{-1})$ where $c\Omega^{-1}$ is the complement of Ω^{-1}.

Note that (2) represents up to factors 2π and i which we shall ignore, the Fourier transform of $\jmath(f)$, that is

$$\phi(t) = \jmath(f).e^{itz} \tag{15}$$

where we regard H as the space of entire functions of z. Since $\jmath(f) \in H'$, the growth of its Fourier transform determines the smallest convex set K such that $\jmath(f)$ can be extended to a continuous linear function on $H(K)$. We write $K = K(f)$ (see FA).

More precisely, if $a(\theta)$ is a convex function then (up to epsilonics)

$$\phi(re^{i\theta}) = O(e^{a(\theta)r}) \tag{16}$$

if and only if K is contained in intersection of the half-planes

$$-\text{Im}(ze^{i\theta}) \leq a(\theta). \tag{17}$$

Thus K is defined by the minimal function $a(\theta)$, which is called the *indicator* of K or of ϕ.

We have noted above that $f(z^{-1})$ defines an element of $H'(c\Omega^{-1})$. Moreover, our construction shows that K is a convex set contained in the convex hull of $c\Omega^{-1}$. In fact, if we wrote more precisely $K = K(f)$ then

$$\bigcup K(f) = \text{convex hull of } c\Omega^{-1}.$$

In any case,

$$K(f) \subset \text{convex hull of } c\Omega^{-1}.$$

By analytic continuation, $f(z^{-1})$ is holomorphic outside $K(f)$ so that $f(z)$ is holomorphic on the complement of $[K(f)]^{-1}$. This is Borel's method of analytic continuation. The Laplace transform (3) then inverts the Fourier transform. This inversion formula is based on the fact that the Γ function, which arises when we expand ϕ in a power series in (3), cancels the denominators $m!$ that appear in (2). This is keeping with the remark made at the end of

Section 1 regarding the relation between the Fourier and Mellin transforms. (The Mellin transform controls the power series.)

As we mentioned in Section 1, there is no direct way of determining the indicator diagram for $K(t)$ from the power series coefficients of f.

The functional analysis approach can deal with more general aspects of the Borel transform. For the usual Borel transform we assume that the function f in (1) is holomorphic in a neighborhood of the origin. This means that

$$| a_m |\leq c^{m+1} \tag{18}$$

for some constant c. Other growth condition can be studied in the same manner. For example, suppose $l \leq 1$ and

$$| a_m |\leq c^{m+2}(lm)!. \tag{19}$$

Of course, the power series (1) does not converge. But the Borel transform ϕ satisfies

$$\begin{aligned} | \phi(z) | &\leq \sum c^{m+1} \frac{(lm)!}{m!} t^m \\ &\sim \sum c_1^{m+1} m^{(l-1)m} t^m \\ &\sim \sum c_2^{m+1} \frac{t^{m/(1-l)}}{m!} \\ &\sim c_3 e^{c_4 t^{1/1-l}}. \end{aligned} \tag{20}$$

For example, if $l = 1/2$ then ϕ is an entire function of order 2. When $l = 0$ then, of course, ϕ is of exponential type.

These classes of functions have been studied extensively in FA. Let us examine how the Borel transform relates to such classes.

For example the Laplace transform of e^{-t^2} is given by (according to (3))

$$F(z) = \int_0^\infty e^{-z^2t^2-t} dt \tag{21}$$

$$F(z) = e^{1/4z^2} \int_0^\infty e^{-(zt+1/2z)^2} dt.$$

When z is real the integral converges, and, as we see from the first integral in (21), $F(z)$ is well defined and bounded on the real axis.

From the viewpoint of real analysis, we evaluate $F^{(n)}(0)$ using the first integral in (21). Clearly

$$F^{(2n)}(0) = (-1)^n \frac{(2n)!}{n!} \int_0^\infty t^{2n} e^{-t} dt \sim (-1)^n (3n)!. \tag{22}$$

(The odd derivatives vanish.) We have used the symbol $\sim$ to indicate equality modulo c^n. This means that

$$| F^{(n)}(0) | \leq c^{n+1}(\frac{3}{2}n)! \tag{23}$$

so F belongs to Gervey 3/2 in accordance with (19).

If we replace e^{-t^2} by e^{-t^p} where p a positive integer, we find that the corresponding $F = F_p$ belongs to Gervey $(2 - 1/p)$. This agrees with (20) since $l = 1 - 1/p$ for Gervey $(2 - 1/p)$.

3 Several Variables

We cannot extend the above ideas as they stand to several complex variables because of Hartogs' theorem (see FA) which asserts that a function which is holomorphic outside a compact set extends holomorphically to the interior. Thus certain modifications must be made.

Let f be holomorphic in a neighborhood of the origin in C^n, then f is holomorphic on a product domain

$$\Omega = \{z_j \in \Omega_j\} \tag{24}$$

all j. Ω has a distinguished boundary so we can again inject f into $\jmath(f) \in H'$ by setting

$$\jmath(f).h = \int f(\frac{1}{z_1}, ..., \frac{1}{z_n})h(z_1, ..., z_n)\frac{dz}{z_1 z_2 ... z_n}. \tag{25}$$

The integral is extended over the distinguished boundary of Ω. (For the theory there is no difficulty in using more general types of distinguished boundaries; we shall discuss this below.)

The Fourier transform of $\jmath(f)$ is defined by

$$\widehat{\jmath(f)}(t) = \jmath(f).e^{it.z} \tag{26}$$

$$\widehat{\jmath(f)}(t) = \sum \frac{a_m}{m!}t^m$$

where

$$f(z) = \sum a_m z^m. \tag{27}$$

Here $m = (m_1, m_2, ..., m_n)$ and $m! = m_1!m_2!...m_n!$.

We can formulate the Fourier transform in a slightly different fashion, which is useful in certain contexts. The usual (Fisher) inner product of polynomials $P(z), Q(z)$ is defined by

$$P \odot Q = \bar{P}(\partial/\partial z)Q(z) \mid_{z=0} . \tag{28}$$

$\bar{P}$ is defined by taking the complex conjugate of the coefficients of P. In particular

$$z^m \odot z^{m'} = m!\delta_{mm'}. \tag{29}$$

We renormalize the Fisher product to

$$z^m * z^{m'} = \delta_{mm'} \tag{30}$$

Thus

$$\widehat{j(f)}(t) = f * e^{itz} \tag{31}$$

In order to relate this to the domain of analyticity of f we apply the Fundamental Principle of FA. As defined, $j(f) \in H'$ is an element of the dual of the kernel of the Cauchy Riemann system $\bar{\partial}$ on R^{2n}. We regard $\widehat{j(f)}$ as an entire function of the complex variable $t \in C^n$ where C^n is identified with the Cauchy-Riemann variety V in C^{2n}.

Precisely, if $x = (x_1, ..., x_{2n})$ is the parameter on R^{2n}, then the Cauchy-Riemann variety V is defined by

$$\hat{x}_j + i\hat{x}_{n+j} = 0. \tag{32}$$

Moreover, $z_j = x_j + ix_{n+j}$ so that, by (32), we have

$$x.\hat{x} = \sum (x_j \hat{x}_j + x_{n+j}\hat{x}_{n+j}) = \sum_1^n \hat{x}_j (x_j + ix_{n+j}) \tag{33}$$

$$\sum (x_j \hat{x}_j + x_{n+j}\hat{x}_{n+j}) = \sum_1^n t_j z_j.$$

We have written

$$t_j = \hat{x}_j \tag{34}$$

for $j = 1, ..., n$. Note that $\hat{x}_1, ..., \hat{x}_n$ parameterizes V.

This shows the relation between the complex Fourier transform on C^n and the real Fourier transform on R^{2n}. The Fundamental Principle goes further; it shows that (up to epsilonics) the indicator diagram of $\widehat{j(f)}$ on V can be realized as the indicator diagram of some extension of $\widehat{j(f)}$ from V to an element $\hat{F} \in \hat{E}'(R^{2n})$. ($\hat{F}$ is not unique.) $\hat{F}$ is the Fourier transform of a distribution F of compact support on R^{2n} for which the convex hull of support F is determined by this indicator function in much the same way as in the case $n = 1$.

Let us make the relation between $\widehat{j(f)}$ and $\hat{F}$ precise. In general the indicator diagram of an element $\hat{G} \in \hat{H}'(C^{2n})$ is a function on the real sphere at infinity in $R^{4n} = C^{2n}$. But for function in $\hat{E}'(R^{2n})$ (which are functions on C^{2n}) the indicator diagram is really a function on I^{2n} which is the imaginary part of C^{2n}. This is the content of Paley-Wiener theory.

From another point of view, the indicator diagram of $\widehat{j(f)}$ lives on the sphere at infinity in V. Since V is defined by $\hat{x} = -i\hat{y}$, the growth at infinity of $\widehat{j(f)}$ on V depends only on the growth in $\hat{y}$. For, on V, the growth conditions of $\hat{H}$ and $\hat{E}$ agree. Thus by the Fundamental Principle of FA, the indicator diagram of $\widehat{\jmath(f)}$ as an element of $\hat{H}'$ corresponds to the indicator diagram of some $\hat{F} \in \hat{E}'(R^{2n})$ which extends $\widehat{\jmath(f)}$ from V to C^{2n}.

By construction $F = \jmath(f)$ as elements in the dual of the kernel of $\hat{\partial}$. In particular,

$$F.z^n = \jmath(f).z^n \tag{35}$$

This means that the restriction of the power series of $\hat{F}$ to a power series in $\{x_j + ix_{n+j}\}$ is the same as the power series of f except that the coefficient a_n is to be divided by $m!$.

It follows from this construction that the indicator diagram of $\hat{F}$ [or of $\widehat{\jmath(f)}$] determines the convex hull of the support of F. (Recall that $F \in E'$ and hence has a well-defined support.) $\jmath(f)$ is an element of H'. Since H is not dense in E, there is no meaning to the support of $\jmath(f)$.

The definition of $\jmath(f)$ as given by (25) represents $\jmath(f)$ as a measure supported on the product of $\{\text{bd}\Omega_j\}$. The convex hull of support F is the smallest convex set $K = K(f)$ in R^{2n} on which we can find a representative on $E'(R^{2n})$ which agrees with $\jmath(f)$ on H.

How does this relate to the analytic continuation of f?

As in the case of $n = 1$, the dual of the space of holomorphic functions on a product domain $\Omega = \{z_j \in \Omega_j\}$ can be identified (modulo boundary conditions) with the space of functions which are holomorphic on the product domain of holomorphy.

$$\Omega^{-1} = \{z_j^{-1} \in \Omega_j\}. \tag{36}$$

Let $K_0(f)$ be a "smallest" product of convex sets containing K. We can assume that (as in case $n = 1$)

$$K_0(f) \subset \prod[\text{convex hull of } c(\Omega_j^{-1})] \tag{37}$$

since the right side of (37) is a convex set containing $c\Omega^{-1}$.

Now supp $F \subset K \subset K_0(f)$. Since $K_0(f)$ is a product then F can be represented in H' by a function F_0 which is holomorphic on the complement of $[K_0(f)]$. $F = \jmath(f)$ as element of H' so $F_0(z) = f(z^{-1})$ for z^{-1} sufficiently small. By analytic continuation, this means that

$$f(z) \text{ is holomorphic on the complement of } [K_0(f)]. \tag{38}$$

(38) can be regarded as Borel continuation for $n \geq 1$.

4 Nonlinear Borel Transform

The Borel (or Fourier) transform uses, as kernel, the exponential of linear functions. In chapter 5 of RT we introduced the idea of using the exponential of polynomials. In that work we dealt with the space $E(R^n)$; we now pass to $H(C^n)$. The use of the nonlinear Borel transform helps us deal with certain nonconvex regions, or, more precisely, regions which are convex in a certain polynomial sense.

Before discussing the nonlinear theory, let us analyze the linear theory in its relation to support. We have, for $z = re^{i\theta}$, $\lambda = \rho e^{i\phi}$

$$| e^{i\lambda z} |= e^{r\rho cos(\theta+\phi+\pi/2)} \tag{39}$$

If we think of λ as fixed, formula (39) describes this growth, of $e^{i\lambda z}$ on the circle at infinity. Conversely, λ can be determined by this growth, more specifically, by the maximal growth, which occurs at $\theta = -\phi - \pi/2$. In fact, the essential structure of the growth of $ae^{i\lambda z}$ is independent of a.

If we have a finite sum $\sum a_\alpha e^{i\lambda_\alpha z}$ then we can determine the convex hull of the points λ_j from the growth of $\sum a_\alpha e^{i\lambda_\alpha z}$. (This will be clarified below.)

The same discussion can be made for points $x \in R^n$ with dual $\hat{x}$. Paley-Wiener theory tells us that we can construct the convex hull L of the support of a function $a(x)$ of compact support from the growth at infinity of its Fourier transform

$$\hat{a}(\hat{x}) = \int a(x)e^{ix.\hat{x}}dx. \tag{40}$$

If L has an interior and we choose the origin in this interior then we can regard the Paley-Wiener theory as establishing a correspondence between the unit sphere in $\{x\}$ and the sphere at imaginary infinity in $\{\hat{x}\}$.

In the complex case we have to consider a as an element of $H'(C^n)$. We cannot apply Paley-Wiener theory directly, but, via the Fundamental Principle of FA, we can produce a measure $\tilde{a}$ on R^{2n} for which the convex hull of its support is determined, as in the real case, by the growth at infinity of the complex Fourier transform of a. (Of course $\tilde{a}$ is not unique.)

This is the underlining current of the ideas of Sections 2 and 3.

There is one further remark which can serve as an introduction to nonlinear convexity. If we fix $\hat{z} = \hat{r}e^{i\hat{\theta}}$ then

$$| e^{iz\hat{z}} |\leq 1 \tag{41}$$

defines the half-plane

$$x\hat{y} + y\hat{x} \geq 0. \tag{42}$$

This is the basis for convexity in the Paley-Wiener theory.

Now, consider the inequality

$$| e^{iz^2\hat{p}} |\leq a. \tag{43}$$

For $\hat{p} = 1$ this is

$$xy \geq \frac{1}{2}\log a \tag{44}$$

which is a hyperbolic region. Thus we might expect that a quadratic Borel transform should lead to conditions of support or analyticity on regions which are convex with respect to such hyperbolas.

The Paley-Wiener theory has a dual side. If $h(z)$ is a holomorphic function on a compact convex set κ which is the closure of its interior, then h admits a Fourier-Borel representation

$$h(z) = \int e^{iz\hat{z}} \frac{d\mu(\hat{z})}{k(\hat{z})} \tag{45}$$

where k grows at infinity faster than the indicator function of κ and μ is a measure on C of bounded variation.

Conversely, any expression (45) for such k defines a holomorphic function on K (see FA).

Let us consider the homogeneous quadratic analog of (45):

$$h(z) = \int e^{iz^2\hat{z}} \frac{d\mu(\hat{z})}{k(\hat{z})}. \tag{46}$$

We want such an expression to represent functions on compact sets κ which are convex with respect to real parts of the complex quadratics $\{\hat{z}_j z^2\}$. If κ is defined by inequalities

$$\text{Re } \alpha_j z^2 \leq c_j \tag{47}$$

where $| \alpha_j |= 1$ then we would suspect that k must satisfy

$$k(\hat{z}) \geq e^{-c_j|\hat{z}|} \tag{48}$$

on the rays

$$\arg i\hat{z} = \arg \alpha_j \tag{49}$$

just as in the linear case. Moreover this quadratic indicator diagram satisfies the same convexity properties as in the linear case.

Indeed, this is the case.

We want to apply a modification of the nonlinear Fourier transform as presented in Chapter 5 of RT. We start with the quadratic nonlinear Fourier transform.

In RT we dealt with the real Fourier transform and functions defined on R^n. In the simplest case we start with a function $u(x)$ of a single variable and form

$$U(\hat{p}, \hat{x}) = \int u(x) e^{i\hat{x}x + i\hat{p}x^2} dx. \tag{50}$$

U satisfies the heat (Schrodinger) equation

$$\left(\frac{\partial}{\partial \hat{p}} + i \frac{\partial^2}{\partial \hat{x}^2} \right) U = 0. \tag{51}$$

U is the (unique) solution of (51) whose initial value (i.e., $\hat{p} = 0$) is $\hat{u}(\hat{x})$. (Actually U is unique only within certain classes of functions defined by growth.)

Conversely, if U is a solution of (51) then, in a general function sense, U has a representation like (50). u is the inverse Fourier transform of $U(0, \hat{x})$. (This idea is developed in FA.)

What is a holomorphic (Borel) version of the nonlinear Fourier transform? Now $u = j(f)$ where f is a holomorphic function, say on the unit disc. Then, a la Borel, we form the Fourier-Borel transform ϕ of f as in (2). We call Φ the solution of the Schrodinger equation with initial value ϕ. Since ϕ is an entire function of exponential type we can define $\hat{\phi}$ in a unique manner by growth conditions.

Precisely, we can write

$$\Phi(\hat{p}, \hat{x}) = \sum (-i)^m \frac{\phi^{(2m)}(\hat{x})\hat{p}^m}{m!}. \tag{52}$$

It is clear that Φ satisfies (50). Moreover, since ϕ is an entire function of exponential type we have

$$| \phi^{(m)}(\hat{x}) | \leq c^{m+1} e^{A|\hat{x}|} \tag{53}$$

for some c, A. Thus the series (52) converges and defines Φ as an an entire function of exponential type.

We want to show how to write the solution Φ of the Schrodinger equation as a nonlinear Fourier transform. $\phi(\hat{x}) = \phi^{(0)}(\hat{x})$ has a representation as the Fourier transform of $j(f) = f(z^{-1})$ considered as an element of H'. Thus

$$\phi(\hat{x}) = \int e^{i\hat{x}z} f(z^{-1}) dz \tag{54}$$

where the integral is taken about a suitable contour. We claim that

$$\Phi(\hat{p}, \hat{x}) = \int e^{i\hat{x}z + i\hat{p}z^2} f(z^{-1}) dz \tag{55}$$

where the integral is taken over the same contour.

It is clear that the value of the integral at $\hat{p} = 0$ is $\phi(\hat{x})$. Moreover, the integral satisfies the Schrodinger equation hence must equal Φ.

We could also expand $e^{i\hat{p}z^2}$ in a power series, yielding

$$\Phi(\hat{p}, \hat{x}) = \sum \frac{i^n \hat{p}^n}{n!} \int z^{2n} e^{i\hat{x}z} f(z^{-1}) dz \tag{56}$$

$$\Phi(\hat{p}, \hat{x}) = \sum \frac{i^n \hat{p}^n (-1)^n \phi^{(2n)}(\hat{x})}{n!}$$

in conformity with (52).

How does nonlinear Borel transform help us? Recall that ϕ is the Fourier transform of $j(f) = f(z^{-1})$ considered as an element of H'. Linear Borel transform provides us with the convex hull of the support of $j(f)$, hence of the complement of the set on which $f(z^{-1})$ is holomorphic. Nonlinear Borel transform gives more. To understand this, suppose that $f(z^{-1})$ is holomorphic on the complement of Figure 1. Linear Borel transform replaces this region by its convex hull.

Suppose we embed Figure 1 into R^3 by adding a variable $t = y^2$. Then Figure 1 becomes Figure 2. The convex hull of Figure 2 clearly differs from the convex hull of Figure 3 which arises by applying $t = y^2$ to Figure 4 which is the convex hull of Figure 1.

It is the difference in the convex hulls which can be perceived by the non-linearity just as in the case of one real variable where the difference between the convex hulls of the quadratic transforms of Figures 5 and 6 namely Figure 7 and 8 provides a distinction.

We have seen that Φ is an entire function of exponential type so the convex hull (in $C^2 = R^4$) of the support of $\jmath(f)$ lifted to the curve $p = z^2$ can be determined from the indicator diagram of Φ.

We can now proceed as in the case of the usual Borel transform to obtain more information about the analytic continuation of f.

Naturally we could use higher degree polynomials in the Borel transform.

PROBLEM. Use the nonlinear Borel transform, or the Oka embedding, to extend the ideas of Section 3 to polynomial convex domains.

References

1. L. Ehrenpreis, *Fourier Analysis in Several Complex Variables*, Wiley-Inter science, 1970. Reprinted by Dover, 2006.
2. L. Ehrenpreis ,*The Universality of the Radon Transform*, Oxford University Press, 2003.
3. L. Ehrenpreis, *Conditionally Convergent Functional Integrals and Partial Differential Equations*, Proc. Int. Congress Math., Stockholm (1962) 337-338.
4. L. Ehrenpreis, *Singularities, Functional Equations, and the Circle Method*, Contemp. Math. 166, AMS, Providence, RI, 35-80.
5. C.J. Howls, T. Kawai, Y. Takai, *Toward the Exact WKB Analysis of Differential Equations, Linear or Nonlinear*, Kyoto University Press, 2000.

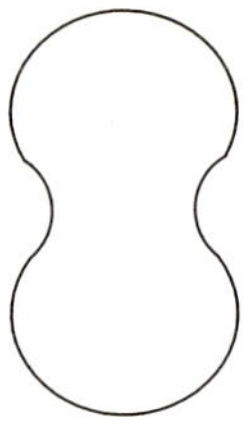

FIGURE 1

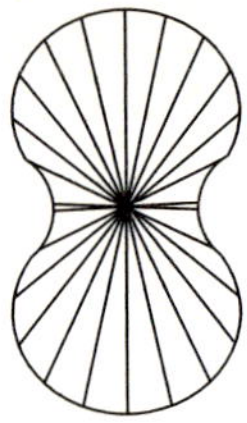

FIGURE 2

FIGURE 3

FIGURE 4

FIGURE 5

FIGURE 6

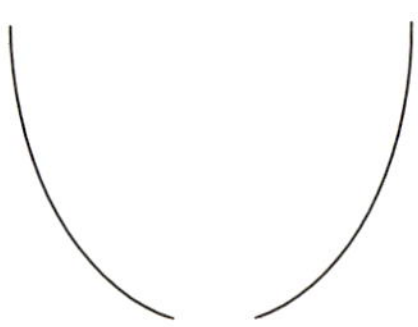

FIGURE 7

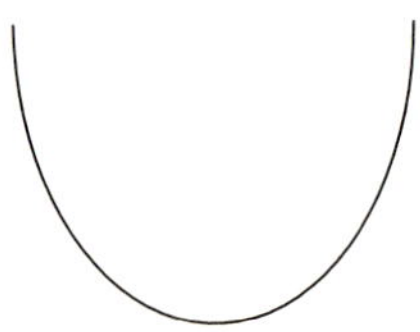

FIGURE 8

On the use of Z-transforms in the summation of transseries for partial differential equations

塵も積もれば山となる

Christopher J. Howls

University of Southampton, U.K.

Summary. In this paper we give examples of the resummation of transseries in partial differential equations using the language of the Z-transform. In so doing we illustrate the relationship between matched and exponential asymptotic approaches to the modelling of shock waves.

1 Introduction

Nonlinear partial differential equations in $\mathcal{C}\times\mathcal{R}$ with an asymptotic parameter ε can generate cascades of exponential scales, leading to transseries of the form

$$u(x,t;\varepsilon) \sim \sum_{n=0}^{\infty} C^n u^{(n)}(x,t;\varepsilon), \qquad 0<\varepsilon\ll 1 \tag{1}$$

where

$$u^{(n)}(x,t;\varepsilon) \sim e^{-nf(x,t)/\varepsilon} \sum_{r=0}^{\infty} a_r^{(n)}(x,t)\varepsilon^r, \qquad n=0,1,2,\cdots. \tag{2}$$

and C does not depend on $x \in \mathcal{C}$ or $t \in \mathcal{R}$.[1] Here we shall consider regions of (x,t) where $\mathrm{Re} f > 0$ so that the successive series in n are exponentially subdominant corrections.

That transseries may be re-summed under certain conditions is well-known, as was demonstrated in [2] for nonlinear ordinary differential equations. In the present context, in its most simple form, we may re-sum the transseries by formally reversing the n and r sums above to get

Received 28 February, 2006. Revised 4 September, 2006. Accepted 4 September, 2006.

[1] The equivalence "$\sim$" denotes a formally exact expansion. For simplicity here we only treat transseries that depend on a single function f.

$$u(x,t;\varepsilon) \sim \sum_{r=0}^{\infty} W^{(r)}(x,t;C,\varepsilon)\varepsilon^{r} \tag{3}$$

where

$$W^{(r)}(x,t;C,\varepsilon) \sim \sum_{n=0}^{\infty} e^{-nf(x,t)/\varepsilon} C^{n} a_{r}^{(n)}(x,t), \qquad r = 0,1,2,\cdots. \tag{4}$$

So that, to leading order in algebraic orders of ε we have

$$u(x,t;\varepsilon) = W^{(0)}(x,t;C,\varepsilon) + \mathcal{O}(\varepsilon) \tag{5}$$

with

$$W^{(0)}(x,t;C,\varepsilon) = \sum_{n=0}^{\infty} e^{-nf(x,t)/\varepsilon} C^{n} a_{0}^{(n)}(x,t) \tag{6}$$

wherever the latter sum converges.

Initially, resummation appears to be of little use. First, the possibility of exponential accuracy in ε has apparently been traded for only accuracy to algebraic order. Secondly, when the asymptotics is presented in this term-by-term "atomic" form, in order to achieve a practical resummation, not only would the general form of $a_0^{(n)}$ as a function of n have to be known, but the user would also have to be fortunate enough to find a closed form for the W−sum.

The first issue can be addressed when the resummation is viewed in the context of multiple scales [2]. As is known from classical asymptotic analysis, a multiple scales approach to nonlinear oscillatory systems can extend the range of validity of an expansion. In addition, as shown in [2], the analytic structure of terms in resummed transseries naturally contains additional information about positions of singularities of the solution.

In [3], we used a trivial resummation technique to locate the viscid singularities associated a Burgers equation in the limit of small viscosity.[2] These singularities are coupled to the existence of a shock, lying close to anti-stokes lines that emanate from caustics (real or virtual) associated with the initial data. Moreover the resummed series provided a transitional approximation across the shock region.

Hence the local exponential accuracy of (1), (2) is actually traded in (5) for a wider range of validity of the expansion. Note that resummation of transseries is closely associated with a variety of existing multiple scales methods, including for example, methods due to Kuzmak [6] or Maslov [8].

The purpose of this paper is to use the language of the Z-transform [5] to illustrate the resummation of transseries in a couple of simple PDE problems. In this way, following [2], we circumvent the need for explicit knowledge of all

[2] We chose not to use the obvious Cole-Hopf method in order to illustrate a more generic approach to nonlinear partial differential equations.

but a few of the individual, complicated, terms in the full expansion (1), (2).

Definition: The Z-transform of a (causal) sequence $a(n)$ is defined [5] to be

$$W(z) = Z\{a(n)\} \equiv \sum_{n=0}^{\infty} a(n) z^{-n}. \tag{7}$$

The Z-transform is used extensively in signals processing. We shall not need the associated inverse transform.

Setting $a(n) = a_0^{(n)}$ we see that the Z-transform of $a_0^{(n)}$ is thus the required leading order transseries sum, provided we set

$$z = \exp(f/\varepsilon)/C \implies W(z) \equiv W^{(0)}(x,t,z;C,\varepsilon) = \sum_{n=0}^{\infty} a_0^{(n)} C^n e^{-nf/\varepsilon} \tag{8}$$

The procedure is straightforward. We simply apply the Z-transform to the recurrence relation for the $a_0^{(n)}$ and obtain an equation for the corresponding Z-transform $W(z)$, this being the leading order of the resummed transseries.

We stress that resummation of transseries should not be viewed as an alternative to a standard multiple scales approach. For practical purposes, in many cases it will be sufficient to carry out a straightforward conventional multiple scales analysis. What this approach illustrates however, is the intimate link between exponential asymptotics and multiple scales analysis, and how the coefficients of expansions on one scale feed directly into those on another.

In the next section we briefly recall the relevant results of [1]. Following that we outline the application of the Z-transform to Burger's equation to illustrate the relationship between a matched asymptotic approach and transseries. We then illustrate the extension of the work to the so-called "modified Burgers equation" discussed in [7]. The latter contains a cubic nonlinearity. Coupled with the Z-transform, the method of [1] can be extended to apply to integer-power or classes of analytic nonlinearity.

2 Burgers equation

We consider Burgers equation

$$u_t + u u_x = \varepsilon u_{xx}, \tag{9}$$

where

$$x \in \mathbb{C}, \qquad t \geqslant 0, \qquad \varepsilon \to 0^+. \tag{10}$$

We consider initial Cauchy conditions

$$u(x,0) = a_0^{(0)}(x,0) \tag{11}$$

where $a_0^{(0)}$ is a given function with suitable decay at infinity. We shall consider $a_0^{(0)}(x,0)$ to have but a single maximum and no minimum at finite real x. As above, we shall further assume that we work in a region in (x,t)-space where $\mathrm{Re}(f) > 0$.

We shall consider a transseries solution of the form (1), (2). Substitution of this ansatz into into (9) and balancing at exponential and algebraic orders in ε shows that the exponential functions $f(x,t)$ satisfy the first order nonlinear equation

$$f_t + a_0^{(0)} f_x + f_x^2 = 0. \tag{12}$$

This may be solved with suitable initial data [3] on complex caustics to give

$$f_j\big(x(x_0,x_j),t(x_0,x_j)\big) = \tfrac{1}{2}\int_{x_0}^{x_j} a_0^{(0)}(z)dz - \tfrac{1}{4}\big(a_0^{(0)}(x_0) + a_0^{(0)}(x_j)\big)(x_j - x_0), \tag{13}$$

where the x_j satisfy the ray equation

$$x = x_0 + a_0^{(0)}(x_j)t. \tag{14}$$

The x_j are the intersection of the rays (14) with the initial $t = 0$ plane [1], [3].

Proceeding to higher order balances we obtain

$$n(n-1)f_x a_0^{(n)} + \sum_{p=1}^{n-1}(n-p)a_0^{(p)}a_0^{(n-p)} = 0. \tag{15}$$

The Z-transform of this equation is, [5],

$$f_x z\frac{d^2W}{dz^2} + (2f_x + a_0^{(0)} - W(z))\frac{dW}{dz} = 0, \tag{16}$$

$$\lim_{z\to\infty} W(z) = a_0^{(0)}, \tag{17}$$

$$\lim_{z\to\infty} z^2 W'(z) = -a_0^{(1)}. \tag{18}$$

Here, of course, $a_0^{(0)}$ and f_x are regarded as constants with respect to z. Note that despite the multidimensional nature of the original differential equation, the equation for W is only ordinary. This follows from the existence of only two effective asymptotic scales. In more complicated equations, with multiple scales behaviour, the equation would be (at least) partial.

A first integral of this equation that satisfies the first of the initial conditions can be written as

$$f_x z\frac{dW}{dz} + (f_x + a_0^{(0)})W - \frac{W^2}{2} = (f_x + a_0^{(0)})a_0^{(0)} - \frac{a_0^{(0)2}}{2} \tag{19}$$

In turn, this may be integrated once more to give

$$W(z) = a_0^{(0)} + \frac{2f_x e^{2f_x C_1}}{e^{2f_x C_1} - z}. \tag{20}$$

The arbitrary constant C_1 can be identified by identifying the coefficient of z^{-1} in the expansion (7) of $W(z)$ as $a_0^{(1)}$. Whence we arrive at

$$-2f_x e^{2f_x C_1} = a_0^{(1)} \tag{21}$$

and thus

$$W(z) = a_0^{(0)} + \frac{2Cf_x a_0^{(1)} e^{-f/\varepsilon}}{2f_x + Ca_0^{(1)} e^{-f/\varepsilon}}. \tag{22}$$

This is the result obtained in [1] by identification of the exact form of the individual $a_0^{(n)}$ and thence from explicit trivial summation of a geometric series. Obviously, the use of the Z-transform has avoided the need to know anything about the explicit form of $a_0^{(n)}$ for $n > 1$.

Note that equation (19) is intimately related to the leading order equation for the matched asymptotic approach as detailed in [1]. To see this, we change variables from z to $X = x - x_s(t)$ with the transformation

$$z = e^{f_x X}, \qquad X = \frac{x - x_s(t)}{\varepsilon}, \qquad W\left(e^{f_x X}\right) = A_0(X). \tag{23}$$

and treat X and t as independent variables. From (23) in (19) we then find

$$\frac{\partial^2 A_0}{\partial X^2} + \left(f_x + a_0^{(0)}(x_0) - A_0\right)\frac{\partial A_0}{\partial X} = 0 \tag{24}$$

On recalling from [1] that

$$f_x = \frac{1}{2}\left(a_0^{(0)}(x_1) - a_0^{(0)}(x_0)\right) \tag{25}$$

we see that (24) becomes

$$\frac{\partial^2 A_0}{\partial X^2} + (\dot{x}_s(t) - A_0)\frac{\partial A_0}{\partial X} = 0. \tag{26}$$

Here, as in [1], we have defined

$$\dot{x}_s(t) = \frac{1}{2}\left(a_0^{(0)}(x_1) + a_0^{(0)}(x_0)\right). \tag{27}$$

This is precisely the Riemann-Hugoniot condition that determines the location of the shock. Equation (26) is then obviously the inner equation as derived in the vicinity of the shock.

The variable X in (23) is just the inner variable in the matched asymptotic approach centred on the related inviscid shock location $x_s(t)$. Hence the quantity $e^{f_x X}$ in (23) is then just the leading order local (exponential) approximation to $e^{f/\varepsilon}$ in (8). A corollary is that we expect the approximations (2) and (22) to be valid outside the inner region of a matched asymptotic approach. Indeed to leading order, a matched asymptoticist might argue that it could be simpler to go straight to the solution of the inner problem and substitute the full f/f_x for X to obtain the exponential asymptotic result.

3 Modified Burgers equation

The so-called "modified Burgers equation"

$$u_t + u^2 u_x = \varepsilon u_{xx}, \tag{28}$$

was studied on a finite range of real x, $t \geq 0$ in [7]. It models, for example, weakly dissipative nonlinear transverse shear waves in an isotropic sold. For general Cauchy initial data it possesses no Bäcklund transformation onto a parabolic equation and so no approach of Cole-Hopf type exists. Its asymptotic treatment thus falls naturally into the realm of transseries. We shall study it on the domain (10) for Cauchy initial data (with suitable decay at ∞) that give rise to solutions of the form (1), (2). A full exponential analysis of the problem using the techniques of [1] will appear elsewhere. Here we focus on the more simple task of transseries summation.

We pose a transseries solution of the form (1), (2)

$$u(x,t) = \sum_{n=0}^{\infty} C^n u^{(n)}(x,t;\varepsilon)$$

where, we assume there is only one active subdominant f and with $\mathrm{Re} f > 0$ in the (x,t) region of interest, for example in advance/behind the first/last shock wave. Substituting (1), (2) in (28) and balancing at $\mathcal{O}(C^n)$ (which is equivalent to balancing at the exponential scales), the $u^{(n)}$ satisfy

$$\frac{\partial u^{(n)}}{\partial t} + \sum_{p=0}^{n}\sum_{r=0}^{p} u^{(r)} u^{(p-r)} \frac{\partial u^{(n-p)}}{\partial x} - \varepsilon \frac{\partial^2 u^{(n)}}{\partial x^2} = 0. \tag{29}$$

We now substitute the expansion of (1) for each $u^{(n)}$ and balance at order $\mathcal{O}(e^{-nf/\varepsilon}/\varepsilon)$. When $n = 0$ we obtain

$$f_t + {a_0^{(0)}}^2 f_x + f_x^2 = 0. \tag{30}$$

Equation (30) is almost identical to the corresponding equation for f in Burgers equation (12). The only difference is that $a_0^{(0)}$ in (12) is replaced with its square. Thus this equation has a solution in terms of coordinates x_j in the $t = 0$ initial plane satisfying

$$x = x_j + a_0^{(0)}(x_j)^2 t \tag{31}$$

where $a_0^{(0)}(x_j) = a_0^{(0)}(x_j, 0) = u(x_j, 0)$. The exact solution of (30) is then

$$f_j\big(x(x_0,x_j), t(x_0,x_j)\big) = \tfrac{1}{2}\int_{x_0}^{x_j} a_0^{(0)}(z)^2 dz - \tfrac{1}{4}\big(a_0^{(0)}(x_0)^2 + a_0^{(0)}(x_j)^2\big)(x_j - x_0). \tag{32}$$

Turning to a fixed value of $n > 0$, equation (32) can be used to simplify the recurrence relation for the $a_0^{(n)}$. This becomes

$$n(n-1)f_x a_0^{(n)} - \sum_{p=1}^{n-1}\sum_{r=0}^{p}(n-p)a_0^{(r)}a_0^{(p-r)}a_0^{(n-p)} = 0 \qquad n > 0. \tag{33}$$

Comparison of (33) with (15) reveals an obvious similarity in structure. Nevertheless it would be rather difficult to deduce the general form of the $a_0^{(n)}$ as a function of n and hence to sum the associated transseries.

To circumvent this problem, we again apply the Z-transform (7) to (33) with definitions (8). The result is

$$f_x z\frac{d^2W}{dz^2} + (2f_x + a_0^{(0)^2} - W(z)^2)\frac{dW}{dz} = 0, \tag{34}$$

$$\lim_{z\to\infty} W(z) = a_0^{(0)}, \tag{35}$$

$$\lim_{z\to\infty} z^2 W'(z) = -a_0^{(1)}. \tag{36}$$

This equation is similar to the corresponding one for Burgers (16). The solution of (34) now generates a transcendental equation for $W(z)$ of the form

$$V(z) \equiv W(z) - a_0^{(0)} \tag{37}$$

$$\left(\frac{zV(z)}{a_0^{(1)}}\right)^{A-B} = \left(1 - \frac{V(z)}{B}\right)^{A}\left(1 - \frac{V(z)}{A}\right)^{-B} \tag{38}$$

$$A = \frac{-3a_0^{(0)} + \sqrt{9a_0^{(0)^2} + 12f_x}}{2} \tag{39}$$

$$B = \frac{-3a_0^{(0)} - \sqrt{9a_0^{(0)^2} + 12f_x}}{2}. \tag{40}$$

For practical calculations, $W(z)$ would obviously have to be extracted numerically for each value of (x, t).

The complexity of this result in a comparatively simple equation demonstrates that for more general problems resummation to obtain even only the first term is likely to result in either a technically complicated differential equation or an involved implicit relation for $W(z)$.

The full analysis of this problem, including an exponential asymptotic treatment of shock interactions will appear elsewhere.

4 Other Riemann-like equations

The above method can be extended to generate resummed transseries associated with the leading order subdominant exponential for a broad class of equations with suitable boundary data.

In particular the techniques of [1] and this paper can be shown formally apply to equations of the form

$$u_t + u^n u_x = \varepsilon u_{xx}, \tag{41}$$

for integer n, and more generally to

$$u_t + F(u)u_x = \varepsilon u_{xx}, \tag{42}$$

where F is a suitable analytic function. These calculations will be outlined at greater length elsewhere [4].

5 Discussion

As we have indicated above, many of these results can be obtained by other means. However the Z-transform is particularly simple to use and, given a transseries template, it avoids some of the intricate details of multiple scales analysis. It can obtain results that appear to be valid in a wider region, extracting information from all exponential orders to generate a locally more appropriate analytical structure.

As has been illustrated here, the Z-transform can provide a transitional approximation across shocks for little effort. In fact, rather than using a Riemann-Hugoniot condition, the shock location emerges naturally from the resummation. The technique is widely applicable to shock systems, wherever transseries occur, subject to the ability to solve the (here) ordinary differential equation satisfied by the Z-transform. However, as the second example shows the resummed function will often be given implicitly. Work is ongoing

Finally, the purpose of the Z-transform (or any initial resummation) approach is well-described by the proverb in the title of the paper: "even dust if stacked up will become a mountain". The specks of dust are the exponentially small terms in the transseries. It is far better to stack them up collectively with, say, a simple, efficient Z-transform as soon as possible, rather than to leave them lying around in the problem to create a mountain of work when one comes to try to explicitly resum them later.

References

[1] Chapman S.J., Howls C.J., King J.R., Olde Daalhuis A.B., *Why is a shock not a caustic? The role of exponential asymptotics in smoothed shock formation*, (2007), Nonlinearity 20, pp. 2425-2452.

[2] Costin O. and Costin R.D., *On the formation of singularities of solutions of nonlinear differential systems in antistokes directions*, Inventiones Mathematicae 145, 3, pp 425-485 (2001)

[3] Howls C.J., *Recent Trends in Exponential Asymptotics*, RIMS Kokyuroku, University of Kyoto, 1424 (2005), 35-52

[4] Howls C.J., *in preparation*

[5] Jury E.I. *Theory and Application of the z-Transform Method*, John Wiley and Son (New York) (1964).

[6] King A.C., Billingham J., Otto S.R., *Differential Equations, Linear, Nonlinear, Ordinary, Partial*, Cambridge University Press (Cambridge, UK).

[7] Lee-Bapty I.P., Crighton D.G., *Nonlinear Wave Motion Governed by the Modified Burgers Equation*, Phil. Trans. Roy Soc. Lond. A. Math. Phys, Sci. **323** (1987), pp. 173-209.

[8] Maslov V.P., Omel'yanov G.A., *Geometrical Asymptotics for Nonlinear PDE. I*, Translations of Mathematical Monographs, vol. 202, American Mathematical Society (Providence, Rhode Island).

Some dynamical aspects of Painlevé VI

Katsunori Iwasaki

Faculty of Mathematics, Kyushu University, 6-10-1 Hakozaki, Higashi-ku, Fukuoka 812-8581, Japan
iwasaki@math.kyushu-u.ac.jp

Summary. We survey some results from our recent studies on the sixth Painlevé equation as a dynamical system. We discuss such topics as phase space and its compactification, Riemann-Hilbert correspondence, Poincaré section, bounded orbits, topological entropy and dynamical degree, and periodic solutions.

Key words: Painlevé VI equation, stable parabolic connection, Riemann-Hilbert correspondence, Klein singularities, Poincaré section, cubic surface, bounded orbits, topological entropy, dynamical degree, periodic solutions

1 Introduction

We survey some results from our recent studies on the sixth Painlevé equation as a dynamical system. Some parts of this article are a résumé of materials which have already been published or accepted for publication, and others are an announcement of new results which will appear elsewhere in full details.

The sixth Painlevé equation in its Hamiltonian form is a rank 2 system of nonlinear ordinary differential equation

$$\frac{dq}{dx} = \frac{\partial H(\kappa)}{\partial p}, \qquad \frac{dp}{dx} = -\frac{\partial H(\kappa)}{\partial q},$$

with an independent variable $x \in X := \mathbb{P}^1 \setminus \{0, 1, \infty\}$ and unknown functions $q = q(x)$, $p = p(x)$ depending on parameters κ in a 4-dimensional affine space

$$\mathcal{K} = \{\kappa = (\kappa_0, \kappa_1, \kappa_2, \kappa_3, \kappa_4) \in \mathbb{C}^5 \;:\; 2\kappa_0 + \kappa_1 + \kappa_2 + \kappa_3 + \kappa_4 = 1\},$$

where the Hamiltonian $H(\kappa) = H(q, p, x; \kappa)$ is given by

$$H(\kappa) = \frac{(q_0 q_1 q_x)p^2 - \{\kappa_1 q_1 q_x + (\kappa_2 - 1) q_0 q_1 + \kappa_3 q_0 q_x\} p + \kappa_0(\kappa_0 + \kappa_4) q_x}{x(x-1)},$$

Received 19 February, 2006. Revised 14 June, 2006. Accepted 14 June, 2006.

with $q_\nu := q - \nu$ for $\nu \in \{0, 1, x\}$.

This rather complicated equation is only a fragmentary appearance of a more substantial entity, which we call the sixth Painlevé dynamical system $P_{VI}(\kappa)$. It is a flow (Painlevé flow) on a certain moduli-theoretical fibration $\pi_\kappa : M(\kappa) \to X$ that is semi-conjugate to an isomonodromic flow on the moduli space of monodromy representations through a Riemann-Hilbert correspondence. For a symmetric description, however, it would be better to lift the phase space to a fibration $\pi_\kappa : \mathcal{M}(\kappa) \to T$ via the cross ratio map

$$T \to X, \quad t = (t_1, t_2, t_3, t_4) \mapsto x = \frac{(t_1 - t_3)(t_2 - t_4)}{(t_1 - t_2)(t_3 - t_4)},$$

where T is the configuration space of distinct ordered 4-points in $\mathbb{P}^1$.

2 Phase Space

We review the construction of phase space $\mathcal{M}(\kappa)$. The media of the Painlevé flow are stable parabolic connections so that the phase space of the Painlevé dynamics $P_{VI}(\kappa)$ is a family of moduli spaces of stable parabolic connections $\pi_\kappa : \mathcal{M}(\kappa) \to T$. More explicitly, given any $(t, \kappa) \in T \times \mathcal{K}$, a (t, κ)-*parabolic connection* is a quadruple of data $Q = (E, \nabla, \psi, l)$ such that

(1) E is a rank 2 algebraic vector bundle of degree -1 over $\mathbb{P}^1$,
(2) $\nabla : E \to E \otimes \Omega^1_{\mathbb{P}^1}(D_t)$ is a Fuchsian connection with pole divisor $D_t = t_1 + t_2 + t_3 + t_4$ and the Riemann scheme as in Table 1,
(3) $\psi : \det E \to \mathcal{O}_{\mathbb{P}^1}(-t_4)$ is a horizontal isomorphism, called a determinantal structure, where the sheaf $\mathcal{O}_{\mathbb{P}^1}(-t_4)$ is equiped with the connection induced from the standard exterior differential $d : \mathcal{O}_{\mathbb{P}^1} \to \Omega^1_{\mathbb{P}^1}$,
(4) $l = (l_1, l_2, l_3, l_4)$ is a *parabolic structure*, where l_i is an eigenline of the residue operator $\mathrm{Res}_{t_i}(\nabla) \in \mathrm{End}(E_{t_i})$ at t_i corresponding to the eigenvalue λ_i (whose negative $-\lambda_i$ is the first exponent at t_i in Table 1).

In Table 1, κ_i stands for the difference of the first exponent at t_i from the second one, so that λ_i is uniquely determined by κ_i. The fourth singular point t_4 is somewhat distinguished. There exists a concept of *stability* for parabolic connections, with which the geometric invariant theory of Mumford can be worked out to establish the following theorem [13, 14].

singularity	t_1	t_2	t_3	t_4
1st exponent	$-\lambda_1$	$-\lambda_2$	$-\lambda_3$	$-\lambda_4$
2nd exponent	λ_1	λ_2	λ_3	$\lambda_4 - 1$
difference	κ_1	κ_2	κ_3	κ_4

Table 1. Riemann scheme of parabolic connections

Theorem 1.

(1) *There is a fine moduli scheme* $\mathcal{M}_t(\kappa)$ *of stable* (t,κ)*-parabolic connections.*
(2) *The moduli space* $\mathcal{M}_t(\kappa)$ *is a smooth, irreducible, quasi-projective surface.*
(3) *Given any* $\kappa \in \mathcal{K}$, *there is a family of moduli spaces* $\pi_\kappa : \mathcal{M}(\kappa) \to T$, *where* π_κ *is a smooth morphism with fiber* $\mathcal{M}_t(\kappa)$ *over* $t \in T$.

The space $\mathcal{M}(\kappa)$ is exactly the phase space of $\mathrm{P_{VI}}(\kappa)$ as a time-dependent Hamiltonian dynamical system. See also the related works [1, 24, 25, 26].

3 Compactification

We proceed to the compactification of moduli spaces. The main idea for our compactifcation is stated as follows: Consider stable parabolic connections as 'matrix-valued Schrödinger operators'. Introduce a 'matrix-valued Planck constant', called a *phi-operator* ϕ. Consider stable parabolic *phi-connections*, where ϕ is allowed to be degenerate, namely, to be 'semi-classical'. Compactify the moduli space by adding those semi-classical objects.

More explicitly, given any $(t,\kappa) \in T \times \mathcal{K}$, a *parabolic phi-connection* is a sextuple of data $Q = (E_1, E_2, \phi, \nabla, \psi, l)$ consisting of

(1) a variant of connection $\nabla : E_1 \to E_2 \otimes \Omega^1_{\mathbb{P}^1}(D_t)$,
(2) an $\mathcal{O}_{\mathbb{P}^1}$-homomorphism $\phi : E_1 \to E_2$ (called a *phi-operator*), which may be *degenerate* or non-isomorphic, satisfying a generalized Leibniz rule

$$\nabla(fs) = \phi(s) \otimes df + f\nabla(s), \qquad (s \in E_1,\ f \in \mathcal{O}_{\mathbb{P}^1}),$$

(3) extra data of a determinantal structure ψ and a parabolic structure l.

There is a concept of *stability* for parabolic phi-connections, with which geometric invariant theory can be worked out to establish the following [13, 14].

Theorem 2.

(1) *There is a coarse moduli scheme* $\overline{\mathcal{M}}_t(\kappa)$ *of stable parabolic phi-connections.*
(2) *The moduli space* $\overline{\mathcal{M}}_t(\kappa)$ *is a smooth, irreducible, projective surface, having a unique effective anti-canonical divisor* $\mathcal{Y}_t(\kappa)$.
(3) *Under the natural embedding* $\mathcal{M}_t(\kappa) \hookrightarrow \overline{\mathcal{M}}_t(\kappa)$ *sending* (E, ∇, ψ, l) *to* $(E, E, \mathrm{id}, \nabla, \psi, l)$, *the space* $\mathcal{M}_t(\kappa)$ *is exactly the locus of* $\overline{\mathcal{M}}_t(\kappa)$ *where the phi-operator* ϕ *is isomorphic, and one has* $\mathcal{M}_t(\kappa) = \overline{\mathcal{M}}_t(\kappa) - \mathcal{Y}_t(\kappa)$.
(4) *As a relative setting over* T, *there is a family of moduli spaces with unique effective anti-canonical divisors,* $\overline{\pi}_\kappa : (\overline{\mathcal{M}}(\kappa), \mathcal{Y}(\kappa)) \to T$, *where* $\overline{\pi}_\kappa$ *is a smooth projective morphism with fiber* $(\overline{\mathcal{M}}_t(\kappa), \mathcal{Y}_t(\kappa))$ *over* $t \in T$.

The divisor $\mathcal{Y}_t(\kappa)$ on $\overline{\mathcal{M}}_t(\kappa)$ is called the *vertical leaves* at time t. We have the following realization of moduli spaces [13, 14] (see Fig. 1).

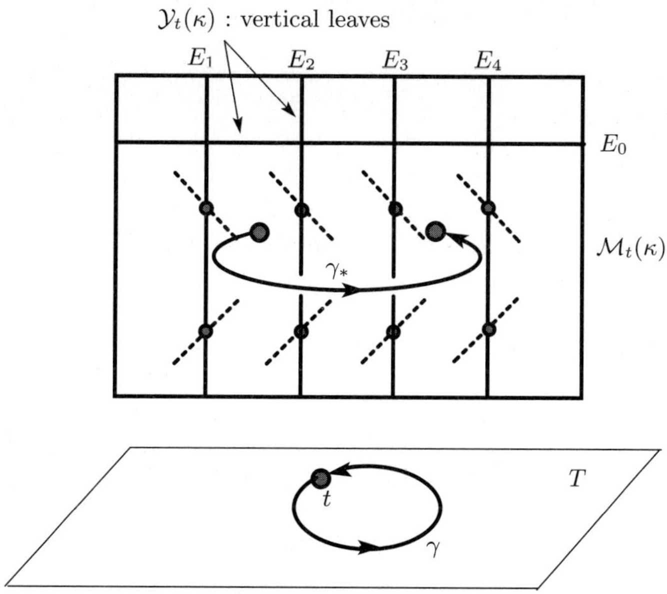

Fig. 1. Poincaré section on the space of initial conditions

Theorem 3.

(1) *The compactified moduli space* $\overline{\mathcal{M}_t}(\kappa)$ *is isomorphic to an* 8*-point blow-up of the Hirzebruch surface* $\Sigma_2 \to \mathbb{P}^1$ *of degree* 2*.*
(2) *The unique effective anti-canonical divisor on* $\overline{\mathcal{M}_t}(\kappa)$ *is given by*

$$\mathcal{Y}_t(\kappa) = 2E_0 + E_1 + E_2 + E_3 + E_4,$$

where E_0 *is the strict transform of the section at infinity of* $\Sigma_2 \to \mathbb{P}^1$*, and* $E_1, \ldots, E_4$ *are the strict transforms of the fibers at* $t_1, \ldots, t_4$*, respectively.*

Now we can talk about our main concern in this article: *Poincaré section* of the Painlevé dynamics. By the Painlevé property of $\mathrm{P}_{\mathrm{VI}}(\kappa)$, each loop $\gamma \in \pi_1(T, t)$ admits unique horizontal lifts along the Painlevé flow and induces an automorphism $\gamma_* : \mathcal{M}_t(\kappa) \to \mathcal{M}_t(\kappa)$, called the Poincaré return map along γ. These transformations altogether define a group homomorphism

$$\mathrm{PS}_t(\kappa) : \pi_1(T, t) \to \mathrm{Aut}\, \mathcal{M}_t(\kappa), \quad \gamma \mapsto \gamma_*.$$

This is exactly what we call the Poincaré section of $\mathrm{P}_{\mathrm{VI}}(\kappa)$, whose global dynamics is particularly interesting to us.

4 Riemann-Hilbert Correspondence

To understand the Poincaré section $\mathrm{PS}_t(\kappa)$, we should proceed to a discrete dynamical system to which it is semi-conjugated by a Riemann-Hilbert correspondence. So let us turn our attension to Riemann-Hilbert correspondence.

First, we start with introducing moduli space of monodromy representations. Given any $a = (a_1, a_2, a_3, a_4) \in A := \mathbb{C}_a^4$, let $\mathcal{R}_t(a)$ be the moduli space of Jordan equivalence classes of representations $\rho : \pi_1(\mathbb{P}^1 \setminus D_t, *) \to SL_2(\mathbb{C})$ with prescribed *local monodromy data*: $a_i = \operatorname{Tr}\rho(\gamma_i)$, $i = 1, 2, 3, 4$, where $\gamma_i \in \pi_1(\mathbb{P}^1 \setminus D_t, *)$ is a loop surrounding t_i in the positive direction, leaving the remaining points t_j $(j \neq i)$ outside.

Any stable parabolic connection $Q = (E, \nabla, \psi, l) \in \mathcal{M}_t(\kappa)$, restricted to $\mathbb{P}^1 \setminus D_t$, induces a flat connection $\nabla|_{\mathbb{P}^1 \setminus D_t} : E|_{\mathbb{P}^1 \setminus D_t} \to E|_{\mathbb{P}^1 \setminus D_t} \otimes \Omega^1_{\mathbb{P}^1 \setminus D_t}$. Let ρ be the Jordan equivalence class of its monodromy representation. Then the *Riemann-Hilbert correspondence* at time $t \in T$ is the map:

$$\mathrm{RH}_{t,\kappa} : \mathcal{M}_t(\kappa) \to \mathcal{R}_t(a), \quad Q \mapsto \rho,$$

together with the correspondence of parameters, $\mathcal{K} \to A$, $\kappa \mapsto a$, given by

$$a_i = \begin{cases} 2\cos \pi\kappa_i & (i = 1, 2, 3), \\ -2\cos \pi\kappa_4 & (i = 4). \end{cases}$$

The moduli space $\mathcal{R}_t(a)$ can be expressed in terms of an affine cubic surfaces. Given any $\theta = (\theta_1, \theta_2, \theta_3, \theta_4) \in \Theta := \mathbb{C}_\theta^4$, let $\mathcal{S}(\theta) = \{x \in \mathbb{C}^3 : f(x, \theta) = 0\}$ be an affine cubic surface with defining function

$$f(x, \theta) = x_1 x_2 x_3 + x_1^2 + x_2^2 + x_3^2 - \theta_1 x_1 - \theta_2 x_2 - \theta_3 x_3 + \theta_4.$$

Then there is an identification of the moduli space $\mathcal{R}_t(a)$ with the cubics $\mathcal{S}(\theta)$ via the isomorphism $\rho \mapsto x = (x_1, x_2, x_3)$ with $x_i = \operatorname{Tr}\rho(\gamma_j \gamma_k)$, $\{i, j, k\} = \{1, 2, 3\}$, under the correspondence of parameters, $A \to \Theta$, $a \mapsto \theta$, where

$$\theta_i = \begin{cases} a_i a_4 + a_j a_k & (i = 1, 2, 3), \\ a_1 a_2 a_3 a_4 + a_1^2 + a_2^2 + a_3^2 + a_4^2 - 4 & (i = 4). \end{cases}$$

Now we have three kinds of parameters and correspondences among them:

$$\begin{array}{ccccc} \text{parameters of } P_{\mathrm{VI}} & & \text{local monodromy data} & & \text{parameters of cubics} \\ \mathcal{K} & \to & A & \to & \Theta \\ \Cup & & \Cup & & \Cup \\ \kappa & \mapsto & a & \mapsto & \theta \end{array}$$

The composition of $\mathcal{K} \to A \to \Theta$ is referred to as the Riemann-Hilbert correspondence in the parameter level and is denoted by $\mathrm{rh} : \mathcal{K} \to \Theta$. Since the moduli space $\mathcal{R}_t(a)$ is identified with the affine cubic surface $\mathcal{S}(\theta)$, the Riemann-Hilbert correspondence $\mathrm{RH}_{t,\kappa}$ is reformulated as the map

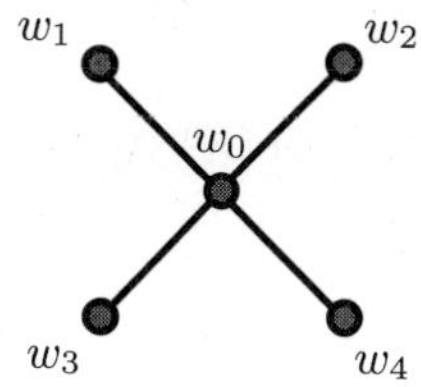

Fig. 2. Dynkin diagram of type $D_4^{(1)}$

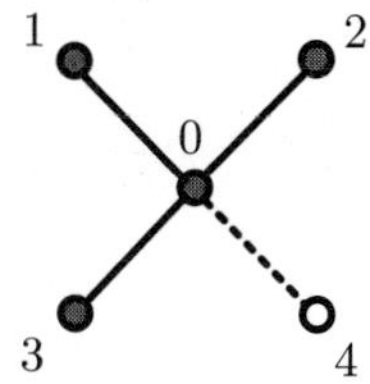

Fig. 3. A stratum of type D_4

$$\mathrm{RH}_{t,\kappa} : \mathcal{M}_t(\kappa) \to \mathcal{S}(\theta) \qquad \text{with} \qquad \theta = \mathrm{rh}(\kappa).$$

To discuss the Riemann-Hilbert correspondence, we have to introduce an affine Weyl group structure on the parameter space $\mathcal{K}$. For each $i \in \{0, 1, 2, 3, 4\}$, let $w_i : \mathcal{K} \to \mathcal{K}$ be the orthogonal reflection with respect to the natural inner product, having $H_i = \{\, \kappa \in \mathcal{K} : \kappa_i = 0 \,\}$ as its reflecting hyperplane. Then $W(D_4^{(1)}) = \langle w_0, w_1, w_2, w_3, w_4 \rangle \curvearrowright \mathcal{K}$ is an *affine Weyl group of type* $D_4^{(1)}$ corresponding to the Dynkin diagram as in Fig. 2. Let **Wall** be the union of all reflecting hyperplanes for the reflection group $W(D_4^{(1)})$. Explicitly, they are given by affine linear relations

$$\kappa_i = m, \qquad \kappa_1 \pm \kappa_2 \pm \kappa_3 \pm \kappa_4 = 2m + 1 \qquad (i \in \{1, 2, 3, 4\}, m \in \mathbb{Z}).$$

It turns out that the Riemann-Hilbert correspondence in the parameter level $\mathrm{rh} : \mathcal{K} \to \Theta$ is a branched $W(D_4^{(1)})$-covering ramified along **Wall** and it maps **Wall** onto the discriminant locus $V = \{\, \theta \in \Theta : \Delta(\theta) = 0 \,\}$ of the family of cubic surfaces $\mathcal{S}(\theta)$, $\theta \in \Theta$. In order to describe the singularities of $\mathcal{S}(\theta)$, let us introduce a stratification into the parameter space $\mathcal{K}$. For any *proper* subset $I \subset \{0, 1, 2, 3, 4\}$ including the empty set $\emptyset$, we put

$$\overline{\mathcal{K}}_I = W(D_4^{(1)})\text{-translates of the subset } \{\kappa_i = 0 \ (i \in I)\},$$
$$D_I = \text{Dynkin subdiagram of } D_4^{(1)} \text{ that has nodes} \bullet \text{ exactly in } I.$$

Let $\mathcal{K}_I$ be the set obtained from $\overline{\mathcal{K}}_I$ by removing the sets $\overline{\mathcal{K}}_J$ with $\#J = \#I + 1$. When $I = \emptyset$ we have the maximal stratum $\mathcal{K}_\emptyset$ called the *big open*, which is just $\mathcal{K} \setminus$ **Wall**. At an opposite extreme the case $I = \{0, 1, 2, 3\}$ gives a minimal stratum of type D_4 as indicated in Fig. 3.

Our solution to the Riemann-Hilbert problem is stated as follows [13, 14].

Theorem 4. *Given any $\kappa \in \mathcal{K}$, put $\theta = \mathrm{rh}(\kappa) \in \Theta$.*

(1) *If $\kappa \in \mathcal{K}_I$, then $\mathcal{S}(\theta)$ has Klein singularities of Dynkin type D_I.*
(2) *The Riemann-Hilbert correspondence $\mathrm{RH}_{t,\kappa} : \mathcal{M}_t(\kappa) \to \mathcal{S}(\theta)$ is a proper surjection that is an analytic minimal resolution of singularities.*

We remark that if κ lies on the big open $\mathcal{K} \setminus$ **Wall**, then $\mathcal{S}(\theta)$ is smooth and the Riemann-Hilbert correspondence $\mathrm{RH}_{t,\kappa}$ is biholomorphic, while if $\kappa = (0, 0, 0, 0, 1)$ and $\theta = (8, 8, 8, 28)$, then $\mathcal{S}(\theta)$ has a Klein singularity of type D_4 and $\mathrm{RH}_{t,\kappa}$ gives an analytic minimal desingularization of it as in Fig. 4.

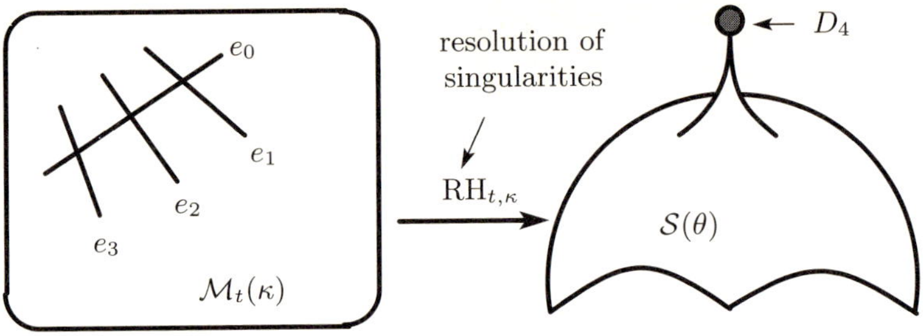

Fig. 4. Analytic minimal desigularization by $\mathrm{RH}_{t,\kappa}$

5 Semi-Conjugacy and Conjugacy

Now we can speak about the semi-conjugacy and conjugacy of the Poincaré section of $\mathrm{P_{VI}}(\kappa)$ in the following manner.

(1) Klein singularities admit *algebraic* minimal resolutions of singularities as constructed by Brieskorn and others. Let $\varpi : \widetilde{\mathcal{S}}(\theta) \to \mathcal{S}(\theta)$ be such a standard algebraic minimal resolution of singularities.
(2) $\mathrm{RH}_{t,\kappa}$ lifts to a biholomorphism $\widetilde{\mathrm{RH}}_{t,\kappa}$ to yield a commutative diagram

$$\begin{array}{ccc} \mathcal{M}_t(\kappa) & \xrightarrow{\widetilde{\mathrm{RH}}_{t,\kappa}} & \widetilde{\mathcal{S}}(\theta) \\ \| & & \downarrow{\scriptstyle\varpi} \\ \mathcal{M}_t(\kappa) & \xrightarrow[\mathrm{RH}_{t,\kappa}]{} & \mathcal{S}(\theta) \end{array}$$

(3) The Poincaré section $\mathrm{PS}_t(\kappa)$ on $\mathcal{M}_t(\kappa)$ is semi-conjugated to a discrete dynamical system on $\mathcal{S}(\theta)$ by $\mathrm{RH}_{t,\kappa}$, and is strictly conjugated to a one on $\widetilde{\mathcal{S}}(\theta)$ by the lifted Riemann-Hilbert correspondence $\widetilde{\mathrm{RH}}_{t,\kappa}$.

Let us describe the semi-conjugate and conjugate dynamical systems on the cubic surface. Since $\mathcal{S}(\theta)$ is a $(2,2,2)$-*surface*, namely, its defining equation $f(x,\theta) = 0$ is a quadratic equation in each variable x_i, $i \in \{1,2,3\}$, the line through a point $x \in \mathcal{S}(\theta)$ parallel to the x_i-axis passes through a unique second point $x' \in \mathcal{S}(\theta)$. This defines an involution

$$\sigma_i : \mathcal{S}(\theta) \to \mathcal{S}(\theta), \; x \mapsto x' \quad \text{with} \quad (x_i', x_j', x_k') = (\theta_i - x_i - x_j x_k, x_j, x_k).$$

Consider the transformation group $G = \langle \sigma_1, \sigma_2, \sigma_3 \rangle$ on $\mathcal{S}(\theta)$ generated by these three involutions, as well as its subgroup $G(2)$ of even words in G. Note that $G(2)$ is an index 2 subgroup of G. Then the desired semi-conjugacy and conjugacy are now descibed as follows.

(1) Via the Riemann-Hilbert correspondence $\mathrm{RH}_{t,\kappa}$, the Poincaré section $\mathrm{PS}_t(\kappa)$ on $\mathcal{M}_t(\kappa)$ is *semi-conjugate* to the group action $G(2) \curvearrowright \mathcal{S}(\theta)$.
(2) The action $G(2) \curvearrowright \mathcal{S}(\theta)$ uniquely lifts to $\widetilde{\mathcal{S}}(\theta)$. Via the lifted Riemann-Hilbert correspondence $\widetilde{\mathrm{RH}}_{t,\kappa}$, the Poincaré section $\mathrm{PS}_t(\kappa)$ is *strictly conjugate* to the lifted action $G(2) \curvearrowright \widetilde{\mathcal{S}}(\theta)$.

These situations are summarized in the following commutative diagram.

$$\begin{array}{ccc} \mathrm{PS}_t(\kappa) \curvearrowright \mathcal{M}_t(\kappa) & \xrightarrow{\widetilde{\mathrm{RH}}_{t,\kappa}} & \widetilde{\mathcal{S}}(\theta) \curvearrowleft G(2) \\ \Big\| & & \Big\downarrow \varpi \\ \mathrm{PS}_t(\kappa) \curvearrowright \mathcal{M}_t(\kappa) & \xrightarrow[\mathrm{RH}_{t,\kappa}]{} & \mathcal{S}(\theta) \curvearrowleft G(2) \end{array}$$

6 Bounded Orbits

Now we can discuss one of the main issues of this article: bounded orbits. A *bounded orbit* of $\mathrm{PS}_t(\kappa)$ is an orbit on $\mathcal{M}_t(\kappa)$ bounded away from the vertical leaves $\mathcal{Y}_t(\kappa)$, which may be thought of as the points at infinity. Through the Riemann-Hilbert correspondence, a bounded orbit on $\mathcal{M}_t(\kappa)$ is converted to a *bounded $G(2)$-orbit* on the cubic surface $\mathcal{S}(\theta)$.

Problem 1. Classify bounded $G(2)$-orbits.

Let us categorize the bounded orbits in the following manner.

(1) *fixed points*, which correspond to the so-called Riccati solutions,
(2) *finite orbits* other than the fixed points; they correspond to finitely many-valued solutions of non-Riccati type, all of which seem to be algebraic solutions,
(3) orbits in the moduli space of $SU(2)$-monodromy representations, and
(4) remaining ones, if any.

We are particularly interested in *infinite bounded orbits*, or in *filled Julia set*. We have also a list of some distinguished finite orbits as in Table 2. A characteristic feature of this table is that orbits of Classes 1, 2, 3, 4 depend continuously on at least one parameters, while those of Classes 5, 6 are not. The family $\mathcal{S} \to \Theta$ of cubic surfaces $\mathcal{S}(\theta)$ parametrized by $\theta \in \Theta$ admits a rather simple action of the group $S_3 \ltimes (\mathbb{Z}/2\mathbb{Z})^2$ and an orbit is said to be of Class $*$ if it is obtained from an orbit of Class $*$ in Table 2 via this action.

What we can say about infinite bounded orbits is the following confinement result which asserts that all those orbits are confined in a very thin set [17].

Theorem 5. *If $\mathcal{S}(\theta)$ admits an infinite bounded $G(2)$-orbit, then the parameters $\theta = (\theta_1, \theta_2, \theta_3, \theta_4)$ must be real with $-8 < \theta_1, \theta_2, \theta_3 < 8$ and every infinite bounded $G(2)$-orbit in $\mathcal{S}(\theta)$ is confined in the real cube $[-2, 2]^3$.*

Class	$\theta = (\theta_1, \theta_2, \theta_3, \theta_4)$	$G(2)$-orbit on $\mathcal{S}(\theta)$	$\sharp$ parameters
Class 1	$\Delta(\theta) = 0$	a fixed point = a singular point	3
Class 2	$(0,\ 0,\ \alpha+\beta,\ \alpha\beta)$	$(0, 0, \alpha)$ $(0, 0, \beta)$	2
Class 3	$(2,\ \alpha,\ \alpha,\ 1)$	$(1, 0, 0)$ $(1, \alpha, 0)$ $(1, 0, \alpha)$	1
Class 4	$(\alpha,\ \alpha,\ \alpha,\ 3\alpha-4)$	$(1, 1, 1)$ $(\alpha-2, 1, 1)$ $(1, \alpha-2, 1)$ $(1, 1, \alpha-2)$	1
Class 5	$(2\sqrt{2},\ 2\sqrt{2},\ 3,\ 4)$	$(0, \sqrt{2}, 1)$ $(0, \sqrt{2}, 2)$ $(\sqrt{2}, 0, 1)$ $(\sqrt{2}, 0, 2)$ $(\sqrt{2}, \sqrt{2}, 0)$ $(\sqrt{2}, \sqrt{2}, 1)$	0
Class 6	$(0,\ 0,\ 0,\ -4)$	$(\sqrt{2}, \sqrt{2}, 0)$ $(\sqrt{2}, -\sqrt{2}, 0)$ $(\sqrt{2}, \sqrt{2}, -2)$ $(\sqrt{2}, -\sqrt{2}, 2)$ $(-\sqrt{2}, \sqrt{2}, 0)$ $(-\sqrt{2}, \sqrt{2}, 2)$ $(-\sqrt{2}, -\sqrt{2}, 0)$ $(-\sqrt{2}, -\sqrt{2}, -2)$	0

Table 2. Some distinguished finite $G(2)$-orbits: $\alpha, \beta \in \mathbb{C}$

So the infinite bounded orbits live on a real 3-dimensional world, which may be pictured by computer. We can also say something about finite orbits [17].

Theorem 6. *If $\mathcal{S}(\theta)$ admits a finite $G(2)$-orbit that is none of Classes 1, 2, 3, 4 in Table 2, then the parameters $\theta = (\theta_1, \theta_2, \theta_3, \theta_4)$ must be real cyclotomic integers with $-8 < \theta_1, \theta_2, \theta_3 < 8$ and the coordinates of every finite $G(2)$-orbit in $\mathcal{S}(\theta)$ must also be real cyclotomic integers.*

Thus all finite orbits but a few exceptions live on a real cyclotomic world.

7 Topological Entropy and Dynamical Degree

If we think of non-integrable aspects of Painlevé dynamics, we will be naturally led to the following rather naïve but genuine questions:

(1) What is the global nature of transcendental trajectories of Painlevé dynamics?
(2) How far or near is Painlevé system from a so-called 'integrable' system?
(3) Does any chaotic phenomenon occur in Painlevé dynamics?

Here, key terminologies for a general dynamical system, not necessarily integrable, are *topological entropy* and *dynamical degree*; usually the former is difficult to handle while the latter is more accessible. Let us recall the definition of topological entropy.

Definition 1. *Let M be a compact topological space, $f : M \circlearrowleft$ a continuous map. Then the topological entropy of f is defined by*

$$h_{\mathrm{top}}(f) := \sup_{\mathcal{U}} H(f, \mathcal{U})$$

where the supremum is taken over all finite open coverings $\mathcal{U}$ of M and

$$\begin{aligned} H(f,\mathcal{U}) &:= \lim_{n\to\infty} \frac{1}{n} H(\mathcal{U} \vee f^{-1}\mathcal{U} \vee \cdots \vee f^{-(n-1)}\mathcal{U}), \\ H(\mathcal{U}) &:= \log \min\{ \sharp\mathcal{U}' : \mathcal{U}' \subset \mathcal{U} \text{ covers } M \}, \\ \mathcal{U} \vee \mathcal{V} &:= \{ U \cap V : U \in \mathcal{U},\, V \in \mathcal{V} \}. \end{aligned}$$

We also review the definition of dynamical degree on complex surfaces.

Definition 2. *Let S be a compact Kähler surface, $f : S \circlearrowleft$ a bimeromorphic map. Even if f is bimeromorphic, an induced map $f^* : H^{1,1}(S) \circlearrowleft$ can be defined appropriately in terms of an induced action f^* on the space $C_+^{1,1}(S)$ of positive closed $(1,1)$-currents on S. Then the first dynamical degree is defined by the limit (which actually exists) :*

$$\lambda(f) := \lim_{n\to\infty} ||(f^n)^*||^{1/n},$$

where $||\cdot||$ is a norm on $H^{1,1}(X)$, which is independent of the norm chosen.

For these concepts there are fundamental results of Gromov [10], Yomdin [27] and Katok [21]. The following are special consequences of them.

Theorem 7 (Gromov-Yomdin). *If S is a compact Kähler surface and $f : S \circlearrowleft$ is a biholomorphic map, then the topological entropy $h_{\rm top}(f)$ is equal to the logarithm of the first dynamical degree $\lambda(f)$, that is, $h_{\rm top}(f) = \log \lambda(f)$.*

Theorem 8 (Katok). *Let S be a compact complex surface, $f : S \circlearrowleft$ a holomorphic map. If f has a positive topological entropy, then there exists an $n \in \mathbb{N}$ such that f^n admits a Smale's horseshoe; this means that*

Positive topological entropy $h_{\rm top}(f) > 0$ implies 'chaos' in dimension 2.

Unfortunately or fortunately, these theorems are not directly applicable to our case, because the semi-conjugate dynamical system $G(2) \curvearrowright \mathcal{S}(\theta)$ of the Poincaré section $\mathrm{PS}_t(\kappa)$ prolongs to the compactification $\mathcal{S}(\theta) \hookrightarrow \overline{\mathcal{S}}(\theta) \subset \mathbb{P}^3$ only *birationally*, not biholomorphically. We should consider to what extent such results as in Theorems 7 and 8 hold for transformations in $G(2)$.

8 Coxeter Transformation

The transformation group $G = \langle \sigma_1, \sigma_2, \sigma_3 \rangle$ may be thought of as a nonlinear reflection group with basic reflections σ_1, σ_2, σ_3, so that the birational map $c := \sigma_1\sigma_2\sigma_3 : \overline{\mathcal{S}}(\theta) \circlearrowleft$ of their product may be regarded as a 'Coxeter' transformation of G. It is worth investigating, since it is expected to be a dominant element controlling the entire G-action. By applying the theory of birational surface dynamics as in [6], the following computation is made in [18].

Theorem 9. *The first dynamical degree* $\lambda = \lambda(c)$ *of* c *is calculated as*

$$\lambda = 2 + \sqrt{5} = 4.23607\cdots.$$

A recent advance in birational surface dynamics [8] shows that a Gromov-Yomdin-type result as in Theorem 7 remains true for a certain class of birational maps. This result applies to our case, yielding the following theorem.

Theorem 10. *The topological entropy* $h_{\mathrm{top}}(c)$ *of* c *is given by*

$$h_{\mathrm{top}}(c) = \log \lambda(c) = 1.44364\cdots > 0.$$

We describe the general strategy to establish this kind of equality.

(1) Given a dynamical system $f : S \circlearrowleft$, there is the notion of *measure-theoretical entropy* $h_\mu(f)$ for any f-invariant measure μ on S.
(2) There are bounds from both sides $h_\mu(f) \le h_{\mathrm{top}}(f) \le \log \lambda(f)$, where the first inequality is the so-called variational principle and the second one follows from a result of [9].
(3) There is the notion of Green currents $T^{\pm} \in C_+^{1,1}(S)$ for $f^{\pm 1}$, which are positive closed $(1,1)$-currents such that $(f^{\pm 1})^* T^{\pm} = \lambda(f) T^{\pm}$.
(4) Define an f-invariant measure μ by $\mu = T^+ \wedge T^-$. Then it turns out that μ is a *maximal entropy measure*, i.e., a measure such that $h_\mu(f) = \log \lambda(f)$.

The results mentioned above are not restricted to the Coxeter transformation. In fact we have an algorithm to compute the dynamical degree $\lambda(g)$ for all elements $g \in G$ and Theorems 9 and 10 extend to those general cases [19]. We can also describe the chaotic behavior of almost all elements of G based on the work of [2]. This means that a large part of the Poincaré return maps of Painlevé VI is a *chaotic* dynamical system. This important topic is fully discussed in [19].

We remark that the Coxeter transformation $c : \overline{\mathcal{S}}(\theta) \circlearrowleft$ is not bimeromorphically conjugate to any holomorphic automorphism of $\overline{\mathcal{S}}(\theta)$. This follows from a theorem of [6]: Let S be a compact Kähler surface, $f : S \circlearrowleft$ an AS bimeromorphic map, λ the first dynamical degree of f and $v \in H^{1,1}(S)$ a λ-eigenvector of f^*. If $\lambda > 1$, then f is bimeromorphically conjugate to a holomorphic automorphism of S iff v has the vanishing self-intersection number. In our case where $S = \overline{\mathcal{S}}(\theta)$ and $f = c$, we can check that c is an AS, $\lambda > 1$, and v has a positive intersection number; hence the assertion.

9 Periodic Solutions

Several authors [3, 4, 7, 11, 12, 22, 23] have been interested in algebraic solutions to $\mathrm{P_{VI}}(\kappa)$, namely, in periodic solutions along *all* loops in $X = \mathbb{P}^1 \setminus \{0, 1, \infty\}$. But little attention has been paid to periodic solutions along a *single* loop chosen particularly. The following theorem [18] is a first result in this direction.

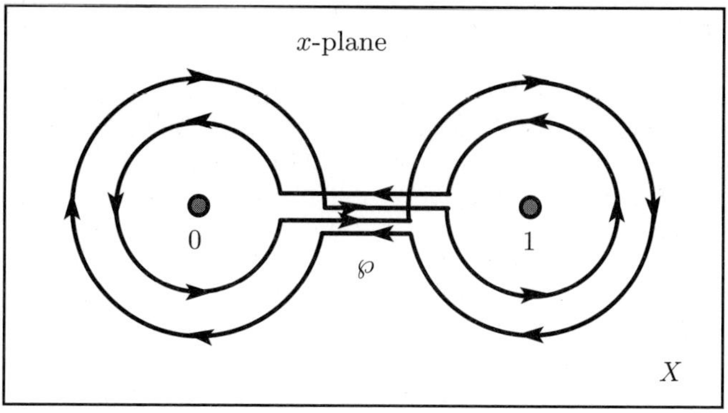

Fig. 5. Pochhammer loop

Theorem 11. *Let* $\mathrm{Per}_N(\kappa) = \{Q \in M_x(\kappa) : \wp_*^N Q = Q\}$ *be the set of all periodic solutions to* $\mathrm{P_{VI}}(\kappa)$ *of period* N *along a Pochhammer loop* $\wp$ *as in Fig.* 5*. For any* $\kappa \in \mathcal{K} \setminus$ **Wall** *the number of* $\mathrm{Per}_N(\kappa)$ *is given by*

$$\#\mathrm{Per}_N(\kappa) = (2+\sqrt{5})^{2N} + (2+\sqrt{5})^{-2N} + 4 \qquad (N = 1, 2, 3, \dots).$$

Note that $2+\sqrt{5}$ is the first dynamical degree of c in Sect. 8. Main ingredients of the proof are the moduli-theoretical formulation of $\mathrm{P_{VI}}(\kappa)$ in Sects. 2-3, the Riemann-Hilbert correspondence in Sect. 4, the birational dynamical systems on cubic surfaces in Sect. 8 and the Lefschetz fixed point formula. Not only for the Pochhammer loop but also for every loop γ in $X = \mathbb{P}^1 \setminus \{0, 1, \infty\}$, we have an algorithm to compute the number of periodic points of a given period along γ. Details are described in [19]. We have also [17]:

Theorem 12. *If* $\kappa_i \pm \kappa_j \notin \mathbb{Z}$ *for every* $1 \le i < j \le 4$ *(this condition might be weakened), then any single-valued solution germ to* $\mathrm{P_{VI}}(\kappa)$ *around each fixed singular point* $x = 0, 1, \infty$ *is necessarily meromorphic around that point, that is, no essential singularity occurs.*

Once this theorem is established, Laurent series expansions of those local meromorphic solutions can be found elementarily. The relevant calculation is made in [5, 20].

These researches make us to realize that the study of Painlevé dynamics is surely at the crossroads of many branches of mathematics:

(1) algebraic geometry: moduli theory; surface theory; Hodge theory,
(2) topology and geometry: symplectic geometry; moduli of monodromy representations; braid group actions, Lefschetz fixed point formula,
(3) analysis: complex differential equations; measure, distributions and currents; pluripotential theory,
(4) interdisciplinary subjects: Riemann-Hilbert correspondence; dynamical system theory; singularity theory.

Note added in proofs. A final form of Theorem 12 is given in [28].

References

1. Arinkin, D., Lysenko, S.: On the moduli of SL(2)-bundles with connections on $\mathbf{P}^1 \setminus \{x_1, \ldots, x_4\}$. Internat. Math. Res. Notices, **1997**, no. 19, 983–999 (1997)
2. Bedford, E., Diller, J.: Energy and invariant measures for birational surface maps. Duke Math. J., **128** (2), 331–368 (2005)
3. Boalch, P.: From Klein to Painlevé via Fourier, Laplace and Jimbo. Proc. London Math. Soc. (3), **90**, 167–208 (2005)
4. Boalch, P.: The fifty-two icosahedral solutions to Painlevé VI. J. Reine Angew. Math., **596**, 183–214 (2006)
5. Bryuno, A.D., Goryuchkina, I.V.: Expansions of solutions of the sixth Painlevé equation. Dokl. Akad. Nauk., **395** (6), 733–737 (2004)
6. Diller, J., Favre, C.: Dynamics of bimeromorphic maps of surfaces. Amer. J. Math., **123**, 1135–1169 (2001)
7. Dubrovin, B., Mazzocco, M.: Monodromy of certain Painlevé-VI transcendents and reflection groups. Invent. Math., **141** (1), 55–147 (2000)
8. Dujardin, R.: Laminar currents and birational dynamics. Duke Math. J., **131** (2), 219–247 (2006)
9. Dinh, T.-C., Sibony, N.: Une borne supérieure pour l'entropie topologique d'une application rationnelle. Ann. of Math. (2) **161** (3), 1637–1644 (2005)
10. Gromov, M.: On the entropy of holomorphic maps. Enseign. Math. (2) **49** (3-4), 217–235 (2003)
11. Hitchin, N.: Poncelet polygons and the Painlevé equations. Geometry and analysis (Bombay, 1992), 151–185, Tata Inst. Fund. Res., Bombay (1995)
12. Hitchin, N.: A lecture on the octahedron. Bull. London Math. Soc., **35** (5), 577–600 (2003)
13. Inaba, M., Iwasaki, K., Saito, M.-H.: Dynamics of the sixth Painlevé equation. Théorie asymptotiques et équations de Painlevé (Angers, juin 2004). Loday, M., Delabaere, E. (Ed.) Séminaires et Congrès. **14**, 103–167, Soc. Math. France, Paris (2006)
14. Inaba, M., Iwasaki, K., Saito, M.-H.: Moduli of stable parabolic connections, Riemann-Hilbert correspondence and geometry of Painlevé equation of type VI. Part I. Publ. Res. Inst. Math. Sci., **42** (4), 987–1089 (2006); Part II. Adv. Stud. Pure Math. **45**, 387–432 (2006)
15. Iwasaki, K.: A modular group action on cubic surfaces and the monodromy of the Painlevé VI equation. Proc. Japan Acad. Ser. A, **78**, 131–135 (2002)
16. Iwasaki, K.: An area-preserving action of the modular group on cubic surfaces and the Painlevé VI equation. Comm. Math. Phys., **242** (1-2), 185–219 (2003)
17. Iwasaki, K.: Bounded trajectories of the sixth Painlevé equation. in preparation.
18. Iwasaki, K., Uehara, T.: Periodic solutions to Painlevé VI and dynamical system on cubic surface. Preprint (arXiv: `math.AG/0512583`)
19. Iwasaki, K., Uehara, T.: An ergodic study of Painlevé VI. Math. Ann., **338** (2), 295–345 (2007)
20. Kaneko, K.: Painlevé VI transcendents which are meromorphic at a fixed singularity. Proc. Japan Acad., Ser. A, Math. Sci., **82**, 71–76 (2006)
21. Katok, A.: Lyapunov exponents, entropy and periodic orbits for diffeomorphisms, Publ. Math. I.H.E.S., **51**, 137–173 (1980)
22. Kitaev, A.V.: Grothendieck's dessins d'enfants, their deformations, and algebraic solutions of the sixth Painlevé and Gauss hypergeometric equations. Algebra i Analiz, **17** (1), 224–275 (2005)

23. Mazzocco, M.: Rational solutions of the Painlevé VI equation. J. Phys. A: Math. Gen., **34**, 2281–2294 (2001)
24. Okamoto, K.: Sur les feuilletages associés aux équations du second ordre à points critiques fixes de P. Painlevé, Espaces des conditions initiales. Japan. J. Math, **5**, 1–79 (1979)
25. Saito, M.-H., Takebe, T., Terajima, H.: Deformation of Okamoto-Painlevé pairs and Painlevé equations. J. Algebraic. Geom, **11** (2), 311–362 (2002)
26. Sakai, H.: Rational surfaces associated with affine root systems and geometry of the Painlevé equations. Comm. Math. Phys, **220**, 165–229 (2001)
27. Yomdin, Y.: Volume growth and entropy. Israel J. Math., **57** (3), 285–300 (1987)
28. Iwasaki, K.: Finite branch solutions to Painlevé VI around a fixed singular point. to appear in Adv. Math. arXiv: 0704.0679 [math.AG]

An algebraic representation for correlation functions in integrable spin chains

Michio Jimbo

Graduate School of Mathematical Sciences, The University of Tokyo, Komaba, Tokyo 153-8914, Japan
jimbomic@ms.u-tokyo.ac.jp

Summary. Exact description of correlation functions in integrable spin chains stands out as one of the most important problems in quantum integrable systems. In this article we give a brief survey of the subject and known results. We then report a recently found 'algebraic' formula for general correlation functions. This is a joint work with Herman Boos, Tetsuji Miwa, Fedor Smirnov and Yoshihiro Takeyama.

Key words: integrable system, spin chain, correlation function, qKZ equation

1 Heisenberg magnet

Consider a quantum mechanical Hamiltonian, known as the XXX model:

$$H = \sum_{j=1}^{L} \left(\sigma_j^x \sigma_{j+1}^x + \sigma_j^y \sigma_{j+1}^y + \sigma_j^z \sigma_{j+1}^z \right) . \tag{1}$$

This is a Hermitian matrix which acts on $V^{\otimes L}$, where $V = \mathbf{C}^2$,

$$\sigma^x = \begin{pmatrix} 0 & 1 \\ 1 & 0 \end{pmatrix}, \quad \sigma^y = \begin{pmatrix} 0 & -i \\ i & 0 \end{pmatrix}, \quad \sigma^z = \begin{pmatrix} 1 & 0 \\ 0 & -1 \end{pmatrix},$$

and σ_j^a $(a = x, y, z)$ signifies the operator which acts as σ^a on the j-th tensor component and as identity elsewhere. For definiteness we consider the periodic boundary condition $\sigma_{L+1}^a = \sigma_1^a$. Let

$$|\psi\rangle_L \in V^{\otimes L}, \quad {}_L\langle\psi| = {}^t\left(|\psi\rangle_L\right) \in V^{*\otimes L}$$

be the normalized eigenvector and the dual eigenvector associated with the smallest eigenvalue of H. By a correlation function, we mean an average value of an operator of the form $\mathcal{O} = \sigma_{i_1}^{a_1} \cdots \sigma_{i_r}^{a_r}$,

Received 22 December, 2005. Revised 1 April, 2006. Accepted 1 April, 2006.

$$\langle \mathcal{O} \rangle := \lim_{L \to \infty} {}_L\langle \psi | \mathcal{O} | \psi \rangle_L,$$

considered in the limit where the length L of the chain is taken to be infinite. It is of great physical and mathematical significance to describe these quantities exactly.

Let us give a few examples. The first simple cases are

$$\langle \sigma_1^z \rangle = 0,$$
$$\langle \sigma_1^z \sigma_2^z \rangle = \frac{1}{3} - \frac{4}{3} \log 2.$$

The next step is already quite non-trivial. In 1977, Takahashi [6] showed that

$$\langle \sigma_1^z \sigma_3^z \rangle = \frac{1}{3} - \frac{16}{3} \log 2 + 3\zeta(3),$$

where $\zeta(s)$ denotes the Riemann zeta function. A recent result of Sakai, Shiroishi, Nishiyama and Takahashi [7] states that

$$\langle \sigma_1^z \sigma_4^z \rangle = \frac{1}{3} - 12 \log 2 + \frac{74}{3}\zeta(3) - \frac{56}{3}\zeta(3) \log 2 - 6\zeta(3)^2 - \frac{125}{6}\zeta(5) + \frac{100}{3}\zeta(5) \log 2.$$

As we go on, the calculation becomes increasingly more difficult, and no *tractable* general formula is known to date. We will come back to this issue later on.

We remark that the Hamiltonian (1) has an integrable anisotropic version, where independent coupling parameters J_a are introduced in front of $\sigma_j^a \sigma_{j+1}^a$ ($a = x, y, z$). This generalized model is called the XYZ model. The special case $J_x = J_y$ is called the XXZ model.

2 A brief history

Systematic study of correlation functions in integrable models has begun relatively recently. Let us flush some relevant works in the chronological order.

~ 1979

The first major results were obtained for the Ising and allied models that can be dealt with using the method of free fermions. For these models, a complete understanding is achieved. For example, $\langle \sigma_1^z \sigma_n^z \rangle$ has a closed expression as a determinant involving hypergeometric functions. In the scaling limit to continuous field theory, correlation functions are described in terms of solutions to the Painlevé differential equation and their generalizations (Wu, McCoy, Tracy, Barouch 1973-76, Sato, Miwa, Jimbo 1977-79).

Beyond the free fermion case, however, very little was known. The only exceptions are Takahashi's $\langle\sigma_1^z\sigma_3^z\rangle$ for the XXX model already mentioned above, and Baxter's $\langle\sigma_1^z\rangle$ for the XXZ model.

∼ 1989

In the mid 80's conformal field theory was given birth (Belavin, Polyakov, Zamolodchikov 1984, and many others). Though applicable only to critical models in the continuum, it is powerful enough to determine all correlation functions in the theory as solutions of *linear* differential equations. Another important discovery made in this period is the corner transfer matrix method, which gives a general way for obtaining one-point functions in lattice models (Baxter 1980–84).

∼ 1998

Combining ideas from conformal field theory and corner transfer matrices, an approach to lattice models was developed in the framework of representation theory of quantum affine algebras. This led to a derivation of a linear difference equation (qKZ equation) for lattice correlation functions, as well as their integral solutions. (Jimbo, Miwa, Nakayashiki 1992–96)

1999 ∼

The turn of the century saw a new development by the group in Lyon, Kitanine, Maillet, Slavnov and Terras (1999–). On the basis of the algebraic Bethe Ansatz, these authors obtained many new results, including (i) a rigorous derivation of integral formulas, (ii) new integral representations, (iii) generalizations to incorporate magnetic field and time, (iv) exact asymptotics for some correlation functions. For a recent review, see [10]. In view of the topic of this conference, we quote from their works [11] one formula for the asymptotics of the quantity called Emptiness Formation Probability (EFP) in the XXZ model:

$$\langle (E_{--})_1 \cdots (E_{--})_n \rangle \sim A n^{-\gamma} e^{-n^2 B} \quad (n \to \infty),$$
$$-B = \log\frac{1}{\nu} + \frac{1}{2}\int_{\mathbf{R}-i0} \frac{d\omega}{\omega} \frac{\sinh\frac{\omega}{2}(1-\nu)\cosh^2\frac{\omega\nu}{2}}{\sinh\frac{\omega}{2}\sinh\frac{\omega\nu}{2}\cosh\omega\nu}.$$

Here $(E_{--})_j = (1-\sigma_j^z)/2$, $0 < \nu < 1$ is a parameter related to the anisotropy parameters of the model as $J_x = J_y = 1, J_z = \cos\pi\nu$, and A, γ are some constants.

3 Principle for solvability

Before proceeding further, let us review some basic features of correlation functions in integrable models [1]. For the integrability of the model, a fundamental building block is afforded by the R matrix

$$R(\lambda) = \frac{\rho(\lambda)}{\lambda+1}(\lambda + P) \quad \in \mathrm{End}(V^{\otimes 2}), \qquad (Pu \otimes v = v \otimes u),$$

satisfying the Yang-Baxter equation

$$R_{12}(\lambda_{12})R_{13}(\lambda_{13})R_{23}(\lambda_{23}) = R_{23}(\lambda_{23})R_{13}(\lambda_{13})R_{12}(\lambda_{12})$$

where $\lambda_{ij} = \lambda_i - \lambda_j$. The factor $\rho(\lambda)$ will be given later (see (8)). In addition, we have

$$R_{12}(\lambda)R_{21}(-\lambda) = I, \quad \sigma_1^y R_{12}(\lambda)^{t_1}(\sigma_1^y)^{-1} = -R_{21}(-\lambda - 1).$$

Introduce the transfer matrix

$$\tau(\lambda) = \mathrm{tr}_0 R_{0L}(\lambda) \cdots R_{01}(\lambda).$$

The properties of the R matrix entail that they constitute a mutually commuting family

$$[\tau(\lambda), \tau(\mu)] = 0 \qquad (\forall \lambda, \mu),$$

and that (up to a constant multiple) the Hamiltonian H is the first member I_1 in the expansion

$$\log \tau(0)^{-1}\tau(\lambda) = I_1\lambda + I_2\lambda^2 + \cdots .$$

In this procedure, one can introduce inhomogeneity parameters λ_j attached independently to each site j. Commutativity of tranfser matrices remains true (though the resulting Hamiltonian H is not local any more). In this case the ground state $|\psi(\lambda_1, \cdots, \lambda_L)\rangle$, and hence correlation functions, depends also on λ_j's. Let $\{E_{\epsilon,\bar{\epsilon}}\}$ be the basis of $\mathrm{End}(V)$ consisting of matrix units. It is convenient to arrange correlation functions into a matrix

$$\begin{aligned} h_n(\lambda_1, \ldots, \lambda_n) = \sum_{\epsilon_1, \ldots, \bar{\epsilon}_n} &\langle (E_{\bar{\epsilon}_1,\epsilon_1})_1 \cdots (E_{\bar{\epsilon}_n,\epsilon_n})_n \rangle \\ &\times E_{\epsilon_1,\bar{\epsilon}_1} \otimes \cdots \otimes E_{\epsilon_n,\bar{\epsilon}_n} \quad \in \mathrm{End}\big(V^{\otimes n}\big). \end{aligned}$$

Here in the expected value $\langle \cdots \rangle$ the infinite lattice limit is assumed. We call h_n density matrix. A particular correlation function is then written as a trace, e.g.,

$$\langle \sigma_1^z \sigma_n^z \rangle = \mathrm{tr}_{V^{\otimes n}} \left(\sigma_1^z \sigma_n^z h_n(\lambda_1, \cdots, \lambda_n) \right).$$

It is a general feature (Z-invariance) that h_n depends only on the parameters $\lambda_1, \cdots, \lambda_n$ relevant to the lattice sites $1, \cdots, n$. We study the matrix h_n as a function of these parameters.

4 Reduced qKZ equation

The fundamental properties of h_n are the following (see e.g.[4]).

$$\begin{aligned}
&\mathrm{tr}_1 h_n(\lambda_1,\cdots,\lambda_n) = h_{n-1}(\lambda_2,\cdots,\lambda_n), && (2)\\
&P_{jj+1}R_{j,j+1}(\lambda_{j,j+1})h_n(\cdots,\lambda_j,\lambda_{j+1},\cdots) && (3)\\
&\quad = h_n(\cdots,\lambda_{j+1},\lambda_j,\cdots)P_{jj+1}R_{j,j+1}(\lambda_{j,j+1}),\\
&h_n(\lambda_1-1,\lambda_2,\cdots,\lambda_n) = A_n(\lambda_1,\cdots,\lambda_n)h_n(\lambda_1,\lambda_2,\cdots,\lambda_n). && (4)
\end{aligned}$$

Here tr_1 means the trace with respect to the first tensor component $V_1 = \mathbf{C}^2$. In the last line (4), we regard h_n as a vector in $V^{\otimes n}\otimes V^{*\otimes n} \simeq \mathrm{End}(V^{\otimes n})$, on which A_n acts as

$$\begin{aligned}
A_n(\lambda_1,\cdots,\lambda_n) = (-1)^n R_{\bar{1}\bar{2}}(\lambda_{12}-1)\cdots R_{\bar{1}\bar{n}}(\lambda_{1n}-1)\\
\times\quad P_{1\bar{1}}R_{1n}(\lambda_{1n})\cdots R_{12}(\lambda_{12}).
\end{aligned}$$

The suffices refer to the tensor components arranged in the order

$$V_1\otimes\cdots\otimes V_n\otimes V_{\bar{n}}^*\otimes\cdots\otimes V_{\bar{1}}^*.$$

The following integral formula is known [8], [9]. Given a set of indices $\{\epsilon_j,\bar{\epsilon}_j\}$, define

$$\begin{aligned}
&A = \{j|\epsilon_j = +\} = \{a_1,\dots,a_r\},\\
&B = \{j|\bar{\epsilon}_j = +\} = \{b_1,\dots,b_s\},\\
&a_1<\cdots<a_r, b_1<\cdots<b_s \quad (r+s=n).
\end{aligned}$$

Arrange integration variables t_a $(a\in A)$, t'_b $(b\in B)$ as

$$(u_1,\dots,u_n) = (t_{a_r},\dots,t_{a_1},t'_{b_1},\dots,t'_{b_s}).$$

Then

$$\begin{aligned}
&h_n(\lambda_1,\dots,\lambda_n)^{\epsilon_1,\dots,\epsilon_n,\bar{\epsilon}_n,\dots,\bar{\epsilon}_1}\\
&= c^{(n)}_{A,B}\prod_{a\in A}\int_{C^+}\frac{dt_a}{t_a-\lambda_a}\prod_{b\in B}\int_{C^-}\frac{dt'_b}{t'_b-\lambda_b}\\
&\quad\times\prod_{\substack{a\in A\\ j<a}}\frac{t_a-\lambda_j-1}{t_a-\lambda_j}\prod_{\substack{b\in B\\ j<b}}\frac{t'_b-\lambda_j+1}{t'_b-\lambda_j}\\
&\quad\times\prod_{j<k}\left(\frac{\sinh\pi i(u_j-u_k)}{u_j-u_k-1}\frac{\sinh\pi i(\lambda_j-\lambda_k)}{\lambda_j-\lambda_k}\right)\\
&\quad\times\prod_{j,k}\frac{u_j-\lambda_k}{\sinh\pi i(u_j-\lambda_k)}.
\end{aligned}$$

Here $c^{(n)}_{A,B}$ is a constant. The integration contour C^+ is parallel to the imaginary axis for $|\mathrm{Im} t_a| \gg 0$, and separates the sequences of the poles of the integrand into the two sets $\lambda_j + \mathbf{Z}_{\le 0}$ and $\lambda_j + \mathbf{Z}_{>0}$. Similarly C^- is the contour separating the poles into $\lambda_j + \mathbf{Z}_{<0}$ and $\lambda_j + \mathbf{Z}_{\ge 0}$.

From the integral representation it is not hard to see the following analyticity and asymptotic properties.

$$\begin{aligned} &h_n(\lambda_1, \cdots, \lambda_n) \text{ is meromorphic in } \lambda_1, \cdots, \lambda_n \\ &\quad \text{with at most simple poles at } \lambda_i - \lambda_j \in \mathbf{Z}\backslash\{0, \pm 1\}, \end{aligned} \tag{5}$$

$$\begin{aligned} &\lim_{\substack{\lambda_1 \to \infty \\ \lambda_1 \in S_\delta}} h_n(\lambda_1, \cdots, \lambda_n) = h_{n-1}(\lambda_2, \cdots, \lambda_n), \\ &\text{where } S_\delta = \{\lambda \in \mathbf{C} \mid \delta < |\arg \lambda| < \pi - \delta\},\ 0 < \forall \delta < \pi. \end{aligned} \tag{6}$$

It turns out that the properties (2), (3), (4), (5), (6), along with $h_0 = h_1 = 1$, uniquely characterize the family of functions $\{h_n\}_{n=0}^{\infty}$.

A remark about the qKZ equation is in order here. In fact, h_n is obtained as a specialization of a function $g_{2n} \in V^{\otimes 2n}$,

$$h_n(\lambda_1, \cdots, \lambda_n) = g_{2n}(\lambda_1, \cdots, \lambda_n, \lambda_n + 1, \cdots, \lambda_1 + 1).$$

The latter satisfies the qKZ equation

$$\begin{aligned} &g_{2n}(\cdots, \lambda_j + k + 2, \cdots) \\ &\quad = R_{j-1\,j}(\lambda_{j-1\,j} + k + 2) \cdots R_{1\,j}(\lambda_{1\,j} + k + 2) \\ &\quad \times R_{jn}(\lambda_{jn}) \cdots R_{j\,j+1}(\lambda_{j\,j+1}) g_{2n}(\cdots, \lambda_j, \cdots). \end{aligned}$$

Here the shift parameter k is called the level. It is $k = -4$ for correlation functions. There is a duality between qKZ equations of level -4 and of level 0, in the sense that their fundamental matrix solutions are transpose inverse to each other. The case of level 0 plays a significant role in the theory of form factors of massive integrable field theory, and is known to have a very special nature: the integral solutions reduce to a linear combination of products of *one*-dimensional integrals. Whether there exists some special feature in the dual case of level -4 has been obscure till recently.

5 Performing integrals

After the integral formula was derived, it took several years before any serious attempt was made to honestly evaluate it. In 2001, Boos and Korepin [2] tackled this problem and computed integrals for the EFP

$$P(n) = \langle (E_{++})_1 \cdots (E_{++})_n \rangle$$

for $n = 3, 4$. Surprisingly, every time the integrals were reduced to one-dimensional ones. They observed that the result can be written in terms of

$\zeta(3), \zeta(5), \cdots$. This analysis was extended by Takahashi and his group [6], who obtained many further exact results from integrals for $n \leq 6$ and including the XXZ case. It was curious why integration can always be reduced to one-dimensional ones. Boos, Korepin and Smirnov [3] explained the reason by utilizing the duality between qKZ equations of level -4 and level 0. They anticipated that the correlation functions have the general form

$$h_n = \sum \omega(\lambda_{i_1 j_1}) \cdots \omega(\lambda_{i_p j_p}) f_{i_1 j_1 \cdots i_p j_p}(\lambda_1, \cdots, \lambda_n)$$

where $\omega(\lambda)$ is a transcendental function (defined in (7) below), and $f_{i_1 j_1 \cdots i_p j_p}$ are some rational functions.

The main difficulty was to describe the complicated rational part.

6 Algebraic formula

All these works indicate that in full generality there exists an alternative 'algebraic' representation for h_n which is essentially free from integrals. We are now in a position to present such a formula.

This representation is based on the following ingredients: a transcendental function $\omega(\lambda)$, the L-operator, and a transfer matrix constructed from it using auxiliary spaces 'of fractional dimension'. Let us explain them.

The function $\omega(\lambda)$ is a logarithmic derivative of the normalizing factor entering the R matrix,

$$\omega(\lambda) = \frac{d}{d\lambda} \log \rho(\lambda) + \frac{1}{2(\lambda^2 - 1)}, \tag{7}$$

$$\rho(\lambda) = \frac{\Gamma\left(1 + \frac{\lambda}{2}\right) \Gamma\left(\frac{1-\lambda}{2}\right)}{\Gamma\left(1 - \frac{\lambda}{2}\right) \Gamma\left(\frac{1+\lambda}{2}\right)}. \tag{8}$$

The L-operator is a relative of the R matrix. Let $\{S_a\}_{a=1}^3$ be a basis of sl_2 such that $[S_a, S_b] = 2i\epsilon_{abc} S_c$. Set

$$L(\lambda) = \frac{\rho(\lambda, d)}{\lambda + \frac{d}{2}} \left(\lambda + \frac{1}{2} + \frac{1}{2} \sum_{a=1}^{3} S_a \otimes \sigma^a \right),$$

where $\sum_{a=1}^3 S_a^2 = d^2 - 1$, $\rho(\lambda, d)$ is a Gamma factor generalizing $\rho(\lambda) = \rho(\lambda, 2)$; it will appear always in the combination

$$\frac{\rho(\lambda, d)}{\lambda + \frac{d}{2}} \frac{\rho(\lambda - 1, d)}{\lambda + \frac{d}{2} - 1} = -\frac{1}{\lambda^2 - \frac{d^2}{4}}.$$

If π_d is the d-dimensional irreducible representation of sl_2, then $(\pi_d \otimes \mathrm{id}) L(\lambda) \in \mathrm{End}(\mathbf{C}^d \otimes V)$ is a solution of the Yang-Baxter equation.

Using L, we consider the 'transfer matrix'

$$\mathrm{Tr}_d T_n(\mu; \lambda_1, \cdots, \lambda_n), \tag{9}$$

where

$$T_n(\mu; \lambda_1, \cdots, \lambda_n) = L_{\bar{1}}(\mu - \lambda_1 - 1) \cdots L_{\bar{n}}(\mu - \lambda_n - 1) \\ \times L_n(\mu - \lambda_n) \cdots L_1(\mu - \lambda_1).$$

We regard d also as a parameter as follows. For each $A \in U(sl_2)$, $\mathrm{tr}_{\mathbf{C}^d}(\pi_d(A))$ depends polynomially on the dimension d. We denote this polynomial by the symbol $\mathrm{Tr}_d(A)$. Hence (9) is a rational function in $\lambda_1, \cdots, \lambda_n$ as well as of the 'dimension' d.

Let $\mathbf{s}_n \in V^{\otimes 2n}$ be the vector corresponding to $\mathrm{id} \in \mathrm{End}(V^{\otimes n})$.

Theorem 1. *[5] We have*

$$h_n(\lambda_1, \cdots, \lambda_n) = 2^{-n} e^{\Omega_n(\lambda_1, \cdots, \lambda_n)} \mathbf{s}_n,$$

where

$$\Omega_n(\lambda_1, \cdots, \lambda_n) = \frac{(-1)^n}{2} \iint \frac{d\mu_1}{2\pi i} \frac{d\mu_2}{2\pi i} \frac{\mu_{12}\, \omega(\mu_{12})}{\mu_{12}^2 - 1} \\ \times \mathrm{Tr}_{\mu_{12}} \left(T_n(\frac{\mu_1 + \mu_2}{2}; \lambda_1, \cdots, \lambda_n) \right) \\ \times \mathrm{Tr}_2 \left(T_n(\mu_1; \lambda_1, \cdots, \lambda_n) T_n(\mu_2 + 1; \lambda_1, \cdots, \lambda_n)^{-1} \right),$$

the integration contour being a small circle around $\lambda_1, \cdots, \lambda_n$.

The integrand is a meromorphic function with poles only at $\mu_i \in \lambda_j + \mathbf{Z}$. The matrix Ω_n is nilpotent,

$$\Omega_n^{[\frac{n}{2}]+1} = 0,$$

and hence the exponential series terminates.

At first glance there seems to be still integrals. However they are just sums of residues and therefore the formula is algebraic. Indeed, if $\lambda_1, \cdots, \lambda_n$ are distinct, then all poles are simple, and

$$\Omega_n(\lambda_1, \cdots, \lambda_n) = \sum_{i<j} \omega(\lambda_{ij}) X_n^{(ij)}(\lambda_1, \cdots, \lambda_n)$$

where $X^{(ij)}$ are rational in $\lambda_1, \cdots, \lambda_n$. These matrices have the properties

$$\left[X_n^{(i,j)}(\lambda_1, \cdots, \lambda_n), X_n^{(k,l)}(\lambda_1, \cdots, \lambda_n)\right] = 0, \qquad (\forall i, j, k, l)$$
$$X_n^{(i,j)}(\lambda_1, \cdots, \lambda_n) X_n^{(k,l)}(\lambda_1, \cdots, \lambda_n) = 0 \quad \text{if } \{i, j\} \cap \{k, l\} \neq \emptyset.$$

From physical point of view, the main interest lies in the homogeneous limit $\lambda_1 = \cdots = \lambda_n = 0$. The formula for Ω_n then becomes

$$\Omega_n = \frac{(-1)^n}{2} \mathrm{res}_{\mu_1,\mu_2=0} \Big\{ \frac{\omega(\mu_{12})}{\mu_{12}^2 - 1} \mathrm{Tr}_{\mu_{12}} T_n\Big(\frac{\mu_1+\mu_2}{2}\Big) \times \mathrm{Tr}_2\big(T_n(\mu_1) T_n(\mu_2+1)^{-1}\big)\Big\}.$$

The expression inside the parenthesis now has poles of high order at $\mu_i = 0$. As noted by Boos et al. [3], this formula readily implies that

$$h_n^{\epsilon_1,\cdots,\epsilon_n,\bar{\epsilon}_n,\cdots,\bar{\epsilon}_1} \in \mathbf{Q}[\log 2, \zeta(3), \zeta(5), \cdots],$$

because the Taylor coefficients of $\omega(\lambda)$ are given by

$$\omega(\lambda) - \frac{1}{2(\lambda^2-1)} = 2\left(\log 2 + \sum_{k=1}^{\infty} \zeta_a(2k+1)\lambda^{2k}\right),$$

$$\zeta_a(s) = (1-2^{1-s})\zeta(s).$$

7 Comments

We have seen that the density matrix for the XXX chain (a solution to the reduced qKZ equation) has a specific structure, written in terms of a single transcendental function with rational function coefficients. The rational part is expressed in terms of a certain transfer matrix of a finite chain. The existing proof of this formula rests on a direct verification of the reduced qKZ equation together with a comparison of poles and asymptotic behavior. It is quite technical and computational. At present, a similar formula is known also for the massive regime of the XXZ model, and as conjectures for the massless XXZ and XYZ models. For the XXZ model two transcendental functions enter the formula, and for XYZ model we have three of them.

Unfortunately, these formulas are not effective for practical computation at all. Although suggestive, we do not understand the nature of the formula well.

References

1. Baxter, R., Exactly solved models in statistical mechanics, Academic Press, London, New York, (1982)
2. H. Boos and V. Korepin, Quantum spin chains and Riemann zeta function with odd arguments, J. Phys. A **34** 5311–5316, (2001)
3. H. Boos, V. Korepin and F. Smirnov, Emptiness formation probability and quantum Knizhnik-Zamolodchikov equation, Nucl. Phys. B **658**, 417–439 (2003)

4. M. Jimbo and T. Miwa, Algebraic analysis of solvable lattice models, Regional Conference Series in Math. **85** AMS, (1994)
5. H. Boos, M. Jimbo, T. Miwa, F. Smirnov and Y. Takeyama, Density matrix of a finite sub-chain of the Heisenberg anti-ferromagnet, Lett. Math. Phys. **75** (2006) 201–208. See also the review by the same authors and references cited therein: Algebraic representation of correlation functions in integrable spin chains, hep-th/0601132, to appear in Annales Henri Poincaré.
6. M. Takahashi, Half-filled Hubbard model at low temperature, J. Phys. C **10** 1298 (1977)
7. K. Sakai, M. Shiroishi, Y. Nishiyama and M. Takahashi, Third neighbor correlators of a one-dimensional spin-1/2 Heisenberg antiferromagnet, Phys. Rev. E **67**, 065101 (2003); for more recent results, see e.g., J. Sato, M. Shiroishi and M. Takahashi, Correlation functions of the spin-1/2 anti-ferromagnetic Heisenberg chain: exact calculation via the generating function, hep-th/0507290.
8. M. Jimbo, T. Miwa, K. Miki and A. Nakayashiki, Correlation functions of the XXZ model for $\Delta < -1$, Phys. Lett. A **168** 256–263, (1992); M. Jimbo and T. Miwa, Quantum Knizhnik-Zamolodchikov equation at $|q| = 1$ and correlation functions of the XXZ model in the gapless regime, J. Phys.**A** **29** 2923–2958, (1996)
9. N. Kitanine, J.-M. Maillet and V. Terras, Correlation functions of the XXZ Heisenberg spin-$\frac{1}{2}$-chain in a magnetic field, Nucl. Phys. B **567** 554–582, (2000)
10. N. Kitanine, J.-M. Maillet, N. Slavnov and V. Terras, On the algebraic Bethe Ansatz approach to the correlation functions of the XXZ spin-1/2 Heisenberg chain, hep-th/0505006
11. N. Kitanine, J.-M. Maillet, N. Slavnov and V. Terras, Large distance asymptotic behavior of the emptiness formation probability of the XXZ spin-1/2 Heisenberg chain, J. Phys. A **35**, L753, (2002)

Inverse image of D-modules and quasi-*b*-functions

Yves Laurent

Institut Fourier Mathématiques, UMR 5582 CNRS/UJF, BP 74, 38402 St Martin d'Hères Cedex, France
Yves.Laurent@ujf-grenoble.fr

Summary. The usual *b*-function of a holonomic $\mathcal{D}$-module is associated to the Euler vector field but the elementary case of a ramification map shows that this Euler vector field is not preserved under inverse image. We define quasi-*b*-functions, that is *b*-functions associated to a quasi-homogeneity and use them to state an inverse image theorem for *b*-functions of holonomic $\mathcal{D}$-modules. We apply this result to an explicit calculation of the usual *b*-function of the Kashiwara-Hotta module on the Grothendieck's simultaneous resolution of a semi-simple Lie algebra.

Key words: D-modules, semi-simple Lie groups, b-function
Mathematics Subject Classification (2000): 35A27, 35D10, 17B15

Introduction

The classical Bernstein-Sato polynomial is defined for a holomorphic (or algebraic) function $f : \mathbb{C}^n \to \mathbb{C}$ by the equation

$$b(s)f(x)^s = P(s, x, D_x)f(x)^{s+1}$$

where b is a polynomial in s and P a differential operator in $(x, D_x = \frac{\partial}{\partial x})$ with polynomial parameter s. More precisely, the *b*-function of f is the generator of the ideal of polynomials satisfying that equation and is denoted by b_f. In [4] Kashiwara proved the following direct image theorem:

Theorem 1. *Let $F : X' \to X$ be a projective holomorphic map, Y a subvariety of X and $Y' = F^{-1}(Y)$. Let $f : X \to \mathbb{C}$ be a holomorphic map such that $Y \subset f^{-1}(0)$ and let $f' = f \circ F$. We assume that F is an isomorphism $X' - Y' \to X - Y$. Then there exists some positive integer N such that $b_f(s)$ divides $b_{f'}(s)b_{f'}(s+1)\ldots b_{f'}(s+N)$.*

Received 26 April, 2006. Revised 11 July, 2006. Accepted 20 July 2006.

The definition of b-function was extended to holonomic $\mathcal{D}_X$-modules in [4],[6] in the following way. Let Y be a submanifold of the complex manifold X. Let $\mathcal{O}_X$ be the sheaf of holomorphic functions on X, $\mathcal{I}_Y$ be the ideal of $\mathcal{O}_X$ defining Y and $\mathcal{D}_X$ be the sheaf of differential operators with coefficients in $\mathcal{O}_X$. Then the V-filtration is defined by:

$$V_k\mathcal{D}_X = \{\, P \in \mathcal{D}_X \mid \forall j \in \mathbb{Z}, P\mathcal{I}_Y^j \subset \mathcal{I}_Y^{j+k} \,\}$$

(with $\mathcal{I}_Y^j = \mathcal{O}_X$ if $j \leq 0$).

Let θ_Y be any vector field such that $\theta_Y(f) - kf \in \mathcal{I}_Y^{k+1}$ if $f \in \mathcal{I}_Y^k$. Then a *b-function* along Y for a section u of a coherent $\mathcal{D}_X$-module $\mathcal{M}$ is a polynomial b such that there exists $P \in V_{-1}\mathcal{D}_X$ and

$$(b(\theta_Y) + P)\, u = 0$$

In local coordinates such that $Y = \{\, (x_1, \ldots, x_{n-d}, t_1, \ldots, t_d) \in X \mid t = 0 \,\}$, we have $\mathcal{I}_Y^j = \left\{ \sum_{|\alpha|=j} f_\alpha(x,t) t^\alpha \right\}$ hence t_i is of order -1 for the V-filtration, D_{t_i} is of order $+1$, x_j, D_{x_j} are of order 0. With $\theta_Y = \sum_{i=1}^d t_i D_{t_i}$ a b-function is given by an equation

$$\Big(b(\theta_Y) + \sum t_i P_i(x,t,D_x,(t_i D_{t_j})_{(i,j)})\Big)\, u = 0$$

and in the case of a hypersurface:

$$(b(tD_t) + tP(x,t,D_x,tD_t))\, u = 0$$

We are interested in the behavior of these b-functions under inverse image of $\mathcal{D}_X$-modules. If we consider the rather elementary case of a map $(x, t_1, \ldots, t_d) \mapsto (x, t_1^{m_1}, \ldots, t_d^{m_d})$, the inverse image of the vector field $\sum_{i=1}^d t_i D_{t_i}$ is $\sum_{i=1}^d 1/m_i t_i D_{t_i}$. So it appears the need of considering b-functions relative to a vector field which is not the Euler vector field $\sum_{i=1}^d t_i D_{t_i}$.

In this paper, we recall the definition of such a "weighted" b-function, which we call a quasi-b-function, and we state an inverse image theorem for them in the case of holonomic $\mathcal{D}$-modules. Then we give an interesting application to some holonomic modules associated to Lie algebra. In a previous paper [2], we have calculated the quasi-b-functions of equivariant $\mathcal{D}$-modules (e.g. the Kashiwara-Hotta module) along the strata of the nilpotent cone. Here, we show that after a resolution of singularities (Grothendieck's simultaneous resolution), we can calculate exactly the usual b-function along the desingularization of the nilpotent cone.

1 Weighted b-functions

Let X be a complex manifold, $\mathcal{O}_X$ the sheaf of holomorphic functions on X and $\mathcal{D}_X$ the sheaf of differential operators with coefficients in $\mathcal{O}_X$. Let

$\varphi = (\varphi_1, \ldots, \varphi_d)$ be a holomorphic map from X to the vector space $W = \mathbb{C}^d$ and $m_1, \ldots, m_d$ be strictly positive and relatively prime integers. We define a filtration on $\mathcal{O}_X$ by:

$$V_k^{\varphi}\mathcal{O}_X = \sum_{<m,\alpha>=-k} \mathcal{O}_X \varphi^{\alpha}$$

with $\alpha \in \mathbb{N}^d$, $<m,\alpha>= \sum m_i\alpha_i$ and $\varphi^{\alpha} = \varphi_1^{\alpha_1} \ldots \varphi_d^{\alpha_d}$. If $k \geq 0$ we set $V_k^{\varphi}\mathcal{O}_X = \mathcal{O}_X$.

This filtration extends to $\mathcal{D}_X$ by:

$$V_k^{\varphi}\mathcal{D}_X = \{\, P \in \mathcal{D}_X \mid \forall l \in \mathbb{Z}, PV_l^{\varphi}\mathcal{O}_X \subset V_{l+k}^{\varphi}\mathcal{O}_X \,\} \tag{1.1}$$

Definition 1.1. A (φ, m)*-weighted Euler vector field* is a vector field η on X such that

- $\eta(\varphi_i) - m_i\varphi_i \in V_{-m_i-1}^{\varphi}\mathcal{O}_X$ for $i = 1 \ldots d$.
- $\eta \in \sum_i \varphi_i \mathcal{V}_X$. ($\mathcal{V}_X$ is the sheaf of vector fields on X.)

Lemma 1.2. *a) Any (φ, m)-weighted Euler vector field is in $V_0^{\varphi}\mathcal{D}_X$.*

b) If η_1 and η_2 are two (φ, m)-weighted Euler vector fields, $\eta_1 - \eta_2$ is in $V_{-1}^{\varphi}\mathcal{D}_X$.

c) A vector field η is a (φ, m)-weighted Euler vector field if and only if

$$\forall k \in \mathbb{Z}, \quad \forall f \in V_k^{\varphi}\mathcal{O}_X, \quad \eta(f) + kf \in V_{k-1}^{\varphi}\mathcal{O}_X$$

d) A vector field η is a (φ, m)-weighted Euler vector field if and only if

$$\forall k \in \mathbb{Z}, \quad \forall P \in V_k^{\varphi}\mathcal{D}_X, \quad [\eta, P] + kP \in V_{k-1}^{\varphi}\mathcal{D}_X$$

Proof. This is a direct consequence of the definitions. □

Definition 1.3. Let u be a section of a coherent $\mathcal{D}_X$-module $\mathcal{M}$. A polynomial b is a quasi-b-function of type (φ, m) for u if there exist a (φ, m)-weighted Euler vector field η and a differential operator Q in $V_{-1}^{\varphi}\mathcal{D}_X$ such that $(b(\eta)+Q)u = 0$.

Lemma 1.2 shows that the quasi-b-function is independent of the vector field η.

The smooth case:

Assume first that the map φ is smooth. Then there exists always, at least locally, a (φ, m)-weighted Euler vector field. Indeed we may always choose local coordinates $(x_1, \ldots, x_{n-d}, t_1, \ldots, t_d)$ such that $\varphi_i(x, t) = t_i$ for $i = 1, \ldots, d$ and take $\eta = \sum_{i=1}^d m_i t_i D_{t_i}$. Definition 1.3 being independent of η, the polynomial b may be defined globally even if η is not.

In [2] we gave another definition of the V-filtration and quasi-b-function. If φ is smooth, the two definition are locally equivalent, as it is clear in local coordinates.

If $m_1 = m_2 = \cdots = m_d = 1$ and φ is smooth, the V-filtration depends only on $Y = \varphi^{-1}(0)$. It is the classical V-filtration of Kashiwara [6] given in the introduction. The graduate $V_0\mathcal{D}_X/V_{-1}\mathcal{D}_X$ is identified to the sheaf $\mathcal{D}_{T_YX}[0]$ of differential operators on T_YX which are homogeneous of degree 0 in the fibers of $T_YX \to Y$ and η is any vector field of $V_0\mathcal{D}_X$ whose image in $\mathcal{D}_{T_YX}[0]$ is the Euler vector field of the fiber bundle T_YX (see [7] for the details).

Remark however that the structure given by the functions φ_i is more precise and will be used in the inverse image theorem.

The non smooth case:

If φ is not smooth, a (φ, m)-weighted Euler vector field does not always exists. Main results as corollary 2.2 are still true with vector fields which satisfy the first but not the second condition of definition 1.1. In theorem 3.5, we shall give an example of such a situation.

Let $\mathcal{M}$ be a coherent $\mathcal{D}_X$-module. A $V^\varphi\mathcal{D}_X$-filtration on $\mathcal{M}$ is a good filtration if it is locally finite.

Definition 1.4. Let $\mathcal{M}$ be a coherent $\mathcal{D}_X$-module and $V^\varphi\mathcal{M}$ a good $V^\varphi\mathcal{D}_X$-filtration. A polynomial b is a quasi-b-function of type (φ, m) for $V^\varphi\mathcal{M}$ if there exists some (φ, m)-weighted Euler vector field η such that:

$$\forall k \in \mathbb{Z}, \quad b(\eta + k)V_k^\varphi\mathcal{M} \subset V_{k-1}^\varphi\mathcal{M}.$$

Definition 1.3 is a special case of definition 1.4 if $\mathcal{D}_X u$ is provided with the filtration induced by the canonical filtration of $\mathcal{D}_X$.

2 Inverse image.

2.1 The main theorem

Let $\varphi : X \to W = \mathbb{C}^d$ and $\varphi' : X' \to W' = \mathbb{C}^{d'}$ be two holomorphic maps, let $m_1, \ldots, m_d$ and $m'_1, \ldots, m'_{d'}$ be strictly positive integers. Let $f : X' \to X$ and $F : W' \to W$ be two holomorphic maps such that $\varphi_\circ f = F_\circ\varphi'$:

$$\begin{array}{ccc} X' & \xrightarrow{f} & X \\ {\scriptstyle\varphi'}\downarrow & & \downarrow{\scriptstyle\varphi} \\ W' & \xrightarrow{F} & W \end{array}$$

We assume that F is a *quasi-homogeneous* polynomial mapping, that is $F = (F_1, \ldots, F_d)$ with $F_i(\lambda^{m'_1}x_1, \ldots, \lambda^{m'_{d'}}x_{d'}) = \lambda^{m_i}F_i(x_1, \ldots, x_{d'})$.

If $\mathcal{N}$ is a $\mathcal{D}_X$-module, its inverse image by f is:

$$f^+\mathcal{N} = \mathcal{O}_{X'} \otimes_{f^{-1}\mathcal{O}_X} f^{-1}\mathcal{N} = \mathcal{D}_{X'\to X} \otimes_{f^{-1}\mathcal{D}_X} f^{-1}\mathcal{N}$$

where $\mathcal{D}_{X'\to X}$ is the $(\mathcal{D}_{X'}, f^{-1}\mathcal{D}_X)$-bimodule $\mathcal{O}_{X'} \otimes_{f^{-1}\mathcal{O}_X} f^{-1}\mathcal{D}_X$.

The filtration on $\mathcal{D}_{X'\to X}$ is defined as:

$$V_k^{\varphi'}\mathcal{D}_{X'\to X} = \sum_{i+j=k} V_i^{\varphi'}\mathcal{O}_{X'} \otimes f^{-1}V_j^{\varphi}\mathcal{D}_X$$

By the hypothesis, $g_\circ f$ is a section of $V_k^{\varphi'}\mathcal{O}_{X'}$ for any section g of $V_k^{\varphi}\mathcal{O}_X$, hence the filtration on $\mathcal{D}_{X'\to X}$ is compatible with the corresponding filtrations on $\mathcal{D}_{X'}$ and $\mathcal{D}_X$.

If a $\mathcal{D}_X$-module $\mathcal{N}$ is provided with a V^{φ}-filtration, this defines a $V^{\varphi'}\mathcal{D}_{X'}$-filtration on $f^+\mathcal{N}$ by

$$V_k^{\varphi'} f^+\mathcal{N} = \sum_{i+j=k} V_i^{\varphi'}\mathcal{O}_{X'} \otimes f^{-1}V_j^{\varphi}\mathcal{N} = \sum_{i+j=k} V_i^{\varphi'}\mathcal{D}_{X'\to X} \otimes f^{-1}V_j^{\varphi}\mathcal{N} \quad (2.1)$$

Let $Y = \varphi^{-1}(0)$ and $Y' = \varphi'^{-1}(0)$, let $p : T_Y X \to X$ and $p' : T_{Y'}X' \to X'$ be the normal bundles, $\widetilde{f} : T_{Y'}X' \to T_Y X$ be the map induced by f.

Theorem 2.1. *We assume that φ' is smooth on X'. Let $\mathcal{N}$ be a holonomic $\mathcal{D}_X$-module provided with a good $V^{\varphi}\mathcal{D}_X$-filtration.*

Then $p'^{-1}gr_{V^{\varphi'}} f^+\mathcal{N}$ is equal to $\widetilde{f}^+p^{-1}gr_{V^{\varphi}}\mathcal{N}$ and isomorphic to the graded module associated with a good $V^{\varphi'}\mathcal{D}_{X'}$-filtration of $f^+\mathcal{N}$.

The coherence of $\mathcal{N}$ does not imply that of $f^+\mathcal{N}$ in general. But if $\mathcal{N}$ is holonomic, then $f^+\mathcal{N}$ is holonomic [5], this is why the theorem is restricted to holonomic modules.

The result was known when f is a submersion, $Y' = f^{-1}(Y)$ and the V-filtrations being the usual V-filtrations along Y and Y' [8]. The introduction of the weights m_i and m_i' allows f to be non submersive and Y' to be a proper subvariety of $f^{-1}(Y)$; the relation between the weights is given by the quasi-homogeneity of F. In this form the theorem was proved in [7, theorem 1.4.1].

Let η' be a vector field on X'. If there exists a vector field η on X such that for any function g on X, $\eta'(g \circ f) = \eta(g) \circ f$, we write $\eta = f_*\eta'$.

Corollary 2.2. *Under the hypothesis of theorem 2.1, we assume that there is a (φ', m')-weighted Euler vector field η' and a (φ, m)-weighted Euler vector field η such that $\eta = f_*\eta'$.*

Let $\mathcal{N}$ be a holonomic $\mathcal{D}_X$-module provided with a good $V^{\varphi}\mathcal{D}_X$-filtration. Then $f^+\mathcal{N}$ is provided with a good $V^{\varphi'}\mathcal{D}_{X'}$-filtration such that a polynomial b is a quasi-b-function of type (φ, m, η) for the filtration of $\mathcal{N}$ if and only if b is a quasi-b-function of type (φ', m', η') for the filtration of $f^+\mathcal{N}$.

Proof. By definition, for any Q in $\mathcal{D}_{X'\to X}$, we have $\eta' Q = Q\eta$ hence for any polynomial $b(\eta')Q = Qb(\eta)$ which shows the corollary. □

Corollary 2.3. *Under the hypothesis of corollary 2.2, if $\mathcal{N}$ is a holonomic $\mathcal{D}_X$-module and u a section of $\mathcal{N}$ with a quasi-b-function of type (φ, m, η), then the section $1_{X' \to X} \otimes u$ of $f^+\mathcal{N}$ has the same polynomial b as a quasi-b-function of type (φ', m', η').*

Proof. Recall that $1_{X' \to X}$ is the canonical section $1 \otimes 1$ in $\mathcal{D}_{X' \to X} = \mathcal{O}_{X'} \otimes_{f^{-1}\mathcal{O}_X} f^{-1}\mathcal{D}_X$. If u is a section of $\mathcal{N}$, we set on $\mathcal{D}_X u$ the filtration induced by the filtration of $\mathcal{D}_X$. Then, by definition of the filtration $V^{\varphi'}[N]f^+\mathcal{N}$ used in the proof of theorem 2.1, $1_{X' \to X} \otimes u$ is of order 0 for this filtration. Then corollary 2.3 is a special case of corollary 2.2. □

Remark 2.4. The proof of these corollaries does not use the fact that η and η' are weighted Euler vector fields.

2.2 Examples

2.2.1 Transversal case

Consider holomorphic maps

$$X' \xrightarrow{\ f\ } X \xrightarrow{\ \varphi\ } W = \mathbb{C}^d$$

with φ smooth. Assume that f is transversal to $Y = \varphi^{-1}(0)$ that is $T^*_{X'}X \cap T^*_Y X = \{0\}$. As usual $T^*_{X'}X$ is the kernel of $T^*X \times_X X' \to T^*X'$.

Then the theorem applies with $\varphi' = \varphi_\circ f$, $W' = W$ and $F = Id_W$.

In particular, if Z and Y are two transversal submanifolds of X, $\mathcal{M}$ a holonomic $\mathcal{D}_X$-module, $\mathcal{M}_Z$ its restriction to Z, then the (usual) b-function of $\mathcal{M}_Z$ along $Y' = Y \cap Z$ is equal to the (usual) b-function of $\mathcal{M}$ along Y.

If f is smooth, that is $T^*_{X'}X = \{0\}$, it is a result of [8].

2.2.2 Ramification map

Let $X' = X = \mathbb{C}^d$ and $f : X' \to X$ given by $f(x_1, \dots, x_d) = (x_1^{m_1}, \dots, x_d^{m_d})$. Let $\varphi = Id_X$. Then the theorem apply to $\varphi' = \varphi$ and $F = f$. If $\mathcal{M}$ is a holonomic $\mathcal{D}_X$-module, the quasi-b-function of $\mathcal{M}$ associated to $\eta = \sum m_i x_i D_{x_i}$ is equal to the usual b-function of $f^+\mathcal{M}$ (i.e. associated to $\eta' = \sum x_i D_{x_i}$).

In the same way, let $X' = \mathbb{C}^{d'}$, $X = \mathbb{C}^d$ and $f : X' \to X$ be a quasi-homogeneous polynomial such that

$$f_i(\lambda^{m'_1} x_1, \dots, \lambda^{m'_{d'}} x_{d'}) = \lambda^{m_i} f_i(x_1, \dots, x_{d'}).$$

The quasi-b-function of $\mathcal{M}$ associated to $\eta = \sum m_i x_i D_{x_i}$ is equal to the quasi-b-function of $f^+\mathcal{M}$ associated to $\eta' = \sum m'_j x'_j D_{x'_j}$.

3 Application to Lie algebras

3.1 Grothendieck's simultaneous resolution

We recall here a few definition about semi-simple Lie algebras, see [1, ch. 3] or [9] for the details.

Let G be a complex semi-simple Lie group with Lie algebra $\mathfrak{g}$. An element X of $\mathfrak{g}$ is said to be semi-simple (resp. nilpotent) if the adjoint map $ad_X : Y \mapsto [X,Y]$ is a semi-simple (resp. nilpotent) endomorphism of the vector space $\mathfrak{g}$. An element X is said to be regular if the dimension of its centralizer $Z_{\mathfrak{g}}(X)$ is minimum (and equal by definition to the rank of $\mathfrak{g}$).

The set of nilpotent elements of $\mathfrak{g}$ is a singular cone of $\mathfrak{g}$ called the nilpotent cone and denoted by $\mathcal{N}$. The number of nilpotent orbits is finite and they define a stratification of the nilpotent cone. The maximal dimensional orbit is also the set of regular nilpotent elements, it is a Zariski open subset of $\mathcal{N}$.

The set of regular semi-simple elements is a Zarisky open set in $\mathfrak{g}$ denoted by $\mathfrak{g}_{rs}$, its complementary is a hypersurface. A Cartan subalgebra of $\mathfrak{g}$ is the centralizer of a regular semi-simple element. A Borel subalgebra is a maximal solvable subalgebra of $\mathfrak{g}$. All Cartan subalgebras are conjugate and so are all Borel subalgebras. The set $\mathfrak{B}$ of all Borel subalgebra of $\mathfrak{g}$ is a projective variety called the flag manifold.

We fix now a Cartan subalgebra $\mathfrak{h}$ of $\mathfrak{g}$ and denote by W the Weyl group $W(\mathfrak{g},\mathfrak{h})$. The algebra of invariant polynomials on $\mathfrak{g}$ under the action of G will be denoted by $\mathcal{O}[\mathfrak{g}]^G$, it is equal to the algebra of polynomials $\mathbb{C}[P_1,\dots,P_d]$ where $(P_1,\dots,P_d)$ are algebraically independent invariant polynomials and d is the rank of $\mathfrak{g}$ (Chevalley's theorem). The algebra of polynomials on $\mathfrak{h}$ invariant under W is $\mathcal{O}[\mathfrak{h}]^W = \mathbb{C}[p_1,\dots,p_d]$ where p_j is the restriction to $\mathfrak{h}$ of P_j. The restriction map $P \mapsto P|_{\mathfrak{h}}$ defines an isomorphism from $\mathcal{O}[\mathfrak{g}]^G$ to $\mathcal{O}[\mathfrak{h}]^W$. The space $\mathfrak{h}/W$ is thus isomorphic to $\mathbb{C}^d$ and will be denoted by V. The functions $P_1,\dots,P_d$ and their restrictions to $\mathfrak{h}$ define two morphisms $\varrho : \mathfrak{g} \to V$ and $\pi : \mathfrak{h} \to V$. The nilpotent cone $\mathcal{N}$ is equal to $\varrho^{-1}(0)$.

Given two Borel subalgebras $\mathfrak{b}$ and $\mathfrak{b}'$, there is a canonical isomorphism

$$\mathfrak{b}/[\mathfrak{b},\mathfrak{b}] \simeq \mathfrak{b}'/[\mathfrak{b}',\mathfrak{b}']$$

If $\mathfrak{h}$ is a Cartan subalgebra contained in $\mathfrak{b}$, the map $\mathfrak{h} \to \mathfrak{b}/[\mathfrak{b},\mathfrak{b}]$ induced by $\mathfrak{h} \to \mathfrak{b}$ is an isomorphism. So there is a canonical morphism from any Borel subalgebra $\mathfrak{b}'$ to $\mathfrak{h}$.

The variety $\widetilde{\mathfrak{g}}$ is defined as:

$$\widetilde{\mathfrak{g}} = \{\, (X,\mathfrak{b}) \in \mathfrak{g} \times \mathfrak{B} \mid X \in \mathfrak{b} \,\}$$

The canonical projection $\widetilde{\mathfrak{g}} \to \mathfrak{g}$ is denoted by μ. We have $\mathcal{N} = \varrho^{-1}(0)$ and we define $\widetilde{\mathcal{N}} = \mu^{-1}(\mathcal{N})$. The map $\nu : \widetilde{\mathfrak{g}} \to \mathfrak{h}$ is defined by the maps $\mathfrak{b}' \to \mathfrak{h}$, that is $\nu(X,\mathfrak{b}) = X \ mod\ [\mathfrak{b},\mathfrak{b}]$ in $\mathfrak{b}/[\mathfrak{b},\mathfrak{b}] \simeq \mathfrak{h}$. In this way we get a commutative diagram, called "Grothendieck's simultaneous resolution":

$$\begin{array}{ccc} \widetilde{\mathfrak{g}} & \xrightarrow{\mu} & \mathfrak{g} \\ \nu \downarrow & & \downarrow \varrho \\ \mathfrak{h} & \xrightarrow{\pi} & V \end{array} \tag{3.1}$$

It has the following properties:

- The varieties $\widetilde{\mathfrak{g}}$ and $\widetilde{\mathcal{N}}$ are smooth manifolds.
- The map μ is proper.
- $\mu : \widetilde{\mathcal{N}} \to \mathcal{N}$ is a resolution of singularities, in particular μ is an isomorphism over the regular points of $\mathcal{N}$.
- For each $X \in \mathfrak{g}_{rs}$, there is a canonical free W-action on $\mu^{-1}(X)$ making $\widetilde{\mathfrak{g}}_{rs} := \mu^{-1}(\mathfrak{g}_{rs})$ a principal W-bundle on $\mathfrak{g}_{rs}$.
- $\nu : \widetilde{\mathfrak{g}} \to \mathfrak{h}$ is smooth

There is an isomorphism $\widetilde{\mathcal{N}} \simeq T^*\mathfrak{B}$ and the map $T^*\mathfrak{B} \to \mathcal{N}$ is called the Springer resolution.

Example 3.1. The $\mathfrak{sl}_n$ case.

If $\mathfrak{g}$ is the algebra $\mathfrak{sl}_n(\mathbb{C})$ of square matrices with null trace, $\mathfrak{g}_{rs}$ is the set of matrices with distinct eigenvalues, a Cartan subalgebra is the set of diagonal matrices and a Borel subalgebra is the set of upper triangular matrices.

The invariant polynomials are the coefficients of the characteristic polynomial of the matrix and the map $\mathfrak{h} \to V = \mathbb{C}^{n-1}$ is given by the symmetric polynomials.

The space $\mathfrak{B}$ is identified with the usual flag variety $\{ F = (F_0 = 0 \subset F_1 \subset \cdots \subset F_n = \mathbb{C}^n) \mid dimF_i = i \}$, $\widetilde{\mathfrak{g}} = \{ (X, F) \in \mathfrak{sl}_n \times \mathfrak{B} \mid X(F_i) \subset F_i \}$ and the map $\nu : \widetilde{\mathfrak{g}} \to \mathfrak{h}$ associates to (X, F) the set of eigenvalues of X ordered by F, that is the eigenvalues of X acting on the 1-dimensional spaces F_i/F_{i-1}.

Example 3.2. $\mathfrak{g} = \mathfrak{sl}_2$

$\mathfrak{g}$ is the set of matrices $\begin{pmatrix} x & y \\ z & -x \end{pmatrix}$ and $\mathcal{N}$ is given by $x^2 + yz = 0$.

$\tilde{\mathfrak{g}} = \{ (x, y, z, (\alpha, \beta)) \in \mathbb{C}^3 \times \mathbb{P}_1(C) \mid \alpha^2 z - 2\alpha\beta x - \beta^2 y = 0 \}$ and the map ν is given by $\nu(x, y, z, (\alpha, \beta)) = z(\alpha/\beta) - x$ if $\beta \neq 0$ and $\nu(x, y, z, (\alpha, \beta)) = x + y(\beta/\alpha)$ if $\alpha \neq 0$.

3.2 Equivariant $\mathcal{D}$-modules

The differential of the adjoint action induces a Lie algebra morphism $\tau : \mathfrak{g} \to \mathrm{Der}\mathcal{O}[\mathfrak{g}]$ by:

$$(\tau(A)f)(x) = \frac{d}{dt} f\left(\exp(-tA).x\right)|_{t=0} \quad \text{for} \quad A \in \mathfrak{g}, f \in \mathcal{O}[\mathfrak{g}], x \in \mathfrak{g}$$

i.e. $\tau(A)$ is the vector field on $\mathfrak{g}$ whose value at $x \in \mathfrak{g}$ is to $[x, A]$. We denote by $\tau(\mathfrak{g})$ the set of all vector fields $\tau(A)$ for $A \in \mathfrak{g}$. It is the set of vector fields on $\mathfrak{g}$ tangent to the orbits of G acting on $\mathfrak{g}$.

The group G acts on $\mathfrak{g}^*$, the dual of $\mathfrak{g}$. We denote by $\mathcal{O}[\mathfrak{g}^*]^G$ the space of invariant polynomials on $\mathfrak{g}^*$ and by $\mathcal{O}_+[\mathfrak{g}^*]^G$ the subspace of polynomials vanishing at $\{0\}$.

Let $\mathcal{D}_{\mathfrak{g}}^G$ be the sheaf of differential operators on $\mathfrak{g}$ invariant under the adjoint action of G. The principal symbol $\sigma(P)$ of such an operator P is a function on $T^*\mathfrak{g} = \mathfrak{g} \times \mathfrak{g}^*$ invariant under the action of G. Examples of such invariant functions are the elements of $\mathcal{O}[\mathfrak{g}^*]^G$ identified to functions on $\mathfrak{g} \times \mathfrak{g}^*$ constant in the variables of $\mathfrak{g}$. If F is a subsheaf of $\mathcal{D}_{\mathfrak{g}}^G$, we denote by $\sigma(F)$ the sheaf of the principal symbols of all elements of F.

Definition 3.3. A subsheaf F of $\mathcal{D}_{\mathfrak{g}}^G$ is of (H-C)-type if $\sigma(F)$ contains a power of $\mathcal{O}_+[\mathfrak{g}^*]^G$. An (H-C)-type $\mathcal{D}_{\mathfrak{g}}$-module is the quotient $\mathcal{M}_F$ of $\mathcal{D}_{\mathfrak{g}}$ by the ideal $\mathcal{I}_F$ generated by $\tau(\mathfrak{g})$ and by F.

The first example of (H-C)-type $\mathcal{D}$-module is the Kashiwara-Hotta module $\mathcal{M}_\lambda^F$ [3], other examples are given in [7]. In the case of the Kashiwara-Hotta module $\mathcal{M}_\lambda^F$, we have $F = \{\, P - P(\lambda) \mid P \in \mathcal{O}[\mathfrak{g}^*]^G \,\}$ for some $\lambda \in \mathfrak{g}^*$ hence $\sigma(F) = \mathcal{O}_+[\mathfrak{g}^*]^G$.

The Killing form of $\mathfrak{g}$ is non-degenerate and defines an isomorphism from $\mathfrak{g}^*$ to $\mathfrak{g}$ and the cotangent bundle $T^*\mathfrak{g} = \mathfrak{g} \times \mathfrak{g}^*$ is identified with $\mathfrak{g} \times \mathfrak{g}$. Then the characteristic variety of an (H-C)-type $\mathcal{D}_{\mathfrak{g}}$-module is a subset of:

$$\{\, (x, y) \in \mathfrak{g} \times \mathfrak{g} \mid [x, y] = 0, y \in \mathcal{N} \,\}$$

so it is a holonomic $\mathcal{D}_{\mathfrak{g}}$-module [3].

In [2], we calculated the quasi-b-function of a (H-C)-type module along each nilpotent orbit of $\mathfrak{g}$ and use this result to show that its solutions are locally integrable functions (Harish-Chandra's theorem). Here we want to show that the situation is much more simple on $\widetilde{\mathfrak{g}}$ and calculate the usual b-function along $\widetilde{\mathcal{N}}$.

Let n be the dimension of $\mathfrak{g}$ (and d its rank). Consider the map $\varrho : \mathfrak{g} \to V$, it is given by the invariant polynomials $(P_1, \ldots, P_d)$ whose degrees $(r_1, \ldots, r_d)$ are called the primitive degrees of $\mathfrak{g}$. The sum of these degrees is $(n + d)/2$ [9]. The map ϱ is smooth on the set of regular points of $\mathfrak{g}$ and in particular on a neighborhood of the regular nilpotent orbit.

Let $\mathcal{M}_F$ be a (H-C)-type $\mathcal{D}$-module. In [2] and [7], we proved that $\mathcal{M}_F$ admits a quasi-b-function of type (ϱ, r) along the regular nilpotent orbit which is exactly the polynomial

$$b(T) = (T - N) \ldots T(T + 1) \ldots (T + \frac{n - d}{2})$$

where $N \geq 0$ and $N = 0$ if $\sigma(F) = \mathcal{O}_+[\mathfrak{g}^*]$.

Let us apply corollary 2.3 to the diagram (3.1) restricted to regular points of $\mathfrak{g}$. The map $\pi : \mathfrak{h} \to V$ is quasi-homogeneous of degrees $(r_1, \ldots, r_d)$, hence we get a b-function with weights $(1, 1, \ldots, 1)$, that is a usual b-function. This means that the usual b-function of $\mu^+\mathcal{M}_F$ along $\widetilde{\mathcal{N}}$ is equal to $b(T)$ on the inverse image by μ of the regular points of $\mathfrak{g}$. In fact, the map μ is a ramification map at regular points and this is a special case of example 2.2.2.

We will now show that this b-function is valid everywhere on $\widetilde{\mathcal{N}}$:

Theorem 3.4. *Let F be an (H-C)-type subsheaf of $\mathcal{D}_{\mathfrak{g}}^G$ and $\mathcal{M}_F$ the associated $\mathcal{D}$-module. The (usual) b-function of the module $\mu^+\mathcal{M}_F$ along $\widetilde{\mathcal{N}}$ is equal to*

$$b(T) = (T - N)\ldots T(T+1)\ldots(T + \frac{n-d}{2})$$

with $N = 0$ if $\sigma(F) = \mathcal{O}_+[\mathfrak{g}^*]$.

Proof. An operator P of F is invariant under G hence define a differential operator $\varrho_*(P)$ on V such that $P(f \circ \varrho) = \varrho_*(P)(f) \circ \varrho$. Let $\mathcal{N}_F$ be the quotient of $\mathcal{D}_V$ by the ideal generated by $\varrho_*(F)$.

Let $1_{\mathfrak{g}\to V}$ be the canonical generator of $\mathcal{D}_{\mathfrak{g}\to V}$ and $u_{\mathfrak{g}\to V}$ its class in $\varrho^+\mathcal{N}_F$. We denote by $\mathcal{M}_F^0$ the $\mathcal{D}_{\mathfrak{g}}$-submodule of $\varrho^+\mathcal{N}_F$ generated by $u_{\mathfrak{g}\to V}$.

It has been proved in [7, th. 2.5.1] that $\mathcal{M}_F$ is equal to $\mathcal{M}_F^0$ hence $\mu^+\mathcal{M}_F$ is a submodule of $\mu^+\varrho^+\mathcal{N}_F$. Moreover, it has been proved in [7, prop. 2.4.3], that the polynomial $b(T) = (T - N)\ldots T(T+1)\ldots(T + \frac{n-d}{2})$ is a quasi-b-function for $\mathcal{N}_F$ of type (Id_V, r) where Id_V is the identity map of V and $r = (r_1, \ldots, r_d)$.

So we may apply corollary 2.3 to the diagram

$$\begin{array}{ccc} \widetilde{\mathfrak{g}} & \xrightarrow{\varrho\circ\mu} & V \\ {\scriptstyle\nu}\downarrow & & \downarrow{\scriptstyle Id_V} \\ \mathfrak{h} & \xrightarrow{\pi} & V \end{array} \tag{3.2}$$

The map ν is smooth and the map π is given by homogeneous functions of degrees $(r_1, \ldots, r_d)$. So we apply the corollary with weights $(r_1, \ldots, r_d)$ on $\mathfrak{g}$ and weights $(1, \ldots, 1)$ on $\widetilde{\mathfrak{g}}$. This shows that the polynomial $b(T) = (T - N)\ldots T(T+1)\ldots(T + \frac{n-d}{2})$ which is a quasi-b-function for $\mathcal{N}_F$ of type (Id_V, r) is a usual b-function for $\mu^+\mathcal{M}_F$ along $\widetilde{\mathcal{N}}$. □

Let E be the Euler vector field of $\mathfrak{g}$, that is the vector field such that $E(f) = kf$ for any homogeneous function f of degree k. With the same method than that of theorem 3.4 we can prove a result in a non smooth case:

Theorem 3.5. *Let F be an (H-C)-type subsheaf of $\mathcal{D}_{\mathfrak{g}}^G$ and $\mathcal{M}_F$ the associated $\mathcal{D}$-module. Let b be the polynomial of theorem 3.4. Let ϱ be the map $\mathfrak{g} \to V$. There exists a good V^ϱ-filtration on $\mathcal{M}_F$ such that:*

$$b(E + k)V^\varrho\mathcal{M}_F \subset V^\varrho_{k-1}\mathcal{M}_F$$

As the map ϱ is defined by homogeneous polynomials, the vector field E satisfy the first condition of definition 1.1. It does not satisfy the second condition but it is well known that on the smooth part of the nilpotent cone, E is equal to a (ϱ, r)-weighted Euler vector field modulo the ideal $\mathcal{I}_F$ [10, Ch 5.6.].

Proof. We consider the diagram

$$\begin{array}{ccc} \mathfrak{g} & \xrightarrow{\varrho} & V \\ {\scriptstyle\varrho}\downarrow & & \downarrow{\scriptstyle Id_V} \\ V & \xrightarrow{Id_V} & V \end{array} \tag{3.3}$$

As ϱ is defined by polynomials which are homogeneous of degree $(r_1, \dots, r_d)$, we have

$$\varrho_*(E) = \sum_{i=1\dots d} r_i t_i D_{t_i}$$

where $(t_1, \dots, t_d)$ are linear coordinates of the vector space V.

So, we may apply corollary 2.3 (with remark 2.4) as in the previous proof. □

References

1. N. Chriss and V. Ginsburg, *Representation theory and complex geometry*, Birkhäuser, 1997.
2. E. Galina and Y. Laurent, *$\mathcal{D}$-modules and characters of semisimple Lie groups*, Duke Mathematical Journal **123** (2004), no. 2, 265–309.
3. R. Hotta and M. Kashiwara, *The invariant holonomic system on a semisimple Lie algebra*, Inv. Math. **75** (1984), 327–358.
4. M. Kashiwara, *B-functions and holonomic systems. Rationality of roots of b-functions*, Invent. Math. **38** (1976), no. 1, 33–53.
5. M. Kashiwara, *On holonomic systems of differential equations II*, Inv. Math. **49** (1978), 121–135.
6. M. Kashiwara, *Vanishing cycles and holonomic systems of differential equations*, Lect. Notes in Math., vol. 1016, Springer, 1983, pp. 134–142.
7. Y. Laurent, *b-functions and integrable solutions of holonomic $\mathcal{D}$-modules*, Volume en l'honneur de Jean-Pierre Ramis, Astérisque, vol. 296, SMF, 2004, 145–165.
8. Y. Laurent and B. Malgrange, *Cycles proches, spécialisation et $\mathcal{D}$-modules*, Ann. Inst. Fourier Grenoble **45** (1995), 1353–1405.
9. V.S. Varadarajan, *Lie groups, lie algebras and their representations*, Prentice-Hall, 1974.
10. V.S. Varadarajan, *Harmonic analysis on real reductive groups*, Lect. Notes in Math., vol. 576, Springer, 1977.

The hypoelliptic Laplacian of J.-M. Bismut

Gilles Lebeau

Département de Mathématiques, Université de Nice Sophia-Antipolis, Parc Valrose, 06108 Nice Cedex 02, France
lebeau@math.unice.fr

Summary. In recents works, J.-M. Bismut has introduced an "hypoelliptic Laplacian" acting on differentials forms on the cotangent bundle T^*X of a Riemannian compact manifold X. This operator is a deformation of the Hodge Laplacian on X. We present here some analytic properties of this new operator.

1 Introduction

J-M. Bismut has introduced in [Bis04c], [Bis04b], [Bis04a], [Bis05] a new operator, the hypoelliptic Laplacian, which acts on differential forms on the cotangent bundle $\Sigma = T^*X$ of a Riemannian manifold X and which is a kinetic deformation of the Hodge Laplacian on X. From an analytic point of view, this operator is a generalisation to the case of non-flat metrics of the kinetic Fokker-Planck equations. The hypoelliptic Laplacian is thus a generalized Kolmogorov equation and so is obviously hypoelliptic. There exist numerous recent works about Fokker-Planck equations on the Euclidian space, both from probabilistic or pde's point of view (see [HN05], [HN04], [HSS05] [DV01] and references therein).
In section 2, we recall the definition of the hypoelliptic Laplacian of J-M. Bismut (see [Bis04a]). In section 3 we state some basic hypoelliptic estimates, and a result on the small time asymptotic of the heat kernel associated to the hypoelliptic Laplacian, and more generally to geometric Fokker-Planck equations (see [Leb05], [Leb06]). Finally, in section 4 we give the result on the convergence of the resolvent of the hypoelliptic Laplacian towards the resolvent of the Hodge Laplacian . This last result is a key ingredient in the study of the Ray-Singer metrics which will appear in [BL06].

We end this introduction by recalling some basic notations in differential geometry. Let (X, g) be a Riemannian, compact, connected n-manifold. Let

Received February, 2006. Revised 17 November, 2006. Accepted 29 December, 2006.

TX and T^*X be the tangent and cotangent bundles on X. We denote by (x,p) points in T^*X, by π the projection of T^*X on X, by g the metric isomorphism from TX to T^*X , by $(.|.)$ the scalar product on TX or T^*X and by $< .,. >$ the duality between T^*X and TX . If $(x_1, ..., x_n)$ are local coordinates defined on a open subset U of X, we will denote by $(x_1, ..., x_n; p_1, ..., p_n)$ the coordinates on $T^*X|_U$ such that $p_j =< p, \frac{\partial}{\partial x_j} >$. For $u = \Sigma u^j \frac{\partial}{\partial x_j} \in T_xX$ one has $|u|^2 = (u|u) = \Sigma g_{i,j}(x)u^iu^j$ and the square of the lenght of $p = \Sigma p_j dx^j \in T^*_xX$ is $|p|^2 = (p|p) = \Sigma g^{i,j}(x)p_ip_j$ with $(g^{i,j}) = g^{-1}$.
Let $T\Sigma$ be the tangent bundle to the cotangent space $\Sigma = T^*X$. The Levi Civita connection on TX induces the canonical splitting of $T\Sigma$

$$T\Sigma = T^H\Sigma \oplus T^V\Sigma \tag{1.1}$$

where $T^V\Sigma$ is the vertical space tangent to the fibration $T^*X \to X$, thus is span by the vectors fields

$$\hat{e}^j = \frac{\partial}{\partial p_j}$$

and the horizontal space $T^H\Sigma$ is span by the vectors fields

$$e_i = \frac{\partial}{\partial x_i} + \Gamma^\alpha_{\beta,i} p_\alpha \frac{\partial}{\partial p_\beta}$$

where the $\Gamma^\alpha_{\beta,i}$ are the Christoffel symbols

$$\Gamma^\alpha_{\beta,i} = \frac{1}{2} g^{\alpha,\mu} [\frac{\partial g_{\mu,\beta}}{\partial x_i} + \frac{\partial g_{i,\mu}}{\partial x_\beta} - \frac{\partial g_{i,\beta}}{\partial x_\mu}]$$

so that the Levi-Civita connection ∇^{LC} on TX is

$$\nabla^{LC}_{\frac{\partial}{\partial x_i}} (\frac{\partial}{\partial x_j}) = \Gamma^k_{i,j} \frac{\partial}{\partial x_k}$$

If $u(x) = \Sigma u^j(x) \frac{\partial}{\partial x_j}$ is a section of TX, then $< p, u >= \Sigma u^j p_j$ is a function on Σ, and one has the identity

$$e_i(< p, u >) =< p, \nabla^{LC}_{\frac{\partial}{\partial x_i}} u >$$

The vectors fields e_i are tangents to the subvarieties $|p|^2 = Cte$, and the Hamiltonian vector field of the function $|p|^2/2$ on the symplectic variety Σ is equal to

$$H_{|p|^2/2} = g^{i,j} p_j e_i \in T^H\Sigma \tag{1.2}$$

We equip $T\Sigma$ with the Riemannian metric such that the splitting (1.1) is orthogonal, the metrics on $T^H\Sigma, T^V\Sigma$ being the canonical metrics associated to the isomorphisms $T^H_{x,p}\Sigma \simeq T_xX$ et $T^V_{x,p}\Sigma \simeq T^*_xX$. We denote by ∇ the associated connection.

Set $\Gamma = \Sigma \Gamma_j dx^j$. The Riemann curvature tensor R is the 2-form with values in $End(TX)$

$$R = R_{j,k} dx^j \wedge dx^k$$

$$R_{j,k} = \frac{\partial \Gamma_k}{\partial x_j} - \frac{\partial \Gamma_j}{\partial x_k} + [\Gamma_j, \Gamma_k]$$

For $u(x) = u^j(x)\frac{\partial}{\partial x_j}, v(x) = v^j(x)\frac{\partial}{\partial x_j}$, one has

$$R(u,v) = R_{j,k} u^j v^k = \nabla_u^{LC} \nabla_v^{LC} - \nabla_v^{LC} \nabla_u^{LC} - \nabla_{[u,v]}^{LC} \in End(TX)$$

and the Riemann tensor R satisfies the following symmetries

$$\begin{aligned} R(u,v) &= -R(v,u) \\ (R(u,v)w|z) &= -(w|R(u,v)z) \end{aligned} \tag{1.3}$$

the second identity being consequence of $(uv - vu - [u,v])(w|z) = 0$ and $u(w|z) = (\nabla_u^{LC} w|z) + (w|\nabla_u^{LC} z)$. The bracket $[e_j, e_k] = e_j e_k - e_k e_j$ of the vector fields e_j, e_k is the vertical vector field on Σ

$$[e_j, e_k] = {}^t R_{j,k}[e_j, e_k] = R^{\alpha}_{j,k,\beta}\, p_\alpha \frac{\partial}{\partial p_\beta} \in T^V \Sigma \tag{1.4}$$

where $p_\alpha \frac{\partial}{\partial p_\beta} \in End(T^*X)$ acts as $p_\alpha \frac{\partial}{\partial p_\beta}(\Sigma p_j \hat{e}^j) = p_\alpha \hat{e}^\beta$.

2 The definition of the Bismut operator on forms

In this section, we recall the construction of the hypoelliptic Laplacian. This operator depends on an deformation parameter $s \in]0, \infty[$. This parameter is a coupling constant related to the probabilistic interpretation of the hypoelliptic Laplacian. The limit $s = 0$ is associated to a deterministic quantization of the geodesic flow on X, and the limit $s = \infty$ is related to the Brownian motion on X. We denote by $\Lambda^p(Y)$ the bundle of differential p-forms on a variety Y and $\Lambda^\cdot(Y) = \oplus_p \Lambda^p(Y)$. The construction of the hypoelliptic Laplacian on T^*X is similar to the one of the standard Hodge Laplacian on X, which involves the De Rham operator on the exterior algebra of X, and the canonical metric on $\Lambda^\cdot(X)$ induced by the Riemannian metric on the 1-forms bundle $\Lambda^1(X)$. However, it differs by the choice of a non degenerate but non symmetric bilinear form on the 1-forms bundle $\Lambda^1(T^*X)$.
Let σ be the canonical symplectic 2-form on $\Lambda^1(\Sigma) = T^*\Sigma$ (dual to the symplectic form on $T\Sigma$) and π the projection from $T^*_{x,p}\Sigma$ onto $T_x X$ (the restriction to the tangent vertical space of Σ at (x,p) of a 1-form on Σ is a linear form on $T^*_x X$ thus a tangent vector to X at x). One defines a bilinear form b_s on $\Lambda^1(\Sigma)$ by setting

$$b_s(\omega, \omega') = \sigma(\omega, \omega') + \sqrt{s}(\pi(\omega)|\pi(\omega')) \tag{2.1}$$

In local coordinates, one has

$$b_s(\omega,\omega') = \beta'^i\alpha_i - \alpha'_i\beta^i + \sqrt{s}g_{i,j}\beta^i\beta'^j$$

$$\omega = \alpha_i dx^i + \beta^i dp_i, \ \omega' = \alpha'_i dx^i + \beta'^i dp_i$$

We denote by $b^{\cdot}{}_s$ the bilinear form on the exterior algebra $\Lambda^{\cdot}(\Sigma)$ such that its restriction on p-forms is the pth power of b_s. Then $b^{\cdot}{}_s$ depends only on the metric g. Note that $b^{\cdot}{}_s$ is non degenerate, but is a sum of a symmetric and a skew-symmetric bilinear form.
If A is a differential operator on $\Lambda^{\cdot}(\Sigma)$, its (right) adjoint with respect to $b^{\cdot}{}_s$ is the operator A^*_b

$$\int b^{\cdot}{}_s(A\omega,\omega')dxdp = \int b^{\cdot}{}_s(\omega, A^*_b\omega')dxdp$$

Let d be the De Rham operator on $\Lambda^{\cdot}(\Sigma)$. In particular, one has

$$d^*_b(\omega_x dx + \omega_p dp) = \frac{\partial\omega_x}{\partial p} - \frac{\partial\omega_p}{\partial x} - \sqrt{s}g_{i,j}\frac{\partial\omega_{p,j}}{\partial p_i}$$

$$(\Omega\wedge)^*_b(\omega_x dx + \omega_p dp) = \omega_p\Omega_x - \omega_x\Omega_p + \sqrt{s}g(\Omega_p,\omega_p)$$

Let h be a function on Σ, and δ_h the conjuguate operator

$$\delta_h = e^{-h}de^h = d + dh\wedge$$

One has $(\delta_h)^*_b = e^h d^*_b e^{-h}$, and we denote by D_h the Dirac operator

$$D_h = \delta_h + (\delta_h)^*_b$$

so that

$$D^2_h = \delta_h(\delta_h)^*_b + (\delta_h)^*_b\delta_h$$

Definition 2.1. The hypoelliptic Laplacian of J.-M. Bismut is the differential operator on $\Lambda^{\cdot}(\Sigma)$

$$\frac{\sqrt{s}}{2}D^2_h = B_s \tag{2.2}$$

with the choice $h = |p|^2/2 = g^{i,j}(x)p_ip_j$.

Remark 2.2. In the above definition , the choice of the right adjoint in the definition of d^*_b is irrelevant; we may as well take the left adjoint: the two operators are related by the symmetry $p \to -p$. However, the Witten twist by the exponantial weight e^h is essential. The choice $h = |p|^2/2$ allows to quantize the geodesic flow on X; one can also replace h by a more general potential $h = |p|^2/2 + V(x)$ in order to study a different dynamics, but we will not do this here.

Observe that the definition 2.1 of the Bismut operator is quite simple and natural. However, the computation of this operator in local coordinates is not so simple. Let us introduce the differential forms on Σ

$$\hat{e}_j = dp_j - \Gamma^{\alpha}_{j,k}\, p_\alpha\, dx^k$$

$$e^i = dx^i$$

Let $\triangle_p$ be the vertical Laplacian associated to the constant coefficients in p metric $g^{-1}(x)$

$$\triangle_p = \Sigma g_{i,j}(x) \frac{\partial^2}{\partial p_i \partial p_j}$$

Let N_V be the number operator on the exterior algebra $\Lambda^{\cdot}(\Sigma)$ which counts the vertical degree in dp, let i_Y be the interior product by the vector field Y, let $\mathcal{L}_Y$ be the Lie derivative in the direction of the vector field Y, and let H_f be the Hamiltonian vector field of the function $f(x,p)$. The following lemma is the Weitzenbock formula for the Bismut operator (see [Bis05])

Lemma 2.3. *The following identity holds true*

$$B_s = \frac{s}{2}[-\triangle_p + |p|^2 + (2N_V - d) - \frac{1}{2} < R(e_i, e_j)e_k, e_l > e^i e^j i_{\hat{e}^k} i_{\hat{e}^l}] - \sqrt{s}\mathcal{L}_{H_{|p|^2/2}} \tag{2.3}$$

By the above formula, B_s is a second order differential operator, partially elliptic in p, involving only first order derivatives in x. Since the vertical derivatives ∂_{p_j} and the brackets $[\partial_{p_k}, \{|p|^2/2, .\}]$ span the tangent space to Σ , the Hörmander theorem implies that B_s is an hypoellic operator, and the same is true for the heat equation $\partial_t + B_s$.

The change of coordinates $q = \sqrt{s}\, p$ allows to rewrite B_s on the form (note that in the following formula, we use $\hat{e}^j(q) = \sqrt{s}\hat{e}^j(p)$ so the operator $i_{\hat{e}^k}(p)$ becomes $\sqrt{s} i_{\hat{e}^k}(q)$)

$$B_s = \frac{1}{2}[-s^2\triangle_q + |q|^2 + s(2N_V - d) - \frac{s^2}{2} < R(e_i, e_j)e_k, e_l > e^i e^j i_{\hat{e}^k} i_{\hat{e}^l}] - \mathcal{L}_{H_{|q|^2/2}} \tag{2.4}$$

In particular, one expect the asymptotics:

$$(A_s - \lambda)^{-1} \sim_{s \to 0} (1/2(-s^2\triangle_q + |q|^2) - \mathcal{L}_{H_{|q|^2/2}} - \lambda)^{-1} \tag{2.5}$$

$$\lim_{s \to \infty} (A_s - \lambda)^{-1} = \Pi_0(\Box_X/2 - \lambda)^{-1}\Pi_0 \tag{2.6}$$

where Π_0 is the L^2-orthogonal projector on the kernel $e^{-\frac{|p|^2}{2}}\Lambda^{\cdot}(X)$ of the harmonic oscillator $-\triangle_p + |p|^2 + (2N_V - d)$ and $\Box_X = (d_X + d_X^*)^2$ is the Hodge Laplacian on X.

In (2.5), the asymptotic is a semi-classical one, with s as a Planck constant. In (2.6), the limit is a limit of operators acting on appropriate spaces. The study of the asymptotic $s \to +\infty$ is done in [BL06]. However, the analysis of the limit $s \to 0$ remains to be done.

3 Hypoellipticity and the heat kernel

In this part, we give some analytical results on the hypoelliptic Laplacian, for s in a fixed compact subset of $]0,\infty[$. For simplicity, we will assume $s = 1$. Let h be a function on Σ, and $Z = H_h$ be the Hamiltonian vector field of h. The Lie derivative $\mathcal{L}_{H_h}$ acting on 1-forms is given by the formula

$$\begin{aligned} \mathcal{L}_{H_h}(\alpha_j dx^j + \beta^j dp_j) &= \alpha'_j dx^j + \beta'^j dp_j \\ \alpha'_j &= \{h, \alpha_j\} + \frac{\partial^2 h}{\partial x_j \partial p_k}\alpha_k - \frac{\partial^2 h}{\partial x_j \partial x_k}\beta^k \\ \beta'^j &= \{h, \beta^j\} + \frac{\partial^2 h}{\partial p_j \partial p_k}\alpha_k - \frac{\partial^2 h}{\partial p_j \partial x_k}\beta^k \end{aligned} \tag{3.1}$$

The matrix

$$\mathcal{N}_h = \begin{pmatrix} \frac{\partial^2 h}{\partial x \partial p} & -\frac{\partial^2 h}{\partial x \partial x} \\ \frac{\partial^2 h}{\partial p \partial p} & -\frac{\partial^2 h}{\partial p \partial x} \end{pmatrix}$$

is skew-adjoint for the symplectic structure. With the choice $h = |p|^2/2$ and the choice of basis $e^i = dx^i, \hat{e}_j = dp_j - \Gamma^{\alpha}_{j,k}\, p_\alpha\, dx^k$, for which e^i is homogeneous of degree 0 in p, and $\hat{e}_j$ homogeneous of degree 1, one gets that $\mathcal{N} = \mathcal{N}_{|p|^2/2}$ has the following homogeneity

$$\begin{aligned} \mathcal{N}(e^i) &= \mathcal{N}^{i,\alpha}_{1,l}(x) p_\alpha e^l + \mathcal{N}^{i,l}_2(x) \hat{e}_l \\ \mathcal{N}(\hat{e}_i) &= \mathcal{N}^{\alpha,\beta}_{3,i,l}(x) p_\alpha p_\beta e^l + \mathcal{N}^{l,\alpha}_{4,i}(x) p_\alpha \hat{e}_l \end{aligned} \tag{3.2}$$

The above formula (3.2) indicates that it is natural to introduce a weight on the exterior algebra so that $\mathcal{N}$ becomes a first order operator in p. If ω is a section of $\Lambda^{\cdot}(\Sigma)$, we write in local coordinates

$$\omega = \Sigma \omega^J_I e^I \hat{e}_J$$

where the $\omega^J_I(x,p)$ are functions on Σ. We thus get

$$N_V(\Sigma \omega^J_I e^I \hat{e}_J) = \Sigma \omega^J_I\ |J|\ e^I \hat{e}_J$$

We introduce the following L^2 structure on the space of sections of $\Lambda^{\cdot}(\Sigma)$

Definition 3.1. Let $dxdp$ be the canonical volume form on Σ, and $< p >= (1+|p|^2)^{1/2}$. For $\omega(x,p) = \Sigma_{0 \le j \le n} \omega_j(x,p)$, $N_V(\omega_j) = j\omega_j$ we define the weight norm

$$|\omega(x,p)|_w^2 = \Sigma |\omega_j|^2(x,p) < p >^{2j}$$

We denote by L^2 et L^2_w the space of sections of $\Lambda^{\cdot}(\Sigma)$ associated to the norms

$$||\omega||^2 = \int |\omega|^2(x,p)\ dxdp < \infty \tag{3.3}$$

$$||\omega||_w^2 = \Sigma \int |\omega_j|_w^2(x,p)\ dxdp < \infty \tag{3.4}$$

Let $M(x,p)$ be a section of $End(\Lambda^{\cdot}(\Sigma))$, and $d \in \mathbb{R}$. Then $M(x,p)$ is a symbol of degree d (resp. a w-symbol of degree d) if for any α, β, there exists $C_{\alpha,\beta}$ such that

$$|\nabla_{e_i}^{\alpha} \nabla_{\hat{e}j}^{\beta} M| \leq C_{\alpha,\beta} < p >^{d-|\beta|} \tag{3.5}$$

resp. in the weight case

$$|\nabla_{e_i}^{\alpha} \nabla_{\hat{e}j}^{\beta} M|_w \leq C_{\alpha,\beta} < p >^{d-|\beta|} \tag{3.6}$$

where $|M|_w$ is the norm of M associated to $|.|_w$.

From (3.2), (3.4), we get that the operator $\mathcal{L}_{H_{|p|^2/2}}$ has the following structure, with $N(x,p) \in End(\Lambda^{\cdot}(\Sigma))$

$$\begin{aligned} &\mathcal{L}_{H_{|p|^2/2}} = \nabla_{\{|p|^2/2,.\}} + N \\ &|| < p >^{-1} N(\omega)||_w \leq C||\omega||_w \end{aligned} \tag{3.7}$$

The vertical derivative ∂_{p_j} is given by

$$\partial_{p_j}(\Sigma \omega_I^J e^I \hat{e}_J) = \Sigma \partial_{p_j}(\omega_I^J)\ e^I \hat{e}_J \tag{3.8}$$

The vertical harmonic oscillator $\mathcal{O}$ is defined by

$$\begin{aligned} &\mathcal{O}(\Sigma \omega_I^J e^I \hat{e}_J) = \Sigma \mathcal{O}(\omega_I^J)\ e^I \hat{e}_J \\ &\mathcal{O} = \frac{1}{2}[-\Delta_p + |p|^2 + (2N_V - n)] \end{aligned} \tag{3.9}$$

Let ρ be the linear map, where $N_V(\omega_j) = j\omega_j$,

$$\rho(\Sigma_{0\leq j \leq n} \omega_j(x,p)) = \Sigma_{0\leq j\leq n} \omega_j(x,p) < p >^j$$

One has $||\rho(\omega)|| = ||\omega||_w$ and by (2.3) et (3.7), the conjuguate operator

$$\rho^{-1} B_1 \rho = B$$

has the following structure

$$\begin{aligned} &B = \mathcal{O} + \nabla_{\{|p|^2/2,.\}} + \mathcal{M} \\ &\mathcal{M} = \Sigma \partial_{p_j} M_0^j + \Sigma p_j M_1^j + M \end{aligned} \tag{3.10}$$

where the matrix $M^j_{0,1}(x,p), M(x,p)$ are symbols of degree 0. In the sequel, we will work with the conjuguate operator $B = \rho^{-1}B_1\rho$, and more generally with any operator B of the form (3.10), and thus we will only use the L^2 standard structure on $\Lambda^{\cdot}(\Sigma)$. One will find in [Leb05] and [Leb06] general results on operators of the form (3.10) (Geometric Fokker Planck equations).

For the analysis of an operator of the form (3.10), we will always use the following rules to evaluate the degree of operators

$$\begin{aligned} &\partial_{x_j} \text{ is of order } 1 \\ &p_j,\ \partial_{p_j} \text{ are of order } 1/2 \end{aligned} \tag{3.11}$$

This rules are the natural ones if one wants to keep the invariance by coordinates changes $y = \varphi(x)$ on X, so that ∂_x and $p\partial_p$ must be of the same degree. Moreover, it is natural to take p and ∂_p of the same degree, in view of the occurence of the harmonic oscillator in (3.10). With these rules, one has

$$\begin{aligned} &\mathcal{O} \text{ is self-adjoint on } L^2 \text{ of order } 1 \\ &\nabla_{\{|p|^2/2,.\}} = \Sigma g^{i,j} p_j \nabla_{e_i} \text{ is of order } 3/2 \\ &\text{and with skew-adjoint principal part on } L^2 \\ &\nabla_{\{|p|^2/2,.\}} + \nabla^*_{\{|p|^2/2,.\}} \text{ and } \mathcal{M} \text{ are at most of order} 1/2 \end{aligned} \tag{3.12}$$

Obviously, when p is bounded, $\mathcal{O}$ is the principal part of B. However, for the global analysis of B on Σ, it turns out that the principal part of B is the sum $\mathcal{O} + \nabla_{\{|p|^2/2,.\}}$, with self-adjoint component $\mathcal{O}$ of degree 1, and skew-adjoint component $\nabla_{\{|p|^2/2,.\}}$ of degree 3/2.

Let $\lambda = \alpha + i\beta \in \mathbb{C}$ be a spectral parameter. One define a scale of Sobolev spaces $\mathcal{H}^\sigma_\lambda$ of sections of $\Lambda^{\cdot}(\Sigma)$, with $\mathcal{H}^0_\lambda = L^2$, by duality and interpolation with the collection of first order operators

$$< p >^2 + |\alpha| + \frac{|\beta|}{< p >},\ \nabla_{e_i},\ < p > \nabla_{p_i} \tag{3.13}$$

Observe that by (3.13), $Re(\lambda) = \alpha$ is of degree 1 and $Im(\lambda) = \beta$ of degree 3/2, which is natural in view of (3.12). Let $\mathcal{S}'$ be the space of tempered sections of $\Lambda^{\cdot}(\Sigma)$. The following result is proved in [Leb06]. For $\delta_0 > 0, \delta_1 > 0$, let U_δ be the open subset of $\mathbb{C}$

$$U_\delta = \{\lambda \in \mathbb{C},\ Re(\lambda) + \delta_0 < \delta_1 |Im(\lambda)|^{1/2}\} \tag{3.14}$$

Theorem 3.2. *The operator B has compact resolvent and there exist $\delta_0 > 0, \delta_1 > 0$ such that its spectrum $\sigma(B)$ satisfies $\sigma(B) \subset \mathbb{C} \setminus U_\delta$. Moreover, for any $\sigma \in \mathbb{R}$ there exists C_σ, such that for all $\lambda \in U_\delta$, and all $u \in \mathcal{S}'$, $(B - \lambda)u \in \mathcal{H}^\sigma$ implies $u \in \mathcal{H}^{\sigma+2/3}$, $\mathcal{O}u \in \mathcal{H}^\sigma$, $\nabla_{\{|p|^2/2,.\}}u \in \mathcal{H}^\sigma$ and the following inequality holds true*

$$\begin{aligned} &||\mathcal{O}u||_{\lambda,\sigma} + ||(\nabla_{\{|p|^2/2,.\}} + iIm(\lambda))u||_{\lambda,\sigma} \\ &+ (|Re(\lambda)| + |Im(\lambda)|^{1/2})||u||_{\lambda,\sigma} + ||u||_{\lambda,\sigma+2/3} \\ &\leq C_\sigma ||(B-\lambda)(u)||_{\lambda,\sigma} \end{aligned} \tag{3.15}$$

Let $P(t,z,z')$, $z=(x,p)$ be the heat kernel associated to B, i.e the Green function of

$$\begin{aligned} &(\partial_t + B)u = 0 \ dans \ t > 0 \\ &u|_{t=0} = v \in L^2 \end{aligned} \tag{3.16}$$

As a byproduct of theorem 3.2, $P(t,z,z')$ is C^∞ in $t>0$, and all its t-derivatives belong to the Schwartz space in z,z'. The following results are proved in [Leb05]. Let $H(z,\zeta)$ be the Hamiltonian on $T^*\Sigma$

$$H(z,\zeta) = \frac{1}{2}(|\zeta^V|^2 - |p|^2) + (p|\zeta^H) \tag{3.17}$$

where ζ^V, ζ^H are the vertical and horizontal components of ζ. Let $\mathcal{L}$ be the action functional of the trajectory $s \in [0,t] \to x(s) \in X$

$$\mathcal{L} = \int_0^t (\frac{|a(s)|^2}{2} + \frac{|v(s)|^2}{2})\ ds \tag{3.18}$$

where $a(s), v(s)$ denotes acceleration and speed. Let $z_0 = (x_0, p_0) \in \Sigma$.

Definition 3.3. We define the large deviation function on $\mathbb{R}_+^* \times \Sigma \times \Sigma$ by

$$\mathcal{D}(t,z,z_0) = min \int_0^t \mathcal{L}(x(s))\ ds \tag{3.19}$$

where the minimum is taken over all trajectories $s \in [0,t] \to x(s)$ such that $(x(0), g(v(0))) = z_0$, $(x(t), g(v(t))) = z$

Theorem 3.4. *i) For all $t>0, z_0, z$, there exists a solution $s \in [0,t] \to z_{opt}(s)$ of the differential equation*

$$\begin{aligned} &-\frac{D}{Dt}\frac{Db}{Dt} + < b|R(v,.)v > + b = 0 \\ &v = \frac{dx}{dt}, \quad b = \frac{Dp}{Dt}, \quad p = g(v) \end{aligned} \tag{3.20}$$

connecting z_0 to z such that

$$\mathcal{D}(t,z,z_0) = \int_0^t \mathcal{L}(x_{opt}(s))\ ds \tag{3.21}$$

The function $t \to \mathcal{D}(t,z,z_0)$ is continuous on $\mathbb{R}_+^*$.
ii)Let $t_0 > 0$ small. There exist $0 < \delta_0 = \delta_1 \leq \delta_2$ such that $\mathcal{D}(t,z,z_0)$ is C^∞ and satisfies the Hamilton Jacobi equation

$$\partial_t \mathcal{D} + \frac{1}{2}((\partial_p \mathcal{D})^2 - |p|^2) + \{|p|^2/2, \mathcal{D}\} = 0 \tag{3.22}$$

in the domain

$$U_{t_0,\delta} = \{(t, z, z_0), t \in]0, t_0], dist(x, x_0) < \delta_0, |tp| < \delta_2, |tp_0| < \delta_1\} \tag{3.23}$$

with $z = (x, p)$, *and* $z_0 = (x_0, p_0)$. *Moreover, in the domain* $U_{t_0,\delta}$, $z \rightarrow \mathcal{D}(t, z, z_0)$ *admits a unique non degenerate minimum* $\gamma(t, z_0)$ *at* $z = Z_*(t, z_0)$. *One has*

$$\begin{aligned} \gamma(t, z_0) = min \int_0^t \mathcal{L}(x(s))\, ds = \\ \frac{t}{2}|p_0|^2 - \frac{t^3}{6}|p_0|^2 + \mathcal{O}(t^3|p_0|^4 + t^5|p_0|^2) \end{aligned} \tag{3.24}$$

where the minimum is taken over all solutions $s \in [0, t] \rightarrow x(s)$ *of* (3.20) *such that* $(x(0), g(v(0))) = z_0$ *with data* $(b_0, b_1 = \frac{Db}{Dt})$ *such that* $t^2|b_0| + t^3|b_1| \leq \delta_2$. *In geodesics coordinates centered at* x_0, *one has*

$$\begin{aligned} Z_*(t, z_0) = exp(tH_{|p|^2/2})(z_0) + \\ (-p_0 t^3/6 + \mathcal{O}(t^5|p_0|), -p_0 t^2/2 + \mathcal{O}(t^4|p_0|)) \end{aligned} \tag{3.25}$$

iii) In geodesics coordinates centered at x_0, *one has for* $(t, z, z_0) \in U_{t_0,\delta}$, *with* $z = Z_*(t, z_0) + (tX, P)$ *and* $|(X, P)|$ *small*

$$\begin{aligned} \mathcal{D}(t, z, z_0) =& \gamma(t, z_0) + \frac{2}{t}(3X^2 - 3PX + P^2) \\ &+ \mathcal{O}(t(1 + |p_0|^2 + (X, P)^2)(X, P)^2) \end{aligned} \tag{3.26}$$

Remark 3.5. It is interesting to developp the analogy between geometric Fokker Planck equations on $\Sigma = T^*X$ and Laplace equations on X, for example $-\frac{\triangle_X}{2}$. For the Laplace equation, the action is

$$\int_0^t \frac{|v|^2}{2}\, ds$$

and the large deviation function $\mathcal{D}_X = \frac{d_X^2(x,x_0)}{2t}$, which satisfies the Hamilton Jacobi equation

$$\partial_t \mathcal{D}_X + \frac{|\partial_x \mathcal{D}_X|^2}{2} = 0,$$

is C^∞ in a neighborhood of x_0 with a zero and non degenerate minimum at $x = x_0$. The differential equation (3.20) is thus a generalisation of the geodesic equation on X, and the local properties near z_0 of $\mathcal{D}$ are similar to the ones of $\mathcal{D}_X$, except that its minimum is shifted by the Hamiltonian of the function $|p|^2/2$.

In order to describe the asymptotics of the heat kernel $P(t, z, z_0)$ for z closed to z_0 and $t \to 0$, we work in geodesic coordinates centered at x_0. Let $Z_*(t, z_0) = (x_{z_0}(t), p_{z_0}(t))$. We introduce the renormalized coordinates (y, q) centered at z_0

$$z = (x, p) = (x_{z_0}(t) + ty, p_{z_0}(t) + q) \tag{3.27}$$

By (3.26), the large deviation function $\mathcal{D}$ satisfies, for some constant $C > 0$, for (y, q) near $(0, 0)$ and $t > 0$ small

$$\mathcal{D} \geq \gamma(t, z_0) + \frac{C}{t}(y^2 + q^2) \tag{3.28}$$

In these coordinates, let $\mathcal{C}^d(\Lambda^{\cdot}(\Sigma))$ be the space of C^∞ functions $f(t, y, q, z_0)$ with values in $f(t, y, q, z_0) \in End(\Lambda^{\cdot}(\Sigma))$, defined for $(t, z_0) \in V_{t_0,\delta_1}$, $|(y, q)| \leq \delta$, and $t_0 > 0, \delta_1 > 0, \delta > 0$ small, and $V_{t_0,\delta_1} = \{t \in [-t_0, t_0],\ |tp_0| \leq \delta_1\}$, such that for all l, α, β, γ, there exists C such that uniformly in (t, z_0, y, q) one has

$$|\partial_t^l \partial_{y,q}^\gamma \nabla_{e_i,z_0}^\alpha \nabla_{\hat{e}^j,z_0}^\beta f| \leq C < p_0 >^{d+l-|\beta|} \tag{3.29}$$

For example, if $M(z_0)$ is a symbol of degree d then

$$f(t, y, q, z_0) = M(x_{z_0}(t) + ty, p_{z_0}(t) + q)$$

belongs to $\mathcal{C}^d(\Lambda^{\cdot}(\Sigma))$, since

$$x_{z_0}(t) + ty \in \mathcal{C}^0, \quad p_{z_0}(t) + q \in \mathcal{C}^1 \tag{3.30}$$

Let $\theta(y, q)$ be a cutoff with support in $|(y, q)| \leq \delta$, equal to 1 near $(0, 0)$. Let $\phi(u)$, $u \in \mathbb{R}$, a cutoff with support in $|u| \leq \delta_1$, equal to 1 in $|u| \leq \delta_1/2$. The following result, proved in [Leb05], gives the small time asymptotic of the heat kernel. Observe that by (3.26), the hypoelliptic diffusion is anisotropic in variables x, p, which is natural for a Kolmogorov type equation.

Theorem 3.6. *For any integer j, there exist*

$$c_j(t, y, q, z_0) \in \mathcal{C}^j(\Lambda^{\cdot}(\Sigma))$$

and for all N, an operator R_N with kernel $R_N(t, z, z_0)$ such that

$$\begin{aligned} P(t, z, z_0) &= P_N(t, z, z_0) + R_N(t, z, z_0) \\ P_N(t, z, z_0) &= t^{-2n}\phi(t|p|)e^{-\mathcal{D}(t,z,z_0)}(\Sigma_{0\leq j\leq N} t^j c_j(t, y, q, z_0))\theta(y, q)\phi(t|p_0|) \end{aligned} \tag{3.31}$$

where the sequence of operators R_N is such that for all $\sigma > 0, M > 0$, there exist N and a constant C such that for $t \in]0, t_0]$ one has

$$||R_N(u)||_\sigma \leq Ct^M\ ||u||_{-\sigma} \tag{3.32}$$

where $||u||_{-\sigma}$ is the norm in the Sobolev space $\mathcal{H}_0^{-\sigma}$.

4 Convergence towards the Hodge Laplacian

In this section, we describe the limit of the hypoelliptic Laplacian when $s \to \infty$. The following results are proven in [BL06]. After an algebraic transformation on the exterior algebra, on is reduced to the study of a familly of operators B_s of the form

$$\begin{aligned} B_s &= s\mathcal{O} + \sqrt{s}\beta + M_2 \\ \mathcal{O} &= \frac{1}{2}[-\Delta_p + |p|^2 + (2N_V - n)] \\ \beta &= -\{|p|^2/2, .\} + \Sigma M_0^j \frac{\partial}{\partial p_j} \end{aligned} \tag{4.1}$$

where the M_k are polynomials in p of degree k. We set $h = \frac{1}{\sqrt{s}}$, so that with the notation $Q_h = \mathcal{O} + h\beta + h^2 M_2$ one has

$$B_s = sQ_h$$

Let Π_0 be the L^2-orthogonal projector on the kernel of the harmonic oscillator $\mathcal{O}$, and $\Pi_\perp = 1 - \Pi_0$. Let $\lambda \in \mathbb{C}$ be the spectral parameter and $\nu = \lambda/s$. Here we will work with the scale of semiclassical Sobolev spaces build on L^2 with the collection of first order operators

$$< p >^2 + \frac{|\nu|}{< p >}, \quad h\nabla_{e_i}, \quad < p > \nabla_{p_i} \tag{4.2}$$

We denote by $||u||_{\nu,sc,\sigma}$ the associated norm, and by $\mathcal{H}^\sigma$ the corresponding Hilbert space (only the norm depends on the parameters ν, h). For any σ, Π_0 maps $\mathcal{H}^\sigma$ in $\mathcal{H}^\sigma$, and there exists C_σ such that

$$||\Pi_0(u)||_{\nu,sc,\sigma} \leq C_\sigma ||u||_{\nu,sc,\sigma} \tag{4.3}$$

Using the orthogonal splitting $L^2 = Ker(\mathcal{O}) \oplus Ker(\mathcal{O})^\perp$, one write the operators on the form of 2×2 matrix

$$\mathcal{O} = \begin{pmatrix} 0 & 0 \\ 0 & \mathcal{O} \end{pmatrix} \qquad \beta = \begin{pmatrix} 0 & \beta_2 \\ \beta_3 & \beta_4 \end{pmatrix} \quad M_2 = \begin{pmatrix} M_{2,1} & M_{2,2} \\ M_{2,3} & M_{2,4} \end{pmatrix} \tag{4.4}$$

so we get

$$B_s = \begin{pmatrix} P_1 & \sqrt{s}P_2 \\ \sqrt{s}P_3 & sP_4 \end{pmatrix}. \tag{4.5}$$

with

$$\begin{aligned} P_1 &= M_{2,1} \\ P_2 &= \beta_2 + hM_{2,2} \\ P_3 &= \beta_3 + hM_{2,3} \\ P_4 &= \mathcal{O} + h\beta_4 + h^2 M_{2,4} \end{aligned} \tag{4.6}$$

Let $\lambda \in \mathbb{C}$; we define the operators Θ, and T (for Θ invertible) by the formulas

$$\begin{aligned} \Theta &= P_4 - \lambda/s \\ T &= P_1 - P_2\Theta^{-1}P_3 \end{aligned} \tag{4.7}$$

Then the resolvent equation

$$(\lambda - B_s)(u_0, u_\perp) = (v_0, v_\perp)$$

is formally equivalent to

$$\begin{pmatrix} u_0 \\ u_\perp \end{pmatrix} = \begin{pmatrix} (\lambda - T)^{-1} & -\frac{1}{\sqrt{s}}(\lambda - T)^{-1}P_2\Theta^{-1} \\ -\frac{1}{\sqrt{s}}\Theta^{-1}P_3(\lambda - T)^{-1} & -\frac{\Theta^{-1}}{s}[1 - P_3(\lambda - T)^{-1}P_2\Theta^{-1}] \end{pmatrix} \begin{pmatrix} v_0 \\ v_\perp \end{pmatrix} \tag{4.8}$$

and it remains to analyse the above formula. We do this in two steps : first, analysis of the inverse of Θ, then analysis of the operator $(\lambda - T)^{-1}$.
The analysis of Θ is technical, and we refer to ([BL06]); it is done by a parametrix construction in a suitable pseudodifferential spaces of operators in x with values in operators in p. As a byproduct, one gets that Θ^{-1} increases the regularity in x by a factor 2/3, and moreover $\Pi_0\Theta^{-1}$ (resp $\Pi_0\Theta^{-1}\Pi_0$) increases the regularity in x by a factor 5/6 (resp 1), the extra regularity in x by a factor 1/6 (resp 1/3) being consequence of an averaging lemma, an usual fact in kinetic theory. The parametrix study allows to get a precise description of T, as stated in the following lemma.

Let $\delta = (\delta_0, \delta_1, \delta_2)$, $\delta_0 \in \mathbb{R}, \delta_1 > 0, \delta_2 > 0$, and let $\mathcal{U}_\delta$ be the domain in $\mathbb{C}$

$$\mathcal{U}_\delta = \{\nu; Re(\nu) \leq \delta_0 + \delta_1 |Im(\nu)|^{\delta_2}\}$$

A symbol $a(x, \xi, h, \nu)$ of degree d associated to a pseudodifferential operator in the class $\mathcal{E}^d_{\delta,h}$ is a function of (x, ξ), holomorphic in $\nu \in \mathcal{U}_\delta$, with parameter $h \in]0, h_0]$, with values in $End(\Lambda^\cdot(T^*X))$, such that for all α, β, there exists $C_{\alpha,\beta}$ independent of h, ν such that

$$|\partial_x^\alpha \partial_\xi^\beta a(x, \xi, h, \nu)| \leq C_{\alpha,\beta}(1 + |\nu| + |\xi|)^{d+1/3|\alpha|-2/3|\beta|} \tag{4.9}$$

In local coordinates and given trivialization, one quantizes a symbol a in an operator $A = Op(a)$ by the usual formula

$$A(x, hD_x, h, \nu)u(x) = (2\pi h)^{-n} \int e^{\frac{i}{h}(x-y)\xi} a(x, \xi, h, \nu)u(y)\, dy d\xi$$

We denote also by $\mathcal{E}^d_{\delta,h,0}$ the set of operators associated to symbols such that the following estimates hold true, instead of (4.9)

$$|\partial_x^\alpha \partial_\xi^\beta a(x, \xi, h, \nu)| \leq C_{\alpha,\beta}(1 + |\nu| + |\xi|)^{d-|\beta|} \tag{4.10}$$

Lemma 4.1. *There exists* δ *with* $\delta_0 > 0, \delta_2 = 1/6$, $A^0_{j,k}, C^0 \in \mathcal{E}^{-1}_{\delta,h,0}$ *and* $A^1_{j,k}, B^1_{k,.}, C^1 \in \mathcal{E}^{-1}_{\delta,h}$ *such that, with* $\nabla_j = \nabla_{\frac{\partial}{\partial x_j}}$, *one has*

$$\begin{aligned} T &= T^0 + hT^1 \\ T^0 &= \Sigma_{j,k} \nabla^*_j A^0_{j,k} \nabla_k + M_{2,1}(x) + C^0 \\ T^1 &= \Sigma \nabla^*_j A^1_{j,k} \nabla_k + \Sigma(B^1_{k,\ell} \nabla_k + \nabla^*_k B^1_{k,r}) + C^1 \end{aligned} \tag{4.11}$$

The operators $A^0_{j,k}$ *have scalar principal symbol* $a_{j,k}(x, \xi, \nu)$, *and the following identity holds true*

$$\Sigma a_{j,k}(x,\xi,\nu)\xi_j\xi_k - \nu = J_0^{-1}(|y|, \nu) \tag{4.12}$$

with $y = \frac{\xi}{\sqrt{2}}$, $|y|^2 = g^{j,k}(x) y_j y_k$ *and*

$$J_0(y,\nu) = \int_0^1 (t_+)^{y^2-\nu-1} \, e^{(1-t)y^2} \, dt$$

Remark 4.2. The estimation $\delta_2 = 1/6$ is not optimal: this comes from the fact that we are not precise enough in the analysis of the remainders terms in the parametrix of Θ. The function $J_0(y,\nu)$ is naturally associated to the calculus with a constant coefficient metric frozen at the given point x_0. Observe that formula (4.11) shows that T looks like a Laplacian, but where the coefficients of the metric are h-pseudodifferential operators of degree -1.

The key point is now the following of J.-M.Bismut (see [Bis05])

Theorem 4.3.

$$T^0|_{\xi=0,\nu=0} = \frac{1}{2}\Box_X \tag{4.13}$$

Let $Spec_X$ be the spectrum of the half Hodge Laplacian $\frac{1}{2}\Box_X$. The following theorem gives the convergence of the resolvent $(T-\lambda)^{-1}$ (take care of the fact that T depends on λ !) towards the resolvent $(\frac{1}{2}\Box_X - \lambda)^{-1}$. Let

$$\mathcal{V}_{\delta,s} = \{\lambda;\ Re(\lambda) \leq s\delta_0 + \delta_1 \ s^{5/6} |Im(\lambda)|^{1/6}\}$$

$$\begin{aligned} \Lambda &= (1 + |\lambda| + \Box_X/2)^{1/2} \\ \Lambda_h &= (1 + |\nu|^2 + h^2\Box_X/2)^{1/2} \\ ||u||_{(\sigma_1,\sigma_2)} &= ||\Lambda^{\sigma_1}\Lambda_h^{\sigma_2} u||_{L^2} \end{aligned} \tag{4.14}$$

For $\lambda \in \mathcal{V}_{\delta,s}$, let $\rho(\lambda)$ be the function

$$\begin{aligned} \rho(\lambda) &= 1 + \frac{1}{|\lambda|} \quad si \ \ Re(\lambda) < 0 \\ \rho(\lambda) &= 1 + \frac{1}{|\lambda|} + \frac{1 + Re(\lambda)}{dist(\lambda, Spec_X)} \quad si \ \ Re(\lambda) \geq 0 \end{aligned} \tag{4.15}$$

The function $\rho(\lambda)$ is equal to $+\infty$ iff $\lambda \in Spec_X$.

Theorem 4.4. *Let $\delta_0 > 0, \delta_1 > 0, h_0 > 0$ small.*
i) There exist C such that for any $\lambda \in \mathcal{V}_{\delta,s} \setminus Spec_X$ and $h \in]0, h_0]$ such that

$$h\rho(\lambda) \leq C \tag{4.16}$$

the "resolvent" $(T-\lambda)^{-1}$ exists as a bounded operator from $\mathcal{H}_X^{\sigma}$ to $\mathcal{H}_X^{\sigma+1}$, for any σ. Moreover, for all (σ_1, σ_2), there exists $C_{(\sigma_1,\sigma_2)}$ such that for all $v \in \mathcal{H}_X^{\sigma_1+\sigma_2+1}$ and all h, λ such that (4.16) holds true, one has

$$||(T-\lambda)^{-1}(\frac{1}{2}\Box_X - \lambda)(v) - v||_{(\sigma_1,\sigma_2)} \leq C_{(\sigma_1,\sigma_2)}\frac{1}{\sqrt{s}}||(\rho(\lambda)+(\Box_X)^{1/2})(v)||_{(\sigma_1,\sigma_2)} \tag{4.17}$$

ii)For all $r > 0$, there exists h_r, such that the "resolvent" $(T-\lambda)^{-1}$ exists for all $h \in]0, h_r]$ and all λ such that

$$\lambda \in \mathcal{V}_{\delta,s}, \quad r(Re(\lambda)+1) \leq |Im(\lambda)| \tag{4.18}$$

Moreover, there exists C_r and for all (σ_1, σ_2), there exists $C_{(\sigma_1,\sigma_2)}$ such that for all $u \in \mathcal{H}_X^{\sigma_1+\sigma_2}$ and all $h \in]0, h_r], \lambda$ such that (4.18) holds true, one has

$$||(T-\lambda)^{-1}(u)||_{(\sigma_1+1,\sigma_2)} \leq C_{(\sigma_1,\sigma_2)}(s^{-1/2} + C_r(1+|\lambda|)^{-1/2})||u||_{(\sigma_1,\sigma_2)} \tag{4.19}$$

By the above theorem, estimates on Θ^{-1}, and the identity (4.11) one gets in particular that the convergence result (2.6) holds true:

Theorem 4.5. *Let K be a compact subset of $\mathbb{C} \setminus Spec_X$. There exist $s_0, C > 0$ such that for $s \geq s_0$ and $\lambda \in K$, the resolvent $(B_s - \lambda)^{-1}$ exists as a bounded operator on L^2, one has $\|(B_s-\lambda)^{-1}\| \leq C$ and*

$$\|(B_s - \lambda)^{-1} - i(\Box_X/2 - \lambda)^{-1}\Pi_0\| \leq \frac{C}{\sqrt{s}} \tag{4.20}$$

References

[Bis04a] J.-M. Bismut. Le Laplacien hypoelliptique. In *Séminaire: Équations aux Dérivées Partielles, 2003–2004*, Sémin. Équ. Dériv. Partielles, pages Exp. No. XXII, 15. École Polytech., Palaiseau, 2004.

[Bis04b] J.-M. Bismut. Le Laplacien hypoelliptique sur le fibré cotangent. *C. R. Math. Acad. Sci. Paris Sér. I*, 338:555–559, 2004.

[Bis04c] J.-M. Bismut. Une déformation de la théorie de Hodge sur le fibré cotangent. *C. R. Acad. Sci. Paris Sér. I*, 338:471–476, 2004.

[Bis05] J.-M. Bismut. The hypoelliptic Laplacian on the cotangent bundle. *To appear in J.A.M.S.*, 2005.

[BL06] J.-M. Bismut and G. Lebeau. The hypoelliptic Laplacian and Ray-Singer metrics. *to appear*, 2006.

[DV01] L. Desvillettes and C. Villani. On the trend to equilibrium in spatially inhomogeneous entropy dissipating systems: the linear Fokker Planck equation. *CPAM*, 54:1–42, 2001.

[HN04] F. Hérau and F. Nier. Isotropic hypoellipticity and trend to equilibrium for Fokker-Planck equations with high degree potential. *Arch. Ration. Mecha. Anal.*, 171(2):151–218, 2004.

[HN05] B. Helffer and F. Nier. Hypoellipticity and spectral theory for Fokker-Planck operators and Witten Laplacians. *Lect. Notes in Math.*, 1862, 2005.

[HSS05] F. Hérau, J. Sjostrand, and C. Stolk. Semiclasical analysis for the Kramers-Fokker-Planck equation. *CPDE*, 30:689–760, 2005.

[Leb05] G. Lebeau. Geometric Fokker Planck equations. *Portugaliae Mathematica*, 62(4), 2005.

[Leb06] G. Lebeau. Equations de Fokker Planck géométriques II: estimations hypoelliptiques maximales. *to appear*, 2006.

Commuting differential operators with regular singularities

Toshio Oshima

Graduate School of Mathematical Sciences, University of Tokyo, 3-8-1, Komaba, Meguro-ku, Tokyo 153-8914, Japan
oshima@ms.u-tokyo.ac.jp

Summary. We study a system of partial differential equations defined by commuting family of differential operators with regular singularities. We construct ideally analytic solutions depending on a holomorphic parameter. We give some explicit examples of differential operators related to $SL(n,\mathbb{R})$ and completely integrable quantum systems.

Key words: differential operators, regular singularity, completely integrable systems

1 Introduction

The invariant differential operators on a semisimple symmetric space have regular singularities along the boundaries of the space which is realized in a compact manifold by [O6]. In the case of a Riemannian symmetric space G/K, the study of such operators in [KO] enables [K–] to have the Poisson integral expression of any simultaneous eigenfunction of the operators. Here G is a connected real semisimple Lie group with finite center and K is its maximal compact subgroup.

In the case of the group manifold G, which is an example of a semisimple symmetric space, Harish-Chandra gives an asymptotic expansion of a right and left K-finite eigenfunction, which plays an important role in the harmonic analysis on G (cf. [Ha]). He uses only the Casimir operator to get the asymptotic expansion, which suggests us that one operator controls other operators together with some geometry.

On the other hand, the Schrödinger operator corresponding to Calogero-Moser-Sutherland system with a trigonometric potential function (cf. [Su]) or a Toda finite chain (cf. [To]) is completely integrable and the integrals with higher orders are uniquely characterized by the Schrödinger operator

Received 2 June, 2006. Revised 26 July, 2006. Accepted 31 July, 2006.

and so are the simultaneous eigenfunctions. These integrals also have regular singularities at infinity.

In this note we study a general commuting system of differential operators with regular singularities by paying attention to the fact that an operator characterizes the system. Our argument used in this note is based on expansions in power series and hence it is rather elementary compared to that in [KO] and [O4] where a microlocal method is used.

In fact we will study matrices of differential operators which may not commute with others in the system but satisfy a certain condition because it is better to do so even in the study of commuting scalar differential operators. Some of its reasons will be revealed in the proof of Theorem 4.1, that of Theorem 6.3, Remark 4.3 ii) etc.

In §2 we study differential operators which commute one operator. We will see that the *symbol map* σ_* plays an important role. In the case of the first example above the map corresponds to Harish-Chandra's isomorphism of the invariant differential operators. In the case of the Schrödinger operator above it corresponds to the commutativity among the integrals with higher orders.

In §3 we construct some of multivalued holomorphic solutions of the system around the singular points which we call *ideally analytic solutions* and then in §4 we study the *induced equations* of other operators, which assures that the solutions automatically satisfy some other differential equations.

In §5 we study the holonomic system of differential equations with constant coefficients holomorphically depending on a parameter, which controls the *leading terms* of the ideally analytic solutions.

In §6 we study a *complete system of differential equations with regular singularities* which means that the system is sufficient to formulate a boundary value problem along the singularities and we describe all the ideally analytic solutions. In particular, when the system has a holomorphic parameter, we construct solutions depending holomorphically on the parameter. It is in fact useful to introduce a parameter for the study of a specific system by holomorphically deforming it to generic simpler ones.

In §7 and §8 we give some explicit examples of the systems related to $SL(n,\mathbb{R})$ and the completely integrable quantum systems with regular singularities at infinity, respectively. Moreover we give Theorem 8.1 in the case of completely integrable quantum systems with two variables.

2 Commuting differential operators with regular singularities

For a positive integer m and a ring R we will denote by $M(m,R)$ the ring of square matrices of size m with components in R and by $R[\xi]$ the ring of polynomials of n indeterminates $\{\xi_1,\dots,\xi_n\}$ if $\xi = (\xi_1,\dots,\xi_n)$. The (i,j)-component of $A \in M(m,R)$ is denoted by A_{ij} and we naturally identify $M(1,R)$ with R.

Let M be an $(n+n')$-dimensional real analytic manifold and let N_i be one-codimensional submanifolds of M such that $N_1, \dots, N_n$ are normally crossing at $N = N_1 \cap \dots \cap N_n$. We assume that M and N are connected. We will fix a local coordinate system $(t, x) = (t_1, \dots, t_n, x_1, \dots, x_{n'})$ around a point $x^o \in N$ so that N_i are defined by the equations $t_i = 0$, respectively.

Let $\mathcal{A}_N$ denote the space of real analytic functions on N and $\mathcal{A}_M$ the space of real analytic functions defined on a neighborhood of N in M. For $\alpha = (\alpha_1, \dots, \alpha_n)$, $\beta = (\beta_1, \dots, \beta_n) \in \mathbb{Z}^n$ we put

$$|\alpha| = \alpha_1 + \dots + \alpha_n,$$
$$\alpha < \beta \Leftrightarrow \alpha_i \le \beta_i \quad \text{for } i = 1, \dots, n \text{ and } \alpha \ne \beta.$$

Let $\mathbb{N}$ be the set of non-negative integers. We will denote

$$\begin{cases} \vartheta_i = t_i \frac{\partial}{\partial t_i}, \quad \partial_x = \left(\frac{\partial}{\partial x_1}, \dots, \frac{\partial}{\partial x_{n'}}\right), \\ \vartheta^\alpha = \vartheta_1^{\alpha_1} \cdots \vartheta_n^{\alpha_n} \quad \text{for } \alpha = (\alpha_1, \dots, \alpha_n) \in \mathbb{N}^n, \\ \partial_x^\beta = \frac{\partial^{|\beta|}}{\partial x_1^{\beta_1} \cdots \partial x_{n'}^{\beta_{n'}}} \quad \text{for } \beta = (\beta_1, \dots, \beta_{n'}) \in \mathbb{N}^{n'}, \\ t^\lambda = t_1^{\lambda_1} \cdots t_n^{\lambda_n} \quad \text{for } \lambda = (\lambda_1, \dots, \lambda_n) \in \mathbb{C}^n. \end{cases}$$

Let $\mathcal{D}_M$ and $\mathcal{D}_N$ denote the rings of differential operators on M and N with coefficients in $\mathcal{A}_M$ and $\mathcal{A}_N$, respectively.

Definition 2.1. *Let $\widetilde{\mathcal{D}}_*$ denote the subring of $\mathcal{D}_M$ whose elements P have the form*

$$P = \sum_{(\alpha,\beta) \in \mathbb{N}^{n+n'}} a_{\alpha,\beta}(t, x) \vartheta^\alpha \partial_x^\beta \quad \textit{with} \quad a_{\alpha,\beta}(t, x) \in \mathcal{A}_M. \tag{1}$$

Here the sum above is finite. Moreover $\mathcal{D}_*$ denotes the subring of $\widetilde{\mathcal{D}}_*$ whose elements P of the form (1) *satisfy*

$$a_{\alpha,\beta}(0, x) = 0 \quad \textit{if} \quad \beta \ne 0. \tag{2}$$

When P is an element of $\mathcal{D}_*$, P is said to have regular singularities in the weak sense along the set of walls $\{N_1, \dots, N_n\}$ with the edge N (cf. [KO]).

Let define a map σ_* of $\widetilde{\mathcal{D}}_*$ to $\mathcal{D}_N[\xi]$ by

$$\sigma_*(P)(x, \xi, \partial_x) := \sum_{\alpha \in \mathbb{N}^n, \, \beta \in \mathbb{N}^{n'}} a_{\alpha,\beta}(0, x) \xi^\alpha \partial_x^\beta$$

for P in (1). *Then*

$$t^{-\lambda} P t^\lambda \phi(t, x) \big|_{t=0} = \sigma_*(P)(x, \lambda, \partial_x) \phi(0, x) \quad \textit{for } \phi \in \mathcal{A}_M \textit{ and } \lambda \in \mathbb{C}^n.$$

Here we note that the condition $P \in \widetilde{\mathcal{D}}_*$ equals

$$t^{-\lambda} P t^\lambda \phi(t, x) \in \mathcal{A}_M \quad \text{for} \quad \forall \phi(t, x) \in \mathcal{A}_M$$

and σ_* is a ring homomorphism of $\widetilde{\mathcal{D}}_*$ to $\mathcal{D}_N[\xi]$ and $\sigma_*(\mathcal{D}_*) = \mathcal{A}_N[\xi]$.

For $k \in \mathbb{N}$ and $P \in \widetilde{\mathcal{D}}_*$ with the form (1) we put

$$\sigma_k(P)(t,x,\xi,\tau) := \sum_{|\alpha|+|\beta|=k} a_{\alpha,\beta}(t,x)\xi^\alpha\tau^\beta$$

and then the order of P, which is denoted by $\operatorname{ord} P$, is the maximal integer k with $\sigma_k(P) \neq 0$.

For $P = \Big(P_{ij}\Big)_{\substack{1\le i\le m\\ 1\le j\le m}} \in M(m, \widetilde{\mathcal{D}}_*)$, the order of P is defined to be the maximal order of the components of P and denoted by $\operatorname{ord} P$. We put

$$\sigma(P) := \Big(\sigma_{\operatorname{ord} P}(P_{ij})\Big)_{\substack{1\le i\le m\\ 1\le j\le m}} \in M(m, \mathcal{A}_M[\xi,\tau]),$$

$$\sigma_*(P) := \Big(\sigma_*(P_{ij})\Big)_{\substack{1\le i\le m\\ 1\le j\le m}} \in M(m, \mathcal{D}_N[\xi]),$$

$$\bar\sigma_*(P) := \sigma(P)(0,x,\xi,\partial_x) \in M(m, \mathcal{D}_N[\xi]).$$

Then as a polynomial of ξ, $\bar\sigma_*(P)$ is the homogeneous part of $\sigma_*(P)$ whose degree equals $\operatorname{ord} P$. For P, $Q \in \widetilde{\mathcal{D}}_*$, we note that $\sigma(PQ) = \sigma(P)\sigma(Q)$ and

$$\begin{aligned}\sigma_{\operatorname{ord} P+\operatorname{ord} Q-1}([P,Q]) &= \sum_{i=1}^{n}\Big(\frac{\partial\sigma(P)}{\partial\xi_i}t_i\frac{\partial\sigma(Q)}{\partial t_i} - \frac{\partial\sigma(Q)}{\partial\xi_i}t_i\frac{\partial\sigma(P)}{\partial t_i}\Big)\\ &\quad+\sum_{j=1}^{n'}\Big(\frac{\partial\sigma(P)}{\partial\tau_j}\frac{\partial\sigma(Q)}{\partial x_j} - \frac{\partial\sigma(Q)}{\partial\tau_j}\frac{\partial\sigma(P)}{\partial x_j}\Big).\end{aligned}$$

Theorem 2.2. *Let P and Q be nonzero elements of $M(m, \widetilde{\mathcal{D}}_*)$ such that $[P,Q]=0$, $P \in M(m,\mathcal{D}_*)$ and $\sigma(P)$ is a scalar matrix satisfying*

$$\sum_{\nu=1}^{n}\gamma_\nu\frac{\partial\bar\sigma_*(P)}{\partial\xi_\nu} \not\equiv 0 \quad \textit{for } \forall\gamma\in\mathbb{N}^n\setminus\{0\}. \tag{3}$$

Here "$\not\equiv 0$" means "not identically zero". Suppose that $\sigma_{\operatorname{ord} P-1}(P)$ or $\sigma(Q)$ is a scalar matrix. Then $[\sigma_*(P),\sigma_*(Q)] = 0$ and $\bar\sigma_*(Q) \neq 0$. Moreover if $\sigma(P)(t,x,\xi,\tau)$ does not depend on t, so does $\sigma(Q)(t,x,\xi,\tau)$.

Proof. Since σ_* is an algebra homomorphism, $[\sigma_*(P),\sigma_*(Q)] = \sigma_*([P,Q]) = 0$.

Put $r_P = \operatorname{ord} P$ and $r_Q = \operatorname{ord} Q$. Fix i and j such that $\sigma_{r_Q}(Q_{ij}) \neq 0$. Note that the assumption implies

$$\sigma_{r_P+r_Q-1}([P,Q]_{ij}) = \sigma_{r_P+r_Q-1}([P_{11},Q_{ij}]).$$

Put

$$\sigma_{r_P}(P_{11}) = \sum_{\substack{\beta,\gamma\\ |\beta|\le r_P}} p_{\beta,\gamma}(x,\xi)t^\gamma\tau^\beta,\quad \sigma_{r_Q}(Q_{ij}) = \sum_{\substack{\beta,\gamma\\ |\beta|\le r_Q}} q_{\beta,\gamma}(x,\xi)t^\gamma\tau^\beta,$$

$$\sigma_{r_P+r_Q-1}([P_{11},Q_{ij}]) = \sum_{\substack{\beta,\gamma \\ |\beta|\le r_P+r_Q-1}} s_{\beta,\gamma}(x,\xi)t^{\gamma}\tau^{\beta}$$

and choose $(\beta^o,\gamma^o)\in\mathbb{N}^{n'+n}$ such that

$$\begin{cases} q_{\beta^o,\gamma^o}\neq 0, & \\ q_{\beta,\gamma}=0 & \text{if } \gamma<\gamma^o, \\ q_{\beta,\gamma^o}=0 & \text{if } \beta>\beta^o. \end{cases}$$

Then

$$s_{\beta^o,\gamma^o}t^{\gamma^o}\tau^{\beta^o} = \Bigl(\sum_{\nu=1}^{n}\frac{\partial p_{0,0}}{\partial\xi_\nu}\gamma^o_\nu\Bigr)\Bigl(q_{\beta^o,\gamma^o}t^{\gamma^o}\tau^{\beta^o}\Bigr), \tag{4}$$

which proves the first claim in the theorem because the condition $[P,Q]=0$ with the assumption of the theorem means $\gamma^o=0$.

Moreover suppose $p_{\beta,\gamma}=0$ for $\gamma\neq 0$. Then (4) is valid for any $\gamma^o\in\mathbb{N}^n$ and $\beta^o\in\mathbb{N}^{n'}$ satisfying $q_{\beta,\gamma^o}=0$ for $\beta>\beta^o$ and hence the condition $[P,Q]=0$ means $q_{\beta^o,\gamma^o}=0$ if $\gamma^o\neq 0$. Thus $q_{\beta,\gamma^o}=0$ if $\gamma^o\neq 0$. □

Corollary 2.3. *Let $P\in M(m,\mathcal{D}_*)$ such that $\sigma(P)$ and $\sigma_{\operatorname{ord}P-1}(P)$ are scalar matrices. Suppose $\bar\sigma_*(P)$ satisfies* (3). *Then the map*

$$\sigma_* : M(m,\widetilde{\mathcal{D}}_*)^P := \{Q\in M(m,\widetilde{\mathcal{D}}_*)\,;\,[P,Q]=0\}\to M(m,\mathcal{D}_N[\xi]),$$
$$Q\mapsto\sigma_*(Q)$$

is an injective algebra homomorphism.

In particular, when $m=1$, $\mathcal{D}_*^P:=\{Q\in\mathcal{D}_*\,;\,[P,Q]=0\}$ is commutative.

Proof. Since σ_* is an algebra homomorphism and the condition $Q_1,Q_2\in M(m,\widetilde{\mathcal{D}}_*)^P$ implies $[Q_1,Q_2]\in M(m,\widetilde{\mathcal{D}}_*)^P$, this corollary is a direct consequence of Theorem 2.2. □

Remark 2.4. i) Retain the notation in Theorem 2.2. Then (3) is valid for $P\in M(m,\mathcal{D}_*)$ if n functions $\frac{\partial\bar\sigma_*(P)}{\partial\xi_1},\dots,\frac{\partial\bar\sigma_*(P)}{\partial\xi_n}$ are linearly independent over $\mathbb{R}$. In particular, if $\operatorname{ord}P=2$ and $\bar\sigma_*(P)$ is a scalar matrix, the condition that

$$\text{the matrix } \Bigl(\frac{\partial^2\bar p}{\partial\xi_i\partial\xi_j}\Bigr)_{\substack{1\le i\le n\\1\le j\le n}} \quad \text{is invertible for generic } x\in N$$

implies (3). Here $\bar p$ is the diagonal element of $\bar\sigma_*(P)$.

ii) The assumption $P\in M(m,\mathcal{D}_*)$ is necessary in Theorem 2.2. For example, $[t\frac{\partial}{\partial t}+x\frac{\partial}{\partial x},t\frac{\partial}{\partial x}]=0$ and $\sigma_*(t\frac{\partial}{\partial x})=0$. Moreover we note that

$$\left[\begin{pmatrix} t\frac{\partial}{\partial t} & 0\\ 0 & t\frac{\partial}{\partial t}+1\end{pmatrix},\begin{pmatrix}0 & t\\ 0 & 0\end{pmatrix}\right]=0.$$

This gives an example such that $\sigma_{\operatorname{ord} P-1}(P)$ and $\sigma(Q)$ are not scalar matrices.

iii) The invariant differential operators on a Riemannian symmetric space G/K of non-compact type have regular singularities along the boundaries of a realization of the space constructed by [O2] and the map σ_* of $\mathcal{D}_*^P$ to $\mathcal{A}_N[\xi]$ in Corollary 2 corresponds to Harish-Chandra isomorphism (cf. [K–]).

The element of the universal enveloping algebra $U(\mathfrak{g})$ of the Lie algebra of G defines a differential operator on the realization of G/K through the infinitesimal action of the left translation by elements of G. Then the differential operator is an element of $\widetilde{\mathcal{D}}_*$.

Moreover the invariant differential operators on a semisimple symmetric space whose rank is larger than its real rank are in $\widetilde{\mathcal{D}}_*$ (cf. [O4]).

The radial parts of the Casimir operator acting on K-finite sections of certain homogeneous vector bundle of G satisfy the assumption of Theorem 2.2 (cf. (26) and (27) for examples).

3 Ideally analytic solutions without logarithmic terms

For a subset Σ of $\mathbb{N}^n$ define

$$\overline{\Sigma} := \{\alpha \in \mathbb{N}^n\,;\ \{\alpha+\gamma\,;\ \gamma\in\mathbb{N}^n\}\cap\Sigma\neq\emptyset\},$$

$$\partial\Sigma := \{\alpha\in\mathbb{N}^n\setminus\overline{\Sigma}\,;\ \text{there exists } \gamma\in\bar{\Sigma} \text{ such that } \sum_{i=1}^n|\alpha_i-\gamma_i|=1\}.$$

Moreover we denote by $\hat{\mathcal{A}}_N$ the ring of formal power series of $t=(t_1,\ldots,t_n)$ with coefficients in $\mathcal{A}_N$.

Theorem 3.1. *Let $P\in M(m,\mathcal{D}_*)$.*

i) *Let Σ be a subset of $\mathbb{N}^n$ such that*

$$\det\bigl(\sigma_*(P)(x,\gamma)\bigr)\not\equiv 0 \quad \textit{for } \forall\gamma\in\mathbb{N}^n\setminus\Sigma.$$

Let $\hat{u}(t,x)=\sum_{\alpha\in\mathbb{N}^n}u_\alpha(x)t^\alpha\in\hat{\mathcal{A}}_M^m$ be a formal solution of $P\hat{u}=0$. Then $\hat{u}=0$ if $u_\alpha=0$ for $\forall\alpha\in\Sigma$.

Hereafter in this theorem suppose

$$\det\bar{\sigma}_*(P)(x,\xi)\neq 0 \quad \textit{for } \forall\xi=(\xi_1,\ldots,\xi_n)\in[0,\infty)^n\setminus\{0\} \textit{ and } \forall x\in N. \tag{5}$$

ii) *If $\hat{u}\in\hat{\mathcal{A}}_M^m$ satisfies $P\hat{u}\in\mathcal{A}_M^m$, then $\hat{u}\in\mathcal{A}_M^m$.*

iii) *Fix $f\in\mathcal{A}_M^m$, a point $x^o\in N$ and a finite subset Σ of $\mathbb{N}^n$ such that*

$$\det\bigl(\sigma_*(P)(x^o,\gamma)\bigr)\neq 0 \quad \textit{for } \forall\gamma\in\mathbb{N}^n\setminus\Sigma.$$

By shrinking $M\ni x^o$ if necessary and denoting

$$\mathcal{A}_M(P^{-1}f):=\{u\in\mathcal{A}_M^m\,;\ Pu=f\},$$

$$\mathcal{A}_M(P^{-1}f)^{\Sigma} := \{\bar{u} = \sum_{\alpha\in\overline{\Sigma}} u_\alpha(x)t^\alpha \in \mathcal{A}_M^m\,;\, P\bar{u} \equiv f \mod \sum_{\beta\in\partial\Sigma} \mathcal{A}_M^m t^\beta\},$$

the natural restriction map

$$\mathcal{A}_M(P^{-1}f) \xrightarrow{\sim} \mathcal{A}_M(P^{-1}f)^{\Sigma}, \quad \sum_{\alpha\in\mathbb{N}^n} u_\alpha(x)t^\alpha \mapsto \sum_{\alpha\in\overline{\Sigma}} u_\alpha(x)t^\alpha$$

is a bijection. Here in particular

$$\mathcal{A}_M(P^{-1}f)^{\{0\}} = \{u \in \mathcal{A}_N^m\,;\, \sigma_*(P)\big(x,0\big)u = f|_{t=0}\}.$$

Proof. The proof proceeds in a similar way as in [O3, Theorem 2.1] where we studies the same problem with $n=1$.

We may assume $x^o = 0$. Expanding functions in convergent power series of (t,x) at $(0,0)$, we will prove the theorem in a neighborhood of $(0,0)$.

Put $r = \operatorname{ord} P$ and

$$P = \sigma_*(P)(x,\vartheta) + \sum_{\substack{(\alpha,\beta)\in\mathbb{N}^{n+n'}\\ |\alpha|+|\beta|\le r}} p_{\alpha,\beta}(t,x)\vartheta^\alpha\partial_x^\beta.$$

Then $p_{\alpha,\beta}(0,x) = 0$. For a finite subset $\Sigma\subset\mathbb{N}^n$ and

$$\hat{u}(t,x) = \sum_{\alpha\in\mathbb{N}^n} \hat{u}_\alpha(x)t^\alpha \in \hat{\mathcal{A}}_M^m,$$

put

$$\bar{u}(t,x) = \sum_{\alpha\in\overline{\Sigma}} \hat{u}_\alpha(x)t^\alpha.$$

Suppose $P\hat{u} \equiv f \mod \sum_{\alpha\in\partial\Sigma}\hat{\mathcal{A}}_M^m t^\alpha$. Put $h = f - P\bar{u}$. Then

$$h = \sum_{\alpha\in\mathbb{N}^n\setminus\overline{\Sigma}} h_\alpha(x)t^\alpha = \sum_{\alpha\in\mathbb{N}^n\setminus\overline{\Sigma},\ \beta\in\mathbb{N}^{n'}} h_{\alpha,\beta}t^\alpha x^\beta \in \mathcal{A}_M^m$$

and

$$P\hat{u} = f \Leftrightarrow Pu = h \quad \text{with } u = \hat{u} - \bar{u}.$$

Then the equation $P\hat{u} = f$ is equal to

$$\sigma_*(P)(x,\vartheta)u = h - \sum_{\substack{(\alpha,\beta)\in\mathbb{N}^{n+n'}\\ |\alpha|+|\beta|\le r}} p_{\alpha,\beta}(t,x)\vartheta^\alpha\partial_x^\beta u.$$

$$u = \sum_{\alpha\in\mathbb{N}^n} u_\alpha(x)t^\alpha \quad \text{with } u_\alpha(x) = \begin{cases} 0 & \text{for } \alpha\in\overline{\Sigma},\\ \hat{u}_\alpha(x) & \text{for } \alpha\in\mathbb{N}^n\setminus\overline{\Sigma},\end{cases}$$

which also equals

$$\begin{aligned}\sigma_*(P)(x,\alpha^o)u_{\alpha^o}(x) &= h_{\alpha^o}(x)\\ &\quad - \mathrm{Coef}(t^{\alpha^o}) \text{ of } \Big(\sum_{\substack{(\alpha,\beta)\in\mathbb{N}^{n+n'}\\|\alpha|+|\beta|\le r}} p_{\alpha,\beta}(t,x)\vartheta^\alpha\partial_x^\beta\Big)\Big(\sum_{\substack{\alpha\in\mathbb{N}^n\\|\alpha|<|\alpha^o|}} u_\alpha(x)t^\alpha\Big) \qquad (6)\end{aligned}$$

for $\forall\alpha^o\in\mathbb{N}^n\setminus\overline{\Sigma}$. Here "Coef($t^{\alpha^o}$)" means "the coefficient of t^{α^o}". Since $\det\sigma_*(P)(x,\gamma)\ne 0$ for $\gamma\in\mathbb{N}^n\setminus\overline{\Sigma}$, $u_{\alpha^o}(x)$ is inductively determined by h.

On the other hand, putting $h=0$, it is clear that the claim i) follows from the induction proving $u_{\alpha^o}=0$ by (6) for $\forall\alpha^o\in\mathbb{N}^n\setminus\Sigma$.

Put

$$u_\alpha(x)=\sum_{\beta\in\mathbb{N}^{n'}} u_{\alpha,\beta}x^\beta \quad\text{with}\quad u_{\alpha,\beta}\in\mathbb{C}.$$

The equation (6) equals

$$\begin{aligned}&\sigma_*(P)(0,\alpha^o)u_{\alpha^o,\beta^o}\\ &= h_{\alpha_o,\beta}+\mathrm{Coef}(x^{\beta^o}) \text{ of } \big(\sigma_*(P)(0,\alpha^o)-\sigma_*(P)(x,\alpha^o)\big)\Big(\sum_{|\beta|<|\beta^o|} u_{\alpha^o,\beta}x^\beta\Big)\\ &\quad - \mathrm{Coef}(t^{\alpha^o}x^{\beta^o}) \text{ of } \Big(\sum_{\substack{(\alpha,\beta)\in\mathbb{N}^{n+n'}\\|\alpha|+|\beta|\le r}} p_{\alpha,\beta}(t,x)\vartheta^\alpha\partial_x^\beta\Big)\Big(\sum_{\substack{(\alpha,\beta)\in\mathbb{N}^{n+n'}\\|\alpha|<|\alpha^o|}} u_{\alpha,\beta}t^\alpha x^\beta\Big)\end{aligned}$$

for any $\alpha^o\in\mathbb{N}^n\setminus\overline{\Sigma}$ and $\beta^o\in\mathbb{N}^{n'}$. Hence the elements u_{α^o,β^o} of $\mathbb{C}^m$ satisfying this equation are uniquely and inductively determined in the lexicographic order of $(|\alpha^o|,|\beta^o|)$. Thus to complete the proof we have only to prove that $\sum u_{\alpha,\beta}t^\alpha x^\beta$ is a convergent power series. Here we may assume $\overline{\Sigma}\ni\{0\}$.

In general, for formal power series $\psi=\sum a_{\alpha,\beta}t^\alpha x^\beta$ and $\phi=\sum b_{\alpha,\beta}t^\alpha x^\beta$ we denote $\psi\ll\phi$ if $|a_{\alpha,\beta}|\le b_{\alpha,\beta}$ for $\forall\alpha$, β and in this case ϕ is called a majorant series of ψ. Note that if ϕ is a convergent power series, so is ψ.

Now assume (5). We note that there exists $\epsilon>0$ such that

$$|\det\bar{\sigma}_*(P)(0,\xi)|\ge\epsilon(\xi_1+\cdots+\xi_n)^{mr}\quad\text{for }\forall\xi\in[0,\infty)^n.$$

As in the proof of [O3, Theorem 2.1], we can choose $C>0$, $c>0$, $M>0$ and $K\ge 1$ so that for $\forall(\alpha,\beta)\in\mathbb{N}^{n+n'}$ and $\forall\gamma\in\mathbb{N}^n\setminus\Sigma$

$$\begin{aligned}cm\big|\big(\sigma_*(P)(0,\gamma)^{-1}\big)_{ij}\big| &\le \prod_{j=0}^{r-1}\big(r|\gamma|-j\big)^{-1},\\ \sigma_*(P)(x,\gamma)_{ij}-\sigma_*(P)(0,\gamma)_{ij} &\le \frac{C(x_1+\cdots+x_{n'})\prod_{j=0}^{r-1}(r|\gamma|-j)}{1-K(x_1+\cdots+x_{n'})},\\ p_{\alpha,\beta}(t,x)_{ij}-p_{\alpha,\beta}(0,x)_{ij} &\ll \frac{C(t_1+\cdots+t_n)}{1-K(t_1+\cdots+t_n+x_1+\cdots+x_{n'})}\\ h(t,x)_i &\ll \frac{M(t_1+\cdots+t_n)}{1-K(t_1+\cdots+t_n+x_1+\cdots+x_{n'})}.\end{aligned}$$

Here i and j represent the indices of square matrices or vectors of size m. Hence the power series $w(s,y)$ of (s,y) satisfying

$$\begin{aligned} c\prod_{j=0}^{r-1}\Big(rs\frac{\partial}{\partial s}-j\Big)w &= \frac{Cmy}{1-Ky}\prod_{j=0}^{r-1}\Big(rs\frac{\partial}{\partial s}-j\Big)w \\ &\quad + \sum_{j+k\le r}\frac{Cm(n+n')^r s}{1-K(s+y)}\Big(s\frac{\partial}{\partial s}\Big)^j\Big(\frac{\partial}{\partial y}\Big)^k w \qquad (7)\\ &\quad + \frac{Ms}{1-K(s+y)}, \\ w(0,y) &= 0 \end{aligned}$$

implies

$$\Big(u(t,x)-\sum_{\alpha\in\mathbb{N}^n\setminus\overline{\Sigma}} u_\alpha(x)t^\alpha\Big)_i \ll w(t_1+\cdots+t_n, x_1+\cdots+x_{n'}) \quad \text{for } 1\le i\le m.$$

Put $s=z^r$. Then (7) changes into

$$\begin{aligned} \Big(c-\frac{Cmy}{1-Ky}\Big)z^r\frac{\partial^r w}{\partial z^r} &= \sum_{j+k\le r}\frac{Cm(n+n')^r z^r}{1-K(z^r+y)}\Big(\frac{z}{r}\frac{\partial}{\partial z}\Big)^j\frac{\partial^k w}{\partial y^k} \\ &\quad + \frac{Mz^r}{1-K(z^r+y)}, \qquad (8)\\ \frac{\partial^j w}{\partial z^j}\Big|_{z=0} &= 0 \quad \text{for} \quad j=0,\dots,r-1. \end{aligned}$$

Since the first equation in the above is equivalent to

$$\Big(c-\frac{Cmy}{1-Ky}\Big)\frac{\partial^r w}{\partial z^r} = \sum_{j+k\le r}\frac{Cm(n+n')^r}{1-K(z^r+y)}\Big(\frac{z}{r}\frac{\partial}{\partial z}\Big)^j\frac{\partial^k w}{\partial y^k} + \frac{M}{1-K(z^r+y)},$$

(8) has a unique solution of power series of (y,z), which is assured to be analytic at the origin by Cauchy-Kowalevsky's theorem. In fact for a sufficiently large positive number L, the solution of the ordinary differential equation

$$\begin{aligned} \Big(c-\frac{Cmt}{1-Kt}\Big)\tilde{w}^{(r)}(t) &= \sum_{j+k\le r}\frac{Cm(n+n')^r L^{-k}}{1-Kt}\Big(\frac{t}{r}\frac{\mathrm{d}}{\mathrm{d}t}\Big)^j\tilde{w}^{(k)}(t) + \frac{M}{1-Kt}, \\ \tilde{w}^{(j)}(0) &= 0 \quad \text{for } j=0,\dots,r-1 \end{aligned}$$

with

$$\begin{cases} t = z+Ly, \\ cL^r > Cm(n+n')^r \end{cases}$$

satisfies $w(z,y)\ll\tilde{w}(z+Ly)$. Hence u is also a convergent power series. □

Let ℓ be a non-negative integer and let U be an open connected neighborhood of a point z^o of $\mathbb{C}^\ell$ and let $\mathcal{O}_U$ be the space of holomorphic functions on U. We denote by ${}_U\mathcal{A}_M$ and ${}_U\mathcal{A}_N$ the space of real analytic functions on M with holomorphic parameter $z \in U$ and that on N with holomorphic parameter $z \in U$, respectively. Moreover we denote by ${}_U\hat{\mathcal{A}}_M$ the space of formal power series of $t = (t_1, \dots, t_n)$ with coefficients in ${}_U\mathcal{A}_N$. Let ${}_U\mathcal{D}_*$ denote the ring of differential operators P of the form

$$\begin{cases} P = \sum_{(\alpha,\beta)\in\mathbb{N}^{n+n'}} a_{\alpha,\beta}(t,x,z)\vartheta^\alpha\partial_x^\beta, \\ a_{\alpha,\beta} \in {}_U\mathcal{A}_N,\ a_{\alpha,\beta}(0,x,z) = 0 \quad \text{if } \beta > 0. \end{cases}$$

Then $\sigma_*(P)(x,z,\xi) := \sum_\alpha p_{\alpha,0}(0,x,z)\xi^\alpha \in {}_U\mathcal{A}_N[\xi]$.

Theorem 3.2. *Let $P \in M(m, {}_U\mathcal{D}_*)$ and $\lambda(z) = \big(\lambda_1(z), \dots, \lambda_n(z)\big) \in \mathcal{O}_U^n$.*

i) *Let Σ be a subset of $\mathbb{N}^n$ such that*

$$\det\big(\sigma_*(P)(x,z,\lambda(z)+\gamma)\big) \not\equiv 0 \quad \textit{for} \quad \forall\gamma \in \mathbb{N}^n \setminus \Sigma.$$

Let $\phi(t,x,z) = \sum_{\alpha\in\mathbb{N}^n} \phi_\alpha(x,z)t^\alpha \in {}_U\hat{\mathcal{A}}_M^m$ satisfying $P\big(t^{\lambda(z)}\phi\big) = 0$. Then $\phi = 0$ if $\phi_\alpha = 0$ for $\forall\alpha \in \Sigma$.

Hereafter in this theorem suppose P satisfies

$$\det\bar{\sigma}_*(P)(x,z,\xi) \neq 0 \quad \textit{for}\ \forall(x,z,\xi) \in N \times U \times \big\{[0,\infty)^n \setminus \{0\}\big\}. \tag{9}$$

ii) *If $\phi(t,x,z) \in {}_U\hat{\mathcal{A}}_M^m$ satisfies $P\big(t^{\lambda(z)}\phi\big) = 0$, then $\phi \in {}_U\mathcal{A}_M^m$.*

iii) *Fix $x^o \in N$. Let Σ be a finite subset Σ of $\mathbb{N}^n$ such that*

$$\det\big(\sigma_*(P)(x^o,z^o,\lambda(z^o)+\gamma)\big) \neq 0 \quad \textit{for} \quad \forall\gamma \in \mathbb{N}^n \setminus \Sigma.$$

Shrinking U and N if necessary and denoting

$$\mathrm{Sol}_U(P;\lambda) := \{u;\ ut^{-\lambda(z)} \in {}_U\mathcal{A}_M^m \textit{ and } Pu = 0\},$$

$$\mathrm{Sol}_U(P;\lambda)^\Sigma := \{\bar{u} = \sum_{\alpha\in\overline{\Sigma}} \phi_\alpha(x,z)t^{\lambda(z)+\alpha};\ \bar{u}t^{-\lambda(z)} \in {}_U\mathcal{A}_M^m \textit{ and}$$
$$P\bar{u} \equiv 0 \mod \sum_{\beta\in\partial\Sigma} {}_U\mathcal{A}_M^m t^{\lambda(z)+\beta}\},$$

we see that the natural restriction map

$$\mathrm{Sol}_U(P;\lambda) \xrightarrow{\sim} \mathrm{Sol}_U(P;\lambda)^\Sigma,$$
$$\sum_{\alpha\in\mathbb{N}^n} \phi_\alpha(x,z)t^{\lambda(z)+\alpha} \mapsto \sum_{\alpha\in\Sigma} \phi_\alpha(x,z)t^{\lambda(z)+\alpha}$$

is a bijection. Here in particular

$$\mathrm{Sol}_U(P;\lambda)^{\{0\}} = \{u \in {}_U\mathcal{A}_N^m;\ \sigma_*(P)\big(x,z,\lambda(z)\big)u = 0\}.$$

Proof. Fix $x^o \in N$. Expanding functions in convergent power series of (t, x, z) at $(0, x^o, z^o)$, we will prove the lemma in a neighborhood of $(0, x^o, z^o)$. Replacing P and the complexification $M_{\mathbb{C}}$ of M by $t^{-\lambda(z)} \circ P \circ t^{\lambda(z)}$ and $M_{\mathbb{C}} \times U$, respectively, we can reduce this theorem to the previous theorem without the parameter z. □

Corollary 3.3. *Retain the notation in the previous theorem. Let $\ell = 1$. Suppose*

$$\sigma_*(P)(x, z, \lambda(z)) = 0 \quad \textit{for } \forall (x, z) \in N \times U$$

and

$$\det\bigl(\sigma_*(P)(x^o, z, \lambda(z) + \gamma)\bigr) \neq 0 \quad \textit{for } \forall \gamma \in \mathbb{N}^n \setminus \{0\} \textit{ and } \forall z \in U \setminus \{z^o\}.$$

Then there exists a non-negative integer k such that the following holds.

The previous theorem assures that for any $\phi_0(x, z) \in {}_U\mathcal{A}_N^m$ and fixed $z \in U \setminus \{z^o\}$ there exists a function $u(t, x, z)$ satisfying

$$\begin{cases} Pu = 0, \\ t^{-\lambda(z)} u \in \mathcal{A}_M^m, \\ t^{-\lambda(z)} u|_{t=0} = \phi_0(x, z). \end{cases}$$

Then $t^{-\lambda(z)} z^k u(x, z)$ extends holomorphically to the point $z = z^o$.

Proof. Since the functions $\det\bigl(\sigma_*(P)(x^o, z, \lambda(z)+\gamma)\bigr)$ have finite order of zeros at $z = z^0$ for $\gamma \in \Sigma \setminus \{0\}$, this corollary follows from the proof of Theorem 3.1 (cf. (6) for $\forall \alpha^o \in \mathbb{N}^n \setminus \{0\}$). In fact it is sufficient to put k the sum of these orders of zeros for $\gamma \in \Sigma \setminus \{0\}$. □

Remark 3.4. It follows from the proves of Theorem 3.1 and Theorem 3.2 that there exist differential operators $P_\alpha^\gamma(x, z, \partial_x)$ such that

$$\phi_\alpha(x, z) = \sum_{\gamma \in \Sigma} P_\alpha^\gamma(x, z, \partial_x) \phi_\gamma(x, z) \quad \text{for } \alpha \in \mathbb{N}^n \setminus \Sigma$$

in Theorem 3.2 iii).

Corollary 3.5. *Fix $(x^o, \lambda^o) \in N \times \mathbb{C}^n$ and let V be a neighborhood of λ^o in $\mathbb{C}^n$. Suppose $P \in M(m, \mathcal{D}_*)$ satisfies* (9) *and*

$$\det\bigl(\sigma_*(P)(x^o, \lambda^o + \gamma) - \sigma_*(P)(x^o, \lambda^o)\bigr) \neq 0 \quad \textit{for } \forall \gamma \in \mathbb{N}^n \setminus \{0\}.$$

Then shrinking N, M and V if necessary, we have a linear bijection

$$\beta_\lambda : \mathrm{Sol}_V(P) := \{u\,;\, ut^{-\lambda} \in {}_V\mathcal{A}_M^m \textit{ and } Pu = \sigma_*(P)(x, \lambda)u\} \xrightarrow{\sim} {}_V\mathcal{A}_N^m, \qquad u \mapsto t^{-\lambda} u|_{t=0}$$

with the coordinate $((t, x), \lambda) \in M \times V$. In particular, we have a bijective map

$$\beta_{\lambda^o} : \mathrm{Sol}_{\lambda^o}(P) := \{u\,;\, ut^{-\lambda^o} \in \mathcal{A}_M^m \textit{ and } Pu = \sigma_*(P)(x, \lambda^o)u\} \xrightarrow{\sim} \mathcal{A}_N^m, \qquad u \mapsto t^{-\lambda^o} u|_{t=0}.$$

Definition 3.6. *The map* β_{λ^o} *of* $\mathrm{Sol}_{\lambda^o}(P)$ *is called the* boundary value map *of the solution space* $\mathrm{Sol}_{\lambda^o}(P)$ *of the differential equation* $Pu = \sigma_*(P)(x, \lambda^o)u$ *with respect to the* characteristic exponent λ^o.

Remark 3.7. When $n = 1$, $u \in \mathrm{Sol}_{\lambda^o}(P)$ is called an *ideally analytic solution* of the equation $Pu = \sigma_*(P)(x, \lambda^o)u$ in [KO].

The following theorem says that $\mathrm{Sol}_V(P)$ and $\sigma_*(P)$ characterize $P \in \mathcal{D}_*$.

Theorem 3.8. *Let* P *be an element of* $M(m, \mathcal{D}_*)$ *satisfying the assumptions in Corollary 3.5. Let* $P' \in M(m, \mathcal{D}_*)$ *with* $\sigma_*(P) = \sigma_*(P')$. *Then the condition* $\mathrm{Sol}_V(P) = \mathrm{Sol}_V(P')$ *implies* $P = P'$.

Proof. Suppose $P \neq P'$. Put

$$P - P' = \sum_{\alpha,\beta,\gamma} r_{\alpha,\beta,\gamma} t^{\gamma} \vartheta^{\alpha} \partial_x^{\beta}.$$

Then we can find $\gamma^o \in \mathbb{N}^{n'} \setminus \{0\}$ such that $\sum_{\alpha,\beta} r_{\alpha,\beta,\gamma^o} t^{\gamma^o} \vartheta^{\alpha} \partial_x^{\beta} \neq 0$ and $r_{\alpha,\beta,\gamma} = 0$ if $\gamma < \gamma^o$. For $v(x) \in \mathcal{A}_N^m$ the coefficients of $t^{\lambda+\gamma^o}$ in $(P - P')\beta_{\lambda}^{-1} v(x)$ show

$$\begin{aligned} 0 &= \bigl(t^{-\lambda} \sum_{\alpha,\beta} r_{\alpha,\beta,\gamma^o} \vartheta^{\alpha} \partial_x^{\beta} t^{\lambda} v(x)\bigr)\big|_{t=0} \\ &= \sum_{\alpha,\beta} r_{\alpha,\beta,\gamma^o} \lambda^{\alpha} \partial_x^{\beta} v(x) \quad \text{for } \forall \lambda \in V \text{ and } \forall v(x) \in \mathcal{A}_N^m, \end{aligned}$$

which means a contradiction. □

4 Induced equations

Retain the notation in the previous section. Moreover we denote by ${}_U\widetilde{\mathcal{D}}_*$ the ring of holomorphic maps of U to $\widetilde{\mathcal{D}}_*$ for a connected open subset U of $\mathbb{C}^{\ell}$.

We recall that the element P of ${}_U\widetilde{\mathcal{D}}_*$ is characterized by the expression

$$P = \sum_{(\alpha,\beta) \in \mathbb{N}^{n+n'}} p_{\alpha,\beta}(t, x, z) \vartheta^{\alpha} \partial_x^{\beta} \tag{10}$$

with $p_{\alpha,\beta}(t, x, z) \in {}_U\mathcal{A}_M$ and

$$\sigma_*(P)(x, z, \xi, \partial_x) = \sum_{\alpha,\beta} p_{\alpha,\beta}(0, x, z) \xi^{\alpha} \partial_x^{\beta}.$$

Theorem 4.1. *Let* $P \in M(m, {}_U\mathcal{D}_*)$ *satisfying the assumption in Theorem 3.2 iii) with* $\Sigma = \{0\}$. *Suppose that* $P_1, \ldots, P_p \in M(m, {}_U\widetilde{\mathcal{D}}_*)$ *satisfy*

$$[P, P_i] = S_i P + \sum_{j=1}^{p} T_{ij} P_j \tag{11}$$

with $S_i \in M(m, {}_U\tilde{\mathcal{D}}_*)$ *and* $T_{ij} \in M(m, {}_U\mathcal{D}_*)$. *Suppose moreover* $\sigma_*(T_{ij}) = 0$. *Then the map*

$$\begin{aligned}\beta_{\lambda(z)} : &\{u\,;\, t^{-\lambda(z)}u \in {}_U\mathcal{A}_M^m \text{ and } Pu = P_i u = 0 \text{ for } i = 1, \dots, p\} \\ &\xrightarrow{\sim} \left\{v \in {}_U\mathcal{A}_N^m\,;\, \begin{cases}\sigma_*(P)\big(x, z, \lambda(z)\big)v = 0, \\ \sigma_*(P_i)\big(x, z, \lambda(z), \partial_x\big)v = 0 \quad (i = 1, \dots, p)\end{cases}\right\}, \\ &u \mapsto t^{-\lambda(z)}u\,\big|_{t=0}\end{aligned} \tag{12}$$

is a bijection.

Proof. Since $(t^{-\lambda(z)}P_j u)|_{t=0} = \sigma_*(P_j)(x, z, \lambda(z), \partial_x)t^{-\lambda(z)}u|_{t=0}$, Theorem 3.2 assures that we have only to prove the surjectivity of the map to get the theorem.

For a given v in the element of the set, we have $u \in t^{\lambda(z)}{}_U\mathcal{A}_M^m$ such that $Pu = 0$ and $t^{-\lambda(z)}u|_{t=0} = v$. Then $PP_i u = \sum_{j=1}^{p} T_{ij} P_j u$, namely,

$$\begin{pmatrix} P - T_{11} & -T_{12} & -T_{13} & \cdots & -T_{1p} \\ -T_{21} & P - T_{22} & -T_{23} & \cdots & -T_{2p} \\ -T_{31} & -T_{32} & P - T_{33} & \cdots & -T_{2p} \\ \vdots & \vdots & \vdots & \ddots & \vdots \\ -T_{p1} & T_{p2} & T_{p3} & \cdots & P - T_{pp} \end{pmatrix} \begin{pmatrix} P_1 u \\ P_2 u \\ P_3 u \\ \vdots \\ P_p u \end{pmatrix} = 0.$$

Since $\sigma_*(T_{ij}) = 0$ and $t^{-\lambda(z)}P_j u|_{t=0} = 0$ for $j = 1, \dots, p$, Theorem 3.2 i) assures $P_j u = 0$. □

Definition 4.2. *The system of differential equations*

$$\sigma_*(P)(x, z, \lambda(z))v = \sigma_*(P_i)(x, z, \lambda(z), \partial_x)v = 0 \quad \textit{for} \quad i = 1, \dots, p$$

in Theorem 4.1 is called the system of induced equations *with respect to the boundary value map* $\beta_{\lambda(z)}$ (cf. (12)).

Remark 4.3. i) Suppose $P \in M(m, {}_U\mathcal{D}_*)$ satisfies the assumption in Theorem 4.1. Let $Q \in M(m, {}_U\mathcal{D}_*)$ such that $[P, Q] = 0$ and $\sigma_*(Q)\big(x, z, \lambda(z)\big) = 0$. Then if $u \in t^{\lambda(z)}{}_U\mathcal{A}_M^m$ satisfies $Pu = 0$, we have $Qu = 0$.

ii) Let p be the rank of an irreducible semisimple symmetric space G/H. The ring of invariant differential operators on G/H is isomorphic to $\mathbb{C}[P_1, \dots, P_p]$, where P_j are algebraically independent and satisfy $[P_i, P_j] = 0$ for $1 \leq i < j \leq p$. Under a suitable coordinate system $(t_1, \dots, t_n, x_1, \dots, x_{n'})$ of a natural realization of G/H constructed by [O6], G/H is defined by $t_1 > 0, \dots, t_n > 0$. Then n is the real rank of G/H and $P_i \in \tilde{\mathcal{D}}_* \setminus \mathcal{D}_*$ if $n < p$. It is shown in [O6] that we can choose $P \in \sum_{j=1}^{p} \mathcal{D}_* P_j$ such that P, $P_1, \dots, P_p$ satisfy the assumption in Theorem 4.1.

5 Holonomic systems of differential equations with constant coefficients

In this section $(\frac{\partial}{\partial y_1}, \dots, \frac{\partial}{\partial y_n})$ is simply denoted by ∂. For $\mu = (\mu_1, \dots, \mu_n) \in \mathbb{C}^n$ and $y = (y_1, \dots, y_n) \in \mathbb{R}^n$, we put

$$\langle \mu, y \rangle = \mu_1 y_1 + \dots + \mu_n y_n.$$

Lemma 5.1. *Let* $\mathrm{Hom}_{\mathbb{C}[\partial]}(\mathcal{M}, \mathcal{N})$ *denote the space of* $\mathbb{C}[\partial]$*-homomorphisms of a* $\mathbb{C}[\partial]$*-module* $\mathcal{M}$ *to a* $\mathbb{C}[\partial]$*-module* $\mathcal{N}$*. Then the space is naturally a* $\mathbb{C}[\partial]$*-module. Let* $\hat{\mathcal{O}}$ *be the space of formal power series of* $y = (y_1, \dots, y_n)$ *and let* $\mathcal{O}(\mathbb{C}^n)$ *be the space of entire functions on* $\mathbb{C}^n \ni y$*. Suppose* $\mathcal{M}$ *is a finite dimensional* $\mathbb{C}[\partial]$*-module. Then*

$$\begin{aligned} \bigoplus_{\lambda \in \mathbb{C}^n} \mathrm{Hom}_{\mathbb{C}[\partial]}\bigl(\mathcal{M}, \mathbb{C}[y]e^{\langle \lambda, y \rangle}\bigr) &\xrightarrow{\sim} \mathrm{Hom}_{\mathbb{C}[\partial]}\bigl(\mathcal{M}, \bigoplus_{\lambda \in \mathbb{C}^n} \mathbb{C}[y]e^{\langle \lambda, y \rangle}\bigr) \\ &\xrightarrow{\sim} \mathrm{Hom}_{\mathbb{C}[\partial]}\bigl(\mathcal{M}, \mathcal{O}(\mathbb{C}^n)\bigr) \\ &\xrightarrow{\sim} \mathrm{Hom}_{\mathbb{C}[\partial]}\bigl(\mathcal{M}, \hat{\mathcal{O}}\bigr), \end{aligned} \tag{13}$$

$$\dim \mathrm{Hom}_{\mathbb{C}[\partial]}\bigl(\mathcal{M}, \mathcal{O}(\mathbb{C}^n)\bigr) = \dim \mathcal{M}. \tag{14}$$

If $\mathcal{M}'$ *is a quotient* $\mathbb{C}[\partial]$*-module of* $\mathcal{M}$ *such that*

$$\mathrm{Hom}_{\mathbb{C}[\partial]}\bigl(\mathcal{M}', \mathcal{O}(\mathbb{C}^n)\bigr) \xrightarrow{\sim} \mathrm{Hom}_{\mathbb{C}[\partial]}\bigl(\mathcal{M}, \mathcal{O}(\mathbb{C}^n)\bigr),$$

then $\mathcal{M} \xrightarrow{\sim} \mathcal{M}'$*.*

Proof. For $\mu = (\mu_1, \dots, \mu_n) \in \mathbb{C}^n$, let $\mathfrak{m}_\mu$ denote the maximal ideal of $\mathbb{C}[\partial]$ generated by $\frac{\partial}{\partial y_i} - \mu_i$ with $i = 1, \dots, n$. Then we have $\mathcal{M} \simeq \mathcal{M}_{\lambda_1} \oplus \dots \oplus \mathcal{M}_{\lambda_m}$ with suitable $\lambda_\nu = (\lambda_{\nu,1}, \dots, \lambda_{\nu,n}) \in \mathbb{C}^n$ and $\mathbb{C}[\partial]$-modules $\mathcal{M}_{\lambda_\nu}$ satisfying $\mathfrak{m}_{\lambda_\nu}^k \mathcal{M}_{\lambda_\nu} = 0$ for a large positive integer k. Hence we have only to prove the lemma for each $\mathcal{M}_{\lambda_\nu}$. By the outer automorphism $\frac{\partial}{\partial y_i} \mapsto \frac{\partial}{\partial y_i} + \lambda_{\nu,i}$ for $i = 1, \dots, n$ which corresponds to the multiplication of the functions in $\mathcal{O}(\mathbb{C}^n)$ or $\hat{\mathcal{O}}$ by $e^{-\langle \lambda_\nu, x \rangle}$ we may assume $\mathfrak{m}_0^k \mathcal{M} = 0$.

Suppose $\mathfrak{m}_0^k \mathcal{M} = 0$. Then $\mathrm{Hom}_{\mathbb{C}[\partial]}\bigl(\mathcal{M}, \mathbb{C}[y]\bigr) \xrightarrow{\sim} \mathrm{Hom}_{\mathbb{C}[\partial]}\bigl(\mathcal{M}, \hat{\mathcal{O}}\bigr)$ and (13) is clear. Since $\hat{\mathcal{O}}$ is the dual space of $\mathbb{C}[\partial]$ by the bilinear form $\langle P(\partial), u \rangle = P(\partial)u|_{x=0}$, (14) is clear. The last statement follows from (14). □

Definition 5.2. *A finite dimensional* $\mathbb{C}[\partial]$*-module* $\mathcal{M}$ *is* semisimple *if*

$$\mathrm{Hom}_{\mathbb{C}[\partial]}\bigl(\mathcal{M}, \bigoplus_{\lambda \in \mathbb{C}^n} \mathbb{C}e^{\langle \lambda, y \rangle}\bigr) \xrightarrow{\sim} \mathrm{Hom}_{\mathbb{C}[\partial]}\bigl(\mathcal{M}, \mathcal{O}(\mathbb{C}^n)\bigr).$$

Let U be a convex open subset of $\mathbb{C}^\ell$, where ℓ is a non-negative integer, and let ${}_U\mathbb{C}[\partial]$ and ${}_U\mathcal{O}(\mathbb{C}^n)$ be the space of holomorphic maps of U to $\mathbb{C}[\partial]$ and that of U to $\mathcal{O}(\mathbb{C}^n)$, respectively.

Proposition 5.3. *Let r be a positive integer and let ${}_U\mathcal{M}$ be a finitely generated ${}_U\mathbb{C}[\partial]$ module with $\dim {}_U\mathcal{M} = r$ for any fixed $z \in U$. Assume that there exist positive integer k and finite number of holomorphic maps λ_i of U to $\mathbb{C}^n$ such that $\left(\prod_{i\in I} \mathfrak{m}^k_{\lambda_i(z)}\right){}_U\mathcal{M} = 0$ for any $z \in U$. Here the indices i run over a finite set I. Then there exist ${}_U\mathbb{C}[\partial]$-homomorphisms $u_1, \ldots, u_r$ of ${}_U\mathcal{M}$ to ${}_U\mathcal{O}(\mathbb{C}^n)$ such that they are linearly independent for any fixed $z \in U$.*

Let $I = I_1 \cup \cdots \cup I_L$ be a decomposition of I such that

$$\lambda_i(z) \neq \lambda_j(z) \quad \text{for } \forall z \in U \quad \text{if } i \in I_\mu \text{ and } j \in I_\nu \text{ and } 1 \leq \mu < \nu \leq L.$$

Then we can choose $\{u_i\,;\, i \in I\}$ such that for each u_i there exists I_ν satisfying

$$u_i \in \operatorname{Hom}_{{}_U\mathbb{C}[\partial]}\Big({}_U\mathcal{M}, \sum_{j\in I_\nu} e^{\langle\lambda_j(z),y\rangle}\mathbb{C}[y]\Big) \quad \text{for any fixed } z \in U. \tag{15}$$

Proof. Let $\{v_1, \ldots, v_m\}$ be a system of generators of ${}_U\mathcal{M}$. We identify the homomorphisms of ${}_U\mathcal{M}$ to ${}_U\mathcal{O}(\mathbb{C}^n)$ with their image of $\{v_1, \ldots, v_m\}$ and hence $u_j(y,z) \in {}_U\mathcal{O}(\mathbb{C}^n)^m$. Note that we can find ${}_U\mathbb{C}[\partial]$-homomorphisms $\tilde{u}_1(y,z), \ldots, \tilde{u}_r(y,z)$ of ${}_U\mathcal{M}$ to ${}_U\mathcal{O}(\mathbb{C}^n)$ if we replace $\mathcal{O}(U)$ by its quotient field.

Fix a point $z^o \in U$. Let $\gamma(t)$ be a holomorphic map of $\{t \in \mathbb{C}\,;\, |t| < 1\}$ to U such that $\gamma(0) = z^o$ and $\tilde{u}_j(y,\gamma(t))$ are holomorphic and linearly independent for $0 < |t| < 1$. Then [OS, Proposition 2.21] assures that there exist meromorphic functions $c_{ij}(t)$ such that the functions $v_i(y,t) = \sum_{j=1}^r c_{ij}(t)\tilde{u}_j(y,\gamma(t))$ are holomorphic at $t = 0$ and that $v_1(y,0), \ldots, v_r(y,0)$ are linearly independent. We can find $P_i \in \mathbb{C}[\partial]^m$ such that $\langle P_i, v_j\rangle = \delta_{ij}$ for $1 \leq i \leq r$ and $1 \leq j \leq r$. Here we put $\langle (Q_1, \ldots, Q_m), (f_1, \ldots, f_m)\rangle := \sum_{\nu=1}^m Q_\nu(f_\nu)(0)$ for $Q_\nu \in \mathbb{C}[\partial]^m$ and $f_\nu \in \mathcal{O}(\mathbb{C}^n)^m$.

Put $A(z) = \Big(\langle P_i, \tilde{u}_j\rangle\Big)_{\substack{1\leq i\leq r\\ 1\leq j\leq r}}$, which is a matrix of meromorphic functions on U and $\det A(z)$ is not identically zero. Let $\tilde{c}_{ij}(z)$ are meromorphic functions on U such that $\langle P_i, u_j\rangle = \delta_{ij}$ by putting $u_i = \sum_{j=1}^r \tilde{c}_{ij}(z)\tilde{u}_j$.

Suppose $u_i(y,z)$ is not holomorphic at $z = z^0$. Then there exist a positive integer L and a holomorphic function $\tilde{\gamma}$ of $\{t \in \mathbb{C}\,;\, |t| < 1\}$ to U such that $\tilde{\gamma}(0) = z^0$ and the function $w(y,t) := t^L u_i(y,\tilde{\gamma}(t))$ is holomorphically extended to the point $t = 0$ and moreover $w(y,0) \neq 0$. Then $w(y,0)$ defines a $\mathbb{C}[\partial]$-homomorphism of ${}_U\mathcal{M}$ to $\mathcal{O}(\mathbb{C}^n)$ at $z = z^o$. But $w(y,0), v_1(y,0), \ldots, v_r(y,0)$ are linearly independent because $\langle P_i, w(y,0)\rangle = 0$ for $i = 1, \ldots, r$, which contradicts to (14).

Hence for any $z^o \in U$ we can construct $u_1(y,z), \ldots, u_r(y,z)$ which are linearly independent and holomorphic in a neighborhood of $z^o \in U$. Then the theorem follows from the theory of holomorphic functions with several variables because U is a convex open subset of $\mathbb{C}^\ell$.

Since we have a decomposition ${}_U\mathcal{M} = {}_U\mathcal{M}_1 \oplus \cdots \oplus {}_U\mathcal{M}_L$ such that $\left(\prod_{i\in I_\nu} \mathfrak{m}^k_{\lambda_i(z)}\right){}_U\mathcal{M}_\nu = 0$ for $\nu = 1, \ldots, L$, we can assume (15). □

Example 5.4. Let W be a finite reflection group on a Euclidean space $\mathbb{R}^n$. Let $\mathbb{C}[p_1,\ldots,p_n]$ be the algebra of W-invariant polynomials on $\mathbb{R}^n$. For example, $p_k(x)=\sum_{1\le i_1<\cdots<i_k\le n} x_{i_1}\cdots x_{i_k}$. Then the system of differential equations

$$\mathcal{M}_\lambda : p_i(\partial)u = p_i(\lambda)u \quad \text{for} \;\; i=1,\ldots,n$$

with $\lambda\in\mathbb{C}^n$ is a fundamental example of a ${}_U\mathbb{C}[\partial]$-module in Proposition 5.3. Here $U=\mathbb{C}^n\ni\lambda$ and $r=\#W$. The system is semisimple if and only if $w\lambda\ne\lambda$ for $\forall w\in W\setminus\{e\}$. When $\lambda=0$, the solutions of this system are called harmonic polynomials for W. In this case, an explicit construction of solutions is given by [O5] such that $u_1(\lambda,y),\ldots,u_r(\lambda,y)$ are entire functions of $(\lambda,y)\in\mathbb{C}^{2n}$ and linearly independent for any fixed $\lambda\in\mathbb{C}^n$.

Remark 5.5. We will apply the result in this section to our original systems with the coordinates $t_i=\mathrm{e}^{-y_i}$ for $i=1,\ldots,n$. Then $\mathbb{C}[\partial]$ and $\mathrm{e}^{\langle\lambda,y\rangle}f(y)$ change into $\mathbb{C}[\vartheta]$ and $t^{-\lambda}f\bigl(-\log t_1,\ldots,-\log t_n\bigr)$, respectively.

6 Ideally analytic solutions for complete systems

In this section we will study the system of differential equations

$$\mathcal{M} : P_i u = 0 \quad \text{for } i=0,1,\ldots,q \tag{16}$$

with $P_i\in M(m,{}_U\mathcal{D}_*)$. Here $z\in U$ is a holomorphic parameter and U is a convex open subset of $\mathbb{C}^\ell$. We assume that $\sigma_*(P_i)$ do not depend on $x\in N$. We moreover assume that $P=P_0$ satisfies (5) and the system

$$\overline{\mathcal{M}} : \sigma_*(P_i)(z,\vartheta)\bar{u} = 0 \quad \text{for} \;\; i=0,1,\ldots,q, \tag{17}$$

which we call *indicial equation*, satisfies the assumption of Proposition 5.3. Then we call $\mathcal{M}$ a *complete system of differential equations with regular singularities along the set of walls* $\{N_1,\ldots,N_n\}$.

For a non-negative integer k let $\mathbb{C}[\log t]_{(k)}$ denote the polynomial function of $(\log t_1,\ldots,\log t_n)$ with degree at most k. Put $\mathbb{C}[\log t]=\bigcup_{k=1}^\infty \mathbb{C}[\log t]_{(k)}$.

Definition 6.1. *A solution $u(t,x,z)$ of $\mathcal{M}$ with the holomorphic parameter z is called an* ideally analytic solution *if $u(t,x,z)\in\bigoplus_{\lambda\in\mathbb{C}} t^\lambda\mathbb{C}[\log t]\mathcal{A}_M^m$ for any fixed $z\in U$.*

First we will examine the system $\mathcal{M}$ without the holomorphic parameter z or U is a point. Then let $\{\bar{u}_i=t^{\lambda_i}v_i(\log t)\,;\, i=1,\ldots,r\}$ be a basis of the solutions of (17). Here $v_i(\xi)\in\mathbb{C}[\xi]$ and these λ_i are called *exponents* of the system $\mathcal{M}$. We define

$$\begin{cases} e(\bar{u}_i) := \lambda_i, \\ \deg(\bar{u}_i) := \deg v_i. \end{cases}$$

We may assume that for any $\lambda \in \mathbb{C}^n$ and $k \in \mathbb{N}$

$$\{\bar{u}_i\,;\,(e(\bar{u}_i), \deg(\bar{u}_i)) = (\lambda, k)\} \text{ is empty}$$
$$\text{or linearly independent in the space } t^\lambda\mathbb{C}[\log t]^m_{(k)}/t^\lambda\mathbb{C}[\log t]^m_{(k-1)}.$$

Definition 6.2. *Let $u(t,x)$ be an ideally analytic solution of $\mathcal{M}$. Then a non-zero function*

$$w(t,x) = \sum_\nu t^\lambda p_\nu(\log t)\phi_\nu(x) \tag{18}$$

with suitable $\lambda \in \mathbb{C}^n$, $p_\nu(\xi) \in \mathbb{C}[\xi]$ and $\phi_\nu(x) \in \mathcal{A}^m_N$ is called a leading term *of $u(t,x)$ if*

$$u(t,x) - w(t,x) \in \sum_{\substack{\mu\in\mathbb{C}^n \\ \lambda-\mu\notin\mathbb{N}^n}} t^\mu\mathbb{C}[\log t]\mathcal{A}^m_M$$

and λ is called a leading exponent *of this leading term. If $\{w_1(t,x), \dots, w_k(t,x)\}$ is the complete set of the leading terms of $u(t,x)$, we say $\sum_{i=1}^k w_i(t,x)$ the* complete leading term *of $u(t,x)$.*

Then we have the following theorem.

Theorem 6.3. *The leading term* (18) *of an ideally analytic solution $u(t,x)$ of $\mathcal{M}$ is a solution of* (17). *Hence there exist $\phi_i(x) \in \mathcal{A}_M$ such that*

$$w(t,x) = \sum_{\lambda_i=\lambda} \bar{u}_i(t)\phi_i(x). \tag{19}$$

In particular, λ is an exponent of $\mathcal{M}$.

Assume

$$\det\sigma_*(P_1)(e(\bar{u}_i)+\gamma) \neq 0 \quad \textit{for} \quad \gamma \in \mathbb{N}^n \setminus \{0\}. \tag{20}$$

Then for any $\phi(x) \in \mathcal{A}_N$ there exists a unique solution of $\mathcal{M}$ in the space $t^{e(\bar{u}_i)}\mathbb{C}[\log t]\mathcal{A}^m_M$ whose leading term equals $\phi(x)\bar{u}_i$. Denoting the solution by $T_{\bar{u}_i}(\phi)$, we have the following bijective isomorphism if (20) *is valid for $1 \le i \le r$.*

$$\mathcal{A}^r_N \xrightarrow{\sim} \{\textit{ideally analytic solutions of } \mathcal{M}\},\ (\phi_i) \mapsto \sum_{i=1}^r T_{\bar{u}_i}(\phi_i). \tag{21}$$

Proof. Examining the equation $Pu(t,x) = 0$ modulo $\sum_{\substack{\mu\in\mathbb{C}^n \\ \lambda-\mu\notin\mathbb{N}^n}} t^\mu\mathbb{C}[\log t]\mathcal{A}^m_M$, we have $\sigma_*(P)(\vartheta)w(t,x) = 0$ and thus (19).

Put $\lambda = e(\bar{u}_i)$. First suppose $\deg(\bar{u}_i) = 0$. Then under the condition (20), Theorem 3.1 assures the unique existence of $\tilde{\phi}(t,x) \in \mathcal{A}^m_M$ such that $P_1 t^\lambda\tilde{\phi}(t,x) = 0$ and $t^{e(\bar{u}_i)}\tilde{\phi}(0,x) = \phi(x)u_i(t)$ and moreover Theorem 4.1 assures $P_j t^\lambda\tilde{\phi}(t,x) = 0$. If there exists another solution $\tilde{u} \in t^\lambda\mathbb{C}[\log t]\mathcal{A}^m_M$

of $\mathcal{M}$ with the same property, the leading exponent λ' of $u - \tilde{u}$ satisfies $\lambda' - e(\bar{u}_i) \in \mathbb{N}^n \setminus \{0\}$, which contradicts to (20). Thus we have proved the required uniqueness of the solution.

Next suppose $u_i = t^\lambda v_i(\log t)$ with $\deg v_i > 0$. Let V be a vector space spanned by the components of elements of $\mathbb{C}[\partial_\xi]v_i(\xi)$ and let $\{w_1(\xi), \dots, w_q(\xi)\}$ be a basis of V. Here we may assume $\mathbb{C}[\partial_\xi]w_k \in \sum_{\nu=1}^k \mathbb{C}w_\nu$ for $k = 1, \dots, q$. Let $\hat{u}$ be the vector of size qm with components $\hat{u}_\nu w_\nu(\log t)$ with $\hat{u}_\nu \in t^\lambda \mathcal{A}_M^m$ for $\nu = 1, \dots, q$. Then the system $\mathcal{M}$ is replaced by a system $\hat{\mathcal{M}}$ with an unknown function $\hat{u}$ where P_i are replaced by suitable $\hat{P}_i \in M(qm, \mathcal{D}_*)$, respectively. We note that $\hat{\mathcal{M}}$ also satisfies the assumption of the theorem because $\det\bigl(\sigma_*(\hat{P}_i)\bigr) = \det\bigl(\sigma_*(P_i)\bigr)^q$. Thus we may only consider the solutions with components in $t^\lambda \mathcal{A}_M$.

For example, if $n = n' = 1$ and $P = (\vartheta - \lambda)^2 + t^2\partial_x^2$, the solution of the equation $Pu = 0$ in the space $t^\lambda \mathcal{A}_M \oplus (t^\lambda \log t)\mathcal{A}_M$ corresponds to the solution of

$$\begin{pmatrix} (\vartheta - \lambda)^2 + t^2\partial_x^2 & 2(\vartheta - \lambda) \\ & (\vartheta - \lambda)^2 + t^2\partial_x^2 \end{pmatrix} \begin{pmatrix} u_1 \\ u_2 \end{pmatrix} = 0$$

in the space $t^\lambda \mathcal{A}_M^2$ by the correspondence $u = u_1 + u_2 \log t$.

To complete the proof of the system we have only to prove that the map (21) is surjective. Let u be any ideally analytic solution of $\mathcal{M}$. Then any leading exponent of u is an exponent of the system $\mathcal{M}$ and therefore we define $\phi_i(x)$ by (19) if $e(\bar{u}_i)$ is a leading exponent of u and by 0 otherwise. Then if $u \neq \sum_i T_{u_i}(\psi)$, any leading exponent of $u - \sum_i T_{\bar{u}_i}(\phi)$ is not in the set $\{e(\bar{u}_i)\}$, which contradicts the first claim in the theorem. □

We will return to the case when (16) is the complete system which has a holomorphic parameter $z \in U \subset \mathbb{C}^\ell$.

First assume that $\overline{\mathcal{M}}$ is semisimple for any $z \in U$ (cf. Definition 5.2) and that the indicial equation $\overline{\mathcal{M}}$ satisfies the assumption in Proposition 5.3 by putting $t_i = e^{-x_i}$ for $1 \leq i \leq n$. Then the proof of the previous theorem implies the following.

Proposition 6.4. *Assume that $\overline{\mathcal{M}}$ is semisimple for any $z \in U$. Let $\{\bar{u}_i(x, z) = t^{\lambda_i(z)} f_i(z)\,;\, i = 1, \dots, r\}$ be a basis of the solutions of* (17) *for any $z \in U$. Here $f_i(z) \in \mathcal{O}(U)^m$. Assume* (20) *for any $z \in U$. Then $T_{\bar{u}_i}(\phi)$ is holomorphic for $z \in U$ under the notation in Theorem 6.3.*

To examine the case without the assumption in this proposition, we study a generic holomorphic curve $t \mapsto z(t)$ through the point $z^o \in U$ where the assumption breaks. Hence we restrict the case when $\ell = 1$.

Suppose $\ell = 1$ and fix $z^o \in U$. For simplicity we put $z^o = 0$. Assume that $\mathcal{M}$ is semisimple (cf. Definition 5.2) for any fixed $z \in U \setminus \{0\}$. We will shrink U if necessary hereafter until the end of the following theorem. Let $\{\bar{u}_1, \dots, \bar{u}_r\}$ be a basis of the solutions of the indicial equation for $\forall z \in U \setminus \{0\}$, where $\bar{u}_i$ are

$$\bar{u}_i(t, z) = t^{\lambda_i(z)} f_i(z) \quad \text{for } i = 1, \dots, r$$

with suitable $f_i \in \mathcal{O}(U)^m$. Then Proposition 5.3 assures that there exist meromorphic functions $c_{ij}(z)$ such that by denoting

$$\bar{w}_i(t,z) = \sum_{j=1}^{r} c_{ij}(z)\bar{u}_j(t,z),$$

$\{\bar{w}_1, \ldots, \bar{w}_r\}$ is a basis of the solutions of the indicial equation for $\forall z \in U$ and $\bar{w}_j(t,z)$ are holomorphic function of $(\log t, z) \in \mathbb{C}^n \times U$. By virtue of (15), we may assume $c_{ij}(z) = 0$ if $\lambda_i(0) \neq \lambda_j(0)$.

Then we have the following theorem which is the main purpose of this note.

Theorem 6.5. *Under the notation above. there exist differential operators* $R_{ij}(x,z,\partial_x)$ *such that for any* $\phi(x,z) \in {}_U\mathcal{A}_M^m$, $\sum_{i=1}^r T_{\bar{u}_i}\big(R_{ij}(x,z,\partial_x)\phi(x,z)\big)$ *is a holomorphic function of* $z \in U$ *and an ideally analytic solution of* $\mathcal{M}$ *with the complete leading term* $\phi(x)\bar{w}_i(t,z)$ *for any fixed* $z \in U$. *Moreover the map*

$$\begin{aligned} \mathcal{A}_N^r &\xrightarrow{\sim} \{\textit{ideally analytic solutions of } \mathcal{M}\}, \\ (\phi_i(x)) &\mapsto \sum_{i,j} T_{\bar{u}_i}\big(R_{ij}(x,z,\partial_x)\phi_j(x)\big) \end{aligned}$$

holomorphically depends on $z \in U$ *and it is bijective for any* $z \in U$. *Here* $R_{ij}(x,z,\partial_x)$ *are holomorphic functions of* $z \in U \setminus \{0\}$ *valued in the space of differential operators on* N *and may have at most poles at* $z = 0$ *and moreover*

$$R_{ij}(x,z,\partial_x) = \begin{cases} 0 & \textit{if } \lambda_i(0) - \lambda_j(0) \notin \mathbb{N}^n, \\ c_{ij}(z) & \textit{if } \lambda_i(0) = \lambda_j(0). \end{cases}$$

Proof. We will inductively construct $R_{ij}(x,z,\partial_x)$ according to the number $L(\lambda_j) = \sum_{\nu=1}^n \Re\lambda_{j,\nu}(0)$. Here $\lambda_j = (\lambda_{j,1}, \ldots, \lambda_{j,n})$ and $\Re\zeta$ denotes the real part of $\zeta \in \mathbb{C}$.

Fix a positive integer k with $k \leq r$. By the hypothesis of the induction we may assume that R_{ij} have been constructed if $L(\lambda_j) > L(\lambda_k)$. Put $R_{jk}^{(0)} = c_{ik}(z)$. We inductively define $R_{ik}^{(\nu)}$ for $\nu = 0, 1, \ldots$ as follows. Put

$$\sum_{i=1}^{r} T_{\bar{u}_i}\big(R_{ik}^{(\nu)}\phi(x,z)\big) = z^{-n_\nu}\phi_{n_\nu}^{(\nu)}(t,x) + \cdots + z^{-1}\phi_1^{(\nu)}(t,x) + \phi_0^{(\nu)}(t,x,z)$$

with $\phi_0^{(\nu)}(t,x,z) \in {}_U\mathcal{A}_M$. Suppose $n_\nu > 0$. By the analytic continuation of $z^{n_\nu}\sum_i T_{\bar{u}_i}\big(R_{ik}^{(\nu)}\phi(x,z)\big)$, it is clear that $\phi_{n_\nu}^{(\nu)}(t,x)$ is a solution of $\mathcal{M}$ at $z = 0$. Any leading exponent μ of $\phi_{n_\nu}^{(\nu)}(t,x)$ satisfies $\mu - \lambda_k(0) \in \mathbb{N}^n \setminus \{0\}$ and hence the complete leading term of $\phi_{n_\nu}^{(\nu)}(t,x)$ is

$$\sum_{\lambda_j(0)\in\lambda_k(0)+(\mathbb{N}^n\setminus\{0\})} \psi_j^{(\nu)}(x)\bar{w}_j(t,0).$$

Note that $\psi_j^{(\nu)}(x) = P_j^{(\nu)}(x, \partial_x)\phi(x)$ for some differential operators which do not depend on $\phi(x)$. Put $P_j^{(\nu)}(x, \partial_x) = 0$ if $\lambda_j(0) - \lambda_k(0) \notin \mathbb{N}^n \setminus \{0\}$. Hence

$$\sum_{i=1}^{r} T_{\bar{u}_i}\big(R_{ik}^{(\nu)}(x, z, \partial_x)\phi(x)\big) - \sum_{i=1}^{r}\sum_{j=1}^{r} z^{-n_\nu} T_{\bar{u}_i}\big(R_{ij}(x, z, \partial_x)P_j^{(\nu)}(x, \partial_x)\phi(x)\big) \tag{22}$$

has a pole of order less than n_k. Defining

$$R_{ik}^{(\nu+1)}(x, z, \partial_x) = R_{ik}^{(\nu)}(x, z, \partial_x) - \sum_{j=0}^{r} z^{-n_\nu} R_{ij}(x, z, \partial_x)P_j^{(\nu)}(x, \partial_x)$$

inductively, we have $R_{ij}(x, z, \partial_x) = R_{ij}^{(\nu)}(x, z, \partial_x)$ for certain ν such that the left hand size of (22) is holomorphic at $z = 0$. □

Remark 6.6. Let $P_i \in \mathcal{D}_*$ for $i = 1, \ldots, n$ satisfies

$$\begin{cases} [P_i, P_j] = \sum_{\nu=1}^{n} R_{ij\nu}P_\nu \quad \text{for} \quad 1 \le i \le j \le n, \\ \sigma_*(P_i) \text{ do not depend on } x \in N, \\ \{\xi \in \mathbb{C}^n \,;\, \bar{\sigma}_*(P_1)(\xi) = \cdots = \bar{\sigma}_*(P_n)(\xi) = 0\} = \{0\} \end{cases}$$

with some $R_{ij\nu} \in \mathcal{D}_*$ satisfying $\sigma_*(R_{ij\nu}) = 0$. Then for a suitable positive integer L there exist $R_i \in \mathbb{C}[\vartheta]$ such that

$$\begin{cases} \operatorname{ord} P_i + \operatorname{ord} R_i = 2L, \\ \sigma_*(P_0) = \xi_1^{2L} + \cdots + \xi_n^{2L} \end{cases}$$

by putting

$$P_0 = \sum_{i=1}^{n} R_i P_i.$$

Then $\{P_0, \ldots, P_n\}$ satisfies (11) with $S = 0$ and $\sigma_*(T_{ij}) = 0$ because

$$[P_0, P_j] = \sum_{i=1}^{n}\Big([P_0, R_j]P_i + \sum_{\nu=1}^{n} R_i R_{ij\nu} P_\nu\Big)$$

and $\sigma_*([P_0, R_i]) = \sigma_*(R_i R_{ij\nu}) = 0$.

In this case let λ^o be an exponent of the system $P_i u = 0$ $(1 \le i \le n)$. Then for a suitable $\rho \in \mathbb{C}^n$ and a positive integer k, the system

$$_U\mathcal{M} : \big(P_i - \sigma_*(P)(\lambda^o + \rho z^k)\big)u = \sum R_i\big(P_i - \sigma_*(P)(\lambda^o + \rho z^k)\big)u = 0$$

satisfies the assumption of Theorem 6.5 for $U = \{z \in \mathbb{C};\, |z| < 1\}$ by changing the lower order terms of R_i if necessary. Hence we can analyze the ideally analytic solutions of $\mathcal{M}$ by the analytic continuation of the parameter z to the origin.

Theorem 6.7. *Retain the notation and the assumption in Theorem 6.5. Let r' be the dimension of the finitely generated $\mathbb{C}[\vartheta]$-module*

$$\bar{\mathcal{M}}^o := \sum_{j=1}^{m} \mathbb{C}[\vartheta]u_j \Big/ \sum_{i=0}^{q}\sum_{k=1}^{m} \mathbb{C}[\vartheta] \sum_{j=1}^{m} \bar{\sigma}_*(P_i)_{kj}(z^o,\vartheta)u_j.$$

Suppose $n' = 0$ and $r' \leq r$. Then $r' = r$ and any solution of $\mathcal{M}$ defined on a small connected neighborhood of $(t^o, x^o) \in M$ with $z = z^o$ and $0 < |t^o_j| \ll 1$ for $j = 1,\ldots,n$ is an ideally analytic solution given in Theorem 6.5. In particular the dimension of space of the solutions equals r.

Proof. Let w_ν for $\nu = 1,\ldots,r'$ be elements of $\sum_{j=1}^{m}\mathbb{C}[\vartheta]u_j$ whose residue classes form a basis of $\bar{\mathcal{M}}^o$. Fix $z = z^o$. Then in a neighborhood of $(0, x^o)$

$$\sum_{j=1}^{r} \mathcal{A}_M[\vartheta]u_j = \sum_{\nu=1}^{r'} \mathcal{A}_M w_\nu + \sum_{i=0}^{q}\sum_{k=1}^{m} \mathcal{A}_M[\vartheta] \sum_{j=1}^{m} (P_i)_{kj}u_j.$$

Let w be a column vector of size r' with components w_ν. Then the system $\mathcal{M}$ implies

$$\mathcal{N} : \vartheta_j w = Q_j(t)w \quad \text{for } j = 1,\ldots,n$$

with suitable $Q_j \in M(r', \mathcal{A}_M)$. Then any solution $w(t)$ of $\mathcal{N}$ on a neighborhood of (t^o, x^o) is analytic and $w = 0$ if $w(t^o) = 0$. Hence the dimension of the space of solutions of $\mathcal{N}$ is smaller than or equals to r'. But we have constructed r linearly independent solutions in Theorem 6.5. Hence we have this theorem. □

Remark 6.8. Retain the notation in Theorem 6.7. Suppose $q = n-1$, $[P_i, P_j] = 0$ for $0 \leq i < j \leq q$, $\bar{\sigma}_*(P_i)$ are diagonal matrices and

$$\{\xi \in \mathbb{C}^n\,;\, \bar{\sigma}_*(P_i)(\xi) = 0 \quad \text{for } i = 0,\ldots,q\} = \{0\}.$$

Then $r' = r$ and $r' = m\prod_{i=0}^{q} \operatorname{ord} P_i$.

7 Examples related to $SL(n, \mathbb{R})$

For a connected real reductive Lie group G and an open subgroup H of the fixed point group of an involutive automorphism σ of G, the homogeneous space G/H is called a *reductive symmetric homogeneous space.* Then in a suitable realization $\widetilde{X}$ of G/H constructed by [O6], the system of differential equations that defines the simultaneous eigenspace of the elements of the ring $\mathbb{D}(G/H)$ of the invariant differential operators on G/H has regular singularities along the boundaries of G/H in this realization. It is an important problem to study the eigenspace. For example, see [K–] in the cases of Riemannian symmetric spaces.

Note that the Lie group G is identified with a symmetric homogeneous space of $G \times G$ with respect to the involutive automorphism σ of G defined by $\sigma(g_1, g_2) = (g_2, g_1)$ for $(g_1, g_2) \in G_1 \times G_2$ and that any irreducible admissible representation of G can be realized in an eigenspace of $\mathbb{D}(G)$.

In this section we will consider differential equations related to the Lie group $G = SL(n, \mathbb{R})$, which give examples of the differential equations we study in this note. The element of the Lie algebra $\mathfrak{sl}(n, \mathbb{R})$ of G is identified with that of $M(n, \mathbb{R})$ whose trace equals 0. Let E_{ij} be the fundamental matrix unit whose (i, j)-component equals 1 and the other components are 0. Then $\mathfrak{sl}(n, \mathbb{R})$ is spanned by the elements $\tilde{E}_{ij} = E_{ij} - \frac{\delta_{ij}}{n}(E_{11} + \cdots + E_{nn})$ with $1 \le i \le j \le n$. For simplicity we put $\tilde{E}_i = \tilde{E}_{ii}$.

We identify $\mathfrak{sl}(n, \mathbb{R})$ with the space of right invariant vector field on G by

$$(Xf)(g) = \frac{\mathrm{d}}{\mathrm{d}t} f(g e^{tX}) \Big|_{t=0} \qquad \text{for } X \in \mathfrak{sl}(n, \mathbb{R}),\ f \in C^\infty(G) \text{ and } g \in G.$$

Here we note that

$$(E_{pq} f)\big((x_{ij})\big) := \frac{\mathrm{d}}{\mathrm{d}t} f\big((x_{ij}) e^{tE_{pq}}\big) \Big|_{t=0} = \Big(\sum_{\nu=1}^{n} x_{\nu p} \frac{\partial f}{\partial x_{\nu q}}\Big)\big((x_{ij})\big)$$

for $g \in C^\infty\big(GL(n, \mathbb{R})\big)$ and $\big(x_{ij}\big)_{\substack{1 \le i \le n \\ 1 \le j \le n}} \in GL(n, \mathbb{R})$ because (i, j)-component of $\big(x_{ij}\big)_{\substack{1 \le i \le n \\ 1 \le j \le n}} E_{pq}$ equals $x_{ip}\delta_{qj}$.

We first review, by examples, that the invariant differential operators of the Riemannian symmetric space G/K has regular singularities along the boundaries of the space in the realization constructed in [O2]. By the Iwasawa decomposition $G = \bar{N}AK$ with

$$\begin{aligned}
K &= SO(n) = \{g \in SL(n, \mathbb{R})\,;\, {}^t g g = I_n\}, \\
A &= \left\{ a = \begin{pmatrix} a_1 & & \\ & \ddots & \\ & & a_n \end{pmatrix} ;\, a_j > 0 \text{ for } 1 \le j \le n \text{ and } a_1 \cdots a_n = 1 \right\}, \qquad (23) \\
\bar{N} &= \left\{ \begin{pmatrix} 1 & & & \\ x_{21} & 1 & & \\ \vdots & \vdots & \ddots & \\ x_{n1} & x_{n2} & \cdots & 1 \end{pmatrix} ;\, x_{ij} \in \mathbb{R} \text{ for } 1 \le j < i \le n \right\}, \\
t_j &:= \frac{a_{j+1}}{a_j} \qquad \text{for } j = 1, \ldots, n-1,
\end{aligned}$$

the Riemannian symmetric space G/K is identified with the product manifold $\bar{N} \times A$ with the coordinate $(t_k, x_{ij}) \in (0, \infty)^{n-1} \times \mathbb{R}^{\frac{n(n-1)}{2}}$. Then the Lie algebra of the solvable group of $\bar{N}A$ is spanned by the elements

$$E_{ij} = \Big(\prod_{\nu=j}^{i-1} t_\nu\Big)\Big(\frac{\partial}{\partial x_{ij}} + \sum_{\nu=i+1}^{n} x_{\nu i}\frac{\partial}{\partial x_{\nu j}}\Big) \quad \text{for } 1 \le j < i \le n,$$

$$\tilde{E}_{ij} = E_{ij} - \frac{\delta_{ij}}{n}(E_{11} + \cdots + E_{nn}) \quad \text{for } 1 \le i \le n \text{ and } 1 \le j \le n,$$

$$E_i := \tilde{E}_{ii} = \vartheta_{i-1} - \vartheta_i \quad \text{for } 1 \le i \le n, \quad \vartheta_0 = \vartheta_{n+1} = 0.$$

The coordinate $(t_k, x_{ij}) \in \mathbb{R}^{\frac{(n+2)(n-1)}{2}}$ can be used for local coordinate of the realization of G/K.

Let $U(\mathfrak{g})$ be the universal enveloping algebra of the complexification $\mathfrak{g}$ of the Lie algebra of G. Then if $G = SL(n,\mathbb{R})$, the ring $\mathbb{D}(G/K)$ is naturally isomorphic to the center $U(\mathfrak{g})^G$ of $U(\mathfrak{g})$ and $U(\mathfrak{g})^G$ is generated by the elements $L_2, \ldots, L_n$ which are given by

$$\det\big(\tilde{E}_{ij} + (\tfrac{n+1}{2} - i - \lambda)\delta_{ij}\big) = L_n - L_{n-1}\lambda + \cdots + (-1)^n\lambda^n$$

for $\lambda \in \mathbb{C}$ (cf. [Ca]). Here $\det(A_{ij}) = \sum_{\sigma\in\mathfrak{S}_n} \operatorname{sgn}(\sigma)A_{\sigma(1)1}\cdots A_{\sigma(n)n}$ and $U(\mathfrak{g})^G$ is generated by the algebraically independent $(n-1)$-elements which are the coefficients of λ^k for $k = 0, 1, \ldots, n-2$.

Let $\mathfrak{k}$ be a Lie algebra of $SO(n)$, which is generated by the elements $E_{ij} - E_{ji}$ for $1 \le i < j \le n$.

Since

$$\begin{aligned}\varDelta_2 &= \det\begin{pmatrix} E_1 + \frac12 & E_{12} \\ E_{21} & E_2 - \frac12 \end{pmatrix} = (E_1 + \tfrac12)(E_2 - \tfrac12) - E_{21}E_{12} \\ &\equiv (E_1 + \tfrac12)(E_2 - \tfrac12) - E_{21}^2 \mod U(\mathfrak{g})\mathfrak{k} \\ &= -(\vartheta - \tfrac12)^2 - t^2\partial_x^2 = -t^2(\partial_t^2 + \partial_x^2) - \tfrac14 \quad \text{with } \vartheta = t\tfrac{\partial}{\partial t},\end{aligned}$$

we see that $\mathbb{D}\big(SL(2,\mathbb{R})/SO(2)\big) = \mathbb{C}[t^2\big(\frac{\partial^2}{\partial t^2} + \frac{\partial^2}{\partial x^2}\big)]$. Here $SL(2,\mathbb{R})/SO(2)$ is realized in the upper half plane $\{x + it\,;\,(t,x) \in (0,\infty)\times\mathbb{R}\}$ and $\varDelta_2$ has regular singularities along the real axis. On the other hand, the explicit form of the vector field L_X defined by the translation $e^{-sX}\cdot p$ for $s \in \mathbb{R}$, $X \in \mathfrak{g}$ and $p \in SL(2,\mathbb{R})/SO(2)$ is given by

$$L_{E_{21}} = -\partial_x,\ L_{E_1} = \vartheta + x\partial_x,\ L_{E_{12}} = 2x\vartheta - (t^2 - x^2)\partial_x.$$

When $G = SL(3,\mathbb{R})$, we have

$$\begin{aligned}\det\begin{pmatrix} E_1 + 1 - \lambda & E_{12} & E_{13} \\ E_{21} & E_2 - \lambda & E_{23} \\ E_{31} & E_{32} & E_3 - 1 - \lambda \end{pmatrix} &= (E_1 + 1 - \lambda)(E_2 - \lambda)(E_3 - 1 - \lambda) \\ &+ E_{21}E_{32}E_{13} + E_{31}E_{12}E_{23} - (E_{11} + 1 - \lambda)E_{32}E_{23} - E_{21}E_{12}(E_3 - 1 - \lambda) \\ &- E_{31}(E_2 - \lambda)E_{13} = \varDelta_3 - \varDelta_2\lambda - \lambda^3\end{aligned}$$

with

$$\begin{aligned}
\Delta_3 &= (E_1+1)E_2(E_3-1) + E_{21}E_{32}E_{13} + E_{31}E_{12}E_{23}\\
&\quad - (E_1+1)E_{32}E_{23} - E_{21}E_{12}(E_3-1) - E_{31}E_2E_{13}\\
&\equiv (E_1+1)E_2(E_3-1) - (E_1+1)E_{32}^2 - (E_3-1)E_{21}^2 - (E_2-1)E_{31}^2\\
&\qquad + 2E_{21}E_{32}E_{31} \mod U(\mathfrak{g})\mathfrak{k}\\
&= -(\vartheta_1-1)(\vartheta_1-\vartheta_2)(\vartheta_2-1) + 2t_1^2t_2^2(\partial_x+y\partial_z)\partial_y\partial_z\\
&\qquad + (\vartheta_1-1)t_2^2\partial_y^2 - (\vartheta_1-\vartheta_2-1)t_1^2t_2^2\partial_z^2 - (\vartheta_2-1)t_1^2(\partial_x+y\partial_z)^2,\\
\Delta_2 &= E_2(E_3-1) + (E_1+1)(E_3-1) + (E_1+1)E_2\\
&\quad - E_{32}E_{23} - E_{21}E_{12} - E_{31}E_{13}\\
&\equiv E_2(E_3-1) + (E_1+1)(E_3-1) + (E_1+1)E_2\\
&\quad - E_{32}^2 - E_{21}^2 - E_{31}^2 \mod U(\mathfrak{g})\mathfrak{k}\\
&= -(\vartheta_1-1)^2 + (\vartheta_1-1)(\vartheta_2-1) - (\vartheta_2-1)^2\\
&\quad - t_2^2\partial_y^2 - t_1^2t_2^2\partial_z^2 - t_1^2(\partial_x+y\partial_z)^2,
\end{aligned}$$

$$x = x_{21},\ y = x_{32} \text{ and } z = x_{31}.$$

Then $\mathbb{D}\big(SL(3,\mathbb{R})/SO(3)\big) = \mathbb{C}[\bar{\Delta}_3, \bar{\Delta}_2]$, where $\bar{\Delta}_3$ and $\bar{\Delta}_2$ are the last expressions of Δ_3 and Δ_2 in the above, respectively. This expression of invariant differential operators on $SL(3,\mathbb{R})/SO(3)$ is given by [O1] to obtain the Poisson integral representation of any simultaneous eigenfunction of the operators on the space, where such representation is first obtained in the space with the rank larger than one. In fact $4\Delta_2$ and $8\Delta_2 + 8\Delta_3$ are explicitly written there under the coordinate (s,t,u,v,w) with $(s,t,u,v,w) = (t_2^2, t_1^2, x, y, z)$, which corresponds to a local coordinate system in the realization given in [OS].

When $G = SL(n,\mathbb{R})$ the second order element L_2 of $U(\mathfrak{g})^G$ is

$$\begin{aligned}
L_2 &= \sum_{1\le i<j\le n} \Big((E_i + \frac{n+1}{2} - i)(E_j + \frac{n+1}{2} - j) - E_{ji}E_{ij}\Big)\\
&\equiv \sum_{1\le i<j\le n} (\tilde{\vartheta}_{i-1} - \tilde{\vartheta}_i)(\tilde{\vartheta}_{j-1} - \tilde{\vartheta}_j)\\
&\qquad - \sum_{1\le i<j\le n} \Big(\prod_{\nu=i}^{j-1} t_\nu^2\Big)\Big(\frac{\partial}{\partial x_{ji}} + \sum_{\nu=j+1}^{n} x_{\nu j}\frac{\partial}{\partial x_{\nu i}}\Big)^2 \mod U(\mathfrak{g})\mathfrak{k},\\
\tilde{\vartheta}_i &= \vartheta_i - \tfrac{i(n-i)}{2}
\end{aligned}$$

and $\mathbb{D}(G/K) = \mathbb{C}[\bar{L}_2, \ldots, \bar{L}_n]$ satisfying

$$\begin{aligned}
\sigma_*(\bar{L}_k) &= \sum_{1\le i_1<i_2<\cdots<i_k\le n} (\tilde{\xi}_{i_1-1} - \tilde{\xi}_{i_1})(\tilde{\xi}_{i_2-1} - \tilde{\xi}_{i_2})\cdots(\tilde{\xi}_{i_k-1} - \tilde{\xi}_{i_k}),\\
\tilde{\xi}_i &= \xi_i - \tfrac{i(n-i)}{2}
\end{aligned}$$

for $k = 2, \ldots, n$.

We will examine more examples. For a in (23) we have

$$\operatorname{Ad}(a^{-1})E_{ij} := aE_{ij}a^{-1} = a_i^{-1}a_jE_{ij} = t_{ij}E_{ij},$$
$$t_{ij} = a_i^{-1}a_j = \begin{cases} t_it_{i+1}\cdots t_{j-1} & \text{if } i \le j,\\ t_j^{-1}t_{j+1}^{-1}\cdots t_{i-1}^{-1} & \text{if } i > j,\end{cases}$$
$$U(\mathfrak{g})\mathfrak{k} = \sum_{1\le i<j\le n} U(\mathfrak{g})(E_{ij} - E_{ji}).$$

Hence

$$\begin{aligned}
&\operatorname{Ad}(a^{-1})(E_{ij} - E_{ji})^2 + (E_{ij} - E_{ji})^2\\
&\qquad - (t_{ij} + t_{ij}^{-1})\operatorname{Ad}(a^{-1})(E_{ij} - E_{ji})\cdot(E_{ij} - E_{ji})\\
&\quad = (t_{ij}^2 - 1)E_{ij}E_{ji} + (t_{ij}^{-2} - 1)E_{ji}E_{ij}\\
&\quad = (t_{ij} - t_{ij}^{-1})^2E_{ji}E_{ij} + (t_{ij}^2 - 1)(E_{ii} - E_{jj}),\\
&\operatorname{Ad}(a^{-1})(E_{ij} - E_{ji})\cdot E_{ij} - t_{ij}E_{ij}^2 = -t_{ij}^{-1}E_{ji}E_{ij}.
\end{aligned}$$

Thus we have

$$\begin{aligned}
E_{ji}E_{ij} &= \frac{t_{ij}^2}{1 - t_{ij}^2}(E_{ii} - E_{jj}) \qquad (24)\\
&\quad + \frac{t_{ij}^2}{(1 - t_{ij}^2)^2}\left(\operatorname{Ad}(a^{-1})(E_{ij} - E_{ji})^2 + (E_{ij} - E_{ji})^2\right)\\
&\quad - \frac{t_{ij}(1 + t_{ij}^2)}{(1 - t_{ij}^2)^2}\operatorname{Ad}(a^{-1})(E_{ij} - E_{ji})\cdot(E_{ij} - E_{ji})\\
&= t_{ij}^2E_{ij}^2 - t_{ij}\operatorname{Ad}(a^{-1})(E_{ij} - E_{ji})\cdot E_{ij}. \qquad (25)
\end{aligned}$$

Let (ϖ, V_ϖ) be a finite dimensional representation of a closed subgroup H of G and $C^\infty(G; V_\varpi)$ denote the space of V_ϖ-valued C^∞-functions on G. Then the space of C^∞-sections $C^\infty(G/H; \varpi)$ of the G-homogeneous bundle associated to ϖ is

$$\{f \in C^\infty(G; V_\varpi);\ f(gh) = \varpi^{-1}(h)f(g) \text{ for } \forall h \in H\}.$$

Consider the case when $H = K$. Because of the decomposition $G = KAK$ the function $f \in C^\infty(G/K; \varpi)$ is determined by its restriction on KA and by the natural map $K \times A \to KA$ the restriction can be considered as a function $\bar{f}$ on $K \times A$. Then the action of the differential operator L_2 to $\bar{f}$ is

$$\begin{aligned}
\bar{L}_2 = \sum_{1\le i<j\le n}\Big(&(\tilde{\vartheta}_{i-1} - \tilde{\vartheta}_i)(\tilde{\vartheta}_{j-1} - \tilde{\vartheta}_j) - \frac{t_{ij}^2}{1 - t_{ij}^2}(\vartheta_{i-1} - \vartheta_i - \vartheta_{j-1} + \vartheta_j)\\
&- \frac{t_{ij}^2}{(1 - t_{ij}^2)^2}\left(\operatorname{Ad}(a^{-1})(E_{ij} - E_{ji})^2 + \varpi(E_{ij} - E_{ji})^2\right)\\
&- \frac{t_{ij}(1 + t_{ij}^2)}{(1 - t_{ij}^2)^2}\operatorname{Ad}(a^{-1})(E_{ij} - E_{ji})\cdot\varpi(E_{ij} - E_{ji})\Big)
\end{aligned}$$

at $(k,a) \in K \times A$, which follows from (24). Here the induced representation of the Lie algebra $\mathfrak{k}$ of K is also denoted by ϖ.

Let (δ, V_δ) be an irreducible representation of K. Then the δ-component of $C^\infty(G/K;\varpi)$ is an element $f \in V \otimes C^\infty(G/K;\varpi)$ which satisfies

$$\frac{\mathrm{d}}{\mathrm{d}t} f(\mathrm{e}^{tX} g)\Big|_{t=0} = \big(\delta(X)f\big)(g)$$

for $X \in \mathfrak{k}$. Hence the function f is determined by its restriction $\bar{f}$ on A and the action of the operator L_2 to $\bar{f}$ is

$$\begin{aligned}\bar{L}_2 = \sum_{1\le i<j\le n} \Big((\tilde{\vartheta}_{i-1} - \tilde{\vartheta}_i)(\tilde{\vartheta}_{j-1} - \tilde{\vartheta}_j) - \frac{t_{ij}^2}{1-t_{ij}^2}(\vartheta_{i-1} - \vartheta_i - \vartheta_{j-1} + \vartheta_j) \\ - \frac{t_{ij}^2}{(1-t_{ij}^2)^2}\big(\delta(E_{ij}-E_{ji})^2 + \varpi(E_{ij}-E_{ji})^2\big) \\ - \frac{t_{ij}(1+t_{ij}^2)}{(1-t_{ij}^2)^2}\delta(E_{ij}-E_{ji}) \otimes \varpi(E_{ij}-E_{ji})\Big).\end{aligned} \tag{26}$$

Note that the operator $P = \bar{L}_2$ satisfies the assumption of Corollary 2.3.

When G is $SL(2,\mathbb{R})$ or its universal covering group and $\bar{f}$ is an eigenfunction of L_2, we can put $\varpi(E_{12} - E_{21}) = \sqrt{-1}k$ and $\delta(E_{12} - E_{21}) = -\sqrt{-1}m$ for certain numbers k and m and

$$\Big(\vartheta^2 + \frac{1}{4} - \frac{1+t^2}{1-t^2}\vartheta + \frac{t(k-mt)(m-kt)}{(1-t^2)^2} - (\lambda + \frac{1}{2})^2\Big)\bar{f} = 0.$$

Put $t = \mathrm{e}^{-x}$ and $u = \bar{f}$. Then $\vartheta = -\frac{\mathrm{d}}{\mathrm{d}x}$ and

$$u'' + \coth x \cdot u' - \frac{(k+m)^2}{4\sinh^2 x}u + \frac{km}{4\sinh^2\frac{x}{2}}u = \lambda(\lambda+1)u,$$

$$\frac{\mathrm{d}^2 v}{\mathrm{d}z^2} - \frac{(k+m-1)(k+m+1)}{\sinh^2 2z}v + \frac{km}{\sinh^2 z}v = (2\lambda+1)^2 v.$$

by denoting $v = \sinh^{\frac{1}{2}} x \cdot u$ and $z = \frac{x}{2}$.

Then for $\tilde{v} = \sinh^m z \cdot \sinh^{-\frac{k+m+1}{2}} 2z \cdot v$ and $w = -\sinh^2 z$ we have

$$w(1-w)\frac{\mathrm{d}^2\tilde{v}}{\mathrm{d}w^2} + \Big(\frac{k-m+2}{2} - (k+2)w\Big)\frac{\mathrm{d}\tilde{v}}{\mathrm{d}w} - \Big(\frac{k}{2} - \lambda\Big)\Big(\frac{k}{2} + \lambda + 1\Big)\tilde{v} = 0$$

and hence $\bar{f}$ is a linear combination of the functions

$$\begin{cases}\sinh^{\frac{k-m}{2}} z \cdot \cosh^{\frac{k+m}{2}} z \cdot F(\frac{k}{2} - \lambda, \frac{k}{2} + \lambda + 1, \frac{k-m}{2} + 1; -\sinh^2 z), \\ \sinh^{\frac{m-k}{2}} z \cdot \cosh^{\frac{k+m}{2}} z \cdot F(\frac{m}{2} - \lambda, \frac{m}{2} + \lambda + 1, \frac{m-k}{2} + 1; -\sinh^2 z).\end{cases}$$

Thus it is clear that the non-zero real analytic solution $\bar{f}$ defined in a neighborhood of the point $z = 0$ exists if and only if $k - m \in 2\mathbb{Z}$. Here $F(\alpha, \beta, \gamma; z)$ denotes the Gauss hypergeometric function (cf. [W]).

Next we assume $H = N$ and ϖ is a character of N. Then there exist complex numbers $c_1, \ldots, c_{n-1}$ such that

$$\varpi(\mathrm{e}^{\sum_{1\le i<j\le n} s_{ij}E_{ij}}) = \mathrm{e}^{\sqrt{-1}(c_1 s_{12}+\cdots+c_{n-1}s_{n-1,n})}.$$

The element $f \in C^\infty(G/N;\varpi)$ is determined by the restriction $\bar{f} = f|_{KA}$ and it follows from (25) that the operation of L_2 to $\bar{f}$ is

$$\sum_{1\le i<j\le n} (\tilde{\vartheta}_{i-1}-\tilde{\vartheta}_i)(\tilde{\vartheta}_{j-1}-\tilde{\vartheta}_j) + \sum_{1\le i<n} \left(c_i^2 t_i^2 + \sqrt{-1}c_i t_i (E_{i,i+1}-E_{i+1,i})\right). \quad (27)$$

Hence if $G = SL(2,\mathbb{R})$, the eigenfunction f of L_2 of the δ-component of $C^\infty(G/N;\varpi)$ with $\delta(E_{12}-E_{21}) = \sqrt{-1}m$ satisfies

$$\Bigl(-(\vartheta-\tfrac{1}{2})^2 + c_1^2 t^2 - c_1 m t + (\lambda+\tfrac{1}{2})^2\Bigr) f|_A = 0$$

and hence

$$\frac{\mathrm{d}^2}{\mathrm{d}t^2}(f|_A) - \Bigl(c_1^2 - \frac{c_1 m}{t} + \frac{\lambda(\lambda+1)}{t^2}\Bigr)(f|_A) = 0.$$

If we put $u(x) = \mathrm{e}^{\frac{x}{2}}\bigl(f|_A(\mathrm{e}^{-x})\bigr)$, then

$$u'' - \bigl(c_1^2 \mathrm{e}^{-2x} - c_1 m \mathrm{e}^{-x}\bigr)u = (\lambda+\tfrac{1}{2})^2 u.$$

Denoting $W(\pm 2c_1 t) = f|_A(t)$, we have the Whittaker equation (cf. [W])

$$W'' + \Bigl(-\frac{1}{4} \pm \frac{m}{2t} + \frac{\frac{1}{4}-(\lambda+\frac{1}{2})^2}{t^2}\Bigr)W = 0.$$

8 Completely integrable quantum systems

A Schrödinger operator

$$P = \sum_{k=1}^{n} \frac{\partial^2}{\partial x_k^2} + R(x_1,\ldots,x_n)$$

of n variables is called *completely integrable* if there exist n algebraically independent differential operators P_k such that

$$[P_i,P_j] = 0 \quad \text{for } 1 \le i < j \le n \quad \text{and} \quad P \in \mathbb{C}[P_1,\ldots,P_n].$$

Under the coordinate system $(t_1,\ldots,t_n)$ with

$$t_1 = \mathrm{e}^{x_1-x_2}, \ldots, t_{n-1} = \mathrm{e}^{x_{n-1}-x_n}, t_n = \mathrm{e}^{x_n},$$

the Schrödinger operators P which belong to $\mathcal{D}_*$ and have elements $Q \in \mathcal{D}_*$ satisfying

$$Q = \sum_{k=1}^{n} \frac{\partial^4}{\partial x_k^4} + Q' \quad \text{with} \quad \operatorname{ord} Q' < 4$$

are classified in [O8] and proved to be completely integrable (cf. [O7] and [O9]). They are reduced to the Schrödinger operators with the potential functions $R(x_1, \ldots, x_n)$ in the following list.

$$\begin{aligned}&\sum_{1 \le i < j \le n} C_1 \bigl(\sinh^{-2} \tfrac{x_i + x_j}{2} + \sinh^{-2} \tfrac{x_i - x_j}{2}\bigr) \\ &\quad + \sum_{k=1}^{n} \bigl(C_2 \sinh^{-2} x_k + C_3 \sinh^{-2} \tfrac{x_k}{2}\bigr),\end{aligned} \qquad \text{(Trig-}BC_n\text{-reg)}$$

$$\sum_{1 \le i < j \le n} C_1 \sinh^{-2} \tfrac{x_i - x_j}{2} + \sum_{k=1}^{n} \bigl(C_2 e^{x_k} + C_3 e^{2x_k}\bigr), \qquad \text{(Trig-}A_{n-1}\text{-bry-reg)}$$

$$C_1 \sum_{i=1}^{n-1} e^{x_i - x_{i+1}} + C_1 e^{x_{n-1} + x_n} + C_2 \sinh^{-2} \tfrac{x_n}{2} + C_3 \sinh^{-2} x_n, \qquad \text{(Toda-}D_n\text{-bry)}$$

$$C_1 \sum_{i=1}^{n-1} e^{x_i - x_{i+1}} + C_2 e^{x_n} + C_3 e^{2x_n}. \qquad \text{(Toda-}BC_n\text{)}$$

Here C_1, C_2 and C_3 are any complex numbers.

We can generalize the Schrödinger operators in terms of root systems (cf. [OP]). Let Σ be an irreducible root system with rank n, Σ^+ a positive system of Σ and $\Psi \subset \Sigma$ a fundamental system of Σ^+. Then Σ is identified with a finite subset of a Euclidean space $\mathbb{R}^n$ and

$$P = \sum_{k=1}^{n} \frac{\partial^2}{\partial x_k^2} + \sum_{\alpha \in \Sigma^+} \frac{C_\alpha}{\sinh^2 \frac{\langle \alpha, x \rangle}{2}} \quad (C_\alpha \in \mathbb{C},\ C_\alpha = C_\beta \text{ if } |\alpha| = |\beta|) \tag{28}$$

and

$$P = \sum_{k=1}^{n} \frac{\partial^2}{\partial x_k^2} + \sum_{\alpha \in \Psi} e^{\langle \alpha, x \rangle} \tag{29}$$

are Schrödinger operators of Heckman-Opdam's hypergeometric system (cf. [HO]) and Toda finite chain (cf. [To]) corresponding to the fundamental system Ψ, respectively. They are in $\mathcal{D}_*$ under the coordinate system

$$t_k = e^{\langle \alpha_k, x \rangle} \quad \text{for} \quad k = 1, \ldots, n$$

with $\Psi = \{\alpha_1, \ldots, \alpha_n\}$ and known to be completely integrable.

If Σ is of type BC_n, then

$$\Sigma^+ = \{e_i - e_j,\ e_k,\ 2e_k\,;\ 1 \le i < j \le n,\ 1 \le k \le n\}$$

and the Schrödinger operators (28) and (29) correspond to (Trig-BC_n-reg) or (Toda-BC_n). If Σ is of other classical type, the operators also correspond to special cases of (Trig-A_{n-1}-bry) or (Toda-D_n-bry) or (Toda-BC_n).

The potential functions $R(x)$ of known completely integral quantum systems which may not have regular singularities at infinity are expressed by functions with one variable. If P_2 and P_3 are operators of order 4 and 6 with the highest order terms $\sum_{k=1}^n \frac{\partial^4}{\partial x_k^4}$ and $\sum_{k=1}^n \frac{\partial^6}{\partial x_k^6}$, respectively, this is proved by [Wa] in general. We will examine this in the case when $n = 2$.

Theorem 8.1. *Let ℓ be a positive integer. Suppose the differential operators*

$$\begin{aligned} P &= \frac{\partial^2}{\partial x^2} + \frac{\partial^2}{\partial y^2} + R(x, y), \\ Q &= \sum_{i=0}^{m} c_i \frac{\partial^m}{\partial x^{m-i}\partial y^i} + \sum_{i+j\le m-2} S_{i,j}(x, y)\frac{\partial^{i+j}}{\partial x^i \partial y^j} \end{aligned} \tag{30}$$

satisfy $[P, Q] = 0$ and $\sigma_m(Q) \notin \mathbb{C}[\sigma(P)]$. Here $R(x, y)$ and $S_{i,j}(x, y)$ are square matrices of size ℓ whose components are functions of (x, y) and $c_i \in \mathbb{C}$. Put

$$\Big(\xi\frac{\partial}{\partial \tau} - \tau\frac{\partial}{\partial \xi}\Big)\sum_{i=0}^{m} c_i \xi^{m-i}\tau^i = \prod_{\nu=1}^{L}(a_\nu\xi - b_\nu\tau)^{m_\nu} \tag{31}$$

with suitable $(a_\nu, b_\nu) \in \mathbb{C}^2 \setminus \{0\}$ satisfying $a_\nu b_\mu \neq a_\mu b_\nu$ for $\mu \neq \nu$. Here m_ν are positive integers and $m_1 + \cdots + m_L = m$. Then

$$R(x, y) = \sum_{\nu=1}^{L}\sum_{i=0}^{m_\nu-1}(b_\nu x + a_\nu y)^i R_{\nu,i}(a_\nu x - b_\nu y) \tag{32}$$

with m square matrices of size ℓ whose components are functions $R_{\nu,i}(t)$ of the one variable t.

Proof. The coefficients of $\frac{\partial^{m+1}}{\partial x^{m-1-j}\partial y^j}$ in the expression $[P, Q]$ for (30) show

$$2\partial_x S_{m-2-j,j} + 2\partial_y S_{m-1-j,j-1} = c_j(m-j)\partial_x R + c_{j+1}(j+1)\partial_y R$$

for $j = 0, \ldots, m-1$. Hence the theorem follows from the following equation.

$$\begin{aligned} 0 &- 2\sum_{j=0}^{m-1}(-1)^j\big(\partial_x^j\partial_y^{m-j}S_{m-2-j,j} + \partial_x^{j+1}\partial_y^{m-1-j}S_{m-1-j,j-1}\big) \\ &= 2\sum_{j=0}^{m-1}(-1)^j\partial_x^j\partial_y^{m-1-j}\big(c_j(m-j)\partial_x R + c_{j+1}(j+1)\partial_y R\big) \\ &= \sum_{j=0}^{m-1}(-1)^j c_j(m-j)\partial_x^{j+1}\partial_y^{m-1-j}R + \sum_{j=0}^{m-1}(-1)^j c_{j+1}(j+1)\partial_x^j\partial_y^{m-j}R \\ &= \Big(\big(\xi\frac{\partial}{\partial\tau} - \tau\frac{\partial}{\partial\xi}\big)\sum_{i=0}^{m}c_i\xi^{m-i}\tau^i\Big)\Big|_{\xi=\partial_y,\,\tau=-\partial_x} R. \end{aligned}$$

□

References

[Ca] Capelli, A.: Üeber die Zurückführung der Cayley'schen Operation Ω auf gewöhnliche Polar-Operationen. Math. Ann. **29**, 331-338(1887)

[Ha] Harish-Chandra: Collected Papers I-IV. Springer(1989)

[HO] Heckman, G.J., Opdam, E.M.: Root system and hypergeometric functions. I. Comp. Math. **64**, 329–352(1987)

[K–] Kashiwara, M., Kowata, A., Minemura, K., Okamoto, K,, Oshima, T., Tanaka, M.: Eigenfunctions of invariant differential operators on a symmetric space. Ann. of Math. **107**, 1–39(1978)

[KO] Kashiwara, M., Oshima, T.: Systems of differential equations with regular singularities and their boundary value problems. Ann. of Math. **106**, 145–200(1977)

[OP] Olshanetsky, M.A., Perelomov, A.M.:, Quantum integrable systems related to Lie algebras. Phys. Rep. **94**, 313–404(1983)

[O1] Oshima, T.: A boundary value problem on a Riemannian symmetric space. SuriKaisekiKenkyusho Koukyuroku **249**, 10–21(1975), in Japanese

[O2] Oshima, T.: A realization of Riemannian symmetric spaces. J. Math. Soc. Japan **53**, 117–132(1978)

[O3] Oshima, T.: A definition of boundary values of solutions of partial differential equations with regular singularities. Publ. RIMS Kyoto Univ. **19**, 1203–1230(1983)

[O4] Oshima, T.: Boundary value problems for systems of linear partial differential equations with regular singularities. Advanced Studies in Pure Math. **4**, 391–432(1984)

[O5] Oshima, T.: Asymptotic behavior of spherical functions on semisimple symmetric spaces. Advanced Studies in Pure Math. **14**, 561–601 (1988)

[O6] Oshima, T.: A realization of semisimple symmetric spaces and construction of boundary value maps. Advanced Studies in Pure Math. **14**, 603–650(1988)

[O7] Oshima, T.: Completely integrable systems with a symmetry in coordinates. Asian Math. J. **2**, 935–956(1998)

[O8] Oshima, T.: A class of completely integrable quantum systems associated with classical root systems. Indag. Mathem. **16**, 655–677(2005)

[O9] Oshima, T.: Completely integrable quantum systems associated with classical root systems. SIGMA **3**, 061, 50pp (2007).

[OS] Oshima, T., Sekiguchi, J.: Eigenspaces of invariant differential operators on an affine symmetric space. Invent. Math. **57**, 1–81(1980)

[Su] Sutherland, B: Exact results for a quantum many-body problem in one dimension II. Phys. Rev. **A5**, 1372–1376(1972)

[To] Toda, M.: Wave propagation in anharmonic lattice. J. Phy. Soc. Japan **23**, 501–506(1967)

[Wa] Wakida, S.: Quantum integrable systems associated with classical Weyl groups. MA Thesis, University of Tokyo, Tokyo(2004)

[W] Whittaker, E.T., Watson, G.N.: A Course of Modern Analysis, Fourth Edition. Cambridge University Press (1927)

The behaviors of singular solutions of some partial differential equations in the complex domain

Sunao Ōuchi

Sophia Univ., Chiyoda-ku, Tokyo 102-8554, Japan
ouchi@mm.sophia.ac.jp

Summary. The aim of this paper is to study behaviors of solutions of nonlinear partial differential equations in $\mathbb{C}^{d+1}$ with singularities on a complex hypersurface K. We introduce a class of differential equations and give asymptotic terms near K with remainder estimates of singular solutions.

Key words: singular solutions, asymptotic behaviors, complex partial differential equations

1 Introduction and Notations

Let $L(u) = 0$ be a partial differential equation in a neighborhood of $z = 0$ in $\mathbb{C}^{d+1}$, $z = (z_0, z_1, \cdots, z_d)$ and K be a complex hypersurface through $z = 0$. We choose a coordinate so that $K = \{z_0 = 0\}$. Suppose that $u(z)$ solves $L(u) = 0$, which is not necessarily holomorphic on K. Our question is *how $u(z)$ behaves as z tends to K*, more concretely,

(1) to obtain more precise growth order of $u(z)$ near singularities,
(2) to obtain asymptotic terms of $u(z)$ as z_0 tends to 0, if possible.

As for linear equations these problems were studied in [3], [4],[5] and [6]. Let $P(z, \partial_z)$ be a linear partial differential operator with holomorphic coefficients. The equation $P(z, \partial_z)u = f(z)$ was studied there, where $f(z)$ is holomorphic or has an asymptotic expansion with respect to z_0 as $z_0 \to 0$. We can define an index $\gamma > 0$ for $P(z, \partial_z)$. The main result in [3] and [5] was as follows:

Suppose that

1. the behavior of $u(z)$ is bounded by some function with at most some exponential growth near K, that is, for any $\varepsilon > 0$ $|u(z)| \le C_\varepsilon \exp(\varepsilon |z_0|^{-\gamma})$,

Received 19 January, 2006. Revised 12 April, 2006. Accepted 17 April, 2006.

2. $f(z)$ has an asymptotic expansion

$$f(z) \sim \sum_{n=0}^{\infty} f_n(z')z_0^n \quad \text{with} \quad |f_n(z')| \le AB^n \Gamma(\frac{n}{\gamma}+1),$$

which is called an asymptotic expansion of Gevrey type with exponent γ^{-1}.

Then under some additional conditions on $P(z,\partial_z)$ and $u(z)$, $u(z)$ has also an asymptotic expansion of the same Gevrey type as $f(z)$. A few results in these papers are stated in section 2 as an introduction to nonlinear equations for the readers. In section 3 we introduce a class of nonlinear partial differential equations and find asymptotic terms of its solution $u(z)$ as z_0 tends to 0,

$$\begin{cases} u(z) \sim \sum_{n=0}^{\infty} u_n(z) \quad z_0 \to 0 \\ u_n(z) = O(|z_0|^{p_n}) \\ p_0 < p_1 < \cdots < p_n < \to +\infty, \end{cases} \tag{1}$$

and obtain remainder estimates of Gevrey type, which is the main result in this article and is a generalization of that in [6] to nonlinear equations. The details of this article will be published elsewhere.

Finally in this section let us give notations used, which are usual. $\mathbb{N} = \{0,1,2,\cdots\}$ is the set of all nonnegative integers. The coordinate is denoted by $z = (z_0, z') = (z_0, z_1, \cdots, z_d) \in \mathbb{C}^{d+1}$ and the norms $|z|$ and $|z'|$ are defined by $|z| = \max_{0\le i\le d} |z_i|$ and $|z'| = \max_{1\le i\le d}|z_i|$ respectively. $\alpha = (\alpha_0, \alpha') = (\alpha_0, \alpha_1, \cdots, \alpha_d) \in \mathbb{N}^{d+1}$ is a multi-index and $|\alpha| = \alpha_0 + |\alpha'| = \sum_{i=0}^d \alpha_i$ is the length of α. As for differentiations $\partial_{z_i} = \dfrac{\partial}{\partial z_i}$, $\vartheta = z_0 \dfrac{\partial}{\partial z_0}$ and $\vartheta^{\alpha_0}\partial^{\alpha'} = (z_0\partial_{z_0})^{\alpha_0}\partial_{z_1}^{\alpha_1}\cdots\partial_{z_d}^{\alpha_d}$ for a multi-index $\alpha \in \mathbb{N}^{d+1}$. $\mathscr{O}(W)$ is the set of all holomorphic functions on a region W. $\mathscr{O}(W)[\lambda]$ is the totality of polynomial in λ with coefficients in $\mathscr{O}(W)$.

In this article we study solutions holomorphic on a sectorial region. Let $U = U_0 \times U'$ be a polydisk with center $z = 0$, where $U_0 = \{z_0; |z_0| < R_0\}$ and $U' = \{z'; |z'| < R\}$. $U_0(\theta) = \{z_0; 0 < |z_0| < R_0, |\arg z_0| < \theta\}$ is a sector and set $U(\theta) = U_0(\theta) \times U'$, which is sectorial with respect to z_0. For open sets V and W, $W \Subset V$ means $\bar{W}$ is compact and $\bar{W} \subset V$. For sectors $T = \{z_0; 0 < |z_0| < r_0, |\arg z_0| < \theta^{**}\}$ and $S = \{z_0; 0 < |z_0| < R_0, |\arg z_0| < \theta^*\}$, $T \Subset S$ means $r_0 < R_0$ and $\theta^{**} < \theta^*$. Solutions considered in this article are in $\mathscr{O}(U(\theta))$ for a polydisk U and a $\theta > 0$.

2 Linear equations

Let us state some results in [3] and [4] concerning behaviors of singular solutions of linear equations. Let $L(z,\partial_z)$ be an m-th order linear partial differential operator with coefficients in $\mathcal{O}(U)$. Suppose that $L(z,\partial_z)$ has the following form:

$$L(z,\partial_z) = A(z,\partial_{z_0}) + B(z,\partial_z),$$

where

$$\begin{cases} A(z,\partial_{z_0}) = \sum_{h=0}^{k} c_h(z')(z_0\partial_{z_0})^h, \\ B(z,\partial_z) = \sum_{|\alpha|\le m} b_\alpha(z)\partial_z^\alpha, \\ \quad b_\alpha(z) = z_0^{j_\alpha} c_\alpha(z), \quad c_\alpha(0,z') \not\equiv 0, \quad j_\alpha \in \mathbb{N}. \end{cases}$$

We put the following assumptions on $L(z,\partial_z)$:

$$\begin{cases} c_k(0) \ne 0, \\ j_\alpha - \alpha_0 > 0. \end{cases} \tag{C}$$

We choose U' so small that $c_k(z') \ne 0$ for $z' \in U'$, if necessary. Hence $A(z,\partial_{z_0})$ is an ordinary differential operator with order k and $z_0 = 0$ is a regular singular surface. Define a polynomial $\chi(\lambda, z')$ induced from $A(z,\partial_{z_0})$ by

$$\chi(\lambda,z') = \sum_{h=0}^{k} c_h(z')\lambda^h \in \mathcal{O}(U')[\lambda] \tag{2}$$

and take constants a_0 and a_1 such that

$$\{\lambda; \chi(\lambda,z') = 0, z' \in U'\} \subset \{\lambda; a_0 \le \mathrm{Re}\lambda \le a_1\}.$$

Define an index γ by

$$\gamma = \begin{cases} \min_{|\alpha|>k} \{\dfrac{j_\alpha - \alpha_0}{|\alpha| - k}\} \\ +\infty, \qquad \text{if } k = m. \end{cases} \tag{3}$$

Before stating theorems, we give an example:
Example -1 Let

$$P(z,\partial_z) = \partial_{z_0}^k + \sum_{\substack{|\alpha|\le m \\ \alpha_0<k}} a_\alpha(z)\partial_z^\alpha. \tag{4}$$

Set $L(z,\partial_z) = z_0^k P(z,\partial_z)$, $A(z,\partial_{z_0}) = z_0^k\partial_{z_0}^k$ and $B(z,\partial_z) = L(z,\partial_z) - A(z,\partial_{z_0})$. Then $L(z,\partial_z)$ satisfies condition (C) and $\chi(\lambda,z') = \lambda(\lambda-1)\cdots(\lambda-k+1)$.

Let us introduce a subspace of $\mathcal{O}(U(\theta))$.

Definition 1. *$\mathcal{O}_{(\kappa)}(U(\theta))$ $(\kappa > 0)$ is a totality of $u(z) \in \mathcal{O}(U(\theta))$ such that for any $\varepsilon > 0$, $0 < \theta' < \theta$ and $V \Subset U$*

$$|u(z)| \le M\exp(\varepsilon|z_0|^{-\kappa}) \quad \text{in } V(\theta') \tag{5}$$

holds for a constant $M = M(\varepsilon,\theta',V)$.

Theorem 1. [4] *Suppose that condition* (C) *holds for* $L(z,\partial_z)$. *Let* $u(z) \in \mathscr{O}_{(\gamma)}(U(\theta))$ *be a solution of* $L(z,\partial_z)u = f(z) \in \mathscr{O}(U)$. *Then there are a constant* c *and a neighborhood* V *of* $z=0$ ($V \Subset U$) *such that for any* θ' $(0<\theta'<\theta)$

$$|u(z)| \le M|z_0|^c \quad \text{in } V(\theta')$$

holds for a constant $M = M(\theta')$.

If $\partial_{z_0}^{\ell} f(0,z') = 0$ for $0 \le \ell \le s-1$, then we can take any c with $c < \min\{a_0, s\}$. Theorem 1 is an answer of the problem (1) in the introduction. As for the problem (2) we have

Theorem 2. [3] *Let* $P(z,\partial_z)$ *be an operator of the form*

$$P(z,\partial_z) = \partial_{z_0}^k + \sum_{|\alpha|\le m, \alpha_0<k} a_\alpha(z)\partial_z^\alpha.$$

Suppose that $u(z) \in \mathscr{O}_{(\gamma)}(U(\theta))$ *solves*

$$P(z,\partial_z)u(z) = f(z) \in \mathscr{O}(U).$$

Then there exist $u_n(z') \in \mathscr{O}(U')$ $(n \in \mathbb{N})$ *and for any* $0<\theta'<\theta$ *and* $V \Subset U$ *there exist* A, B *such that*

$$|u(z) - \sum_{n=0}^{N-1} u_n(z')z_0^n| \le AB^N \Gamma(\frac{N}{\gamma}+1)|z_0|^N \tag{6}$$

holds for any $N \in \mathbb{N}$ *and* $z \in V(\theta')$.

As for behaviors of singular solutions to linear equations, especially a generalization of Theorem 2, we refer to [3] and [5].

3 Nonlinear equations

Let us proceed to study nonlinear equations. For this purpose we introduce some notations, which are not always usual. $\boldsymbol{\Delta}_m$ is a subset of $\mathbb{N}^{d+1}$ such that $\boldsymbol{\Delta}_m = \{\alpha \in \mathbb{N}^{d+1}; |\alpha| \le m\}$ and set $M = \#\boldsymbol{\Delta}_m$. For $A = (A_\alpha; \alpha \in \boldsymbol{\Delta}_m) \in \mathbb{N}^M$ and $Z = (Z_\alpha; \alpha \in \boldsymbol{\Delta}_m)$

$$|A| = \sum_{\alpha\in\boldsymbol{\Delta}_m} A_\alpha, \; Z^A = \prod_{\alpha\in\boldsymbol{\Delta}_m} Z_\alpha^{A_\alpha}.$$

Z^A is a monomial in (Z_α) with degree $|A|$. Set $\mathbb{N}^{M*} := \mathbb{N}^M - \{0\} = \{A \in \mathbb{N}^M; |A| \ge 1\}$. For $A \in \mathbb{N}^{M*}$

$$m_A = \max\{|\alpha|; A_\alpha \ne 0\}. \tag{7}$$

Let

$$L(u) := L(z, \vartheta^{\alpha_0}\partial^{\alpha'} u) \tag{8}$$

be a nonlinear partial differential operator with order m, where $L(z,Z)$ is holomorphic in $(z,Z) \in U \times \Omega$, $\mathbb{C}^M \supset \Omega \ni 0$:

$$L(z,Z) = \sum_{A \in \mathbb{N}^{M*}} c_A(z) Z^A + f(z). \tag{9}$$

Hence

$$L(z, \vartheta^{\alpha_0}\partial^{\alpha'} u) = \sum_{A \in \mathbb{N}^{M*}} c_A(z) \prod_{\alpha \in \Delta(m)} (\vartheta^{\alpha_0}\partial^{\alpha'} u)^{A_\alpha} + f(z). \tag{10}$$

Let $S = \{z_0 \in U_0; |\arg z_0| < \theta^*\}$ $(= U_0(\theta^*))$. The aim in this section is to study the behavior of $u(z) \in \mathcal{O}(S \times U')$ that solves $L(u) = 0$ with

$$\sup_{z' \in U'} |u(z)| \le C|z_0|^{\nu_0} \qquad \nu_0 > 0. \tag{11}$$

$L(z,Z)$ is defined only in a neighborhood of $(z,Z) = (0,0)$, so we assume the growth condition (11) on $u(z)$, which differs from the growth condition for linear equations (see Theorems 1 and 2).

For our purpose let us define a class of functions represented by Mellin type integral.

Definition 2. *Let $\varphi(\lambda, z') \in \mathcal{O}(U')[\lambda]$ with nonvanishing leading coefficient and $\widetilde{\mathbb{C} - \{0\}}$ be the universal covering space of $\mathbb{C} - \{0\}$. $\mathcal{M}_\varphi(U')$ is the set of all $w(z) \in \mathcal{O}(\widetilde{\mathbb{C} - \{0\}} \times U')$ represented by*

$$w(z) = \frac{1}{2\pi i}\int_C z_0^{-\lambda}\frac{\psi(\lambda,z')}{\varphi(\lambda,z')}d\lambda, \tag{12}$$

where C is a Jordan curve enclosing all the zeros of $\varphi(\lambda, z')$ and $\psi(\lambda, z') \in \mathcal{O}(U')[\lambda]$ such that $\deg.\psi(\lambda,z') < \deg.\varphi(\lambda,z')$.

Set $\mathcal{M}_{rat}(U') = \bigcup_\varphi \mathcal{M}_\varphi(U')$. If all zeros $\{a_j(z')\}_{j=1}^p$ of $\varphi(\lambda, z')$ are distinct in (12), then

$$w(z) = \sum_{j=1}^{p} z_0^{-a_j(z')} A_j(z').$$

Now let us return to (10). We represent $L(u)$ in another form:

$$\begin{aligned} L(u) :=& L(z, \vartheta^{\alpha_0}\partial^{\alpha'} u; |\alpha| \le m) \\ =& P(u) + Q(u) + f(z), \end{aligned} \tag{13}$$

where

$$\begin{aligned} P(u) &= \sum_{h=0}^{k} c_h(z')\vartheta^h u \\ Q(u) &= \sum_{A\in\mathbb{N}^{M*}} c_A(z) \prod_{\alpha\in\Delta(m)} (\vartheta^{\alpha_0}\partial^{\alpha'} u)^{A_\alpha}. \end{aligned} \tag{14}$$

Let $e_A \in \mathbb{N}$ such that $c_A(z) = z_0^{e_A} b_A(z)$, $b_A(0,z') \not\equiv 0$ for $c_A(z) \not\equiv 0$. We assume

$$c_k(0) \neq 0 \tag{C$_0$}$$

and

$$e_A > 0 \quad \text{for } A \in \mathbb{N}^{M*} \text{ with } |A| = 1. \tag{C$_1$}$$

Suppose that $u(z) \in \mathcal{O}(S \times U')$ solves $L(u) = 0$ with growth property (11). Define an index γ by

$$\gamma = \begin{cases} \min_{\{A;m_A>k\}} \{\dfrac{e_A + \nu_0(|A|-1)}{m_A - k}\}, \\ \\ +\infty, \qquad \text{if } k = m. \end{cases} \tag{15}$$

In the case of linear equations, the index γ is determined by operators, however, γ defined by (15) for nonlinear operator depends on ν_0 in (11). It follows from (C$_0$) that

$$\chi(\lambda, z') = \sum_{h=0}^{k} c_h(z')\lambda^h$$

has k-roots $\{\lambda_i(z')\}_{1\le i\le k}$ in a small neighborhood of U' of $z' = 0$. For simplicity we assume for $z' \in U'$

$$\chi(\lambda, z') \neq 0 \quad \text{on } \mathrm{Re}\lambda = \nu_0. \tag{C$_2$}$$

So

$$\begin{aligned} Re\lambda_i(z') &> \nu_0 \quad \text{for } 1 \le i \le k', \\ Re\lambda_i(z') &< \nu_0 \quad \text{for } k' < i \le k. \end{aligned} \tag{16}$$

Set $\lambda_0 = 1$. Define the discrete sets $\Lambda(z')$ and $\Lambda_{-\nu_0}(z')$ by

$$\begin{cases} \Lambda(z') = \{\lambda = -\sum_{i=0}^{k'} n_i\lambda_i(z'); (n_0, n_1, \cdots, n_{k'}) \in \mathbb{N}^{k'+1}\}, \\ \Lambda_{-\nu_0}(z') = \Lambda(z') \bigcap \{\mathrm{Re}\lambda \le -\nu_0\}. \end{cases} \tag{17}$$

Now we give a result concerning behavior of a solution $u(z)$ of (10). Asymptotic terms of $u(z)$ are given by functions in $\mathcal{M}_\varphi(U')$, where the poles of $\varphi(\lambda, z')$ are contained in $\Lambda_{-\nu_0}(z')$.

Theorem 3. *Suppose that $L(u)$ satisfies conditions* (C_0), (C_1) *and* (C_2), *and* $u(z) \in \mathscr{O}(S \times U')$ *with*

$$|u(z)| \leq C|z_0|^{\nu_0} \quad (\nu_0 > 0)$$

solves $L(u) = 0$. *Then there exist* $\varphi_n(\lambda, z') \in \mathscr{O}(W')[\lambda]$ *and* $u_n(z) \in \mathcal{M}_{\varphi_n}(W')$ $(n \in \mathbb{N})$ *in a neighborhood* W' *of* $z' = 0$ *such that for any sector* $T \Subset S$

$$|u(z) - \sum_{n=0}^{N-1} u_n(z)| \leq AB^N |z_0|^{N+\nu_0} \Gamma(\frac{N}{\gamma} + 1) \tag{18}$$

holds for $z \in T \times W'$ *and* $N \in \mathbb{N}$. *Here as for the zeros of* $\varphi_n(\lambda, z') \in \mathscr{O}(W')[\lambda]$, *there exist* $a, b > 0$ *such that*

$$\{\lambda; \varphi_n(\lambda, z') = 0\} \subset \Lambda_{-\nu_0}(z') \cap \{\mathrm{Re}\lambda > -(an + b)\}.$$

Outline of the proof is given in section 4.
Remark (1). If $L(u) = 0$ is linear, the similar result is obtained in [6]. As stated in the introduction, Theorem 3 is a generalization to nonlinear equations.
(2). Suppose $k = m$. Then it follows from $\gamma = +\infty$ that $u(z) = \sum_{n=0}^{+\infty} u_n(z)$ converges. In this case in [1] and [2] linear equations, which are called equations of Fuchsian type, were studied and in [7] nonlinear equations were studied. Theorem 3 is a generalization of these results to equations with $k \leq m$.
(3) The order of singularities of $u_n(z)$ is characterized by the zeros of $\varphi_n(\lambda, z')$ and that of $u(z)$ may better than ν_0. Hence the constant $\gamma > 0$ in the remainder estimate (18) will be improved.

4 Outline of the proof of Theorem 3

As stated in the introduction, the proof of Theorem 3 is given in the forthcoming paper. We sketch it. It consists of four parts.
(1) In order to study singularity of $u(z)$ we use the Mellin transform of $u(z)$ with respect to z_0,

$$(\mathcal{M}u)(\lambda, z') = \int_0^T z_0^{\lambda-1} u(z_0, z') dz_0.$$

It follows from (11) that $(\mathcal{M}u)(\lambda, z')$ is holomorphic on $\{\lambda; \mathrm{Re}\lambda > -\nu_0\}$. Since $u(z)$ satisfies $L(u) = 0$, we can show that $(\mathcal{M}u)(\lambda, z')$ is meromorphically extensible to a wider domain. The following holds.

Proposition 1. $(\mathcal{M}u)(\lambda, z')$ $(z \in W')$ *is meromorphically extensible in* λ *to the whole plane. The poles of* $(\mathcal{M}u)(\lambda, z')$ *are contained in* $\Lambda_{-\nu_0}(z')$.

Let us sketch the proof Proposition 1. First, by integration by parts, we note that $(\mathcal{M}u)(\lambda, z')$ satisfies a nonlinear functional equation

$$(\mathcal{M}u)(\lambda, z') = -\frac{T^{\lambda}H(\lambda, z') + (\mathcal{M}Q(u))(\lambda, z') + (\mathcal{M}f)(\lambda, z')}{\chi(-\lambda, z')}, \tag{19}$$

where $H(\lambda, z')$ is a polynomial in λ. It follows from (C_1) and (11) that $(\mathcal{M}Q(u))(\lambda, z')$ is holomorphic in $\{\lambda; \mathrm{Re}\lambda > -\nu_0 - \min\{\nu_0, 1\}\}$. Since $f(z)$ is holomorphic, the singularities of $(\mathcal{M}f)(\lambda, z')$ are at most poles at $\lambda \in -\mathbb{N}$. Therefore the numerator of the right hand side of (19) is meromorphic in $\{\lambda; \mathrm{Re}\lambda > -\nu_0 - \min\{\nu_0, 1\}\}$, hence $(\mathcal{M}u)(\lambda, z')$ is also meromorphic there from (19). The zeros of $\chi(-\lambda, z')$ may be poles of $(\mathcal{M}u)(\lambda, z')$. Since $(\mathcal{M}u)(\lambda, z')$ is holomorphic in $\{\lambda; \mathrm{Re}\lambda > -\nu_0\}$, the zeros of $\chi(-\lambda, z')$ contained in $\{\mathrm{Re}\lambda \le -\nu_0\}$ may be poles, however, others are not. By modifying and repeating the above process, we can show that $(\mathcal{M}u)(\lambda, z')$ is meromorphically extensible to larger domains step by step, and consequently to the whole plane and obtain the location of its poles.

It follows from Proposition 1 that $(\mathcal{M}u)(\lambda, z')$ is an infinite sum of $\mathcal{M}_{rat}(W')$, hence from which we can obtain asymptotic terms of $u(z)$ as z_0 approaches to 0 by the inverse Mellin transform. However our purpose is to obtain remainder estimates. In order to do so we need further steps.

(2) Next step is a modification of the equation $L(u) = 0$. Let $\nu > 0$ be a constant such that

$$\nu \ge \nu_0 \quad \text{and} \quad \nu > \max_{1\le i\le k} \sup_{z'\in W'} \mathrm{Re}\ \lambda_i(z'). \tag{20}$$

We can choose $v(z) \in \mathcal{O}(S \times W')$ and $w(z) \in \mathcal{M}_{rat}(W')$ such that $u(z) = v(z) + w(z)$,

$$|w(z)| \le C|z_0|^{\nu_0} \quad \text{and} \quad |v(z)| \le C|z_0|^{\nu}. \tag{21}$$

Let us consider $v(z)$ as an unknown function instead of $u(z)$. It follows from $\nu \ge \nu_0$ that the behavior of singularities of $v(z)$ is milder than that of the original unknown $u(z)$. The poles of $(\mathcal{M}v)(\lambda, z')$ is contained in $\Lambda(z') \cap \{\mathrm{Re}\ \lambda \le -\nu\}$. From $L(u) = P(v+w) + Q(v+w) + f(z) = 0$ we have an equation with unknown $v(z)$:

$$\begin{cases} P(v) + Q(w; v) + f(w; v) = 0 \\ Q(w; v) = Q(v+w) - Q(w) \\ f(w; z) = P(w) + Q(w) + f(z). \end{cases}$$

$$Q(w; v) = \sum_{A\in\mathbb{N}^{M_*}} c_A(w; z) \prod_{\alpha\in\Delta(m)} (\vartheta^{\alpha_0}\partial^{\alpha'} v)^{A_\alpha},$$

$$c_A(w; z) \in \mathcal{O}(\widetilde{W_0 - \{0\}} \times W').$$

$Q(w; v)$ depends on $w(z)$, so coefficients $\{c_A(w; z)\}_{A\in\mathbb{N}^{M_*}}$ are holomorphic in a neighborhood of $z = 0$ except on $z_0 = 0$.

(3) By representing $\{c_A(w;z); A \in \mathbb{N}^{M*}\}$ and $f(w;z)$ by coefficients of $L(u)$ and $w(z)$, we can decompose them as an infinite sum of functions in $\mathcal{M}_{rat}(W')$

$$\begin{cases} c_A(w;z) = \sum_{j=0}^{+\infty} c^w_{A,j}(z) \\ f(w;z) = \sum_{j=0}^{+\infty} f^w_j(z) \end{cases} \tag{22}$$

such that the poles of $(\mathcal{M}c^w_{A,j})(\lambda, z')$ and $(\mathcal{M}f^w_j)(\lambda, z')$ are contained in $\Lambda_{-\nu_0}(z')$, and it holds that

$$|c^w_{A,j}(z)| \le C^{|A|+j+e^w_A}|z_0|^{j+e^w_A}, \quad |f^w_j(z)| \le C^{j+1}|z_0|^{j+\nu}$$

for some constants $e^w_A \ge 0$ and $C > 0$.

Set

$$L^w(v) := P(v) + Q(w;v) + f(w;z). \tag{23}$$

Thus the equation to study is $L^w(v) = 0$.

(4) Finally we study asymptotic behavior of $v(z)$ that solves

$$L^w(v) = 0 \quad \text{with} \quad |v(z)| \le C|z_0|^\nu. \tag{24}$$

Let us find asymptotic terms of $v(z)$, that is, $\{v_n(z)\}_{n=0}^\infty \subset \mathcal{M}_{rat}(W')$ such that $v(z) \sim \sum_{n=0}^\infty v_n(z)$ and remainder estimate of Gevrey type holds. By using (22), we define $v_n(z) \in \mathcal{M}_{rat}(W')$ successively as follows:

$$\begin{aligned} &P(z',\vartheta)v_0(z) + f^w_0(z) = 0 \\ &P(z',\vartheta)v_n(z) + Q_n(w; v_0, v_1, \cdots, v_{n-1}) + f^w_n(z) = 0. \end{aligned} \tag{25}$$

Here $Q_n(w; v_0, v_1, \cdots, v_{n-1}) \in \mathcal{M}_{rat}(W)$ is determined by $\{v_i(z)\}_{i=1}^{n-1}$ and of slightly complicated form. It is a finite sum of terms

$$c^w_{A,j}(z) \prod_{i=1}^{|A|} \vartheta^{\alpha_0(i)} \partial^{\alpha'(i)} v_{n_i}(z),$$

which will be given more concretely in the forthcoming paper.

It follows from the condition (20) that $\{v_n(z)\}_{n\in\mathbb{N}}$ with $|v_n(z)| \le C|z_0|^\nu$ are uniquely determined and it holds that the poles of $(\mathcal{M}Q_n)(\lambda, z')$ are in $\Lambda_{-\nu_0}(z')$, so those of $(\mathcal{M}v_n)(\lambda, z')$ are also in $\Lambda_{-\nu_0}(z')$. Here we give a remark about constructing of $\{v_n(z)\}_{n\in\mathbb{N}}$: it follows from (25) that the formal sum $\hat{v}(z) = \sum_{n=0}^{+\infty} v_n(z)$ satisfies formally

$$P(z',\vartheta)\hat{v}(z) + \sum_{n=1}^{+\infty} Q_n(w; v_0, v_1, \cdots, v_{n-1}) + f(w;z) = 0.$$

Hence in order that we have $P(z',\vartheta)\hat{v}(z)+Q(w;\hat{v})+f(w;z)=0$ (see (23)) in some sense, it will hold that $Q(w;\hat{v})=\sum_{n=1}^{+\infty}Q_n(w;v_0,v_1,\cdots,v_{n-1})$ not rigrously but formally. Set

$$r_n(z):=u(z)-(w(z)+\sum_{i=0}^{n-1}v_i(z)).$$

By considering how to determine $\{v_n(z)\}_{n=0}^{\infty}$ carefully (see (25)), we estimate the remainder $r_n(z)$ by majorant method and obtain Theorem 3.

Let us give an example.

Example -2. $z=(z_0,z_1)\in\mathbb{C}^2$

$$L(u)=z_0\frac{\partial u}{\partial z_0}-a(z)u+c_0(z)\frac{\partial u}{\partial z_1}+c_1(z)\prod_{i=1}^{\ell}(\frac{\partial^{m_i}u}{\partial z_1^{m_i}})+f(z), \qquad (26)$$

where $a(z),c_0(z),c_1(z)$ and $f(z)$ are holomorphic in a neighborhood of $z=0$, $c_0(0,z_1)=0,\ \ c_1(z)=z_0^pb_1(z)$ with $b_1(0,z')\not\equiv 0$ and $\ell,m_i\geq 2$. Let $m=\max_{1\leq i\leq\ell}m_i$. Now let $u(z)\in\mathscr{O}(S\times U')$ be a solution of $L(u)=0$ such that

$$|u(z)|\leq C|z_0|^{\nu_0}\quad \nu_0>0.$$

Then $\gamma=\dfrac{p+\nu_0(\ell-1)}{m-1}$. In the following T is a sector such that $T\Subset S$ and W' is a small neighborhood of $z'=0$.

(1) Suppose $\nu_0>\mathrm{Re}\ a(0)$. Then

$$u(z)\sim\sum_{n\in\mathbb{N},n\geq\nu_0}u_n(z_1)z_0^n$$

and the remainder estimate is obtained by Theorem 3. Let $\nu'=\min\{n;u_n(z_1)\not\equiv 0\}$. Then $|u(z)|\leq C|z_0|^{\nu'}$ holds in $T\times W'$. Set $\gamma'=\dfrac{p+\nu'(\ell-1)}{m-1}$. Hence $\gamma\leq\gamma'$, which means that the remainder estimate is improved, as stated in Remark in section 3.

(2) Suppose $\nu_0<\mathrm{Re}\ a(0)$. The asymptotic terms of $u(z)$ are of more complicated form

$$u(z)\sim\frac{1}{2\pi i}\int_C z_0^{-\lambda}\frac{\psi(\lambda,z_1)}{\varphi(\lambda,z_1)}d\lambda$$

given by $\mathcal{M}_\varphi(W')$ with a polynomial $\varphi(\lambda,z_1)$ such that $\{\lambda;\varphi(\lambda,z_1)=0\}\subset\{-n_0-n_1a(0,z_1);n_0,n_1\in\mathbb{N}\}$.
Suppose $\nu_0<\mathrm{Re}\ a(0)<1$. Then there is a $\psi(z_1)\in\mathscr{O}(W')$ such that

$$u(z)\sim\frac{1}{2\pi i}\int_C z_0^{-\lambda}\frac{\psi(z_1)}{\lambda+a(0,z_1)}d\lambda=\psi(z_1)z_0^{a(0,z')}$$

and for some ν' with $\nu_0 < \nu' \leq \text{Re } a(0)$, $|u(z)| \leq C|z_0|^{\nu'}$ holds in $T \times W'$.
Suppose $\nu_0 < 1 < \text{Re } a(0)$. Then

$$u(z) \sim \frac{1}{2\pi i} \int_C z_0^{-\lambda} \frac{f_1(z_1)}{(a(0, z_1) - 1)(\lambda + 1)} d\lambda = \frac{f_1(z_1)}{a(0, z_1) - 1} z_0$$

and $|u(z)| \leq C|z_0|$ holds in $T \times W'$.
Suppose $\nu_0 < \text{Re } a(0) = 1$. Then

$$u(z) \sim \int_C z_0^{-\lambda} \left(\frac{\psi(z_1)}{\lambda + a(0, z_1)} + \frac{f_1(z_1)}{(\lambda + a(0, z_1))(\lambda + 1)} \right) d\lambda$$

and for ν' with $\nu_0 < \nu' < 1$, $|u(z)| \leq C|z_0|^{\nu'}$ holds in $T \times W'$.

We give only the main term of asymptotic expansion of $u(z)$, and can also calculate the asymptotic terms following the main term, which are given by functions in $\mathcal{M}_{rat}(W')$ by Theorem 3.

References

1. R. Gérard, H. Tahara, Singular Nonlinear Partial Differential Equations, Aspects of Mathematics E28, Vieweg, (1996).
2. T. Mandai, The method of Frobenius to Fuchsian partial differential equations, *J. Math. Soc. Japan,* **52** (2000), 561-582.
3. S. Ōuchi, Singular solutions with asymptotic expansion of linear partial differential equations in the complex domain, *Publ. RIMS Kyoto Univ.,* **34** (1998), 291-311.
4. S. Ōuchi, Growth property and slowly increasing behavior of singular solutions of linear partial differential equations in the complex domain, *J. Math. Soc. Japan,* **52** (2000), 767-792.
5. S. Ōuchi, Asymptotic expansion of singular solutions and the characteristic polygon of linear partial differential equations in the complex domain, *Publ. RIMS Kyoto Univ.,* **36** (2000), 457-482.
6. S. Ōuchi, The behaviors of singular solutions of partial differential equations in some class in the complex domain, Partial Differential Equations and Mathematical Physics, In Memory of Jean Leray, ed. by K.Kajitani and J. Vaillant, Progress in Nonlinear Differential Equations and Their Applications vol 52, Birkhauser (2003) 177-194.
7. H. Tahara, H. Yamazawa, Structure of Solutions of Nonlinear Partial Differential Equations of Gérard-Tahara Type, *Publ. RIMS Kyoto Univ.,* **41** (2005), 339-373.

Observations on the JWKB treatment of the quadratic barrier

Hujun Shen[1] and Harris J. Silverstone[2]

[1] Department of Chemistry, The Johns Hopkins University, 3400 N. Charles St., Baltimore, MD 21218, USA

[2] Department of Chemistry, The Johns Hopkins University, 3400 N. Charles St., Baltimore, MD 21218, USA

hjsilverstone@jhu.edu

Summary. Historically, the "lowest-order" JWKB "parabolic" connection formula between the left and right classically allowed regions for tunneling through a parabolic barrier involved a rather non-JWKB-like square-root $\sqrt{1+\mathrm{e}^{-2\pi(-E)/\hbar}}$ and a non-JWKB phase factor that had been extracted from the asymptotic expansion of the parabolic cylinder function. Generalization to higher order was not obvious. We show how the usual JWKB connection formulas at the linear turning points, combined with matching in a common Stokes region when $\hbar$ is complex, lead to the historical formula and its generalization. The limit of real $\hbar$ is tricky, because Stokes lines coalesce, and the common Stokes region that joins the two turning points disappears. Certain Stokes lines are thus "doubled," but which ones depend on the sign of $\arg\hbar \to \pm 0$. The square root and phase arise from the Borel sum of the "normalization factors" $\mathrm{e}^{\pm\mathrm{i}\sum_{n=1}^{\infty} S_R^{(n)}(\infty)\hbar^{2n-1}}$, which as conjectured by Sato are summed by a gamma function. Real $\hbar$ is a Stokes line for these factors, causing the Borel sums of a single JWKB wave function even in the classically allowed region to be different for $\arg\hbar \to \pm 0$. A proof of Sato's conjecture is given in an Addendum.

Key words: JWKB; connection formula; Borel sum; asymptotic expansion; semi-classical; tunneling

1 Introduction

Miller and Good [1], Ford, Hill, Wakano, and Wheeler [2], Child [3], and Connor [4] established the lowest-order Jeffreys-Wentzel-Kramers-Brillouin (JWKB) connection formula for transmission from $x < -x_0$ to $x > x_0$ through a parabolic barrier $V = -\frac{1}{2}kx^2$, with energy $E < 0$:

$$\mathrm{i}\left(-p^{-1/2}\mathrm{e}^{\frac{\mathrm{i}\pi}{4}+\frac{\mathrm{i}}{\hbar}\int_x^{-x_0} p dx} + \mathrm{e}^{\mathrm{i}\phi}\sqrt{1+\mathrm{e}^{-2\pi(-E)/\hbar}}\, p^{-1/2}\mathrm{e}^{-\frac{\mathrm{i}\pi}{4}-\frac{\mathrm{i}}{\hbar}\int_x^{-x_0} p dx}\right)$$
$$\longleftrightarrow\ \mathrm{e}^{-\frac{\pi(-E)}{\hbar}} p^{-1/2}\mathrm{e}^{\frac{\mathrm{i}\pi}{4}+\frac{\mathrm{i}}{\hbar}\int_{x_0}^{x} p dx}, \quad (1)$$

Received 23 February, 2006. Revised 14 May, 2006, 23 June, 2006. Accepted 11 July, 2006.

where $\pm x_0 = \pm\sqrt{-2E}$ are the classical turning points, $p = +\sqrt{x^2 - (-2E)}$, $m = k = 1$, and ϕ is a phase obtained from the large-x asymptotic expansion of the parabolic cylinder function, and which, according to Child [3], "cannot be [determined] by purely phase integral methods."

In the mathematics literature, the two-turning-point problem has been dealt with in some detail by Fedoryuk [5] and in theoretical generality from the point of view of Borel summability by Voros [6], by Aoki, Kawai, and Takei [7],[9], by Kawai and Takei [8], and by Delabaere, Dillinger, and Pham [10],[11]. Complexification with respect to $\hbar$ disentangles coalesced Stokes lines, and the JWKB wave functions are proven to be Borel summable, and therefore exact.

This paper aims to make the mathematical developments more accessible to practitioners via a concrete example, the complete JWKB expansion for the quadratic barrier potential. The JWKB wave function can be found term-by-term in simple closed form to high order; the exact solution is the known parabolic cylinder function. From the application point of view, Eq. (1) raises two immediate questions: (i) Where does the square root in $\sqrt{1 + \mathrm{e}^{-2\pi(-E)/\hbar}}$ come from in the JWKB context? (ii) Is ϕ implicit in the JWKB expansion? More puzzling in the context of physical applications is (iii) the conceptual dilemma inside the barrier ($-x_0 < x < +x_0$): if the JWKB wave function that is exponentially decreasing from $-x_0$ smoothly joins the exponentially increasing wave function from $+x_0$, what happens to the Stokes line of the increasing function that is irrelevant to the decreasing function? Our aim is to look in detail at an example simple enough to be solvable but rich enough to illustrate some of the subtle twists and turns of the exact JWKB method.

The answers, most of which can be found in the paper of Kawai and Takei [8], come from complexification with respect to $\hbar$, Borel summation, the Stokes lines of Stirling's formula for the gamma function, and the disappearance of a domain and coalescence of Stokes lines in the limit of real $\hbar$. The gamma-function Stokes lines are crucial to evaluating the series that converts between "normalization at a turning point" and "normalization at infinity."

In addition, we give in an Addendum an elementary proof of Sato's conjecture about the normalization constant.

2 JWKB wave function for the quadratic barrier

The JWKB solution of the Schrödinger equation for a particle scattered by an inverted parabolic potential (Fig. 1)

$$\left(-\frac{\hbar^2}{2}\frac{d^2}{dx^2} - \frac{1}{2}x^2 - E\right)\psi(x) = 0 \tag{2}$$

has the form

$$\psi_{\mathrm{JWKB}\pm} \sim \dot{S}^{-1/2}\mathrm{e}^{\pm\mathrm{i}(S/\hbar + \pi/4)}\,, \tag{3}$$

$$S(x,\hbar) = \textstyle\sum_{n=0}^{\infty} S^{(n)}(x)\hbar^{2n}, \qquad \dot{S} \equiv dS/dx\,. \tag{4}$$

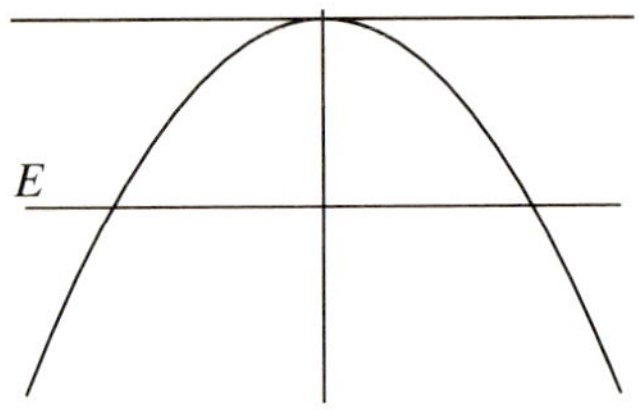

Fig. 1. Inverted-parabola potential

We explicitly take $E < 0$, but there is no difficulty to treat $E > 0$. For notational clarity, the dependence of $S^{(n)}(x)$ on E is suppressed.

The derivatives $\dot{S}^{(n)}(x)$ can be found recursively:

$$\dot{S}^{(0)}(x) = \pm\sqrt{x^2 - (-2E)}\,, \tag{5}$$

$$\dot{S}^{(n)}(x) = \frac{1}{2\dot{S}^{(0)}}\left[\left(\frac{3}{4}\frac{\ddot{S}^2}{\dot{S}^2} - \frac{1}{2}\frac{\dddot{S}}{\dot{S}}\right)^{\text{“}(n-1)\text{”}} - \sum_{k=1}^{n-1}\dot{S}^{(k)}\dot{S}^{(n-k)}\right], \tag{6}$$

$$= \pm(-1)^{n-1}16^{-n}[x^2 - (-2E)]^{-2n+1/2}P_n\left(\frac{1}{x^2/(-2E) - 1}\right). \tag{7}$$

Equation (7) corresponds to the first equation of Lemma A.4 of [8]. The superscript “(n)”, as in $\dot{S}^{(n)}$, denotes the coefficient of $\hbar^{2n}$ in the expansion of $\dot{S}$ in powers of $\hbar^2$. As calculated with *Mathematica* [12] to order $n = 50$, the polynomials $P_n(y)$ have positive, integer coefficients. The first three are

$$P_1(y) = 6 + 10y\,, \tag{8}$$

$$P_2(y) = 594 + 2652y + 2210y^2\,, \tag{9}$$

$$P_3(y) = 200556 + 1545948y + 2981700y^2 + 1656500y^3\,. \tag{10}$$

When integrating $\dot{S}^{(n)}(x)$ to get $S^{(n)}(x)$, we choose the “integration constant” to preserve the square-root branch point at the classical turning point, and we choose the sign of the square root that makes $S^{(0)}(x)$ positive where x is classically allowed. Thus $S^{(n)}(x)$ is tied to a particular turning point. For the classically allowed region to the right of the right-hand turning point (labeled by the subscript “R”) one finds

$$S_R^{(0)}(x) = +\frac{1}{2}x\sqrt{x^2 - (-2E)} + \frac{-E}{2}\log\left(\frac{x - \sqrt{x^2 - (-2E)}}{x + \sqrt{x^2 - (-2E)}}\right), \tag{11}$$

$$S_R^{(n)}(x) = +x\frac{16^{-n}(-2E)^{-2n+1}}{\sqrt{x^2 - 2(-E)}}T_{3n-2}\left(\frac{1}{x^2/(-2E) - 1}\right), \tag{12}$$

where $T_{3n-2}(y)$ is a polynomial of degree $3n - 2$. The first three are

$$T_1(y) = \frac{2}{3} - \frac{10}{3}y\,, \tag{13}$$

$$T_4(y) = \frac{112}{45} - \frac{56}{45}y + \frac{14}{15}y^2 + \frac{884}{9}y^3 + \frac{2210}{9}y^4 \,, \tag{14}$$

$$T_7(y) = \frac{15872}{315} - \frac{7936y}{315} + \frac{1984y^2}{105} - \frac{992y^3}{63} + \frac{124y^4}{9} \\ -20068y^5 - \frac{331300y^6}{3} - \frac{331300y^7}{3} \,. \tag{15}$$

One can straightforwardly obtain $S_R^{(n)}(x)$ to high order n.

We note that what is called Darwin's expansion [13],[14] for the parabolic cylinder function is essentially the JWKB expansion.

3 Stokes regions; doubled-Stokes lines

At Stokes curves the Borel sum of the dominant JWKB wave function changes discontinuously. If $\arg\hbar = \pm\epsilon \neq 0$, the Stokes curves – on which $iS^{(0)}(x)$ is real – divide the x plane into five regions; one is shared by both turning points [6],[8],[10],[11]. We number the local regions around the two turning points sequentially. The shared region flips with the sign of $\arg\hbar$, as illustrated in Fig. 2, and disappears when $\hbar$ is real; the region that disappears depends on the sign of $\arg\hbar$. Figure 2 is the "inner part" of Figs. 2.1 and 2.2 of [8], which deals with the more complicated four-turning-point double-well oscillator. One

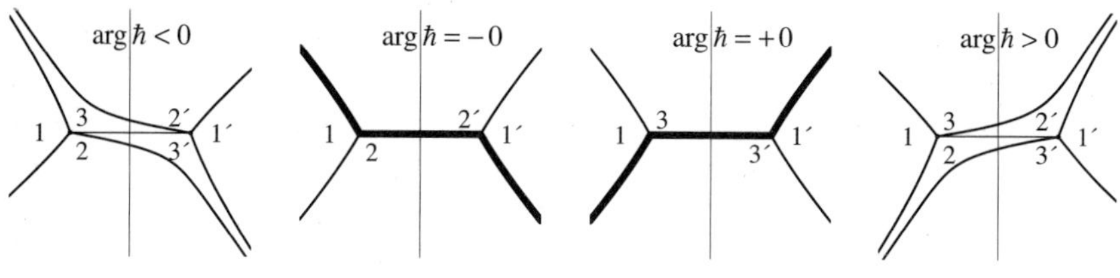

Fig. 2. Stokes lines when $\arg\hbar < 0$, $\arg\hbar = \mp 0$, and $\arg\hbar > 0$, and the specification of Regions 1–3, $1'$–$3'$. The thick, coalesced Stokes lines in the $\arg\hbar = \mp 0$ graphs behave as *doubled-Stokes* lines.

glance is enough to see that the JWKB descriptions of the wave function should be formally different for $|\pm\arg\hbar| > 0$, mirror images, distinct even in the limit of real $\hbar$ [10],[11].

- The overall left-right, 1–$1'$ connection formula will depend on the sign of $\arg\hbar$, because different Stokes lines are crossed.
- The Borel sum of a JWKB wave function in Region $1'$, may depend discontinuously on the sign of $\arg\hbar$, because Region $1'$ lies on different sides of the Stokes line emanating from the left turning point towards the right.
- In the limit that $\hbar$ is real, for each turning point the long-range influence of the other turning point squeezes out one of its local Stokes sectors. As

a consequence, moving from the classically allowed sector in the direction of the collapsed sector requires crossing two Stokes lines. We regard the coalesced, thick Stokes lines in the $\arg \hbar = \mp 0$ portions of Fig. 2 as *doubled-Stokes* lines.

- The puzzle of how an exponentially decreasing, Stokes-line-continuous solution can smoothly join an exponentially increasing, Stokes-line-discontinuous solution from the other turning point is explained by Fig. 2: the match is made on only one side of the increasing solution's Stokes line; the limit of real $\hbar$ obliterates the common region from the decreasing solution and forces it to cross an additional (nonlocal) Stokes line.

4 JWKB connection formulas when $\arg \hbar \neq 0$

To discuss the connection formulas some details are unavoidable: notation for the JWKB wave functions in each region; the basic connection formulas among the regions; the implications of different normalizations; and the differences among the connection formulas that depend on normalization and on the sign of $\arg \hbar$. This section gives these details.

4.1 Notational elaboration

We extend our notation so that the JWKB wave functions are transparently imaginary- or real-exponential of a function positive on the physical axes as in (1), and also in the choice of sign for the square root in (11). To keep the connection formula local, each of the exponentiated functions has a square-root branch point at its associated turning point (a choice of integration constant). Accordingly, we define in addition to S_R, the related Q_R, S_L, and Q_L, which have square-root branch points at either the right or left turning points as labeled, and which in zeroth order are positive in the adjacent classically allowed regions (S) or adjacent classically forbidden regions (Q).

The localized versions of the so-called action functions are defined by

$$\dot{S}_L^{(0)}(x) = -\sqrt{x^2-(-2E)} = -\dot{S}_R^{(0)}(x)\,, \tag{16}$$

$$S_L^{(0)}(x) = -S_R^{(0)}(x)\,, \tag{17}$$

$$\dot{Q}_L^{(0)}(x) = +\sqrt{(-2E)-x^2} = -\dot{Q}_R^{(0)}(x)\,, \tag{18}$$

$$Q_L^{(0)}(x) = \frac{1}{2}x\sqrt{(-2E)-x^2} - (-E)\arccos\frac{x}{\sqrt{-2E}} + (-E)\pi\,, \tag{19}$$

$$Q_R^{(0)}(x) = (-E)\pi - Q_L^{(0)}(x)\,, \tag{20}$$

$$S_L^{(n)}(x) = -S_R^{(n)}(x) = S_R^{(n)}(-x), \qquad (n \geq 0)\,, \tag{21}$$

$$Q_L^{(n)}(x) = -x\frac{16^{-n}(-2E)^{-2n+1}}{\sqrt{(-2E)-x^2}}T_{3n-2}\left(\frac{1}{x^2/(-2E)-1}\right)\,, \tag{22}$$

$$Q_R^{(n)}(x) = -Q_L^{(n)}(x) = Q_L^{(n)}(-x), \qquad (n \geq 1)\,. \tag{23}$$

The formal JWKB pairs of wave functions are then

$$\psi_{L,\pm S}(x,\hbar) = (-\dot{S}_L)^{-1/2} \mathrm{e}^{\pm \mathrm{i}(S_L/\hbar+\pi/4)} , \tag{24}$$

$$\psi_{L,\pm Q}(x,\hbar) = \dot{Q}_L^{-1/2} \mathrm{e}^{\pm Q_L/\hbar} , \tag{25}$$

$$\psi_{R,\pm Q}(x,\hbar) = (-\dot{Q}_R)^{-1/2} \mathrm{e}^{\pm Q_R/\hbar} , \tag{26}$$

$$\psi_{R,\pm S}(x,\hbar) = \dot{S}_R^{-1/2} \mathrm{e}^{\pm \mathrm{i}(S_R/\hbar+\pi/4)} , \tag{27}$$

Note that

$$\psi_{R,\pm Q}(x,\hbar) = \mathrm{e}^{\pm\frac{(-E)\pi}{\hbar}} \psi_{L,\mp Q}(x,\hbar) . \tag{28}$$

4.2 Connection formulas

The connection formulas are obtained by applying the standard JWKB connection formulas [6],[7],[8],[9],[15],[16],[17] at each linear turning point, $x = \pm\sqrt{-2E}$, and by using the identity (28) to match the left and right functions in their common domain: 2 and 2′ for $\arg\hbar > 0$; 3 and 3′ for $\arg\hbar < 0$. We fix a linear combination of the imaginary-exponential solutions on the right in Region 1′, and give the corresponding linear combinations in the other regions in Table 1. The formulas depend on whether $\arg\hbar > 0$ or $\arg\hbar < 0$. The common region in each case has been repeated to list the JWKB wave function in both the left-turning-point and right-turning-point notations.

Table 1. Connection formula when $\arg h \neq 0$

Region	JWKB wave function
1′	$d_+\psi_{R,+S}(x,\hbar) + d_-\psi_{R,-S}(x,\hbar)$
2′	$(d_+ + d_-)\psi_{R,+Q}(x,\hbar) - \mathrm{i}d_-\psi_{R,-Q}(x,\hbar)$
3′	$(d_+ + d_-)\psi_{R,+Q}(x,\hbar) + \mathrm{i}d_+\psi_{R,-Q}(x,\hbar)$
	$\arg\hbar > 0$
2 (= 2′)	$(d_+ + d_-)\mathrm{e}^{\frac{(-E)\pi}{\hbar}}\psi_{L,-Q}(x,\hbar) - \mathrm{i}d_-\mathrm{e}^{-\frac{(-E)\pi}{\hbar}}\psi_{L,+Q}(x,\hbar)$
3	$[d_+ + d_-(1+\mathrm{e}^{-2\frac{(-E)\pi}{\hbar}})]\mathrm{e}^{\frac{(-E)\pi}{\hbar}}\psi_{L,-Q}(x,\hbar) - \mathrm{i}d_-\mathrm{e}^{-\frac{(-E)\pi}{\hbar}}\psi_{L,+Q}(x,\hbar)$
1	$-\mathrm{i}[d_+ + d_-(1+\mathrm{e}^{-2\frac{(-E)\pi}{\hbar}})]\mathrm{e}^{\frac{(-E)\pi}{\hbar}}\psi_{L,+S}(x,\hbar) + \mathrm{i}(d_+ + d_-)\mathrm{e}^{\frac{(-E)\pi}{\hbar}}\psi_{L,-S}(x,\hbar)$
	$\arg\hbar < 0$
3 (= 3′)	$(d_+ + d_-)\mathrm{e}^{\frac{(-E)\pi}{\hbar}}\psi_{L,-Q}(x,\hbar) + \mathrm{i}d_+\mathrm{e}^{-\frac{(-E)\pi}{\hbar}}\psi_{L,+Q}(x,\hbar)$
2	$[d_+(1+\mathrm{e}^{-2\frac{(-E)\pi}{\hbar}}) + d_-]\mathrm{e}^{\frac{(-E)\pi}{\hbar}}\psi_{L,-Q}(x,\hbar) + \mathrm{i}d_+\mathrm{e}^{-\frac{(-E)\pi}{\hbar}}\psi_{L,+Q}(x,\hbar)$
1	$-\mathrm{i}(d_+ + d_-)\mathrm{e}^{\frac{(-E)\pi}{\hbar}}\psi_{L,+S}(x,\hbar) + \mathrm{i}[d_+(1+\mathrm{e}^{-2\frac{(-E)\pi}{\hbar}}) + d_-]\mathrm{e}^{\frac{(-E)\pi}{\hbar}}\psi_{L,-S}(x,\hbar)$

For a quick comparison with (1), set $d_+ = 1, d_- = 0$ in the Region-1 and -1′ rows of Table 1. The exponential ratio $\mathrm{e}^{\frac{(-E)\pi}{\hbar}}$ of the left to right is present, but there is yet no square root. To find the square root, it is necessary to extract the implicit relative normalization factor.

4.3 "Relative normalization factor at ∞"

One can see from (7) that the large-x behavior of $S_R^{(n)}(x)$ is

$$S_R^{(n)}(x) = S_R^{(n)}(\infty) + O(|x|^{-4n+2}), \quad \text{as } |x| \to \infty, \text{ for } n \geq 1 . \tag{29}$$

The integration constants $S_R^{(n)}(\infty)$ are determined indirectly by the square-root-branch-point requirement. With the assistance of *Mathematica* [12] to solve (6), (7), (12), and (29), we find that the "squared normalization factor" $\mathrm{e}^{+2\mathrm{i}\sum_1^\infty S_R^{(n)}(\infty)\hbar^{2n-1}}$ is given "experimentally" through order 50 by

$$N(\hbar)^2 \equiv \mathrm{e}^{+2\mathrm{i}\sum_1^\infty S_R^{(n)}(\infty)\hbar^{2n-1}} , \tag{30}$$

$$= 1 + \mathrm{i}\frac{\hbar(-E)^{-1}}{24} - \frac{\hbar^2(-E)^{-2}}{1152} + \mathrm{i}\frac{1003\hbar^3(-E)^{-3}}{414720} + \cdots . \tag{31}$$

That (31) is the Stirling series associated with $\Gamma(\frac{1}{2} + \frac{\mathrm{i}(-E)}{\hbar})$ is a conjecture attributed by Kawai and Takei [8] to Sato, with reference also to Voros [6]. (See also [14]. We give a short "theoretical" proof in the Addendum.) At first glance, $N(\hbar)$ would appear to have magnitude 1. However, $\arg\hbar = 0$ corresponds to a Stokes line [8],[18] of the gamma-function series [with respect to $\frac{1}{2} + \mathrm{i}(-E/\hbar)$]. Kawai and Takei (Prop. 2.2 of [8]) further show that

$$N^2 \overset{(B)}{=} \hat{N}^2, (\arg\hbar > 0) , \tag{32}$$

$$\overset{(B)}{=} \hat{N}^2 \left(1 + \mathrm{e}^{-\frac{\pi(-2E)}{\hbar}}\right), (\arg\hbar < 0) , \tag{33}$$

$$\hat{N}^2 \equiv \frac{1}{(2\pi)^{1/2}} \left(\frac{\mathrm{e}\hbar}{-E}\right)^{\frac{\mathrm{i}(-E)}{\hbar}} \mathrm{e}^{\frac{\pi(-E)}{2\hbar}} \Gamma\left(\frac{1}{2} + \frac{\mathrm{i}(-E)}{\hbar}\right) , \tag{34}$$

where by the symbol $\overset{(B)}{=}$ we mean "equality in the sense of Borel sum," and by $\hat{N}^2$, the right side of (34).

In the limits $\arg\hbar \to \pm 0$, the series for N^2 is evaluated at its Stokes line where it is both discontinuous and not of magnitude 1. The gamma-function reflection formula [13] implies that (for real E),

$$\left|\Gamma(\frac{1}{2} \pm \frac{\mathrm{i}(-E)}{\hbar})\right| = \mathrm{e}^{-\frac{\pi(-E)}{2\hbar}} \sqrt{\frac{2\pi}{1 + \mathrm{e}^{-\frac{\pi(-2E)}{\hbar}}}}, \quad (\arg\hbar = 0) , \tag{35}$$

from which the magnitudes of N^2, N^{-2}, and $\hat{N}^2$, for $\arg\hbar = \pm 0$, follow:

$$|N^2| = \left|\mathrm{e}^{+2\mathrm{i}\sum_{n=1}^\infty S_R^{(n)}(\infty)\hbar^{2n-1}}\right| \overset{(B)}{=} \left(1 + \mathrm{e}^{-\frac{\pi(-2E)}{\hbar}}\right)^{\mp 1/2}, (\arg\hbar = \pm 0) , \tag{36}$$

$$|N^{-2}| = \left|\mathrm{e}^{-2\mathrm{i}\sum_{n=1}^\infty S_R^{(n)}(\infty)\hbar^{2n-1}}\right| \overset{(B)}{=} \left(1 + \mathrm{e}^{-\frac{\pi(-2E)}{\hbar}}\right)^{\pm 1/2}, (\arg\hbar = \pm 0) , \tag{37}$$

$$\left|\hat{N}^2\right| = \left(1 + \mathrm{e}^{-\frac{\pi(-2E)}{\hbar}}\right)^{-1/2}, (\arg\hbar = 0) . \tag{38}$$

The JWKB origins of the square-root factor in (1) are thus revealed in (36) – (38). We sharpen this in the next subsection.

4.4 Connection formulas for functions "well-normalized at ∞"

Instead of "square-root" normalization at the turning point, consider instead a normalization by which $S^{(n)}_{R,L}(\infty) = 0$. We attach a superscript $^{[\infty]}$ and define

$$S_L^{(n)[\infty]}(x) = S_L^{(n)}(x) - S_L^{(n)}(-\infty), (n \geq 1)\,, \tag{39}$$

$$S_R^{(n)[\infty]}(x) = S_R^{(n)}(x) - S_R^{(n)}(+\infty), (n \geq 1)\,, \tag{40}$$

$$\psi^{[\infty]}_{R,\pm S}(x,\hbar) = N^{\mp 1}\psi_{R,\pm S}(x,\hbar), \quad \psi^{[\infty]}_{L,\pm S}(x,\hbar) = N^{\mp 1}\psi_{L,\pm S}(x,\hbar)\,. \tag{41}$$

Since $S_L^{(n)}(-\infty) = S_R^{(n)}(+\infty)$, it is not necessary to distinguish left from right normalization factors N. This is the normalization used by Delabaere, Dillinger and Pham [10],[11] who call such JWKB wave functions "well-normalized at ∞." The connection formulas for the "∞-normalized" JWKB wave functions are given in Table 2. It is particularly significant that when the relative normalization factor N is replaced by its Borel-summed explicit evaluation in terms of $\hat{N}$, there is no $\arg\hbar = 0$ Stokes line in the 1–1′ connection formula. (See the discussion in Sect. 5.)

Table 2. 1-1′ connection formula for "∞-normalized" JWKB functions

Region	JWKB wave function
1′	$c_+ \mathrm{e}^{-\frac{(-E)\pi}{\hbar}}\psi^{[\infty]}_{R,+S}(x,\hbar) + c_- \mathrm{e}^{-\frac{(-E)\pi}{\hbar}}\psi^{[\infty]}_{R,-S}(x,\hbar)$
	$\arg\hbar > 0$
1	$-\mathrm{i}[c_+ + c_- N^2(1+\mathrm{e}^{-2\frac{(-E)\pi}{\hbar}})]\psi^{[\infty]}_{L,+S}(x,\hbar) + \mathrm{i}(N^{-2}c_+ + c_-)\psi^{[\infty]}_{L,-S}(x,\hbar)$
	$\arg\hbar < 0$
1	$-\mathrm{i}(c_+ + c_- N^2)\psi^{[\infty]}_{L,+S}(x,\hbar) + \mathrm{i}[c_- + c_+ N^{-2}(1+\mathrm{e}^{-2\frac{(-E)\pi}{\hbar}})]\psi^{[\infty]}_{L,-S}(x,\hbar)$
	independent of the sign of $\arg\hbar$
1	$-\mathrm{i}(c_+ + c_- \hat{N}^2)\psi^{[\infty]}_{L,+S}(x,\hbar) + \mathrm{i}(c_- + c_+ \hat{N}^{-2})\psi^{[\infty]}_{L,-S}(x,\hbar)$

4.5 Preliminary comment on the details: generalization of (1)

Formally the connection formulas that connect Region 1′ to 2, 3, and especially 1 are *different*, depending on the sign of $\arg\hbar$. In Table 2, the factor $(1+\mathrm{e}^{-2(-E)\pi/\hbar})$ shifts places; in Table 1 there are even more changes. The differences are a manifestation of the paired-picture view of Fig. 2 and will be discussed in detail in the next section. Here we note that the cleanest generalization of (1) is already contained in Table 2 when $c_+ = 1$ and $c_- = 0$:

$$\mathrm{i}\left(-\psi^{[\infty]}_{L,+S}(x,\hbar) + \hat{N}^{-2}\psi^{[\infty]}_{L,-S}(x,\hbar)\right) \longleftrightarrow \mathrm{e}^{-\frac{(-E)\pi}{\hbar}}\psi^{[\infty]}_{R,+S}(x,\hbar)\,. \tag{42}$$

The $\hat{N}^{-2}$ in (42) by virtue of (34) and (38) is exactly the $e^{i\phi}\sqrt{1+e^{-2\pi(-E)/\hbar}}$ of (1) when $\hbar$ is real, but with its JWKB origins clear from (30)–(33). At the same time, the lowest-order JWKB functions in (1) generalize to the the complete "∞-normalized" JWKB functions.

5 Discussion of the connection formula differences and the analytic functions the JWKB functions represent

5.1 Borel sums of $\psi_{R,\pm S}(x,\hbar)$ and of $\psi^{[\infty]}_{R,\pm S}(x,\hbar)$

An advantage of the quadratic barrier is that the exact solutions are the known parabolic cylinder functions, to which the JWKB functions $\psi_{R,\pm S}(x,\hbar) = \dot{S}_R^{-1/2} e^{\pm i(S_R/\hbar+\pi/4)}$ are Borel summable (similarly for $\psi_{L,\pm S}$). The proportionality constants can be found by matching asymptotic expansions. Let

$$\eta = e^{\frac{i\pi}{8}} \left(\frac{-E}{e\hbar}\right)^{\frac{i(-E)}{2\hbar}}, \tag{43}$$

and let $D_\nu(z)$ denote Whittaker's principal parabolic cylinder function (Eq. 19.3.7 of [13]). The ∞-normalized functions have no real-$\hbar$ Stokes line:

$$\psi^{[\infty]}_{R,+S}(x,\hbar) \overset{(B)}{=} \eta \left(\frac{2}{\hbar}\right)^{\frac{1}{4}} e^{\frac{\pi(-E)}{4\hbar}} D_{-i(\frac{-E}{\hbar})-\frac{1}{2}} \left(e^{-\frac{\pi i}{4}} x\sqrt{\frac{2}{\hbar}}\right), \tag{44}$$

$$\psi^{[\infty]}_{R,-S}(x,\hbar) \overset{(B)}{=} \eta^{-1} \left(\frac{2}{\hbar}\right)^{\frac{1}{4}} e^{\frac{\pi(-E)}{4\hbar}} D_{+i(\frac{-E}{\hbar})-\frac{1}{2}} \left(e^{+\frac{\pi i}{4}} x\sqrt{\frac{2}{\hbar}}\right). \tag{45}$$

The turning-point-normalized functions do have a real-$\hbar$ Stokes line:

$$\psi_{R,+S}(x,\hbar)$$

$$\overset{(B)}{=} \hat{N}\eta \left(\frac{2}{\hbar}\right)^{\frac{1}{4}} e^{\frac{\pi(-E)}{4\hbar}} D_{-i(\frac{-E}{\hbar})-\frac{1}{2}} \left(e^{-\frac{\pi i}{4}} x\sqrt{\frac{2}{\hbar}}\right), (\arg\hbar > 0), \tag{46}$$

$$\overset{(B)}{=} \hat{N}\eta \left(\frac{2}{\hbar}\right)^{\frac{1}{4}} e^{\frac{\pi(-E)}{4\hbar}} D_{-i(\frac{-E}{\hbar})-\frac{1}{2}} \left(e^{-\frac{\pi i}{4}} x\sqrt{\frac{2}{\hbar}}\right) \sqrt{1+e^{-\frac{\pi(-2E)}{\hbar}}}, \quad (\arg\hbar < 0), \tag{47}$$

$$\psi_{R,-S}(x,\hbar)$$

$$\overset{(B)}{=} (\hat{N}\eta)^{-1} \left(\frac{2}{\hbar}\right)^{\frac{1}{4}} e^{\frac{\pi(-E)}{4\hbar}} D_{+i(\frac{-E}{\hbar})-\frac{1}{2}} \left(e^{+\frac{\pi i}{4}} x\sqrt{\frac{2}{\hbar}}\right) \sqrt{1+e^{-\frac{\pi(-2E)}{\hbar}}}, \quad (\arg\hbar > 0), \tag{48}$$

$$\overset{(B)}{=} (\hat{N}\eta)^{-1} \left(\frac{2}{\hbar}\right)^{\frac{1}{4}} e^{\frac{\pi(-E)}{4\hbar}} D_{+i(\frac{-E}{\hbar})-\frac{1}{2}} \left(e^{+\frac{\pi i}{4}} x\sqrt{\frac{2}{\hbar}}\right), (\arg\hbar < 0). \tag{49}$$

The value of $\hat{N}$ is the square root of (34).

5.2 Current densities

Take $\hbar$ to be real and compare (46) and (47):

$$\psi_{R,+S}(x,\hbar - \mathrm{i}0) \overset{(B)}{=} \sqrt{1 + \mathrm{e}^{-\frac{\pi(-2E)}{\hbar}}}\, \psi_{R,+S}(x,\hbar + \mathrm{i}0)\,. \tag{50}$$

The implication is that the current density represented by the outgoing-wave JWKB wave function $\psi_{R,+S}$ when $\mathrm{Im}\hbar = -0$ is exactly $(1 + \mathrm{e}^{-\frac{\pi(-2E)}{\hbar}})$ times the current density represented by the same formal JWKB wave function when $\mathrm{Im}\hbar = +0$. With (48) and (49), the further implication is that the incoming-wave current density represented by $\psi_{R,-S}$ is $(1 + \mathrm{e}^{-\frac{\pi(-2E)}{\hbar}})$ times larger than the outgoing-wave current density represented by $\psi_{R,+S}$ when $\mathrm{Im}\hbar = +0$, but that the reverse is true when $\mathrm{Im}\hbar = -0$. Both these observations are counter-intuitive, in part because the notation for the JWKB wave functions does not indicate that evaluation is taking place on a Stokes line (for $\hbar$) where "real" may Borel-sum to "complex."

In the classically allowed region to the left, the outgoing-wave JWKB function is $\psi_{L,+S}$. By virtue of (21), it has the same current-density "normalization" as $\psi_{R,+S}$, and similarly for $\psi_{L,-S}$ and $\psi_{R,-S}$.

The $\hbar$-Stokes line comes from the implicit normalization factors. The ∞-normalized wave functions do not have a Stokes line with respect to $\hbar$, and the implication from (44) and (45) is that all four of $\psi^{[\infty]}_{L,\pm S}$ and $\psi^{[\infty]}_{R,\pm S}$ share a common current-density normalization.

5.3 Transmission through the barrier

Transmission through the barrier from left to right is specified, in the notations of Tables 1 and 2, by setting $d_- = c_- = 0, d_+ = c_+ = 1$ and taking $\hbar$ real, which yields the following versions of the 1–$1'$ connection formula:

$$\mathrm{i}\,[-\psi_{L,+S}(x,\hbar + \mathrm{i}0) + \psi_{L,-S}(x,\hbar + \mathrm{i}0)] \longleftrightarrow \mathrm{e}^{-(-E)\pi/\hbar}\psi_{R,+S}(x,\hbar + \mathrm{i}0)\,, \tag{51}$$

$$\mathrm{i}\left[-\psi_{L,+S}(x,\hbar - \mathrm{i}0) + (1 + \mathrm{e}^{-2(-E)\pi/\hbar})\psi_{L,-S}(x,\hbar - \mathrm{i}0)\right] \\ \longleftrightarrow \mathrm{e}^{-(-E)\pi/\hbar}\psi_{R,+S}(x,\hbar - \mathrm{i}0)\,, \tag{52}$$

$$\mathrm{i}\left[-\psi^{[\infty]}_{L,+S}(x,\hbar) + \frac{\psi^{[\infty]}_{L,-S}(x,\hbar)}{N(\hbar + \mathrm{i}0)^2}\right] \longleftrightarrow \mathrm{e}^{-\frac{(-E)\pi}{\hbar}}\psi^{[\infty]}_{R,+S}(x,\hbar)\,, \tag{53}$$

$$\mathrm{i}\left[-\psi^{[\infty]}_{L,+S}(x,\hbar) + \frac{(1 + \mathrm{e}^{-2(-E)\pi/\hbar})\psi^{[\infty]}_{L,-S}(x,\hbar)}{N(\hbar - \mathrm{i}0)^2}\right] \longleftrightarrow \mathrm{e}^{-\frac{(-E)\pi}{\hbar}}\psi^{[\infty]}_{R,+S}(x,\hbar)\,, \tag{54}$$

$$\mathrm{i}\left[-\psi^{[\infty]}_{L,+S}(x,\hbar) + \mathrm{e}^{\mathrm{i}\phi}\sqrt{1 + \mathrm{e}^{-2\pi(-E)/\hbar}}\,\psi^{[\infty]}_{L,-S}(x,\hbar)\right] \longleftrightarrow \mathrm{e}^{-\frac{(-E)\pi}{\hbar}}\psi^{[\infty]}_{R,+S}(x,\hbar)\,. \tag{55}$$

All five equations, although different, lead by an elementary argument to the same transmission and reflection coefficients: since the outgoing waves are in each case identically normalized, the ratio of the transmission to reflection coefficients is exactly

$$T/R = \mathrm{e}^{-2(-E)\pi/\hbar} . \tag{56}$$

From $T + R = 1$, it follows that

$$T = \frac{\mathrm{e}^{-2(-E)\pi/\hbar}}{1 + \mathrm{e}^{-2(-E)\pi/\hbar}} , \qquad R = \frac{1}{1 + \mathrm{e}^{-2(-E)\pi/\hbar}} . \tag{57}$$

The interesting observation is why the coefficient of the incident wave $\psi_{L,-S}(x,\hbar)$ or $\psi^{[\infty]}_{L,-S}(x,\hbar)$ depends of the sign of $\arg\hbar$. The answer from the preceding subsection is that the relative normalization factor $N(\hbar)$ has a Stokes line for real $\hbar$, and that incoming and outgoing turning-point-normalized JWKB wave functions have different current-density normalizations depending on the sign of $\arg\hbar$. The ∞-normalized functions have a common current-density normalization, but it is still necessary to evaluate (one way or another [19]) the normalization factor $N(\hbar)$.

5.4 Conceptualization of the connection formula for real $\hbar$: doubled-Stokes lines

What is the connection formula at the right-hand turning point in the limit that $\arg\hbar = +0$? Region 1′ to Region 3′ is normal, but Region 2′ has disappeared and has been replaced by Region 3 (see Fig. 2). The limit is shown in Table 3. $\psi_{R,+Q}(x,\hbar)$ in Region 3 (new Region 2′) has an extra term when compared to the old Region 2′. It would seem descriptive in the light of Fig. 2 to refer to this situation as a doubled-Stokes line. Which Stokes lines are doubled depend on the sign of $\arg\hbar = \pm 0$.

Table 3. Connection formula at the right-hand turning point when $\arg\hbar = +0$ for turning-point-normalized JWKB wave functions

Region	JWKB wave function, ($\arg\hbar = +0$)
1′	$d_+\psi_{R,+S}(x,\hbar) + d_-\psi_{R,-S}(x,\hbar)$
old 2′ (empty)	$(d_+ + d_-)\psi_{R,+Q}(x,\hbar) - \mathrm{i}d_-\psi_{R,-Q}(x,\hbar)$
3 (new 2′)	$[d_+ + d_-(1 + \mathrm{e}^{-2(-E)\pi/\hbar})]\psi_{R,+Q}(x,\hbar) - \mathrm{i}d_-\psi_{R,-Q}(x,\hbar)$
3′ (normal)	$(d_+ + d_-)\psi_{R,+Q}(x,\hbar) + \mathrm{i}d_+\psi_{R,-Q}(x,\hbar)$

6 Conclusions

The quadratic barrier provides a solvable illustration of the exact JWKB method. With complexification of $\hbar$, the connection formulas follow directly from the connection formulas at the two linear turning points and the matching of solutions in their common Stokes region, as pictured in Fig. 2. The square root in (1) comes from explicit Borel summation of the implicit normalization factors in the JWKB wave function through the Stirling series to a gamma function. A Stokes line in this $\hbar$-series leads to formulas, including current densities, that depend on the sign of $\arg\hbar = \pm 0$. The understanding of the limiting case of real $\hbar$, and especially the disappearance of the common Stokes region where the left and right functions were matched, is aided by the concept of doubled-Stokes lines.

Acknowledgement

The authors wish to thank Professor Gabriel Álvarez for most helpful comments and suggestions.

Addendum: a straightforward proof of Sato's conjecture

With the stimulation and encouragement of Professors Kawai and Takei, we add this short, straightforward, informal proof of the conjecture by Sato mentioned above.

We adapt the notation of Eq. (1.19) of Kawai and Takei [8] and denote the constant ratios of the Weber and JWKB functions that correspond at large x by c_1 and c_2,

$$c_1 = \frac{D_{-\mathrm{i}(\frac{-E}{\hbar})-\frac{1}{2}}\left(\mathrm{e}^{-\frac{\pi\mathrm{i}}{4}}x\sqrt{\frac{2}{\hbar}}\right)}{\psi_{R,+S}(x,\hbar)}\,, \tag{58}$$

$$c_2 = \frac{D_{+\mathrm{i}(\frac{-E}{\hbar})-\frac{1}{2}}\left(\mathrm{e}^{+\frac{\pi\mathrm{i}}{4}}x\sqrt{\frac{2}{\hbar}}\right)}{\psi_{R,-S}(x,\hbar)}\,. \tag{59}$$

The c_2 here differs slightly from C_2 in [8]. Computation of c_1 and c_2 at large x yields [cf. (30), (31), and (43)–(45)]

$$c_1 = (N\eta)^{-1}(\hbar/2)^{1/4}\mathrm{e}^{-\pi(-E)/4\hbar}\,, \tag{60}$$

$$c_2 = N\eta(\hbar/2)^{1/4}\mathrm{e}^{-\pi(-E)/4\hbar}\,. \tag{61}$$

The ratio c_2/c_1 is proportional to the square of the relative normalization factor N:

$$\frac{c_2}{c_1} = N^2\eta^2\,, \tag{62}$$

$$= \mathrm{e}^{\frac{\pi\mathrm{i}}{4}}\left(\frac{-E}{\mathrm{e}\hbar}\right)^{\frac{\mathrm{i}(-E)}{\hbar}} N^2\,. \tag{63}$$

Note that N, $\hbar^{-1/4}c_1$, and $\hbar^{-1/4}c_2$ [(30), (58)–(63)] are all asymptotic expansions in $\hbar/(-E)$ whose Borel sums turn out to have a Stokes discontinuity on $\arg\hbar = 0$. In the present notation, Sato's conjecture takes the form,

$$\frac{c_2}{c_1} = N^2 \mathrm{e}^{\frac{\pi \mathrm{i}}{4}} \left(\frac{-E}{\mathrm{e}\hbar}\right)^{\frac{\mathrm{i}(-E)}{\hbar}} = \frac{\mathrm{e}^{\frac{\pi\mathrm{i}}{4}}\mathrm{e}^{\frac{(-E)\pi}{2\hbar}}}{\sqrt{2\pi}} \Gamma\left(\frac{1}{2} + \frac{\mathrm{i}(-E)}{\hbar}\right), \quad (\arg\hbar > 0)\,, \tag{64}$$

the important content of which is that the square of the JWKB normalization series (30) is term-by-term exactly the same as the asymptotic power series obtained from the gamma function.

Remark 1 Equation (64) is essential to the connection formulas for the ∞-normalized JWKB wave functions (Table 2) and was obtained from an analysis of the Jost function by Voros [6].

Remark 2 Although N arose here as the relative normalization factor (41) between $\psi_{R,+S}(x,\hbar)$ and $\psi^{[\infty]}_{R,+S}(x,\hbar)$, N *also* carries the ratio of $\psi_{R,+S}(x,\hbar)$ to the Weber function [(58) and (60)]. It is *this fact* that we exploit to evaluate N. Our derivation uses values of the Weber function and its derivative at 0.

Remark 3 The calculation is unexpectedly straightforward because of a most fortuitous result: that both $\psi_{R,+S}(0,\hbar)$ and $\mathrm{d}\psi_{R,+S}(x,\hbar)/\mathrm{d}x|_{x=0}$ can be calculated from the single asymptotic power series $\dot{Q}_R(0)$, which in turn can be "Borel-summed" via the equality of the logarithmic derivatives of the JWKB and Weber [13] functions.

Sketch of proof By (60) and (61) it is sufficient to calculate c_1, since

$$c_2 = \sqrt{\hbar/2}\,\mathrm{e}^{-\pi(-E)/2\hbar}/c_1\,. \tag{65}$$

We specify that $\arg\hbar > 0$, so that there is no Stokes discontinuity in the Borel sum of $\psi_{R,+S}(x,\hbar)$ when x is continued from $x = +\infty$ to $x = 0$ (the point at which the potential is stationary). That is, $\psi_{R,+S}(x,\hbar) \leftrightarrow \psi_{R,+Q}(x,\hbar)$. [Cf. Table 1 and Fig. 2; the relevant regions are $1'$ and $2'$. We do not need $\psi_{R,-S}(x,\hbar) \leftrightarrow \psi_{R,+Q}(x,\hbar) - \mathrm{i}\psi_{R,-Q}(x,\hbar)$.] The essential fortunate fact is that $Q_R(x,\hbar) - (-E)\pi/2$ is an odd function of x (also all even derivatives are odd; all odd derivatives are even—see Secs. 2 and 4.1), which leads to

$$\psi_{R,+Q}(0,\hbar) = [-\dot{Q}_R(0,\hbar)]^{-1/2}\mathrm{e}^{(-E)\pi/2\hbar}\,, \tag{66}$$

$$\dot{\psi}_{R,+Q}(0,\hbar) \equiv \mathrm{d}\psi_{R,+Q}(x,\hbar)/\mathrm{d}x|_{x=0} = -\frac{1}{\hbar}[-\dot{Q}_R(0,\hbar)]^{+1/2}\mathrm{e}^{(-E)\pi/2\hbar}\,, \tag{67}$$

$$\dot{Q}_R(0,\hbar) = \hbar\frac{\dot{\psi}_{R,+Q}(0,\hbar)}{\psi_{R,+Q}(0,\hbar)}\,, \tag{68}$$

$$= \hbar\frac{\frac{\mathrm{d}}{\mathrm{d}x}D_{-\mathrm{i}(\frac{-E}{\hbar})-\frac{1}{2}}\left(\mathrm{e}^{-\frac{\pi\mathrm{i}}{4}}x\sqrt{\frac{2}{\hbar}}\right)_{x=0}}{D_{-\mathrm{i}(\frac{-E}{\hbar})-\frac{1}{2}}(0)}\,, \quad (\arg\hbar > 0)\,, \tag{69}$$

$$= -2\mathrm{e}^{-\pi\mathrm{i}/4}\sqrt{\hbar}\frac{\Gamma\left(\frac{3}{4} + \frac{1}{2}\mathrm{i}\frac{(-E)}{\hbar}\right)}{\Gamma\left(\frac{1}{4} + \frac{1}{2}\mathrm{i}\frac{(-E)}{\hbar}\right)}\,, \quad (\arg\hbar > 0)\,. \tag{70}$$

c_1 follows from evaluation of the wave-function ratio at $x = 0$ and the gamma-function duplication formula [13]:

$$c_1 = \frac{D_{-\mathrm{i}(\frac{-E}{\hbar})-\frac{1}{2}}(0)}{\psi_{R,+S}(0,\hbar)}, \tag{71}$$

$$= \frac{\mathrm{e}^{-\pi\mathrm{i}/8}\pi^{1/4}\hbar^{1/4}\mathrm{e}^{-(-E)\pi/2\hbar}}{\sqrt{\Gamma\left(\frac{1}{2}+\mathrm{i}\frac{(-E)}{\hbar}\right)}}, \quad (\arg\hbar > 0)\,. \tag{72}$$

Equations (72) and (65) give (64).

References

1. Miller, S.C., Good, R.H.: Phys. Rev., **91**, 174 (1953).
2. K. W. Ford, D. L. Hill, M. Wakano and J. A. Wheeler, Ann. Phys. (NY), **7**, 239 (1959).
3. M. S. Child, *Semiclassical Mechanics with Molecular Applications* (Oxford University Press, Oxford, 1991).
4. J. N. Connor, Molec. Phys., **15**, 37 (1968).
5. M. V. Fedoryuk, *Asymptotic Analysis* (Springer-Verlag, Berlin, 1993).
6. A. Voros, Ann. Inst. Henri Poincaré, **39**, 211 (1983).
7. T. Aoki, T. Kawai and Y. Takei, in *Special Functions,ICM-90 Satellite Conference Proceedings*, edited by M. Kashiwara and T. Miwa (Springer-Verlag, Berlin, 1991), p. 1.
8. T. Kawai and Y. Takei, in *Analyse Algébrique des Perturbations Singulière. I: Méthodes Résurgentes*, edited by L. Boutet de Monvel (Hermann, Paris, 1994), p. 85.
9. T. Aoki, T. Kawai and Y. Takei, Sugaku Expositions, **8**, 217 (1995).
10. E. Delabaere, H. Dillinger and F. Pham, J. Math. Phys., **38**, 6126 (1997).
11. E. Delabaere and F. Pham, Ann. Inst. Henri Poincaré Phys. Théor., **71**, 1 (1999).
12. S. Wolfram, *The Mathematica Book* (Cambridge University Press and Wolfram Media, Champaign, 1999).
13. M. Abramowitz and I. A. Stegun, *Handbook of Mathematical Functions* (National Bureau of Standards, Washington, D.C., 1964).
14. C. G. Darwin, Quart. J. Mech. Appl. Math., **2**, 311 (1949).
15. H. J. Silverstone, Phys. Rev. Lett., **55**, 2523 (1985).
16. H. Shen and H. J. Silverstone, Int. J. Quantum Chem., **99**, 336 (2004).
17. M. V. Berry and K. E. Mount, Rep. Prog. Phys., **35**, 315 (1972).
18. R. B. Paris and A. D. Wood, J. Comp. and Appl. Math., **41**, 135 (1992).
19. G. Álvarez, V. Martín-Mayor and J. J. Ruiz-Lorenzo, J. Phys. A, **33**, 841 (2000).

A role of virtual turning points and new Stokes curves in Stokes geometry of the quantum Hénon map

Akira Shudo

Department of Physics, Tokyo Metropolitan University, Minami-Ohsawa, Hachioji, Tokyo 192-0397, Japan
shudo@phys.metro-u.ac.jp

Summary. A role of virtual turning points and new Stokes curves, that have been proposed as entirely new notions appearing only in higher-order differential equations, is studied in the Stokes geometry of the quantum Hénon map. Characteristics of the Stokes geometry in multi-steps are particularly focused on and generic bifurcation patterns of the Stokes geometry are listed up, which is intended to develop a "pruning theory of Stokes geometry".

1 Introduction

The WKB method has broad applicability in various branches of mathematical physics. It was originally developed to obtain approximate solutions of the Schrödinger equation, but it has soon been extended to much wider settings. The expansion with respect to a small parameter (or the reciprocal of a large parameter) is usually divergent and it is an asymptotic expansion at most since it is essentially a kind of a singular perturbation. Nevertheless, the validity of approximation is satisfactory and so it is widely used.

Alternative utility of the WKB method recently receiving much attentions particularly in the field of quantum chaos is to use it as a language connecting quantum and classical mechanics. The Gutzwiller trace formula has extensively been studied, not because it is a quantitatively efficient to evaluate correct density of states for chaotic systems, but because it reveals how quantum mechanics links to classical one in an appropriate asymptotic limit [Gutz, Les].

The topic of the present article is along the line of such an investigation, that is to study quantum manifestation of classical chaos through the WKB method. However, as explained below, since what we focus on here is tunneling phenomena in quantum mechanics [SI1], the analysis inevitably involves the

Received 27 February, 2006. Revised 29 May, 2006. Accepted 29 May, 2006.

WKB analysis in the complex domain and one should go one step further from the arguments made so far in the conventional WKB analysis. The aim of the article is to stress that new ingredients introduced by the RIMS group within the exact WKB framework [AKT1, AKT2, AKT3] play essential roles to understand the quantum tunneling effect in the system whose classical counterpart generates chaos.

2 Quantum Tunneling in the Hénon Map

The simplest possible choice of the chaotic system to apply the complex semiclassical method would be polynomial mappings defined on $\mathbb{R}^2$. Torus automorphism or piecewise linear mappings on the cylinder are often used in the semiclassical analysis of chaotic systems, but they are not suitable here because discontinuity of the map makes it difficult to extend the map into the complex plane.

A classification theorem of Friedland and Milnor tells us that a canonical form of non-trivial polynomial diffeomorphism is expressed as

$$f_a : \begin{pmatrix} x \\ y \end{pmatrix} \mapsto \begin{pmatrix} y \\ y^2 - x + a \end{pmatrix}, \tag{1}$$

and diffeomorphism not reducible to the above form is classified either into an affine or elementary map, both of which show simple behaviors and do not generate chaos [FM]. The map (1) is called the Hénon map [Hen] (here we limit ourselves only to the area-preseving case) and it was also shown that non-trivial polynomial diffeomorphism with any order is essentially a composition of the Hénon map [FM]. An alternative familiar form is obtained via an affine change of variables as $(p, q) = (y - x, x - 1)$ together with a parameter $c = 1 - a$,

$$\mathcal{F} : \begin{pmatrix} q \\ p \end{pmatrix} \mapsto \begin{pmatrix} q + p \\ p - V'(q + p) \end{pmatrix}, \qquad V(q) = -\frac{q^3}{3} - cq. \tag{2}$$

Here we construct quantum mechanics of the area-preserving map by introducing a discrete analog of the Feynman-type path integral:

$$< q_n|U^n|q_0 >= \int_{-\infty}^{\infty} \cdots \int_{-\infty}^{\infty} dq_1 \cdots dq_{n-1} \exp\Big[\frac{i}{\hbar} S(q_0, \cdots, q_n)\Big]. \tag{3}$$

The function $S(q_0, \cdots, q_n)$ represents the discretized Lagrangian or the action functional given by

$$S(q_0, \cdots, q_n) = \sum_{j=0}^{n-1} \frac{1}{2}(q_{j+1} - q_j)^2 - \sum_{j=1}^{n-1} V(q_j). \tag{4}$$

The action functional is derived so that applying the variational principle to $S(q_0, \cdots, q_n)$ recovers the area preseving map (2).

Applying the saddle point method, we get the semiclassical formula that is expressed as a sum over contributions of classical trajectories connecting the initial and final states:

$$< q_n|U^n|q_0 > \approx \sum_\gamma A_\gamma(q_0, q_n) \exp\{\frac{i}{\hbar} S_\gamma(q_0, q_n) - i\mu_\gamma \frac{\pi}{2}\}, \tag{5}$$

where $A_\gamma(q_0, q_n)$ stands for the amplitude factor for a classical orbit γ, and $S_\gamma(q_0, q_n)$ is given by putting the data of the corresponding classical path γ into the action functional $S(q_0, \cdots, q_n)$. μ_γ represents the Maslov index. The summation is taken over such classical orbits that are located initially on the manifold $q_0 = \alpha$, and finally on $q_n = \beta$, where both α and β should take real values since they are observables in the representation under consideration. Note that the number of classical orbits satisfying the saddle point condition is 2^{n-1}.

For given $q_0 = \alpha \in \mathbb{R}$ and $q_n = \beta \in \mathbb{R}$, if all the solutions $q_1^{(i)}$ $(1 \le i \le 2^{n-1})$ satisfying

$$\mathcal{F}^{n-1}(q_0 = \alpha, q_1) = (q_{n-1}, q_n = \beta) \tag{6}$$

are real-valued, itinerary of each trajectory $(q_0, q_1^{(i)}, q_2^{(i)}, \cdots, q_{n-1}^{(i)}, q_n^{(i)})$ is also real-valued, meaning that the initial and final states are connected only via the real dynamics. On the other hand, complex-valued q_1 generates itinerary in the complex plane, so it is described by the complex dynamics. As shown in Fig. 1, in case of $n = 2$ if the final position q_2 is put outside the region covered by the real Lagrangian manifold $(q_2(q_0 = \alpha, q_1), q_1)$ $(-\infty < q_1 < \infty)$, q_1 should be complex and the transition from the initial $q_0 = \alpha$ to the final state $q_2 = \beta$ occurs via complex classical dynamics. Therefore we call the part of such complex Lagrangian manifold the *tunneling branch*. Lagrangian manifold for $n = 2$ is exactly parabola and the connection of WKB solutions between the tunneling branch and real parabola is described by the Airy function.

As the map is iterated, the Hénon map generates stretching and folding dynamics. For example, in case of $n = 3$, the manifold has three folding points from each of which a tunneling branch emanates as shown Fig. 1. At each folding point, we expect that the connection is also described by the Airy-type at least locally. The question we will ask here is whether the local treatment is sufficient or not.

3 Ingredients and Recipes for Stokes Geometry

Definition of Virtual Turning Points and New Stokes Curves

A central problem in the complex WKB description of quantum propagator is to find a proper prescription to treat the Stokes phenomenon appearing in

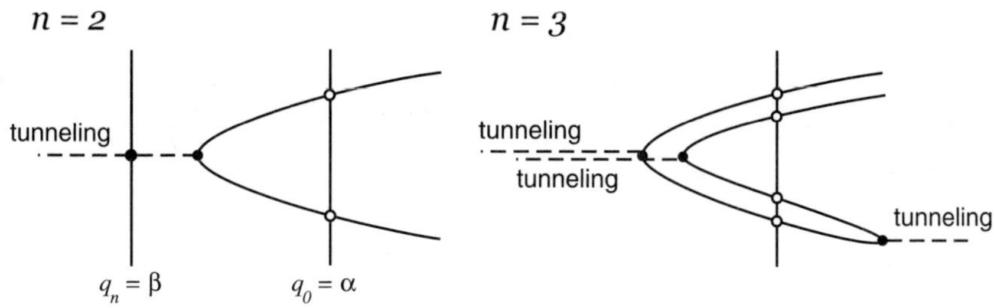

Fig. 1. Iterated Lagrangian manifold in (q_n, q_{n-1}) plane together with tunneling branches attaching to the real one.

multiple integrals (3). In what follows, we fix the initial coordinate $q_0 = \alpha$ and regard the quantum propagator (3) as a function of the final coordinate q_n. We therefore use the notation $I(q_n) \equiv < q_n|U^n|q_0 >$.

An annoying issue for us when we only cares the local rule around folding points on the manifold was the priority problem: which connection is preferentially operated [SI2]. A very crude and phenomenological way of treating it was tested in [SI2], but later it was realized that such a problem is originates from that Stokes curves of the integral $I(q_n) =< q_n|U|q_0 >$ cross each other. The crossing problem was already discussed [BNR], in which the authors pointed out the necessity of introducing *new Stokes curves.* Later, within the exact WKB framework, it was asserted that not only new Stokes curves but also *virtual turning points* should be introduced to construct a complete Stokes geometry [AKT1, AKT2]. Their argument is based on general theory on the propagation of singularities of differential equations [SKK], and virtual turning points are defined as self-intersection points of bicharacteristic curves for the Borel transformed differential equation [AKT1]. A new Stokes curves is obtained as the one emanating from a virtual turning point. It was emphasized that there is no special distinction between ordinary Stokes curves emanating from the ordinary turning point and new Stokes curves. An explicit algorithm to take into account these newly introduced ingredients is described in [AKSST], and also a recent work [Hon]. We should note the work of Howls [How] and the one with his coworkers [HLO], in which hyperasymptotic expansions are employed to analyze the Riemann sheet structure of multiple integrals, and to obtain implications of crossing of Stokes curves, that are exactly the objects of study in this report.

We can apply their scheme in order to treat Stokes phenomena in the saddle point evaluation of integral $I(q_n)$. To this end, though it might be redundant, we first derive a set of differential equations acting on $I(q_n)$ and then consider bicharacteristic equations for their Borel transform [S1]. The principal symbol so derived is

$$\hat{P}(q_n, S, \xi, \eta) \equiv -\eta \hat{S}(q_1, q_2) \Bigg|_{\substack{q_1 = q_1(q_{n-1}, q_n) \\ q_2 = q_2(q_{n-1}, q_n)}} \Bigg|_{q_{n-1} = q_n - \xi\eta^{-1}} \tag{7}$$

where

$$\hat{S}(q_1, q_2) \equiv \frac{\partial}{\partial q_1} S(q_0, q_1, \cdots, q_n)\big|_{q_0=0} = 2q_1 + q_1^2 + q_2 - c \ ,$$

and $\xi \equiv \eta p_n = \eta(q_n - q_{n-1})$. A set of equation describing the bicharacteristic strip is therefore

$$\frac{dq_n}{dq_1} = \frac{\partial \hat{P}(q_n, S, \xi, \eta)}{\partial \xi}, \qquad \frac{d\xi}{dq_1} = -\frac{\partial \hat{P}(q_n, S, \xi, \eta)}{\partial q_n},$$
$$\frac{dS}{dq_1} = \frac{\partial \hat{P}(q_n, S, \xi, \eta)}{\partial \eta}, \qquad \frac{d\eta}{dq_1} = -\frac{\partial \hat{P}(q_n, S, \xi, \eta)}{\partial S}.$$

Here, (q_n, ξ) and (X, η) respectively form canonical conjugate pairs in Hamiltonian equations. An interesting fact is that the initial condition q_1 plays the role of time in bicharacteristic equations. This means that expressing $q_n(q_0, q_1)$ or $S(q_0, \cdots q_n)$ as a function of q_1, which can be easily done by iterating the Hénon map itself, is nothing but solving the bicharacteristic equation. The above set of equations gives the bicharacteristic strip and bicharacteristic curves, the projection of the bicharacteristic strip onto (q_n, S) plane, is then obtained. Hence, by applying the definition proposed in [AKT1], we have all necessary ingredients. That is, for $q_1^{(i)} \neq q_1^{(j)}$, q_n^T is a virutal turning point if

$$q_n^T(q_0, q_1^{(i)}) = q_n^T(q_0, q_1^{(j)}),$$
$$S(q_0, q_1^{(i)}, \cdots, q_n^T(q_0, q_1^{(i)})) = S(q_0, q_1^{(j)} \cdots, q_n^T(q_0, q_1^{(j)})).$$

We say curves emanating from a turning point q_n^T with the following relation are Stokes curves:

$$\operatorname{Im} S(q_0, q_1^{(i)}, \cdots, q_{n-1}^{(i)}, q_n^T) = \operatorname{Im} S(q_0, q_1^{(j)}, \cdots, q_{n-1}^{(j)}, q_n^T). \tag{8}$$

Formal definitions of ordinary and new Stokes curves are the same, and only difference is the starting point of each curve. The procedure to derive differential equations and the principal symbol associated with them can straightforwardly be extended to generic polynomial maps.

Virtual Turning Points : How Observable ?

Before going to the Hénon map, we examine whether or not the virtual turning points are observable objects by taking a one-step twice folding map. A simple example is an integral

$$I(q_2, q_0) = \int_{-\infty}^{\infty} \exp\left[\frac{i}{\hbar} S(q_0, q_1, q_2)\right] dq_1, \tag{9}$$

where

$$S(q_0, q_1, q_2) = \frac{1}{2}(q_2 - q_1)^2 + \frac{1}{2}(q_1 - q_0)^2 + \frac{1}{4}q_1^4 - \frac{5}{2}q_1^2. \tag{10}$$

Here, q_0 is fixed and considered as a parameter as before. The action (10) generates a twice folding classical map that has been investigated in [DM]. Note that the integral (9) is analogous to the so-called Pearcy function. As shown in Fig. 2(a), one-step iteration generates twice folded Lagrangian manifold on which two folding points appear. Figure 2(b) shows the corresponding bicharacteristic curve. As shown in Fig. 2(c), Stokes geometry is the same as a 3rd order differential equation studied by [BNR]: there are two ordinary turning points and one virtual turning point. Ordinary turning points appear as a cusp-type point on the bicharacteristic curve, and a new turning point as a crossing point as expected. We stress that in the theory of dynamical systems, observing the Lagrangian manifold in (q_n, q_{n-1})-plane as Fig. 3(a), that is nothing but phase space, is the most standard way to understand the classical dynamics, but the bicharacteristic curve, just the same manifold drawn on (q_n, S)-plane, is more informative and carrying global information of the Borel plane. This point will be discussed later.

As shown in Fig. 3(d), ordinary turning points are observed as divergent points of the amplitude factor A_γ in (5) in leading order evaluation of $|I(q_2)|^2$. On the other hand, at a virtual turning point, there is no divergent behavior in the amplitude, however, actions of two different branches are degenerate, thus constructive interference necessarily occurs there irrespective of η. If many saddles (or classical orbits) appear in the sum (5), such degeneracy may be hidden among other contributions, but in case of the integral (9), interference can be indeed observed for appropriate choice of $\hbar$. In Fig. 2(d), a hump around $q_2 = 0$ is a result of constructive interference between two degenerate actions.

Two-Step Once Folding Map and One-Step Three Times Folding Map

As seen above, if the degree of the action function is greater than 3, one has to take into account virtual turning points and new Stokes curves. The increase of the degree leads to the increase of folding points, and so ordinary and new Stokes curves. Another way of causing the increase of folding points is the iteration of the map as seen from Fig. 1. In order to characterize the Stokes geometry for the integral associated with the multi-step chaotic map more explicitly, we further examine the difference between Stokes geometry of single step propagator with multiple folding points and that of multistep propagator with once folding dynamics.

We here compare; (a) two-step once folding case:

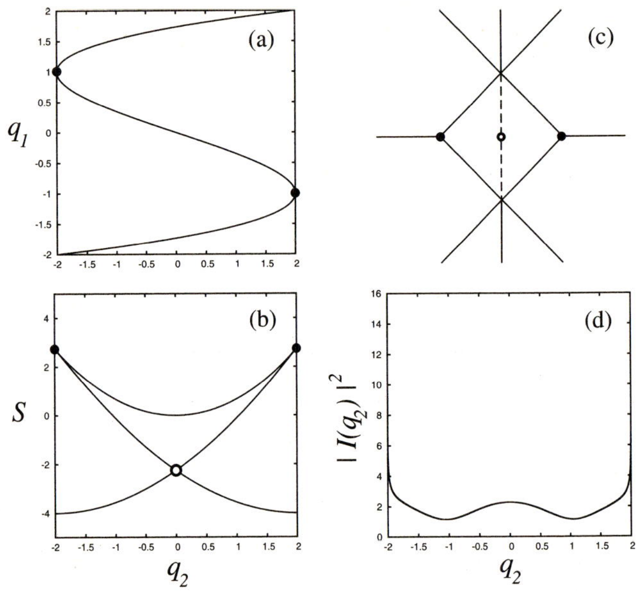

Fig. 2. (a) Lagrangian manifold $(\mathcal{F}(q_1, q_0), q_1)$ generated by the map which is induced by the action (10). (b) Bicharacteristic curve in which filled in dots denote ordinary turning points, an open dot virtual turning point, respectively. (c) Stokes geometry. (d) Leading order saddle point evaluation of integral $I(q_2)$.

$$I(q_3) = \int_{\infty}^{\infty} \int_{-\infty}^{\infty} dq_1 dq_2 \exp\Big[i\eta S(q_0, q_1, q_2, q_3)\Big], \tag{11}$$

where

$$S(q_0, q_1, q_2 \cdot q_3) = \frac{1}{2}(q_1 - q_0)^2 + \frac{1}{2}(q_2 - q_1)^2 + \frac{1}{2}(q_3 - q_2)^2 + \frac{1}{3}q_1^3 + \frac{1}{3}q_2^3 + cq_1 + cq_2, \tag{12}$$

and; (b) one-step three times folding case:

$$I(q_2) = \int_{-\infty}^{\infty} dq_1 \exp\Big[i\eta S(q_0, q_1, q_2)\Big], \tag{13}$$

where

$$S(q_0, q_1, q_2) = \frac{1}{2}(q_1 - q_0)^2 + \frac{1}{2}(q_2 - q_1)^2 + \frac{1}{5}q_1^5 + \frac{a}{3}q_1^2 + bq_1^2. \tag{14}$$

Here a, b, c are some parameters. In appropriate parameter loci, both maps have genuine horseshoe structures with nice symbolic codings.

As shown in Fig. 3, both maps create three folding points B_1, B_2, C_1 that correspond to three ordinary turning points. However, in case of (b), the six

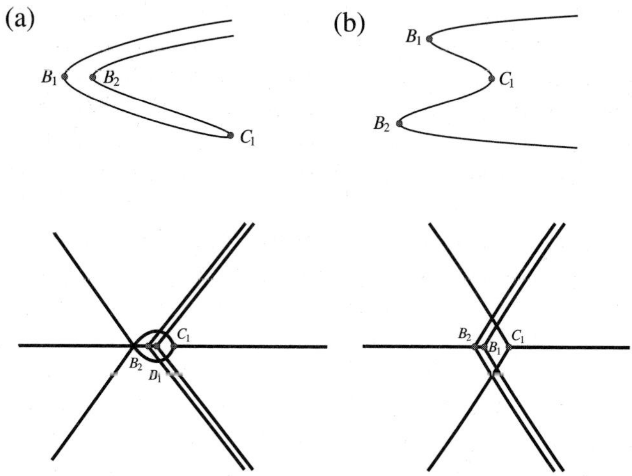

Fig. 3. Lagrangian manifolds (upper) and the ordinary Stokes curves (lower) for (a) two-step once folding case and (b) one-step three times folding case.

ordinary Stokes curves extending to $m\pi/3$ ($m = 1, 2, 4, 5$) directions monotonically go to infinity, whereas in case of (a), they turn their directions and cross $\mathrm{Re}\, q_3$ axis. As will be proved in [SI3], even when the degree of polynomial becomes larger, ordinary Stokes curves for one-step multiply times folding maps do not intersect the real axis. On the other hand, as indeed presented in an example in [S2], the number of turn of Stokes curves increases with step n.

As a result, virtual turning points and new Stokes curves, though not illustrated in Fig. 3, behave differently: in (a) some virtual turning points are located on complex plane and the others are not, but in (b) all the virtual turning points are on the $\mathrm{Re}\, q_3$ axis, thereby new Stokes curves have no effects on the connection along the $\mathrm{Re}\, q_3$ axis. This is similar to the Stokes geometry of non-adiabatic transition problems analyzed in [AKT4]. Therefore, we expect that multiple folding procedure or horseshoe dynamics, which is a fundamental mechanism generating chaos, can be a source that activates new ingredients of Stokes geometry predicted in the exact WKB analysis.

4 Bifurcation of Stokes Geometry: Strategy to Analyze Chaotic Maps

An Idea of Pruning

The Hénon map (1) has a nonlinear paremeter a (or c in the form (2)) controlling the classical dynamics qualitatively. When $a \gg 1$, the so-called horseshoe condition is satisfied and the mapping is conjugate to the symbolic dynamics with binary full shift. All the stable and unstable manifolds for periodic orbits intersect transversally when the horseshoe is realized, and the system keeps hyperbolicity. Non-wondering set forms fractal repeller on the real plane. A sufficient condition for the horseshoe was given in [DN], a recent result using techniques developed in the theory of higher-dimensional complex dynamical systems revealed that hyperbolic structure keeps exactly up to the first tangency point at which transversality between stable and unstable manifold is first broken [BS].

In the horseshoe regime, as shown in Fig. 1, folding points appear on the real plane. As the parameter a decreases some of folding points fall into complex plane. Such events occur as a result of coalescence of turning points. Similarly, some periodic orbits of the map (1) also become complex. If the horseshoe condition is satisfied, each point on the the nonwandering set corresponds to a point of the symbol plane that represents a doubly infinite sequence. As the system parameter is varied, complete horseshoe is destroyed and a certain set of points in the symbol plane may lose the corresponding orbits in the dynamics. Some regions in the symbol plane are *pruned* in this way, and the border, which divides the symbol plane into admissible and non-admissible regions, is called the *pruning front* [CGP]. Clearly, the pruning front is executed to play a similar role as a critical point in kneading theory [MT]. The latter captures all topological information in one-dimensional maps. *Pruning front theory* is therefore two-dimensional version of kneading theory and an idea controlling topological aspects of dynamics.

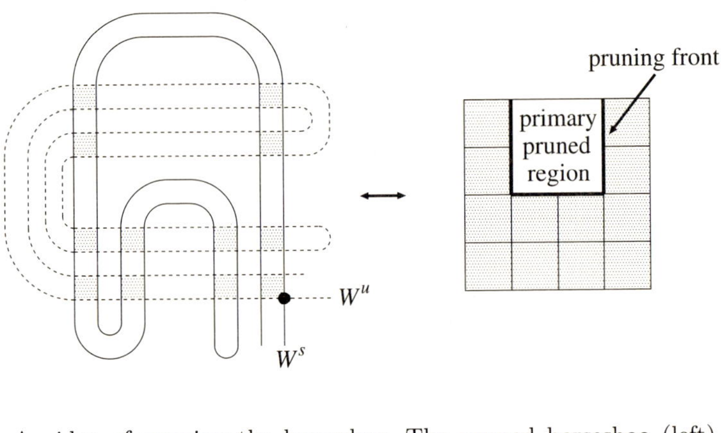

Fig. 4. An idea of pruning the horseshoe. The pruned horseshoe (left) and the corresponding symbolic plane.

Stokes Graph in the Horseshoe Limit

If we perform an analogous program as pruning front theory, our first task is to understand Stokes geometry in the horseshoe limit. Fortunately, reflecting a simple structure in the corresponding classical dynamics, we can predict how ordinary Stokes curves behave as a function of time step n [S3]. Even at $n = 4$ the ordinary Stokes curves look complicated as depicted in Fig. 5(b). Stokes curves of the oldest generation, meaning that they appear at the first iteration, behave simply, but as the generation becomes newer, Stokes curves rotate several times around the origin.

Not observing the Stokes geometry on q_n-plane but q_1-plane makes it easy to find a definite rule for ordinary Stokes curves [S3], and the rotation number for each generation can be derived analytically. Thus, we may say that at least the behavior of ordinary Stokes curves is controllable. The location of virtual turning points and the behavior of new Stokes curves, on the other hand, look not following so simple rule that directly reflects the horseshoe structure. However, numerical observations strongly suggest that types of virtual turning points and new Stokes curves are quite restricted, so it is hopeful to approach it in a similar way as the ordinary ones [S3].

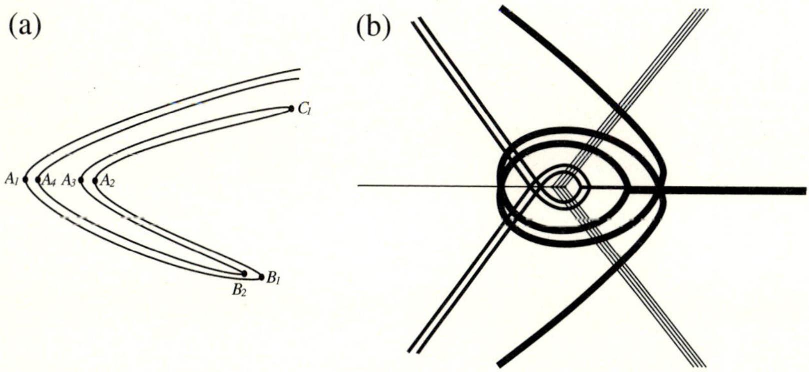

Fig. 5. (a) The Lagrangian manifold of 3 step Hénon map in the horseshoe parameter region (schemactic) and (b) the corresponding ordinary Stokes curves.

Classification of Bifurcation Patterns

Analogously, since the folding point exactly corresponds to the ordinary turning point in Stokes geometry, all the ordinary turning points are located on the real axis in the horseshoe parameter locus. With the decrease of a, some turning points fall into complex plane after the coalescence as well. Then one natural strategy to organize complicated Stokes geometry in more generic paramteter regions is then to trace successive bifurcations of Stokes graphs, starting at the horseshoe limit. Taking into account the symmetry with respect to Re q_n axis, which is rather specific in our integral, we can list up three types of bifurcation, each of which is depicted in Fig. 6.

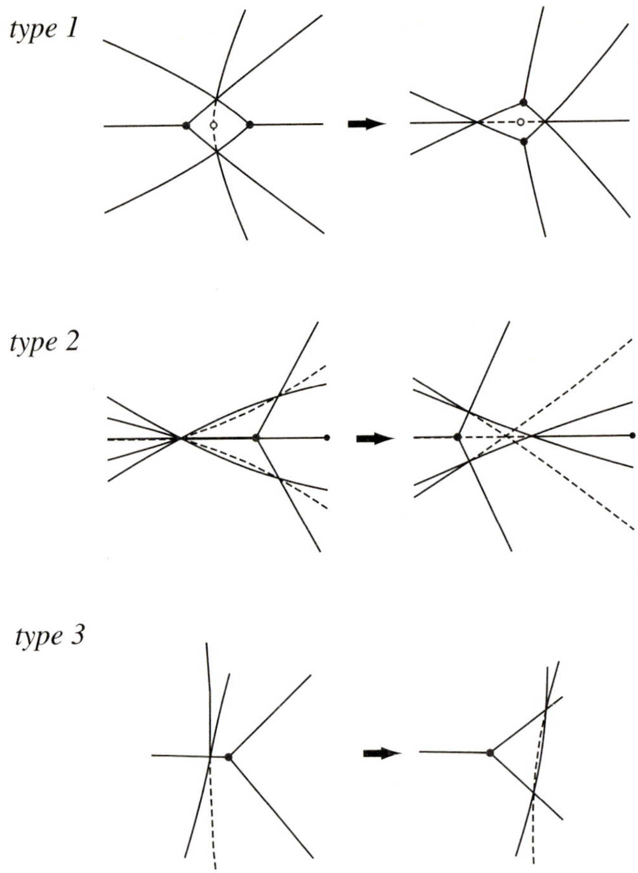

Fig. 6. Three types of bifurcation that can occur in Stokes geometry. On the broken potions, no connection occurs. The filled and open dots represent ordinary and virtual turning points, respectively.

In *Type 1* bifurcation, two ordinary turning points, both located on $\mathrm{Re}\, q_n$-axis, and a virtual turning point are degenerated and ordinary turning points form a complex conjugate pair after bifurcation. It is important to note that roles of Stokes curves are exchanged before and after bifurcation: before bifurcation, a new Stokes curve emanating from the virtual turning point runs vertically, but it runs in the horizontal direction after bifurcation. Correspondingly, ordinary Stokes curves running along $\mathrm{Re}\, q_n$-axis before bifurcation run in the vertical direction after bifurcation. As is stated in [AKSST], the exchange of the role of Stokes curves is one piece of evidence showing that there is no essential distinction between ordinary and new Stokes curves.

Type 2 bifurcation in Fig. 6 also occurs as a result of symmetry of this system. An ordinary turning point located also on the $\mathrm{Re}\, q_n$ axis moves to the left and passes through a degenerated crossing point at which six Stokes curves cross to each other. Note that two Stokes curves, each of which are Stokes curves between different pairs, say i, i' and j, j', run along the $\mathrm{Re}\, q_n$ axis. In order that the univaluedness condition is fulfilled around the crossing point six Stokes curves must cross. Such degenerated crossing is a generic event in Stokes geometry having symmetry. After bifurcation, this degenerated crossing point is split into two distinct simple crossing point as shown in Fig. 6. We again observe that the roles of some ordinary and new Stokes curves are exchanged through the bifurcation process.

Type 3 bifurcation is most generic and can be observed in Stokes geometry of other systems [AKSST]. In the present case, an ordinary turning point is located in complex plane, not on $\mathrm{Re}\, q_n$ axis as the former two types, and it passes through a simple crossing point. The exchange of the role of Stokes curves occurs as well.

These are local patterns of bifurcation which can occur with the change of parameter. We should note that all these indeed occur in Stokes geometry for the Hénon map, though not shown here, even at the smallest non-trivial time step, that is at $n = 3$.

5 Concluding Remarks

There are many physical problems to which the WKB analysis can be applied. Since WKB analysis treats exponentially small objects, it is particularly suitable for phenomena in which such small contributions play a central role. Non-adiabatic transition problem is a typical example in which exponentially small transition amplitude is in question. Quantum tunneling is exactly exponentially small process for which no classical counterpart exists. The WKB, especially complex WKB method, can well cope with such subjects.

However, as the WKB method gives asymptotic expansion at most, a precise control of exponentially small terms is not so easy task. As stressed in introduction of the paper by Voros [V], it is important to notice that identifying exponentially small subdominant contributions cannot be achieved without specifying the precise meaning of divergent series. An idea of employing the Borel resummation is introduced to fix ambiguity necessarily caused in asymptotic expansions [BW, DDP, V, Z]. It would be a plausible approach to understand divergent objects, and as a necessary outcome of such a direction of research, combined with a outcome of microlocal analyses [SKK], Aoki, Kawai and Takei noticed that extra ingredients should be introduced in order to complete the Stokes geometry in higher-order differential equations. The present paper is one piece of evidence supporting that their program is so reasonable and works quite well in the WKB analysis of chaotic maps. Although our approach is still on the way to our final goal and we can sketch only an outline of our program in this paper, we are confident that quantum tunneling in multi dimensions or especially in the presence of chaos cannot be understood without these newly introduced ideas.

It is a great honor for the author to dedicate the present article to Professor T. Kawai's sixtieth birthday. The author greatly appreciate T. Aoki, T. Kawai, T. Koike, and Y. Takei for their many valuable suggestions and fruitful comments on this work. He also thank K.S. Ikeda for collaborations with him, which will be published as [SI3].

References

[AKSST] Aoki, T., Kawai, T., Kokie, T., Sasaki, S., Shudo, A., Takei, T. : A background story and some know-how of virtual turning points. RIMS Kôkyûroku, **1424**, 52–63 (2005)

[AKSST] Aoki, T., Kawai, T., Sasaki, S., Shudo, A., Takei, T. : Virtual turning points and bifurcation of Stokes curves for higher order ordinary differential equations. J. Phys. A, **38**, 3317–3336 (2005)

[AKT1] Aoki, T., Kawai, T., Takei, T. : New turning points in the exact WKB analysis for higher-order ordinary differential equations. In: Boutet de Monvel, L. (ed) Analyse algébrique des perturbationssingulières. I. Hermann, pp. 69–84 (1994)

[AKT2] Aoki, T., Kawai, T., Takei, T. : On the exact WKB analysis for the third order ordinary differential equations with a large parameter. Asian J. Math., **2**, 625–640 (1998)

[AKT3] Aoki, T., Kawai, T., Takei, T. : On the exact steepest descent method. J. Math. Phys., **42**, 3691–3713 (2001)

[AKT4] Aoki, T., Kawai, T., Takei, T. : Exact WKB analysis of nonadiabatic transition probabilities for three levels. J. Phys. A, **35**, 2401–2430 (2002)

[BNR] Berk, H.L., Nevins, M.W., Roberts, K.V. : J. Math. Phys., **23**, 988–1002(1982)

[BS] Bedford, E., Smillie, J. : Real polynomial diffeomorphism with maximal entropy: Tangencies. Ann. Math., **160**, 1–26 (2004)

[BW] Bender, C.M., Wu, T.T. : Anharmonic oscillator, Phys. Rev., **184**, 1231–1260 (1969)

[CGP] Cvitanović, P., Gunaratne, G,. Procaccia I. : Topological and metric properties of Hénon type strange attractors. Phys. Rev. A, **38**, 1503–20 (1988)

[DDP] Delabaere, E., Dillinger, H., Pham., P: Eact semi-classical expansions for on dimensional quantum oscillators. J. Math. Phys., **38**, 6166–6184 (1997)

[DM] Dullin, H.R., Meiss, J.D. : Generalized Hénon maps: the cubic diffeomorphisms of the plane. Physica D, **143**, 262–289 (2000)

[DN] Devaney, R., Nitecki, Z. : Shift automorphisms in the Héon mapping. Commun. Math. Phys., **67**, 137–146 (1979)

[FM] Friedland, S., Milnor J. : Dynamical properties of plane polynomial automorphisms. Ergod. Theor. Dyn. Sys., **9**, 67–99 (1989)

[Gutz] Gutzwiller, M.C. : Chaos in Classical and Quantum Mechanics. Springer, New York (1990)

[HLO] Howls, C.J., Langman, P.J., Olde Daalhuis, A.B. : On the higher-order Stokes phenomeon. Proc. R. Soc. London, **460**, 2285–2303 (2004)

[Hon] Honda, N. : Toward the complete description of the Stokes geometry. In press.

[How] Howls, C.J. : Proc. R. Soc. London, **453**, 2271–2294 (1997)

[Hen] Hénon, M. : A two-dimensional mapping with a strange attractor. Commun. Math. Phys., **50**, 69–77 (1976)

[Les] Giannoni, M.-J., Voros, A., Zinn-Justin, J. (ed): Chaos and quamtum Physics. Les Houches Summer School, Session LII, North-Holland, Amsterdam (1989).

[MT] Milnor, J., Thurston, W., : On iterated maps of the interval. In: Dynamical Systems (College Park, MD, 1986–7). Springer, Berlin, pp. 465-563 (1988)

[S1] Shudo, A. : A recipe for finding Stokes geomtry in the quantized Hénon map. RIMS Kôkyûroku, **1433**, 110–119 (2005)

[S2] Shudo, A. : Stokes geometry for the quantized Hénon map. RIMS Kôkyûroku, **1424**, 184–199 (2005)

[S3] Shudo, A. : Stokes geometry for the quantized Hénon map in the horseshoe regime. RIMS Kôkyûroku, **1431**, 107–115 (2005)

[SI1] Shudo, A., Ikeda, K.S. : Complex Classical Trrajectories and Chaotic Tunneling. Phys. Rev. Lett., **74**, 682–685 (1995)

[SI2] Shudo, A., Ikeda, K.S. : Stokes Phenomenon in Chaotic Systems: Pruning Trees of Complex Paths with Principle of Exponential Dominance. Phys. Rev. Lett., **76**, 4151–5154 (1997)

[SI3] Shudo, A., Ikeda, K.S.: to be submitted.

[SKK] Sato, M., Kawai, T., Kashiwara, M. : Microfunctions and pseudo-differential equations. In: Lecture Notes in Math., **287**, 265–529 (1973)

[V] Voros, A. : The return of the quartic oscillator: The complex WKB method. Poincaré Phys. Théor., **39**, 211–338 (1983)

[Z] Zinn-Justin, J. : Instantons in quantum mechanics: numerical evidence for a conjecture. J. Math. Phys., **25**, 549–555 (1984).

Spectral instability for non-selfadjoint operators

Johannes Sjöstrand

CMLS, Ecole Polytechnique, FR 91120 Palaiseau, France
johannes@math.polytechnique.fr

Summary. We describe a recent result of M. Hager, stating roughly that for non-selfadjoint ordinary differential operators with a small random perturbation we have a Weyl law for the distribution of eigenvalues with a probability very close to 1.

Key words: Non-selfadjoint, eigenvalue, random perturbation
Mathematics Subject Classification (2000): 34L20, 47B80

1 Introduction and some history

In this talk we discuss a recent result by M. Hager [9] which is part of her thesis. Some of the basic ideas in that work have their roots in the general theory of partial differential equations. Let us recall that H. Lewy [11] gave an example of an operator in $\mathbf{R}^3$ which is not locally solvable near any point,

$$P = D_{x_1} - x_2 D_{x_3} + i(D_{x_2} + x_1 D_{x_3}). \tag{1.1}$$

Hörmander [10] discovered the role of the Poisson brackets in this context and gave a very general result on non-local solvability of linear partial differential equations saying that a differential operator with smooth coefficients is non locally solvable if the Poisson bracket $\frac{1}{2i}\{p, \overline{p}\}$ is not identically zero on the characteristic set $p = 0$, where p denotes the principal symbol of the operator. The simplest model of a non locally solvable operator is perhaps the the Mizohata operator in two or more dimensions ([12])

$$P = D_{x_2} + i x_2 D_{x_1}. \tag{1.2}$$

It is not locally solvable near any point of the hyperplane $x_2 = 0$. Since then, there have been great developments in the solvability theory, but we shall here follow a slightly different historical line.

Received 22 October, 2005. Accepted 30 December, 2005.

We first recall that the proof in [10] is based on what nowadays is called a quasi-mode construction for the adjoint operator: If this adjoint is denoted by P and its principal symbol by p, let $(x_0, \xi_0) \in \mathbf{R}_x^n \times (\mathbf{R}_\xi^n \setminus \{0\})$ be a point in the cotangent space where $p = 0$ and $\frac{1}{2i}\{p, \overline{p}\} > 0$. Then there exists a smooth function $\phi \in C^\infty(\mathrm{neigh}\,(x_0))$ with $\phi'(x_0) = \xi_0$, $\mathrm{Im}\,\phi''(x_0) > 0$, $\phi(x_0) = 0$ and a classical symbol

$$a \sim a_0(x) + h a_1(x) + ... \text{ in } C^\infty(\mathrm{neigh}\,(x_0)),\ a_0(x_0) \neq 0,$$

such that

$$P(x, D_x)(a(x;h)e^{\frac{i}{h}\phi(x)}) = \mathcal{O}(h^\infty),\ h \to 0,\ \text{in } C^\infty(\mathrm{neigh}\,(x_0)). \tag{1.3}$$

Here $\mathcal{O}(h^\infty)$ stands for "$\mathcal{O}_N(h^N)$ for every $N \geq 0$". When P has analytic coefficients, we can replace $\mathcal{O}(h^\infty)$ by $\mathcal{O}(e^{-1/(C_0 h)})$ for some constant $C_0 > 0$. The latter fact follows from the work of Sato-Kawai-Kashiwara [14], but the result is at least partially older (L. Boutet de Monvel, P. Kree [1]).

Following work of Yu. Egorov and V. Kondratiev on the oblique derivative problem ([6]), L. Hörmander asked me to make a more complete study of pseudodifferential operators P on a (para-)compact manifold X, with

$$\{p, \overline{p}\}(x,\xi) \neq 0 \text{ whenever } p(x,\xi) = 0,\ \xi \neq 0. \tag{1.4}$$

Under some additional assumptions (in order to get a global result), I obtained in my thesis [15] that a certain operator

$$\mathcal{P} = \begin{pmatrix} P & R_- \\ R_+ & 0 \end{pmatrix} : \mathcal{D}(P) \times \mathcal{H}_- \to L^2(X) \times \mathcal{H}_+ \tag{1.5}$$

has an inverse

$$\mathcal{E} = \begin{pmatrix} E & E_+ \\ E_- & E_{-+} \end{pmatrix}$$

modulo smoothing operators, where E_{-+} is smoothing. Here $\mathcal{H}_\pm$ are Sobolev spaces on manifolds $\Gamma_\pm$ with $\dim \Gamma_\pm = \dim\,(X) - 1$. This result implies that the null-space of

$$P : \mathcal{D}'(X)/C^\infty(X) \to \mathcal{D}'(X)/C^\infty(X)$$

is equal to

$$\{E_+ v_+;\ v_+ \in \mathcal{D}'(\Gamma_+)\}/C^\infty(X),$$

which is a big space, and a similar statement can be made about the cokernel of P in $\mathcal{D}'(X)/C^\infty(X)$.

At about the same time, Sato-Kawai-Kashiwara [14] made a complete study in the analytic category, and they showed among many other things that operators satisfying (1.4) can be reduced to the Mizohata operator.

In September 1972 I met T. Kawai for the first time and we had a very interesting discussion. He pointed out to me that the results of my thesis do not imply that the space of local solutions to the exact equation $Pu = 0$ is large, while in the analytic category the corresponding results ([14]) do so, thanks to the Cauchy-Kowalewski theorem. Indeed, it was showed by L. Nirenberg [13] for perturbations of the H. Lewy operator that there may be a radical difference between the analytic and the C^∞ case. (See also [17], [16] for related results for perturbations of the Mizohata operator.)

This talk deals with closely related problems for eigenvalues and we will establish an "opposite result": By destroying analyticity, the spectral properties "improve" with high probability.

Before describing the precise result in the next section, we end this introduction with an extremely quick review of the notion of pseudospectrum which has been developped by L.N. Trefethen and other specialists in numerical analysis and which then migrated towards PDE thanks to the efforts of E.B. Davies and M. Zworski. See [7, 2, 18]. Let $P : \mathcal{H} \to \mathcal{H}$ be a closed densely defined operator, where $\mathcal{H}$ is a complex Hilbert space. For $\epsilon > 0$, we define the ϵ-pseudospectrum by

$$\sigma_\epsilon(P) = \sigma(P) \cup \{z \in \mathbf{C} \setminus \sigma(P);\ \|(z-P)^{-1}\| > \frac{1}{\epsilon}\}, \tag{1.6}$$

where $\sigma(P)$ denotes the spectrum of P. We have

$$\sigma_\epsilon(P) = \bigcup_{\substack{Q \in \mathcal{L}(\mathcal{H}),\\ \|Q\| < \epsilon}} \sigma(P+Q), \tag{1.7}$$

which shows that σ_ϵ is a region of spectral instability.

When P is selfadjoint or more generally normal, we have

$$\sigma_\epsilon(P) = \{z \in \mathbf{C};\ \mathrm{dist}\,(z, \sigma(P)) < \epsilon\},$$

but in general, $\sigma_\epsilon(P)$ is much larger than the right hand side in the above equation.

Example 1.1. Consider a Jordan block

$$J = \begin{pmatrix} 0\ 1\ 0 \ldots 0 \\ 0\ 0\ 1 \ldots 0 \\ \ldots\ldots\ldots\ldots \\ 0\ 0\ 0 \ldots 1 \\ 0 \ldots\ldots\ldots 0 \end{pmatrix} : \mathbf{C}^N \to \mathbf{C}^N,\ N \gg 1.$$

Then $\|(z-J)^{-1}\| \geq |z|^{-N}$, so $\sigma_\epsilon(J) \supset D(0, \epsilon^{1/N})$, while $\sigma(J) = \{0\}$.

E.B. Davies [3] considered a general Schrödinger operator

$$P = -h^2 \frac{d^2}{dx^2} + V(x) \tag{1.8}$$

on $\mathbf{R}$, where the potential $V(x)$ is smooth and complex-valued. Here we are interested in the semi-classical limit $h \to 0$. The associated semi-classical symbol is $p(x,\xi) = \xi^2 + V(x)$. Davies observed that if

$$z = \xi_0^2 + V(x_0), \tag{1.9}$$

with $\xi_0 \neq 0$, $\operatorname{Im} V'(x_0) \neq 0$, then there exists a function

$$u(x) = u(x;h) = a(x;h)e^{i\phi(x)/h}$$

with $\phi(x_0) = 0$, $\phi'(x_0) = \pm\xi_0$, $\operatorname{Im}\phi''(x_0) > 0$, $a(x;h) \sim h^{-1/4}(a_0(x)+ha_1(x)+ ...)$ in C^∞, such that

$$\|u\| = 1,\ \ \|(P-z)u\| = \mathcal{O}(h^\infty).$$

This implies that

$$\text{either } z \in \sigma(P) \text{ or } \|(z-P)^{-1}\| \geq \frac{1}{\mathcal{O}(h^\infty)}.$$

In many cases when $V(x)$ is analytic, the spectrum of V is confined to a union of curves, much smaller then the set of values in (1.9).

M. Zworski [18] observed that this was a rediscovery of the old Hörmander construction. Then with N. Dencker and Zworski [4], we established

Proposition 1.2. *Let $P(x,hD_x;h)$ be an h-pseudodifferential operator on $\mathbf{R}^n$ with symbol*

$$P(x,\xi;h) \sim p(x,\xi) + hp_{-1}(x,\xi) + ... \in \ \textit{a suitable class.}$$

Assume

$$z = p(x_0,\xi_0),\ \frac{1}{2i}\{p,\overline{p}\}(x_0,\xi_0) > 0.$$

Then there exists a function $u = u_h = a(x;h)e^{i\phi(x)/h}$ with the same properties as in and around (1.3) such that $\|u_h\| = 1$, $\|(P-z)u_h\| = \mathcal{O}(h^\infty)$. When P is analytic (in a suitable class) we may replace $\mathcal{O}(h^\infty)$ by $\mathcal{O}(e^{-1/(Ch)})$ for some $C > 0$.

In the Schrödinger case $p = \xi^2 + V(x)$ we have $\frac{1}{2i}\{p,\overline{p}\}(x,\xi) = -\xi \cdot \operatorname{Im} V'(x)$.

2 The result of M. Hager

We work in $L^2(\mathbf{R})$. Let $P = p^w(x,hD_x)$ be the Weyl quantization of $p(x,h\xi)$. Assume that $p(x,\xi)$ is holomorphic in a tubular neighborhood of $\mathbf{R}^2$, and

$$p(x,\xi) = \mathcal{O}(m(\operatorname{Re}(x,\xi))), \tag{2.1}$$

where $1 \leq m$ is an order function on $\mathbf{R}^2$, in the sense that

$$m(\rho) \leq C_0 \langle \rho - \mu \rangle^{N_0} m(\mu), \ \rho, \mu \in \mathbf{R}^2, \tag{2.2}$$

where $\langle \rho - \mu \rangle = \sqrt{1 + |\rho - \mu|^2}$. We may assume without loss of generality that m belongs to its own symbol class: $\partial^\alpha m = \mathcal{O}(m)$ for every $\alpha \in \mathbf{N}^2$. Then for $h > 0$ small enough, $P : L^2(\mathbf{R}) \to L^2(\mathbf{R})$ is a closed densely defined operator with domain $\mathcal{D}(P) = H(m) := (m^w(x, hD))^{-1}(L^2(\mathbf{R}))$.

Let

$$\Sigma = \overline{p(\mathbf{R}^2)},$$

and let Σ_∞ denote the set of accumulation points of p at $(x, \xi) = \infty$. Let $\widetilde{\Omega} \subset\subset \mathbf{C} \setminus \Sigma_\infty$ be a connected open set not entirely contained in Σ. Assume that

$$|p(x, \xi) - z_0| \geq \frac{m(x, \xi)}{C_0}, \ (x, \xi) \in \mathbf{R}^2,$$

for some $z_0 \in \widetilde{\Omega} \setminus \Sigma$ and some constant $C_0 > 0$. If $\widetilde{K} \subset \mathbf{C} \setminus \Sigma$ is compact, then $\sigma(P) \cap \widetilde{K} = \emptyset$ for h small enough. Moreover, the spectrum of P in $\widetilde{\Omega}$ is discrete when $h > 0$ is small enough.

Assume that

$$\xi \mapsto p(x, \xi) \text{ is even.} \tag{2.3}$$

Let $\Omega \subset\subset \widetilde{\Omega}$ be open and simply connected. Assume

$$p(x, \xi) \in \overline{\Omega} \Rightarrow \{p, \overline{p}\}(x, \xi) \neq 0. \tag{2.4}$$

Then for $z \in \overline{\Omega}$:

$$\begin{aligned} p^{-1}(z) &= \{\rho_j^+(z), \rho_j^-(z); \ j = 1, 2, .., n\}, \\ \rho_j^\pm &= (x_j, \pm \xi_j), \ \pm \frac{1}{2i}\{p, \overline{p}\}(\rho_j^\pm) > 0. \end{aligned} \tag{2.5}$$

Assume for simplicity that $x_j \neq x_k$ for $j \neq k$.

We now add a small random perturbation $\delta q_\omega(x)$ where ω denotes the random parameter and δ is a small parameter satisfying

$$e^{-\frac{1}{D_0 h}} < \delta < \frac{1}{C_0} h^{\frac{3}{2}}, \ D_0, C_0 \gg 1. \tag{2.6}$$

q_ω will be a random linear combination of eigenfunctions of an auxiliary operator

$$\begin{aligned} \widetilde{P} &= \widetilde{p}^w, \ \text{where } \partial^\alpha \widetilde{p} = \mathcal{O}(\widetilde{m}), \ \widetilde{p} \geq \frac{\widetilde{m}}{C}, \\ \langle \rho \rangle^{k_0} &\leq \widetilde{m}(\rho) \leq C_0 \langle \rho - \mu \rangle^{N_0} \widetilde{m}(\mu), \ k_0 > 0. \end{aligned} \tag{2.7}$$

Let $q_1, q_2, \ldots$ be an orthonormal basis of eigenfunctions of $\widetilde{P}$ corresponding to the eigenvalues $E_1 \leq E_2 \leq \ldots \to +\infty$. Let $N = C/h$ for $C \gg 1$ and put

$$q_\omega(x) = \sum_{\ell \le N} \alpha_\ell(\omega) q_\ell(x), \tag{2.8}$$

where α_ℓ are independent identically distributed complex random variables with $\langle \alpha_\ell \rangle = 0$, variance $\sigma = \delta^{2/n}$ and distribution:

$$\frac{1}{\pi\sigma^2} e^{-|\alpha|^2/\sigma^2} d(\operatorname{Re}\alpha) d(\operatorname{Im}\alpha).$$

Theorem 2.1. **(M. Hager)** *Let $\Gamma \subset\subset \Omega$ be open with smooth boundary. There exist $C, K, D > 0$ such that for $h > 0$ sufficiently small, we have with probability $\ge 1 - C\delta^{\frac{1}{2n}} h^{-K}$ that*

$$\left|\#(\sigma(P + \delta q_\omega) \cap \Gamma) - \frac{1}{2\pi h} \operatorname{vol}(p^{-1}(\Gamma))\right| \le D\left(\frac{\ln \frac{1}{\delta}}{h}\right)^{\frac{1}{2}}. \tag{2.9}$$

Hager also has a similar theorem saying that with a probability very close to 1, the Weyl asymptotics (2.9) holds simultaneously for all Γ varying in a class of sets that satisfy the above assumptions uniformly.

3 Quick outline of the proof

For more details, see [9]. Let e_j^+, e_j^- be normalized Davies quasimodes associated to $(P - z, \rho_j^+)$ and $((P - z)^*, \rho_j^-)$ respectively with exponentially small remainders as in the last part of Proposition 1.2. Consider

$$R_+ : H(m) \to \mathbf{C}^n, \; (R_+ u)(j) = (u|e_j^+), \tag{3.1}$$

$$R_- : \mathbf{C}^n \to L^2(\mathbf{R}), \; R_- u_- = \sum_1^n u_-(j) e_j^- .$$

With high probability we have $\|q_\omega\|_{L^\infty} \le 1$, and then

$$\mathcal{P}^\delta(z) = \begin{pmatrix} P + \delta q_\omega - z & R_-(z) \\ R_+(z) & 0 \end{pmatrix} = H(m) \times \mathbf{C}^n \to L^2 \times \mathbf{C}^n \tag{3.2}$$

is invertible with inverse

$$\mathcal{E}^\delta(z) = \begin{pmatrix} E^\delta & E_+^\delta \\ E_-^\delta & E_{-+}^\delta \end{pmatrix}, \tag{3.3}$$

where

$$E_{-+}^\delta = E_{-+}^0 + \delta E_{-+}^{(1)} + \mathcal{O}(\frac{\delta^2}{\sqrt{h}}) \tag{3.4}$$

$$= \delta E_{-+}^{(1)} + \mathcal{O}(\frac{\delta^2}{\sqrt{h}}),$$

$$(E_{-+}^{(1)})_{j,k} = -(q_\omega e_+^k | e_-^j) + \mathcal{O}(e^{-\frac{1}{Ch}}). \tag{3.5}$$

Now,

$$z \in \sigma(P + \delta q_\omega) \Leftrightarrow \det E_{-+}^\delta = 0.$$

We can show that there exists a function

$$\ell^\delta(z) = \ell^0(z) + \mathcal{O}(\frac{\delta}{\sqrt{h}}),\ \ell^0 \in C^\infty(\Omega),$$

such that

$$F^\delta(z) = e^{\frac{\ell^\delta(z)}{h}} \det E_{-+}^\delta(z) \tag{3.6}$$

is holomorphic. Moreover,

$$(\Delta \mathrm{Re}\, \ell_0(z) + \mathcal{O}(h)) d(\mathrm{Re}\, z) d(\mathrm{Im}\, z) = \sum_j (d\xi_j^-(z) \wedge dx_j^-(z) - d\xi_j^+(z) \wedge dx_j^+(z)) \tag{3.7}$$

so

$$\iint_\Gamma \Delta(\mathrm{Re}\, \ell_0) d(\mathrm{Re}\, z) d(\mathrm{Im}\, z) = \mathrm{vol}\,(p^{-1}(\Gamma)) + \mathcal{O}(h). \tag{3.8}$$

Now,

$$|F^\delta(z)| \le e^{\frac{\mathrm{Re}\, \ell_0(z)}{h}}, \tag{3.9}$$

and for every $z \in \Omega$ we have with a high probability that

$$|F^\delta(z)| \ge e^{\frac{\mathrm{Re}\, \ell_0(z)}{h} - \frac{\epsilon}{h}}. \tag{3.10}$$

Here $\epsilon \ll 1$ should be suitably chosen, possibly depending on h. It then suffices to apply

Proposition 3.1. *Let Γ and Ω be as above, $\phi \in C^\infty(\Omega; \mathbf{R})$. Let f be holomorphic in Ω with*

$$|f(z;h)| \le e^{\phi(z)/h},\ z \in \Omega.$$

Assume there exist $\epsilon \ll 1$, $z_k \in \Omega$, $k \in J$, such that

$$\partial\Gamma \subset \bigcup_{k \in J} D(z_k, \sqrt{\epsilon}),\ \#J = \mathcal{O}(\frac{1}{\sqrt{\epsilon}}),$$

$$|f(z_k;h)| \ge e^{\frac{1}{h}(\phi(z_k) - \epsilon)},\ k \in J.$$

Then,

$$\#(f^{-1}(0) \cap \Gamma) = \frac{1}{2\pi h} \iint_\Gamma (\Delta\phi) d(\mathrm{Re}\, z) d(\mathrm{Im}\, z) + \mathcal{O}(\frac{\sqrt{\epsilon}}{h}).$$

4 Prospects

- Extension to higher dimensions. This works ([Hager-Sj] in preparation).
- More general random perturbations. In higher dimensions we run into questions about random matrices.
- Weyl asymptotics for large eigenvalues in the non-semiclassical case. Here one would like to have results with probabality 1. This is under investigation.
- It would be interesting to see wether one could get similar results about resonances.

References

[1] L. Boutet de Monvel, P. Kree, *Pseudo-differential operators and Gevrey classes,* Ann. Inst. Fourier, 17(1)(1967), 295–323.

[2] E.B. Davies, *Semi-classical analysis and pseudospectra,* J. Diff. Eq. 216(1)(2005), 153–187.

[3] E.B. Davies, *Semi-classical states for non-self-adjoint Schrödinger operators,* Comm. Math. Phys., 200(1999), 35–41.

[4] N. Dencker, J. Sjöstrand, M. Zworski, *Pseudospectra of (pseudo)differential operators,* Comm. Pure Appl. Math., 57(2004), 384–415.

[5] M. Dimassi, J. Sjöstrand, *Spectral asymptotics in the semi-classical limit,* London Math. Soc. Lecture Note Series, 268(1999).

[6] Yu.V. Egorov, V.A. Kondratiev, *The oblique derivative problem,* Mat. sb. (N.S.) 78(120)(1969), 148–176.

[7] M. Embree, L.N. Trefethen, *Spectra and pseudospectra, the behaviour of non-normal matrices and operators,* Princeton Univ. Press, Princeton NJ (2005).

[8] M. Hager, *Instabilité spectrale semiclassique pour des opérateurs non-autoadjoints I: un modèle,* préprint 2004, http://hal.ccsd.cnrs.fr/ccsd-00001594, Annales de la Faculté des Sciences de Toulouse, to appear,

[9] M. Hager, *Instabilité spectrale semiclassique d'opérateurs non-autoadjoints II:* http://hal.ccsd.cnrs.fr/ccsd-00004677.

[10] L. Hörmander, *Differential equations without solutions,* Math. Ann. 140(1960), 169-173.

[11] H. Lewy, *An example of a smooth linear partial differential equation without solution,* Ann. of Math., 66(2)(1957), 155–158.

[12] S. Mizohata, *Solutions nulles et solutions non analytiques,* J. Math. Kyoto Univ. 1(1961/1962), 271–302.

[13] L. Nirenberg, *Lectures on linear partial differential equations,* CBMS Regional Conference Series in Mathematics, No. 17. American Mathematical Society, Providence, R.I., 1973.

[14] M. Sato, T. Kawai, M. Kashiwara, *Microfunctions and pseudo-differential equations.* Hyperfunctions and pseudo-differential equations. Lecture Notes in Math., Vol. 287, 265–529 Springer, Berlin, 1973.

[15] J. Sjöstrand, *Operators of principal type with interior boundary conditions,* Acta Math., 130(1973), 1–51.

[16] J. Sjöstrand, *Note on a paper of F. Treves concerning Mizohata type operators,* Duke Math. J., 47(3)(1980), 601-608.

[17] F. Treves, *Remarks about certain first-order linear PDE in two variables,* Comm. PDE 5(4)(1980), 381–425.

[18] M. Zworski, *A remark on a paper of E. B Davies: "Semi-classical states for non-self-adjoint Schrödinger operators",* Comm. Math. Phys. 200(1)(1999), 35–41.

Boundary and lens rigidity, tensor tomography and analytic microlocal analysis

Plamen Stefanov[1*] and Gunther Uhlmann[2**]

[1] Department of Mathematics, Purdue University, West Lafayette, IN 47907, USA
stefanov@math.purdue.edu

[2] Department of Mathematics, University of Washington, Seattle, WA 98195, USA
gunther@math.washington.edu

Summary. The boundary rigidity problem consists of determining a compact, Riemannian manifold with boundary, up to isometry, by knowing the boundary distance function between boundary points. Lens rigidity consists of determining the manifold, by knowing the scattering relation which measures, besides the travel times, the point and direction of exit of a geodesic from the manifold if one knows its point and direction of entrance. Tensor tomography is the linearization of boundary rigidity and length rigidity. It consists of determining a symmetric tensor of order two from its integral along geodesics. In this paper we survey some recent results obtained on these problems using methods from microlocal analysis, in particular analytic microlocal analysis. Although we use the distribution version of analytic microlocal analysis, many of the ideas were based on the pioneer work of the Sato school of microlocal analysis of which Professor Kawai was a very important member.

1 Boundary Rigidity and Tensor Tomography

Let $(M, \partial M, g)$ be a compact Riemannian manifold with boundary. Denote by ρ_g the distance function in the metric g. The boundary rigidity problem consists of whether $\rho_g(x, y)$, known for all x, y on ∂M, determines the metric uniquely. It is clear that any isometry which is the identity at the boundary will give rise to the same distance functions on the boundary. Therefore, the natural question is whether this is the only obstruction to unique identifiability of the metric. The boundary distance function only takes into account the shortest paths and it is easy to find counterexamples where ρ_g does not carry any information about certain open subset of M, so one needs to pose some restrictions on the metric. One such condition is simplicity of the metric.

Received 13 June, 2006. Revised 26 August, 2006. Accepted 6 September, 2006.

* The first author was supported in part by NSF Grant DMS-0400869.

** The second author was supported in part by NSF Grant DMS-0245414.

Definition 1. *We say that the Riemannian metric g is* simple *in M, if ∂M is strictly convex w.r.t. g, and for any $x \in M$, the exponential map $\exp_x : \exp_x^{-1}(M) \to M$ is a diffeomorphism.*

By strictly convex we mean convex (any two points are connected by a unique minimizing geodesic), and the second fundamental form on the boundary is positive.

Michel [18] conjectured that a *simple* metric g is uniquely determined, up to an action of a diffeomorphism fixing the boundary, by the boundary distance function $\rho_g(x, y)$ known for all x and y on ∂M.

This problem also arose in geophysics in an attempt to determine the inner structure of the Earth by measuring the travel times of seismic waves. It goes back to Herglotz [15] and Wiechert and Zoeppritz [38] who considered the case of a radial metric conformal to the Euclidean metric. Although the emphasis has been in the case that the medium is isotropic, the anisotropic case has been of interest in geophysics since the Earth is anisotropic. It has been found that even the inner core of the Earth exhibits anisotropic behavior [8].

Unique recovery of g (up to an action of a diffeomorphism fixing the boundary) from the boundary distance function is known for simple metrics conformal to each other [10], [6], [19], [20], [21], [4], for flat metrics [13], for locally symmetric spaces of negative curvature [5]. In two dimensions it was known for simple metrics with negative curvature [9] and [22], and recently it was shown in [24] for simple metrics with no restrictions on the curvature. In [29], the authors proved a local result for metrics in a small neighborhood of the Euclidean one. This result was used in [17] to prove a semiglobal solvability result assuming that one metric is close to the Euclidean and the other has bounded curvature. Burago and Ivanov have recently extended the latter result; they show that metrics close to the Euclidean metric are boundary rigid [7].

It is known [25], that a linearization of the boundary rigidity problem near a simple metric g is given by the following integral geometry problem: recover a symmetric tensor of order 2, which in any coordinates is given by $f = (f_{ij})$, by the geodesic X-ray transform

$$I_g f(\gamma) = \int f_{ij}(\gamma(t))\dot\gamma^i(t)\dot\gamma^j(t)\, dt$$

known for all geodesics γ in M. It can be easily seen that $I_g dv = 0$ for any vector field v with $v|_{\partial M} = 0$, where dv denotes the symmetric differential

$$[dv]_{ij} = \frac{1}{2}\left(\nabla_i v_j + \nabla_j v_i\right), \tag{1.1}$$

and $\nabla_k v$ denote the covariant derivatives of the vector field v. This is the linear version of the fact that ρ_g does not change on $(\partial M)^2 := \partial M \times \partial M$ under an action of a diffeomorphism as above. The natural formulation of the linearized problem is therefore that $I_g f = 0$ implies $f = dv$ with v vanishing on the boundary. We will refer to this property as *s-injectivity* of I_g. More precisely, we have.

Definition 2. *We say that I_g is* s-injective *in M, if $I_g f = 0$ and $f \in L^2(M)$ imply $f = dv$ with some vector field $v \in H_0^1(M)$.*

Any symmetric tensor $f \in L^2(M)$ admits an orthogonal decomposition $f = f^s + dv$ into a *solenoidal* and *potential* parts with $v \in H_0^1(M)$, and f^s divergence free, i.e., $\delta f^s = 0$, where δ is the adjoint operator to $-d$ given by $[\delta f]_i = g^{jk}\nabla_k f_{ij}$ [25]. Therefore, I_g is s-injective, if it is injective on the space of solenoidal tensors.

The inversion of I_g is a problem of independent interest in integral geometry, also called *tensor tomography*. We first survey the recent results on this problem. S-injectivity, respectively injectivity for 1-tensors (1-forms) and functions is known, see [25] for references. S-injectivity of I_g was proved in [23] for metrics with negative curvature, in [25] for metrics with small curvature and in [27] for Riemannian surfaces with no focal points. A conditional and non-sharp stability estimate for metrics with small curvature is also established in [25]. In [30], we proved stability estimates for s-injective metrics (see (1.5) below) and sharp estimates about the recovery of a 1-form $f = f_j dx^j$ and a function f from the associated $I_g f$ which is defined by

$$I_g f(\gamma) = \int f_i(\gamma(t))\dot{\gamma}^i(t))\ dt.$$

The stability estimates proven in [30] were used to prove local uniqueness for the boundary rigidity problem near any simple metric g with s-injective I_g.

Similarly to [36], we say that f is analytic in the set K (not necessarily open), if it is real analytic in some neighborhood of K.

The results that follow in this section are based on [32]. The first main result we discuss is about s-injectivity for simple analytic metrics.

Theorem 1. *Let g be a simple, real analytic metric in M. Then I_g is s-injective.*

Sketch of the proof. Note that a simple metric g in M can be extended to a simple metric in some M_1 with $M \Subset M_1$. A simple manifold is diffeomorphic to a (strictly convex) domain $\Omega \subset \mathbf{R}^n$ with the Euclidean coordinates x in a neighborhood of Ω and a metric $g(x)$ there. For this reason, it is enough to prove the results of this section for domains Ω in $\mathbf{R}^n$ provided with a Riemannian metric g.

The proof of Theorem 1 is based on the following. For smooth metrics, the normal operator $N_g = I_g^* I_g$ is a pseudodifferential operator with a non-trivial null space which is given by

$$(N_g f)_{kl}(x) = \frac{2}{\sqrt{\det g}} \int \frac{f^{ij}(y)}{\rho_g(x,y)^{n-1}} \frac{\partial \rho_g}{\partial y^i} \frac{\partial \rho_g}{\partial y^j} \frac{\partial \rho_g}{\partial x^k} \frac{\partial \rho_g}{\partial x^l} \det \frac{\partial^2(\rho_g^2/2)}{\partial x \partial y}\, dy, \quad x \in \Omega. \tag{1.2}$$

In the case that the metric g is real-analytic, N_g is an analytic pseudodifferential operator with a non-trivial kernel. We construct an analytic parametrix, using the analytic pseudodifferential calculus in [36], that allows us to reconstruct the solenoidal part of a tensor field from its geodesic X-ray transform, up to a term that is analytic near Ω. If $I_g f = 0$, we show that for some v vanishing on $\partial\Omega$, $\tilde f := f - dv$ must be flat at $\partial\Omega$ and analytic in $\bar\Omega$, hence $\tilde f = 0$. This is similar to the known argument that an analytic elliptic pseudodifferential operator resolves the analytic singularities, hence cannot have compactly supported functions in its kernel. In our case we have a non-trivial kernel, and complications due to the presence of a boundary, in particular a lost of one derivative. For more details see [32]. □

As shown in [30], the s-injectivity of I_g for analytic simple g implies a stability estimate for I_g. In next theorem we show something more, namely that we have a stability estimate for g in a neighborhood of each analytic metric, which leads to stability estimates for generic metrics.

As above, let $M_1 \Supset M$ be a compact manifold which is a neighborhood of M and g extends as a simple metric there. We always assume that our tensors are extended as zero outside M, which may create jumps at ∂M. Define the *normal operator* $N_g = I_g^* I_g$, where I_g^* denotes the operator adjoint to I_g with respect to an appropriate measure. We showed in [30] that N_g is a pseudo-differential operator in M_1 of order -1.

We introduce the norm $\|\cdot\|_{\tilde H^2(M_1)}$ of $N_g f$ in $M_1 \supset M$ in the following way. Choose $\chi \in C_0^\infty$ equal to 1 near ∂M and supported in a small neighborhood of ∂M and let $\chi = \sum_{j=1}^J \chi_j$ be a partition of χ such that for each j, on $\operatorname{supp}\chi_j$ we have coordinates (x'_j, x_j^n), with x_j^n a normal coordinate. Set

$$\|f\|_{\tilde H^1}^2 = \int \sum_{j=1}^J \chi_j \left(\sum_{i=1}^{n-1} |\partial_{x_j^i} f|^2 + |x_j^n \partial_{x_j^n} f|^2 + |f|^2 \right) \mathrm{d}x, \tag{1.3}$$

$$\|N_g f\|_{\tilde H^2(M_1)} = \sum_{i=1}^n \|\partial_{x^i} N_g f\|_{\tilde H^1} + \|N_g f\|_{H^1(M_1)}. \tag{1.4}$$

In other words, in addition to derivatives up to order 1, $\|N_g f\|_{\tilde H^2(M_1)}$ includes also second derivatives near ∂M but they are realized as first derivatives of $\nabla N_g f$ tangent to ∂M.

The reason to use the $\tilde H^2(M_1)$ norm, instead of the stronger $H^2(M_1)$ one, is that this allows us to work with $f \in H^1(M)$, not only with $f \in H_0^1(M)$, since for such f, extended as 0 outside M, we still have that $N_g f \in \tilde H^2(M_1)$, see [30]. On the other hand, $f \in H^1(M)$ implies $N_g f \in \tilde H^2(M_1)$ despite the possible jump of f at ∂M.

Our stability estimate for the linearized inverse problem is as follows:

Theorem 2. *There exists k_0 such that for each $k \geq k_0$, the set $\mathcal{G}^k(M)$ of simple $C^k(M)$ metrics in M for which I_g is s-injective is open and dense in the $C^k(M)$ topology. Moreover, for any $g \in \mathcal{G}^k(M)$,*

$$\|f^s\|_{L^2(M)} \leq C\|N_g f\|_{\tilde{H}^2(M_1)}, \quad \forall f \in H^1(M), \tag{1.5}$$

with a constant $C > 0$ *that can be chosen locally uniform in* $\mathcal{G}^k(M)$ *in the* $C^k(M)$ *topology.*

Of course, $\mathcal{G}^k(M)$ includes all real analytic simple metrics in M, according to Theorem 1.

Sketch of the proof. The proof of the basic estimate (1.5) is based on the following ideas. For g of finite smoothness, one can still construct a parametrix Q_g of N_g as above that allows us to reconstruct f^s from $N_g f$ up to smoothing operator terms. This is done in a way similar to that in [30] in two steps: first we invert N_g modulo smoothing operators in a neighborhood M_1 of M, and that gives us $f^s_{M_1}$, i.e., the solenoidal projection of f but associated to the manifold M_1. Next, we compare $f^s_{M_1}$ and f^s and show that one can get the latter from the former by an operator that loses one derivative. This is the same construction as in the proof of Theorem 1 above but the metric is only C^k, $k \gg 1$.

After applying the parametrix Q_g, the equation for recovering f^s from $N_g f$ is reduced to solving the Fredholm equation

$$(\mathcal{S}_g + K_g)f = Q_g N_g f, \quad f \in \mathcal{S}_g L^2(M) \tag{1.6}$$

where $\mathcal{S}_g$ is the projection to solenoidal tensors, similarly we denote by $\mathcal{P}_g$ the projection onto potential tensors. Here, K_g is a compact operator on $\mathcal{S}_g L^2(M)$. We can write this as an equation in the whole $L^2(M)$ by adding $\mathcal{P}_g f$ to both sides above to get

$$(I + K_g)f = (Q_g N_g + \mathcal{P}_g)f. \tag{1.7}$$

Then the solenoidal projection of the solution of (1.7) solves (1.6). A finite rank modification of K_g above can guarantee that $I + K_g$ has a trivial kernel, and therefore is invertible, if and only if N_g is s-injective. The problem then reduces to that of invertibility of $I + K_g$. The operators above depend continuously on $g \in C^k$, $k \gg 1$. Since for g analytic, $I + K_g$ is invertible by Theorem 1, it would still be invertible in a neighborhood of any analytic g, and estimate (1.5) is true with a locally uniform constant. Analytic (simple) metrics are dense in the set of all simple metrics, and this completes the sketch of the proof of Theorem 2. For more details see [32]. □

The analysis of I_g can also be carried out for symmetric tensors of any order, see e.g., [25] and [26]. Since we are motivated by the boundary rigidity problem, and to simplify the exposition, we study only tensors of order 2.

Theorem 2 and especially estimate (1.5) allow us to prove the following local generic uniqueness result for the non-linear boundary rigidity problem.

Theorem 3. *Let k_0 and $\mathcal{G}^k(M)$ be as in Theorem 2. There exists $k \geq k_0$, such that for any $g_0 \in \mathcal{G}^k$, there is $\varepsilon > 0$, such that for any two metrics g_1, g_2 with $\|g_m - g_0\|_{C^k(M)} \leq \varepsilon$, $m = 1, 2$, we have the following:*

$$\rho_{g_1} = \rho_{g_2} \text{ on } (\partial M)^2 \quad \text{implies } g_2 = \psi_* g_1 \tag{1.8}$$

with some $C^{k+1}(M)$-diffeomorphism $\psi : M \to M$ fixing the boundary.

Sketch of the proof. We prove Theorem 3 by linearizing and using Theorem 2, and especially (1.5), see also [30]. This requires first to pass to special semi-geodesic coordinates related to each metric in which $g_{in} = \delta_{in}$, $\forall i$. We denote the corresponding pull-backs by g_1, g_2 again. Then we show that if g_1 and g_2 have the same distance on the boundary, then $g_1 = g_2$ on the boundary with all derivatives. As a result, for $f := g_1 - g_2$ we get that $f \in C_0^l(\bar{\Omega})$ with $l \gg 1$, if $k \gg 1$; and $f_{in} = 0$, $\forall i$. Then we linearize to get

$$\|N_{g_1} f\|_{L^\infty(\Omega_1)} \leq C\|f\|_{C^1}^2,$$

where $\Omega_1 \supset \bar{\Omega}$ is as above. Combine this with (1.5) and interpolation estimates, to get $\forall \mu < 1$,

$$\|f^s\|_{L^2} \leq C\|f\|_{L^2}^{1+\mu}.$$

One can show that tensors satisfying $f_{in} = 0$ also satisfy $\|f\|_{L^2} \leq C\|f^s\|_{H^2}$, and using this, and interpolation again, we get

$$\|f\|_{L^2} \leq C\|f\|_{L^2}^{1+\mu'}, \quad \mu' > 0.$$

This implies $f = 0$ for $\|f\| \ll 1$. Note that the condition $f \in C_0^l(\bar{\Omega})$ is used to make sure that f, extended as zero in $\Omega_1 \setminus \Omega$, is in $H_0^l(\Omega)$, and then use this fact in the interpolation estimates. Again, for more details see [32]. □

Finally, in [32] it is proven a conditional stability estimate of Hölder type. A similar estimate near the Euclidean metric was proven in [37] based on the approach in [29].

Theorem 4. *Let k_0 and $\mathcal{G}^k(M)$ be as in Theorem 2. Then for any $\mu < 1$, there exits $k \geq k_0$ such that for any $g_0 \in \mathcal{G}^k$, there is an $\varepsilon_0 > 0$ and $C > 0$ with the property that that for any two metrics g_1, g_2 with $\|g_m - g_0\|_{C(M)} \leq \varepsilon_0$, and $\|g_m\|_{C^k(M)} \leq A$, $m = 1, 2$, with some $A > 0$, we have the following stability estimate*

$$\|g_2 - \psi_* g_1\|_{C^2(M)} \leq C(A)\|\rho_{g_1} - \rho_{g_2}\|_{C(\partial M \times \partial M)}^{\mu}$$

with some diffeomorphism $\psi : M \to M$ fixing the boundary.

Sketch of the proof. To prove Theorem 4, we basically follow the uniqueness proof sketched above by showing that each step is stable. The analysis is more delicate near pairs of points too close to each other. An important ingredient of the proof is stability at the boundary, that is also of independent interest:

Theorem 5. *Let g_0 and g_1 be two simple metrics in Ω, and $\Gamma \subset\subset \Gamma' \subset \partial\Omega$ be two sufficiently small open subsets of the boundary. Then for some diffeomorphism ψ fixing the boundary,*

$$\|\partial_{x^n}^k(\psi_* g_1 - g_0)\|_{C^m(\bar{\Gamma})} \leq C_{k,m}\|\rho_{g_1}^2 - \rho_{g_0}^2\|_{C^{m+2k+2}(\overline{\Gamma' \times \Gamma'})},$$

where $C_{k,m}$ depends only on Ω and on a upper bound of g_0, g_1 in $C^{m+2k+5}(\bar{\Omega})$.

Theorem 4 can be used to obtain stability near generic simple metrics for the inverse problem of recovering g from the hyperbolic *Dirichlet-to-Neumann map* Λ_g. It is known that g can be recovered uniquely from Λ_g, up to a diffeomorphism as above, see e.g. [3]. This result however relies on a unique continuation theorem by Tataru [35] and it is unlikely to provide Hölder type of stability estimate as above. By using the fact that ρ_g is related to the leading singularities in the kernel of Λ_g, we proved a Hölder stability estimate under the assumptions above, relating g and Λ_g. We refer to [31] for details. □

2 Lens rigidity for a class of non-simple manifolds

Suppose we have a Riemannian metric in Euclidean space which is the Euclidean metric outside a compact set. The inverse scattering problem for metrics is to determine the Riemannian metric by measuring the scattering operator (see [14]). A similar obstruction to the boundary rigidity problem occurs in this case with the diffeomorphism ψ equal to the identity outside a compact set. It was proven in [14] that from the wave front set of the scattering operator, one can determine, under some conditions on the metric including non-trapping, the *scattering relation* on the boundary of a large ball. This uses high frequency information of the scattering operator. In the semiclassical setting Alexandrova [1], [2] has shown for a large class of operators that the scattering operator associated to potential and metric perturbations of the Euclidean Laplacian is a semiclassical Fourier integral operator that quantizes the scattering relation. The scattering relation maps the point and direction of a geodesic entering the manifold to the point and direction of exit of the geodesic. As mentioned in the previous section, the boundary rigidity problem only takes into account the shortest paths. For non-simple manifolds in particular, if we have conjugate points or the boundary is not strictly convex, we need to look at the behavior of all the geodesics and the scattering relation encodes this information. We proceed to define in more detail the scattering relation and the lens rigidity problem and state our results. We note that we also consider the case of incomplete data, that is when we don't have information about all the geodesics entering the manifold. More details can be found in [33], [34].

Denote by $SM = \{(x,\xi) \in TM;\ |\xi| = 1\}$ the unit sphere bundle and set

$$\partial_\pm SM = \{(x,\xi) \in \partial SM;\ \pm\langle \nu, \xi\rangle < 0\}, \tag{2.1}$$

where ν is the unit interior normal, $\langle \cdot, \cdot \rangle$ and stands for the inner product. The scattering relation

$$\Sigma : \partial_- SM \to \overline{\partial_+ SM} \tag{2.2}$$

is defined by $\Sigma(x,\xi) = (y,\eta) = \Phi^\ell(x,\xi)$, where Φ^t is the geodesic flow, and $\ell > 0$ is the first moment, at which the unit speed geodesic through (x,ξ) hits ∂M again. If such an ℓ does not exists, we formally set $\ell = \infty$ and we call the corresponding initial condition and the corresponding geodesic *trapping*. This defines also $\ell(x,\xi)$ as a function $\ell : \partial_- SM \to [0,\infty]$. Note that Σ and ℓ are not necessarily continuous.

It is convenient to think of Σ and ℓ as defined on the whole ∂SM with $\Sigma = \mathrm{Id}$ and $\ell = 0$ on $\overline{\partial_+ SM}$.

We parametrize the scattering relation in a way that makes it independent of pulling it back by diffeomorphisms fixing ∂M pointwise. Let $\kappa_\pm : \partial_\pm SM \to B(\partial M)$ be the orthogonal projection onto the (open) unit ball tangent bundle that extends continuously to the closure of $\partial_\pm SM$. Then $\kappa_\pm$ are homeomorphisms, and we set

$$\sigma = \kappa_+ \circ \Sigma \circ \kappa_-^{-1} : \overline{B(\partial M)} \longrightarrow \overline{B(\partial M)}. \tag{2.3}$$

According to our convention, $\sigma = \mathrm{Id}$ on $\partial(\overline{B(\partial M)}) = S(\partial M)$. We equip $\overline{B(\partial M)}$ with the relative topology induced by $T(\partial M)$, where neighborhoods of boundary points (those in $S(\partial M)$) are given by half-neighborhoods, i.e., by neighborhoods in $T(\partial M)$ intersected with $\overline{B(\partial M)}$.

It is possible to define σ in a way that does not require knowledge of $g|_{T(\partial M)}$ by thinking of any boundary vector ξ as characterized by its angle with ∂M and the direction of its tangential projection. Let $\mathcal{D}$ be an open subset of $\overline{B(\partial M)}$. A priori, the latter depends on $g|_{T(\partial M)}$. By the remark above, we can think of it as independent of $g|_{T(\partial M)}$ however.

The lens rigidity question we study is the following:

Given M and $\mathcal{D}$, do σ and ℓ, restricted to $\mathcal{D}$, determine g uniquely, up to a pull back of a diffeomorphism that is identity on ∂M?

The answer to this question, even when $\mathcal{D} = B(\partial M)$, is negative, see [12]. The known counter-examples are trapping manifolds.

The boundary rigidity problem and the lens rigidity one are equivalent for simple metrics.

2.1 Main assumptions

Definition 3. *We say that $\mathcal{D}$ is* **complete** *for the metric g, if for any $(z,\zeta) \in T^*M$ there exists a maximal in M, finite length unit speed geodesic $\gamma : [0,l] \to M$ through z, normal to ζ, such that*

$$\{(\gamma(t),\dot\gamma(t));\ 0\le t\le l\}\cap S(\partial M)\subset \mathcal{D}, \tag{2.4}$$

$$\textit{there are no conjugate points on } \gamma. \tag{2.5}$$

We call the C^k metric g **regular**, *if a complete set $\mathcal{D}$ exists, i.e., if $\overline{B(\partial M)}$ is complete.*

If $z \in \partial M$ and ζ is conormal to ∂M, then γ may reduce to one point.

Topological Condition (T): Any path in M connecting two boundary points is homotopic to a polygon $c_1 \cup \gamma_1 \cup c_2 \cup \gamma_2 \cup \cdots \cup \gamma_k \cup c_{k+1}$ with the properties that for any j,
(i) c_j is a path on ∂M;
(ii) $\gamma_j : [0, l_j] \to M$ is a geodesic lying in M^{int} with the exception of its endpoints and is transversal to ∂M at both ends; moreover, $\kappa_-(\gamma_j(0), \dot\gamma_j(0)) \in \mathcal{D}$;

Notice that (T) is an open condition w.r.t. g, i.e., it is preserved under small C^2 perturbations of g. To define the $C^K(M)$ norm below in a unique way, we choose and fix a finite atlas on M.

2.2 Results about the linear problem

We refer to [33] for more details about the results in this subsection. It turns out that a linearization of the lens rigidity problem is again the problem of s-injectivity of the ray transform I. Here and below we sometimes drop the subscript g. Given $\mathcal{D}$ as above, we denote by $I_{\mathcal{D}}$ (or $I_{g,\mathcal{D}}$) the ray transform I restricted to the maximal geodesics issued from $(x,\xi) \in \kappa_-^{-1}(\mathcal{D})$.

The first result of this subsection generalizes Theorem 1.

Theorem 6. *Let g be an analytic, regular metric on M. Let $\mathcal{D}$ be complete and open. Then $I_{\mathcal{D}}$ is s-injective.*

Sketch of the proof. Since we know integrals over a subset of geodesics only, this creates difficulties with cut-offs in the phase variable that cannot be analytic. For this reason, the proof of Theorem 6 is different than that of Theorem 1.

Let g be an analytic regular metrics in M, and let $M_1 \Supset M$ be the manifold where g is extended analytically. There is an analytic atlas in M, and ∂M can be assumed to be analytic, too. In other words, now $(M, \partial M, g)$ is a real analytic manifold with boundary. We denote by $\mathcal{A}(M)$ (respectively $\mathcal{A}(M_1)$) the set of analytic functions on M (respectively M_1). Next, $f^s_{M_1}$ denotes the solenoidal part of the tensor f, extended as zero to M_1, in the manifold M_1.

The main step is to show that $I_{\mathcal{D}} f = 0$ implies $f^s \in \mathcal{A}(M)$. In order to do that one shows that $f^s_{M_1} \in \mathcal{A}(M_1)$. Let us first notice, that in $M_1 \setminus M$, $f^s_{M_1} = -dv_{M_1}$, where v_{M_1} satisfies $\delta dv_{M_1} = 0$ in $M_1 \setminus M$, $v|_{\partial M_1} = 0$ since $f = 0$ in $M_1 \setminus M$. Therefore, v_{M_1} is analytic up to ∂M_1. Therefore, we only need to show that $f^s_{M_1}$ is analytic in the interior of M_1. Below, $\mathrm{WF_A}(f)$ stands for the analytic wave front set of f, see [28, 36].

The crucial point is the following microlocal analytic regularity result.

Proposition 1. *Let γ_0 be a fixed maximal geodesic in M with endpoints on ∂M, without conjugate points, and let $I_g f(\gamma) = 0$ for $\gamma \in$ neigh(γ_0). Let g be analytic in* neigh(γ_0). *Then*

$$N^*\gamma_0 \cap \mathrm{WF_A}(f^s_{M_1}) = \emptyset. \tag{2.6}$$

Sketch of the proof. Let U_ε be a tubular neighborhood of γ_0, and $x = (x', x^n)$ be semigeodesic coordinates in it such that $x' = 0$ on γ_0. Fix $x_0 \in \gamma_0 \cap M$. We can assume that $x_0 = 0$ and $g_{ij}(0) = \delta_{ij}$. Then we can assume that $U_\varepsilon = \{-l_1 - \varepsilon < x^n < l_2 + \varepsilon, |x'| < \varepsilon\}$ with the part of γ_0 corresponding to $x^n \notin [-l_1, l_2]$ outside M.

Fix $\xi^0 = ((\xi^0)', 0)$ with $\xi^0_n = 0$. We will show that

$$(0, \xi^0) \notin \mathrm{WF_A}(f). \tag{2.7}$$

We choose a local chart for the geodesics close to γ_0. Set first $Z = \{x^n = 0;\ |x'| < 7\varepsilon/8\}$, and denote the x' variable on Z by z'. Then z', θ' (with $|\theta'| \ll 1$) are local coordinates in neigh(γ_0) determined by $(z', \theta') \to \gamma_{(z',0),(\theta',1)}$ where the latter denotes the geodesic through the point $(z', 0)$ in the direction $(\theta', 1)$. Let $\chi_N(z')$ be a smooth cut-off function equal to 1 for $|z'| \le 3\varepsilon/4$ and supported in Z, also satisfying $|\partial^\alpha \chi_N| \le (CN)^{|\alpha|}$, $|\alpha| \le N$. Set $\theta = (\theta', 1)$, $|\theta'| \ll 1$, and multiply

$$If\left(\gamma_{(z',0),\theta}\right) = 0$$

by $\chi_N(z') e^{\mathrm{i}\lambda z' \cdot \xi'}$, where $\lambda > 0$, ξ' is in a complex neighborhood of $(\xi^0)'$, and integrate w.r.t. z' to get

$$\iint e^{\mathrm{i}\lambda z' \cdot \xi'} \chi_N(z') f_{ij}\left(\gamma_{(z',0),\theta}(t)\right) \dot\gamma^i_{(z',0),\theta}(t) \dot\gamma^j_{(z',0),\theta}(t)\, \mathrm{d}t\, \mathrm{d}z' = 0. \tag{2.8}$$

Set $x = \gamma_{(z',0),\theta}(t)$. If $\theta' = 0$, we have $x = (z', t)$. By a perturbation argument, for θ' fixed and small enough, (t, z') are analytic local coordinates, depending analytically on θ'. In particular, $x = (z' + t\theta', t) + O(|\theta'|)$ but this expansion is not enough for the analysis below. Performing a change of variables in (2.8), we get

$$\int e^{\mathrm{i}\lambda z'(x,\theta') \cdot \xi'} a_N(x, \theta') f_{ij}(x) b^i(x, \theta') b^j(x, \theta')\, \mathrm{d}x = 0 \tag{2.9}$$

for $|\theta'| \ll 1$, $\forall\lambda$, $\forall\xi'$, where, for $|\theta'| \ll 1$, the function $(x, \theta') \mapsto a_N$ is positive for x in a neighborhood of γ_0, vanishing for $x \notin U_\varepsilon$, and satisfies the same estimate as χ_N. The vector field b is analytic, and $b(0, \theta') = \theta$, $a_N(0, \theta') = 1$.

To clarify the approach, note that if g is Euclidean in neigh(γ_0), then (2.9) reduces to

$$\int e^{\mathrm{i}\lambda(\xi',-\theta'\cdot\xi')\cdot x}\chi f_{ij}(x)\theta^i\theta^j\,\mathrm{d}x = 0,$$

where $\chi = \chi(x' - x^n\theta')$. Then $\xi = (\xi', -\theta'\cdot\xi')$ is perpendicular to $\theta = (\theta', 1)$. This implies that

$$\int e^{\mathrm{i}\lambda\xi\cdot x}\chi f_{ij}(x)\theta^i(\xi)\theta^j(\xi)\,\mathrm{d}x = 0 \tag{2.10}$$

for any function $\theta(\xi)$ defined near ξ^0, such that $\theta(\xi)\cdot\xi = 0$. This has been noticed and used before if g is close to the Euclidean metric (with $\chi = 1$), see e.g., [29]. We will assume that $\theta(\xi)$ is analytic. A simple argument (see e.g. [25, 29]) shows that a constant symmetric tensor f_{ij} is uniquely determined by the numbers $f_{ij}\theta^i\theta^j$ for finitely many θ's (actually, for $N' = (n+1)n/2$ θ's); and in any open set on the unit sphere, there are such θ's. On the other hand, f is solenoidal. To simplify the argument, assume for a moment that f vanishes on ∂M. Then $\xi^i\hat{f}_{ij}(\xi) = 0$. Therefore, combining this with (2.10), we need to choose $N = n(n-1)/2$ vectors $\theta(\xi)$, perpendicular to ξ, that would uniquely determine the tensor $\hat{f}$ on the plane perpendicular to ξ. To this end, it is enough to know that this choice can be made for $\xi = \xi^0$, then it would be true for $\xi \in \text{neigh}(\xi^0)$. This way, $\xi^i\hat{f}_{ij}(\xi) = 0$ and the N equations (2.10) with the so chosen $\theta_p(\xi)$, $p = 1,\dots,N$, form a system with a tensor-valued symbol elliptic near $\xi = \xi^0$. The C^∞ ΨDO calculus easily implies the statement of the lemma in the C^∞ category, and the complex stationary phase method below, or the analytic ΨDO calculus in [36] with appropriate cut-offs in ξ, implies the lemma in this special case (g locally Euclidean).

The general case is considered in [33], and is based on an application of a complex stationary phase argument [28] to (2.9) as in [16]. □

Proposition 1 makes it possible to prove that $f^s \in \mathcal{A}(M)$. We combine this with a boundary determination theorem for tensors, a linear version of Theorem 10 below, to conclude that then $f = 0$. □

Next, we formulate a stability estimate in the spirit of Theorem 2. We need first to parametrize (a complete subset of) the geodesics issued from $\mathcal{D}$ in a different way that would make them a manifold. The parametrization provided by $\mathcal{D}$ is inconvenient near the directions tangent to ∂M.

Let H_m be a finite collection of smooth hypersurfaces in M_1^{int}. Let $\mathcal{H}_m$ be an open subset of $\{(z,\theta) \in SM_1;\ z \in H_m, \theta \notin T_zH_m\}$, and let $\pm l_m^\pm(z,\theta) \geq 0$ be two continuous functions. Let $\Gamma(\mathcal{H}_m)$ be the set of geodesics

$$\Gamma(\mathcal{H}_m) = \left\{\gamma_{z,\theta}(t);\ l_m^-(z,\theta) \leq t \leq l_m^+(z,\theta),\ (z,\theta) \in \mathcal{H}_m\right\}, \tag{2.11}$$

that, depending on the context, is considered either as a family of curves, or as a point set. We also assume that each $\gamma \in \Gamma(\mathcal{H}_m)$ is a simple geodesic (no conjugate points).

If g is simple, then one can take a single $H = \partial M_1$ with $l^- = 0$ and an appropriate $l^+(z,\theta)$. If g is regular only, and Γ is any complete set of geodesics, then any small enough neighborhood of a simple geodesic in Γ has the properties listed in the paragraph above and by a compactness argument one can choose a finite complete set of such $\Gamma(\mathcal{H}_m)$'s, that is included in the original Γ.

Given $\mathcal{H} = \{\mathcal{H}_m\}$ as above, we consider an open set $\mathcal{H}' = \{\mathcal{H}'_m\}$, such that $\mathcal{H}'_m \Subset \mathcal{H}_m$, and let $\Gamma(\mathcal{H}'_m)$ be the associated set of geodesics defined as in (2.11), with the same $l^\pm_m$. Set $\Gamma(\mathcal{H}) = \cup\Gamma(\mathcal{H}_m)$, $\Gamma(\mathcal{H}') = \cup\Gamma(\mathcal{H}'_m)$.

The restriction $\gamma \in \Gamma(\mathcal{H}'_m) \subset \Gamma(\mathcal{H}_m)$ can be modeled by introducing a weight function α_m in $\mathcal{H}_m$, such that $\alpha_m = 1$ on $\mathcal{H}'_m$, and $\alpha_m = 0$ otherwise. More generally, we allow α_m to be smooth but still supported in $\mathcal{H}_m$. We then write $\alpha = \{\alpha_m\}$, and we say that $\alpha \in C^k(\mathcal{H})$, if $\alpha_m \in C^k(\mathcal{H}_m)$, $\forall m$.

We consider $I_{\alpha_m} = \alpha_m I$, or more precisely, in the coordinates $(z,\theta) \in \mathcal{H}_m$,

$$I_{\alpha_m} f = \alpha_m(z,\theta) \int_0^{l_m(z,\theta)} \langle f(\gamma_{z,\theta}), \dot\gamma^2_{z,\theta}\rangle \, \mathrm{d}t, \quad (z,\theta) \in \mathcal{H}_m. \tag{2.12}$$

Next, we set

$$I_\alpha = \{I_{\alpha_m}\}, \quad N_{\alpha_m} = I^*_{\alpha_m} I_{\alpha_m} = I^*|\alpha_m|^2 I, \quad N_\alpha = \sum N_{\alpha_m}, \tag{2.13}$$

where the adjoint is taken w.r.t. the measure $\mathrm{d}\mu := |\langle \nu(z), \theta\rangle| \, \mathrm{d}S_z \, \mathrm{d}\theta$ on $\mathcal{H}_m$, $\mathrm{d}S_z \, \mathrm{d}\theta$ being the induced measure on SM, and $\nu(z)$ being a unit normal to H_m.

S-injectivity of N_α is equivalent to s-injectivity for I_α, which in turn is equivalent to s-injectivity of I restricted to $\operatorname{supp}\alpha$.

Theorem 7.

(a) Let $g = g_0 \in C^k$, $k \gg 1$ be regular, and let $\mathcal{H}' \Subset \mathcal{H}$ be as above with $\Gamma(\mathcal{H}')$ complete. Fix $\alpha = \{\alpha_m\} \in C^\infty$ with $\mathcal{H}'_m \subset \operatorname{supp}\alpha_m \subset \mathcal{H}_m$. Then if I_α is s-injective, we have

$$\|f^s\|_{L^2(M)} \le C\|N_\alpha f\|_{\tilde H^2(M_1)}. \tag{2.14}$$

(b) Assume that $\alpha = \alpha_g$ in (a) depends on $g \in C^k$, so that $C^k(M_1) \ni g \to C^l(\mathcal{H}) \ni \alpha_g$ is continuous with $l \gg 1$, $k \gg 1$. Assume that $I_{g_0,\alpha_{g_0}}$ is s-injective. Then estimate (2.14) remains true for g in a small enough neighborhood of g_0 in $C^k(M_1)$ with a uniform constant $C > 0$.

The theorem above allows us to formulate a generic result:

Theorem 8. *Let $\mathcal{G} \subset C^k(M)$ be an open set of regular Riemannian metrics on M such that (T) is satisfied for each one of them. Let the set $\mathcal{D}' \subset \partial SM$ be open and complete for each $g \in \mathcal{G}$. Then there exists an open and dense subset $\mathcal{G}_s$ of $\mathcal{G}$ such that $I_{g,\mathcal{D}'}$ is s-injective for any $g \in \mathcal{G}_s$.*

Of course, the set $\mathcal{G}_s$ includes all real analytic metrics in $\mathcal{G}$.

Corollary 1. *Let $\mathcal{R}(M)$ be the set of all regular C^k metrics on M satisfying (T) equipped with the $C^k(M_1)$ topology. Then for $k \gg 1$, the subset of metrics for which the X-ray transform I over all simple geodesics through all points in M is s-injective, is open and dense in $\mathcal{R}(M)$.*

2.3 Results about the non-linear lens rigidity problem

Using the results above, we prove the following about the lens rigidity problem on manifolds satisfying the assumptions in Section 2.1. More details can be found in [34].

Theorem 9 below says, loosely speaking, that for the classes of manifolds and metrics we study, the uniqueness question for the non-linear lens rigidity problem can be answered locally by linearization. This is a non-trivial implicit function type of theorem however because our success heavily depends on the a priori stability estimate that the s-injectivity of $I_{\mathcal{D}}$ implies; see Theorem 7; and the latter is based on the hypoelliptic properties of $I_{\mathcal{D}}$. We work with two metrics g and $\hat{g}$; and will denote objects related to $\hat{g}$ by $\hat{\sigma}$, $\hat{\ell}$, etc.

Theorem 9. *Let (M, g_0) satisfy the topological assumption (T), with $g_0 \in C^k(M)$ a regular Riemannian metric with $k \gg 1$. Let $\mathcal{D}$ be open and complete for g_0, and assume that there exists $\mathcal{D}' \Subset \mathcal{D}$ so that $I_{g_0,\mathcal{D}'}$ is s-injective. Then there exists $\varepsilon > 0$, such that for any two metrics g, $\hat{g}$ satisfying*

$$\|g - g_0\|_{C^k(M)} + \|\hat{g} - g_0\|_{C^k(M)} \leq \varepsilon, \tag{2.15}$$

the relations

$$\sigma = \hat{\sigma}, \quad \ell = \hat{\ell} \quad \text{on } \mathcal{D}$$

imply that there is a C^{k+1} diffeomorphism $\psi : M \to M$ fixing the boundary such that

$$\hat{g} = \psi^* g.$$

By Theorem 8, the requirement that $I_{g_0,\mathcal{D}'}$ is s-injective is a generic one for g_0. Therefore, Theorems 9 and 8 combined imply that there is local uniqueness, up to isometry, near a generic set of regular metrics.

Corollary 2. *Let $\mathcal{D}' \Subset \mathcal{D}$, $\mathcal{G}$, $\mathcal{G}_s$ be as in Theorem 8. Then the conclusion of Theorem 9 holds for any $g_0 \in \mathcal{G}_s$.*

2.4 Boundary determination of the jet of g

The first step of the proof of Theorem 9 is to determine all derivatives of g on ∂M. The following theorem is interesting by itself. Notice that g below does not need to be analytic or generic.

Theorem 10. *Let (M, g) be a compact Riemannian manifold with boundary. Let $(x_0, \xi_0) \in S(\partial M)$ be such that the maximal geodesic γ_{x_0,ξ_0} through it is of finite length, and assume that x_0 is not conjugate to any point in $\gamma_{x_0,\xi_0} \cap \partial M$. If σ and ℓ are known on some neighborhood of (x_0, ξ_0), then the jet of g at x_0 in boundary normal coordinates is determined uniquely.*

Sketch of the proof of Theorem 10. To make the arguments below more transparent, assume that the geodesic γ_0 issued from (x_0, ξ_0) hits ∂M for the first time transversally at $\gamma_0(l_0) = y_0$, $l_0 > 0$. Then y_0 is the only point on ∂M reachable from (x_0, ξ_0), and x_0, y_0 are not conjugate points on γ_0 by assumption. Assume also that γ_0 is tangent of finite order at x_0. Then there is a half neighborhood V of x_0 on ∂M visible from y_0. The latter is not always true if γ_0 is tangent to ∂M of infinite order at x_0.

Choose local boundary normal coordinates near x_0 and y_0, and let g_0 be the Euclidean metric in each of them w.r.t. to the so chosen coordinates. We can then consider a representation of Σ, denoted by $\Sigma^\sharp$ below, defined locally on $\mathbf{R}^{n-1} \times S^{n-1}$, with values on another copy of the same space. If $(x, \theta) \in \mathbf{R}^{n-1} \times S^{n-1}$, then the associated vector at $x \in \partial M$ is $\xi = \theta/|\theta|_g$; and $\Sigma^\sharp(x, \theta) = \Sigma(x, \xi)$. The same applies to the second component of $\Sigma^\sharp(x, \theta)$. Namely, if $(y, \eta) = \Sigma(x, \xi)$, then we set $\omega = \eta/|\eta|_{g_0}$, then $\Sigma^\sharp : (x, \theta) \mapsto (y, \omega)$. Similarly, we set $\ell^\sharp(x, \theta) = \ell(x, \xi)$. Let also θ_0 and ω_0 correspond to ξ_0 and η_0, respectively, where $\Sigma(x_0, \xi_0) = (y_0, \eta_0)$.

Set $\tau(x) := \tau(x, y_0)$, where τ is the smooth travel time function localized near $x = x_0$ such that $\tau(x_0, y_0) = l_0$. Then τ is well defined in a small neighborhood of x_0 by the implicit function theorem and the assumption that x_0 and y_0 are not conjugate on γ_0. In the normal boundary coordinates $x = (x', x^n)$ near x_0, $g_{in} = \delta_{in}$, $\forall i$. Since x_0 and y_0 are not conjugate, for $\eta \in S_{y_0}M$ close enough to η_0, the map $\eta \mapsto x \in \partial M$ is a local diffeomorphism as long as the geodesic connecting x and y_0 is not tangent to ∂M at x. Moreover, that map is known, being the inverse of Σ. Similarly, the map $S^{n-1} \ni \omega \mapsto x$ is a local diffeomorphism and is also known. Then we know $(x, -\theta) = \Sigma^\sharp(y_0, -\omega)$, and we know $\ell^\sharp(y_0, -\omega) = \ell^\sharp(x, \theta) = \tau(x)$. Then we can recover $\operatorname{grad}' \tau = -\theta'/|\theta|_g$, where the prime stands for tangential projection as usual. Taking the limit $\omega \to \omega_0$, we recover $|\theta_0|_g^2 = g_{\alpha\beta}\theta_0^\alpha\theta_0^\beta$. We use again the fact that a symmetric $n \times n$ tensor f_{ij} can be recovered by knowledge of $f_{ij}p_k^i p_k^j$ for $N = n(n+1)/2$ "generic" vectors p_k, $k = 1, \dots, N$; and such N vectors exist in any open set on S^{n-1}, see e.g. [34]. Thus choosing appropriate $n(n-1)/2$ perturbations of θ_0's, we recover $g(x_0)$. Thus, we recover g in a neighborhood of x_0 as well; we can assume that V covers that neighborhood.

Note that we know all tangential derivatives of g in $V \ni x_0$. Then τ solves the eikonal equation

$$g^{\alpha\beta}\tau_{x^\alpha}\tau_{x^\beta} + \tau_{x^n}^2 = 1. \tag{2.16}$$

Next, in V, we know τ_{x^α}, $\alpha \le n-1$, we know g, therefore by (2.16), we get $\tau_{x^n}^2$. It is easy to see that $\tau_{x^n} \le 0$ on the visible part, so we recover τ_{x^n} there. We therefore know the tangential derivatives of τ_{x^n} on ∂M near x_0.

Differentiate (2.16) w.r.t. x^n at $x = x_0$ to get

$$\left[\frac{dg^{\alpha\beta}}{dx^n}\tau_{x^\alpha}\tau_{x^\beta} + 2g^{\alpha\beta}\tau_{x^\alpha x^n}\tau_{x^\beta} + 2\tau_{x^n x^n}\tau_{x^n}\right]\Big|_{x=x_0} = 0. \tag{2.17}$$

Since γ_0 is tangent to ∂M at x_0, we have $\tau_{x^n}(x_0) = 0$ by (2.16). The third term in the r.h.s. of (2.17) therefore vanishes. Therefore the only unknown term in (2.17) is $\gamma^{\alpha\beta} := dg^{\alpha\beta}/dx^n$ at $x = x_0$. Since $\tau_{x^\alpha}(x_0) = -\xi_0$, using the fact that $\operatorname{grad}\tau(x_0) = -\xi_0$ again, we get that we have to determine $\gamma^{\alpha\beta}$ from $\gamma_{\alpha\beta}\xi_0^\alpha\xi_0^\beta$. This is possible if as above, we repeat the construction and replace ξ_0 by a finite number of vectors, close enough to ξ_0. So we get an explicit formula for $\partial g/\partial x^i|_{\partial M}$ in fact.

Next, for $x \in V$ but not on ∂V, we can recover $\tau_{x^n x^n}(x)$ by (2.17) because $\tau_{x^n}(x) < 0$. By continuity, we recover $\tau_{x^n x^n}(x_0)$, therefore we know $\tau_{x^n x^n}$ near x_0, and all tangential derivatives of the latter.

We differentiate (2.17) w.r.t. x^n again, and as above, recover $d^2g/d(x^n)^2|_{\partial M}$ near x_0. Then we recover $d^3\tau/d(x^n)^3$, etc.

In the general case, we repeat those arguments with ξ_0 replaced by $\xi_0 + \varepsilon\nu$, where ν is the interior unit normal, and take the limit $\varepsilon \to 0$. □

Sketch of the proof of Theorem 9. We first find suitable metric $\hat{g}_1$ isometric to $\hat{g}$, and then we show that $\hat{g}_1 = g$. First, we can always assume that g and $\hat{g}$ have the same boundary normal coordinates near ∂M. By [11], there is a metric h isometric to $\hat{g}$ so that h is solenoidal w.r.t. g. Moreover, $h = \hat{g} + O(\varepsilon)$. By a standard argument, by a diffeomorphism that identifies normal coordinates near ∂M for h and g, and is identity away from some neighborhood of the boundary, we find a third $\hat{g}_1$ isometric to h (and therefore to $\hat{g}$), so that $\hat{g}_1 = \hat{g}$ near ∂M, and $\hat{g}_1 = h$ away from some neighborhood of ∂M (and there is a region that $\hat{g}_1$ is neither). Then $\hat{g}_1 - h$ is as small as $g - h$, more precisely,

$$\|\hat{g}_1 - h\|_{C^{k-3}} \le C\|g - h\|_{C^{k-1}}, \quad k \gg 1. \tag{2.18}$$

Set

$$f = h - g, \quad \tilde{f} = \hat{g}_1 - g. \tag{2.19}$$

Estimate (2.18) implies

$$\|\tilde{f} - f\|_{C^{l-3}} \le C\|f\|_{C^{l-1}}, \quad \forall l \le k. \tag{2.20}$$

By (2.15), (2.20),

$$\|f\|_{C^{k-1}} \le C\varepsilon, \quad \|\tilde{f}\|_{C^{k-3}} \le C\varepsilon. \tag{2.21}$$

By Theorem 10,

$$\partial^\alpha \tilde{f} = 0 \quad \text{on } \partial M \text{ for } |\alpha| \le k - 5. \tag{2.22}$$

It is known [25] that $2dv$ is the linearization of $\psi_\tau^* g$ at $\tau = 0$, where ψ_τ is a smooth family of diffeomorphisms, and $v = d\psi_\tau/d\tau$ at $\tau = 0$. Next proposition is therefore a version of Taylor's expansion:

Proposition 2. *Let $\hat{g}$ and g be in C^k, $k \ge 2$ and isometric, i.e.,*

$$\hat{g} = \psi^* g$$

for some diffeomorphism ψ fixing ∂M. Set $f = \hat{g} - g$. Then there exists v vanishing on ∂M, so that

$$f = 2dv + f_2,$$

and for g belonging to any bounded set U in C^k, there exists $C(U) > 0$, such that

$$\|f_2\|_{C^{k-2}} \le C(U)\|\psi - \mathrm{Id}\|^2_{C^{k-1}}, \quad \|v\|_{C^{k-1}} \le C(U)\|\psi - \mathrm{Id}\|_{C^{k-1}}.$$

We will sketch now the rest of the proof of Theorem 9. We apply Proposition 2 to h and $\hat{g}_1$ to get

$$\tilde{f} = f + 2dv + f_2, \quad \|f_2\|_{C^{l-3}} \le C\|f\|^2_{C^{l-1}}, \quad \forall l \le k. \tag{2.23}$$

In other words, $\tilde{f}^s = f$ up to $O(\|f\|^2)$.

We can assume that g is extended smoothly on $M_1 \Supset M$. Next, with g extended as above, we extend $\hat{g}_1$ so that $\hat{g}_1 = g$ outside M. This can be done in a smooth way by Theorem 10.

The next step is to reparametrize the scattering relation. We show that one can extend the maximal geodesics of g, respectively $\hat{g}_1$, outside M (where $g = \hat{g}_1$), and since the two metrics have the same scattering relation and travel times, they will still have the same scattering relation and travel times if we locally push ∂M a bit outside M. Then we can arrange that the new pieces of ∂M are transversal to the geodesics close to a fixed one, which provides a smooth parametrization. By a compactness argument, one can do this near finitely many geodesics issued from point on $\mathcal{D}$, and still have a complete set. This puts as in the situation of Theorem 7, where the set of geodesics is parametrized by $\alpha = \{\alpha_j\}$.

Next, we linearize the energy functional near each geodesic (in our set of data) related to g. Using the assumption that g and $\hat{g}_1$ have the same scattering relation and travel times, we deduct

$$\|N_{\alpha_j}\tilde{f}\|_{L^\infty} \le C\|\tilde{f}\|^2_{C^1}, \quad \forall j. \tag{2.24}$$

Using interpolation inequalities, and the fact that the extension of $\tilde{f}$ outside M is smooth enough across ∂M as a consequence of the boundary recovery, we get by (2.24), and (2.20),

$$\|N_\alpha\tilde{f}\|_{\tilde{H}^2(M_1)} \le C\|\tilde{f}\|^{3/2}_{C^3} \le C'\|f\|^{3/2}_{C^5}. \tag{2.25}$$

Since $I_{g_0,\mathcal{D}'}$ is s-injective, so is N_α, related to g_0, by the support properties of α. Now, since g is close enough to g_0 with s-injective N_α by (2.15), N_α (the one related to g) is s-injective as well by Theorem 7. Therefore, by (2.25) and (2.14),

$$\|f^s\|_{L^2(M)} \le C\|N_\alpha\tilde{f}\|_{\tilde{H}^2} \le C'\|f\|^{3/2}_{C^5}. \tag{2.26}$$

A decisive moment of the proof is that by Proposition 2, see (2.23), $\tilde{f}^s = f + f_2^s$, the latter being the solenoidal projection of f_2. Therefore,

$$\|\tilde{f}^s\|_{L^2(M)} \geq \|f\|_{L^2(M)} - C\|f\|_{C^2}^2.$$

Together with (2.26), this yields

$$\|f\|_{L^2(M)} \leq C\left(\|f\|_{C^2}^2 + \|f\|_{C^5}^{3/2}\right) \leq C'\|f\|_{C^5}^{3/2}$$

because the C^5 norm of f is uniformly bounded when $\varepsilon \leq 1$. Using interpolation again, we easily deduct $\|f\|_{L^2(M)} \geq 1/C$ if $f \neq 0$. This contradicts (2.21) if $\varepsilon \ll 1$.

Now, $f = 0$ implies $h = g$, therefore, g and $\hat{g}$ are isometric.

This concludes the sketch proof of Theorem 9. □

References

1. I. Alexandrova, Semi-Classical wavefront set and Fourier integral operators, to appear in *Can. J. Math.*
2. ———, Structure of the Semi-Classical Amplitude for General Scattering Relations, *Comm. PDE* **30**(2005), 1505-1535.
3. M. Belishev and Y. Kurylev, *To the reconstruction of a Riemannian manifold via its boundary spectral data (BC-method)*, Comm. PDE **17**(1992), 767–804.
4. I. N. Bernstein and M. L. Gerver, *Conditions on distinguishability of metrics by hodographs*, Methods and Algorithms of Interpretation of Seismological Information, Computerized Seismology **13**, Nauka, Moscow, 50–73 (in Russian).
5. G. Besson, G. Courtois, and S. Gallot, *Entropies et rigidités des espaces localement symétriques de courbure strictment négative*, Geom. Funct. Anal., **5**(1995), 731-799.
6. G. Beylkin, *Stability and uniqueness of the solution of the inverse kinematic problem in the multidimensional case*, J. Soviet Math. **21**(1983), 251–254.
7. D. Burago and S. Ivanov, *Boundary rigidity and filling volume minimality of metrics close to a flat one*, manuscript, 2005.
8. K. C. Creager, *Anisotropy of the inner core from differential travel times of the phases PKP and PKIPK*, Nature, **356**(1992), 309–314.
9. C. Croke, *Rigidity for surfaces of non-positive curvature*, Comment. Math. Helv., **65**(1990), 150–169.
10. ———, *Rigidity and the distance between boundary points*, J. Differential Geom., **33**(1991), no. 2, 445–464.
11. C. Croke, N. Dairbekov, V. Sharafutdinov, *Local boundary rigidity of a compact Riemannian manifold with curvature bounded above*, Trans. Amer. Math. Soc. **352**(2000), no. 9, 3937–3956.
12. C. Croke and B. Kleiner, Conjugacy and Rigidity for Manifolds with a Parallel Vector Field, *J. Diff. Geom.* 39(1994), 659–680.
13. M. Gromov, *Filling Riemannian manifolds*, J. Diff. Geometry **18**(1983), no. 1, 1–148.
14. V. Guillemin, *Sojourn times and asymptotic properties of the scattering matrix*, Proceedings of the Oji Seminar on Algebraic Analysis and the RIMS Symposium on Algebraic Analysis (Kyoto Univ., Kyoto, 1976). *Publ. Res. Inst. Math. Sci.* **12**(1976/77), supplement, 69–88.

15. G. Herglotz, Uber die Elastizitaet der Erde bei Beruecksichtigung ihrer variablen Dichte, *Zeitschr. fur Math. Phys.*, **52**(1905), 275–299.
16. C. Kenig, J. Sjöstrand and G. Uhlmann, The Calderón Problem with partial data, to appear in *Ann. Math.*
17. M. Lassas, V. Sharafutdinov and G. Uhlmann, *Semiglobal boundary rigidity for Riemannian metrics*, Math. Ann., **325**(2003), 767–793.
18. R. Michel, *Sur la rigidité imposée par la longueur des géodésiques*, Invent. Math. **65**(1981), 71–83.
19. R. G. Mukhometov, *The reconstruction problem of a two-dimensional Riemannian metric, and integral geometry (Russian)*, Dokl. Akad. Nauk SSSR **232**(1977), no. 1, 32–35.
20. R. G. Mukhometov, *On a problem of reconstructing Riemannian metrics* Siberian Math. J. **22**(1982), no. 3, 420–433.
21. R. G. Mukhometov and V. G. Romanov, *On the problem of finding an isotropic Riemannian metric in an n-dimensional space (Russian)*, Dokl. Akad. Nauk SSSR **243**(1978), no. 1, 41–44.
22. J. P. Otal, *Sur les longuer des géodésiques d'une métrique a courbure négative dans le disque*, Comment. Math. Helv. **65**(1990), 334–347.
23. L. Pestov, V. Sharafutdinov, *Integral geometry of tensor fields on a manifold of negative curvature (Russian)* Sibirsk. Mat. Zh. **29**(1988), no. 3, 114–130; translation in Siberian Math. J. **29**(1988), no. 3, 427–441.
24. L. Pestov and G. Uhlmann, *Two dimensional simple compact manifolds with boundary are boundary rigid*, *Annals of Math.* **161**(2005), 1089-1106.
25. V. Sharafutdinov, *Integral geometry of tensor fields*, VSP, Utrech, the Netherlands, 1994.
26. V. Sharafutdinov, M. Skokan, and G. Uhlmann, *Regularity of ghosts in tensor tomography, Journal of Geometric Analysis* textbf15(2005), 517-560.
27. V. Sharafutdinov and G. Uhlmann, *On deformation boundary rigidity and spectral rigidity for Riemannian surfaces with no focal points*, Journal of Differential Geometry, **56** (2001), 93–110.
28. J. Sjöstrand, Singularités analytiques microlocales, Astérique **95**(1982), 1–166.
29. P. Stefanov and G.Uhlmann, *Rigidity for metrics with the same lengths of geodesics*, Math. Res. Lett. **5**(1998), 83–96.
30. ———, *Stability estimates for the X-ray transform of tensor fields and boundary rigidity*, Duke Math. J. **123**(2004), 445–467.
31. ———, *Stable determination of generic simple metrics from the hyperbolic Dirichlet-to-Neumann map*, IMRN **17**(2005), 1047–1061.
32. ———, *Boundary rigidity and stability for generic simple metrics*, Journal Amer. Math. Soc. **18**(2005), 975–1003.
33. ———, *Integral geometry of tensor fields on a class of non-simple Riemannian manifolds*, arXiv:math.DG/0601178, to appear in Amer. J. Math.
34. ———, *Local lens rigidity with incomplete data for a class of non-simple Riemannian manifolds*, in progress.
35. D. Tataru, *Unique continuation for solutions to PDE's; between Hörmander's theorem and Holmgren's theorem*, Comm. P.D.E. **20**(1995), 855–884.
36. F. Treves, *Introduction to Pseudodifferential and Fourier Integral Operators, Vol. 1. Pseudodifferential Operators.* The University Series in Mathematics, Plenum Press, New York–London, 1980.
37. J. Wang, *Stability for the reconstruction of a Riemannian metric by boundary measurements*, Inverse Probl. **15**(1999), 1177–1192.

38. E. Wiechert E and K. Zoeppritz, *Uber Erdbebenwellen*, Nachr. Koenigl. Geselschaft Wiss, Goettingen **4**(1907), 415–549.

Coupling of two partial differential equations and its application

Hidetoshi Tahara

Department of Mathematics, Sophia University, Kioicho, Chiyoda-ku, Tokyo 102-8554, Japan
h-tahara@hoffman.cc.sophia.ac.jp

Summary. The paper considers the coupling of the following two nonlinear partial differential equations

$$\frac{\partial u}{\partial t} = F\Big(t, x, u, \frac{\partial u}{\partial x}\Big) \quad \text{and} \quad \frac{\partial w}{\partial t} = G\Big(t, x, w, \frac{\partial w}{\partial x}\Big),$$

and establishes the equivalence of them. The result is applied to the problem of analytic continuation of the solution.

Key words: coupling equation, equivalence of two PDEs, analytic continuation

1 Introduction

In this paper, I will present a new approach to the study of nonlinear partial differential equations in the complex domain. Since my reasearch is still in the first stage, as a model study I will discuss only the following partial differential equations:

$$\text{(A)} \qquad \frac{\partial u}{\partial t} = F\Big(t, x, u, \frac{\partial u}{\partial x}\Big)$$

(where $(t, x) \in \mathbb{C}^2$ are variables and $u = u(t, x)$ is the unknown function) and

$$\text{(B)} \qquad \frac{\partial w}{\partial t} = G\Big(t, x, w, \frac{\partial w}{\partial x}\Big)$$

(where $(t, x) \in \mathbb{C}^2$ are variables and $w = w(t, x)$ is the unknown function). For simplcity we suppose that $F(t, x, u_0, u_1)$ (resp. $G(t, x, w_0, w_1)$) is a holomorphic function defined in a neighborhood of the origin of $\mathbb{C}_t \times \mathbb{C}_x \times \mathbb{C}_{u_0} \times \mathbb{C}_{u_1}$ (resp. $\mathbb{C}_t \times \mathbb{C}_x \times \mathbb{C}_{w_0} \times \mathbb{C}_{w_1}$).

My basic question is

Received 3 December, 2005. Revised 8 March, 2006. Accepted 8 March, 2006.

Question. *When can we say that the two equations* (A) *and* (B) *are equivalent? or when can we transform* (A) *into* (B) (*or* (B) *into* (A))*?*

One way to treat these questions is to consider the coupling of (A) and (B), and solve their coupling equation: I will do this in sections 3 and 4.

In the case of ordinary differential equations, the theory of the coupling of two differential equations was discussed in section 4.1 of Gérard–Tahara [1]; it will be summarized in the next section 2.

2 Coupling of two ordinary differential equations

First, let us consider the following two differential equations:

$$\text{(a)} \qquad \frac{du}{dt} = f(t,u),$$

$$\text{(b)} \qquad \frac{dw}{dt} = g(t,w).$$

Definition 2.1. *The coupling of* (a) *and* (b) *means that we consider the following partial differential equation* (2.1) *or* (2.2):

$$\frac{\partial \phi}{\partial t} + f(t,u)\frac{\partial \phi}{\partial u} = g(t,\phi) \tag{2.1}$$

(*where* (t,u) *are variables and* $\phi = \phi(t,u)$ *is the unknown function*), *or*

$$\frac{\partial \psi}{\partial t} + g(t,w)\frac{\partial \psi}{\partial w} = f(t,\psi) \tag{2.2}$$

(*where* (t,w) *are variables and* $\psi = \psi(t,w)$ *is the unknown function*). *We call* (2.1) *or* (2.2) *the coupling equation of* (a) *and* (b).

The convenience of considering the coupling equation lies in the following proposition.

Proposition 2.2. (1) *Let* $\phi(t,u)$ *be a solution of* (2.1). *If* $u(t)$ *is a solution of* (a) *then* $w(t) = \phi(t,u(t))$ *is a solution of* (b).

(2) *Let* $\psi(t,w)$ *be a solution of* (2.2). *If* $w(t)$ *is a solution of* (b) *then* $u(t) = \psi(t,w(t))$ *is a solution of* (a).

Proof. We will prove only (1). Let $\phi(t,u)$ be a solution of (2.1) and let $u(t)$ be a solution of (a). Set $w(t) = \phi(t,u(t))$. Then we have

$$\begin{aligned}\frac{dw(t)}{dt} &= \frac{d}{dt}\phi(t,u(t)) = \frac{\partial \phi}{\partial t}(t,u(t)) + \frac{\partial \phi}{\partial u}(t,u(t))\frac{du(t)}{dt} \\ &= \frac{\partial \phi}{\partial t}(t,u(t)) + f(t,u(t))\frac{\partial \phi}{\partial u}(t,u(t)) = g(t,\phi(t,u(t))) = g(t,w(t)). \qquad \square\end{aligned}$$

Next, let us give a relation between two coupling equations (2.1) and (2.2). We have

Proposition 2.3. (1) *Let $\phi(t,u)$ be a solution of* (2.1) *and suppose that the relation $w=\phi(t,u)$ is equivalent to $u=\psi(t,w)$ for some function $\psi(t,w)$, then $\psi(t,w)$ is a solution of* (2.2).

(2) *Let $\psi(t,w)$ be a solution of* (2.2) *and suppose that the relation $u=\psi(t,w)$ is equivalent to $w=\phi(t,u)$ for some function $\phi(t,u)$, then $\phi(t,u)$ is a solution of* (2.1).

Proof. We will show only the part (1). Since $w=\phi(t,u)$ is equivalent to $u=\psi(t,w)$, we get $u\equiv\psi(t,\phi(t,u))$. By derivating this with respect to t and u we get

$$0\equiv\frac{\partial\psi}{\partial t}(t,\phi(t,u))+\frac{\partial\psi}{\partial w}(t,\phi(t,u))\frac{\partial\phi}{\partial t}(t,u),$$
$$1\equiv\frac{\partial\psi}{\partial w}(t,\phi(t,u))\frac{\partial\phi}{\partial u}(t,u).$$

By using these relations we obtain

$$\begin{aligned}\Big(\frac{\partial\psi}{\partial t}+g(t,w)\frac{\partial\psi}{\partial w}\Big)\Big|_{w=\phi(t,u)}&=\frac{\partial\psi}{\partial t}(t,\phi(t,u))+g(t,\phi(t,u))\frac{\partial\psi}{\partial w}(t,\phi(t,u))\\ &=-\frac{\partial\psi}{\partial w}(t,\phi(t,u))\frac{\partial\phi}{\partial t}(t,u)+g(t,\phi(t,u))\frac{\partial\psi}{\partial w}(t,\phi(t,u))\\ &=\frac{\partial\psi}{\partial w}(t,\phi(t,u))\Big(-\frac{\partial\phi}{\partial t}(t,u)+g(t,\phi(t,u))\Big)\\ &=\frac{\partial\psi}{\partial w}(t,\phi(t,u))f(t,u)\frac{\partial\phi}{\partial u}(t,u)=f(t,u).\end{aligned}$$

Since $w=\phi(t,u)$ is equivalent to $u=\psi(t,w)$, this proves the result (1). □

Now, let us discuss the equivalence of two differential equations. Let $f(t,z)$ and $g(t,z)$ be functions of (t,z) in a neighborhood of (0,0), and let us consider the following two equations:

[a] $$\frac{du}{dt}=f(t,u),\quad u(t)\longrightarrow 0\ \ (\text{as } t\longrightarrow 0),$$

[b] $$\frac{dw}{dt}=g(t,w),\quad w(t)\longrightarrow 0\ \ (\text{as } t\longrightarrow 0).$$

Denote by $\mathcal{S}_a$ (resp. $\mathcal{S}_b$) the set of all solutions of [a] (resp. [b]) in a suitable sectorial neighborhood of $t=0$. If $\phi(t,u)$ is a solution of (2.1) satisfying $\phi(0,0)=0$ and if $u(t)\in\mathcal{S}_a$, then we have $\phi(t,u(t))\longrightarrow\phi(0,0)=0$ (as $t\longrightarrow 0$) and therefore the mapping

$$\Phi\ :\ \mathcal{S}_a\ni u(t)\longrightarrow\phi(t,u(t))\in\mathcal{S}_b$$

is well defined. Hence, if $\Phi:\mathcal{S}_a\longrightarrow\mathcal{S}_b$ is bijective, solving [a] is equivalent to solving [b]. Thus:

Definition 2.4. *If $\Phi:\mathcal{S}_a\longrightarrow\mathcal{S}_b$ is well defined and bijective, then we say that the two equations* [a] *and* [b] *are equivalent.*

The following result gives a sufficient condition for Φ to be bijective.

Theorem 2.5. *If the coupling equation* (2.1) *has a solution* $\phi(t,u)$ *in a neighborhood of* $(0,0)$ *satisfying* $\phi(0,0)=0$ *and* $(\partial\phi/\partial u)(0,0)\neq 0$, *then the mapping* Φ *is bijective and so the two equations* [a] *and* [b] *are equivalent.*

Proof. Let us show that the mapping $\Phi : \mathcal{S}_a \longrightarrow \mathcal{S}_b$ is bijective. Since $(\partial\phi/\partial u)(0,0)\neq 0$, by the implicit function theorem we can solve the equation $w=\phi(t,u)$ with respect to u and obtain $u=\psi(t,w)$ for some function $\psi(t,w)$ satisfying $\psi(0,0)=0$. Moreover we know that the above $\psi(t,w)$ is a solution of (2.2). Therefore, by (2) of Proposition 2.2 we can define the mapping

$$\Psi \ : \ \mathcal{S}_b \ni w(t) \longrightarrow \psi(t,w(t)) \in \mathcal{S}_a.$$

Since $w=\phi(t,u)$ is equivalent to $u=\psi(t,w)$, $w(t)=\phi(t,u(t))$ is also equivalent to $u(t)=\psi(t,w(t))$; this implies that $w(t)=\Phi(u(t))$ is equivalent to $u(t)=\Psi(w(t))$. Thus we can conclude that $\Phi : \mathcal{S}_a \longrightarrow \mathcal{S}_b$ is bijective and Ψ is the inverse mapping of Φ. □

3 Coupling of two partial differential equations

In this section we will present a formal theory of the coupling of the following two nonlinear partial differential equations:

$$\text{(A)} \qquad \frac{\partial u}{\partial t} = F\Big(t,x,u,\frac{\partial u}{\partial x}\Big)$$

(where $(t,x)\in\mathbb{C}^2$ are variables and $u=u(t,x)$ is the unknown function) and

$$\text{(B)} \qquad \frac{\partial w}{\partial t} = G\Big(t,x,w,\frac{\partial w}{\partial x}\Big)$$

(where $(t,x)\in\mathbb{C}^2$ are variables and $w=w(t,x)$ is the unknown function). For simplcity we suppose that $F(t,x,u_0,u_1)$ (resp. $G(t,x,w_0,w_1)$) is a holomorphic function defined in a neighborhood of the origin of $\mathbb{C}_t\times\mathbb{C}_x\times\mathbb{C}_{u_0}\times\mathbb{C}_{u_1}$ (resp. $\mathbb{C}_t\times\mathbb{C}_x\times\mathbb{C}_{w_0}\times\mathbb{C}_{w_1}$).

In order to express the coupling equation, let us introduce a vector field D with infinitely many variables $(x,u_0,u_1,\ldots)$ (resp. $(x,w_0,w_1,\ldots)$):

$$D=\frac{\partial}{\partial x}+\sum_{i\geq 0}u_{i+1}\frac{\partial}{\partial u_i} \quad \Big(\text{resp. } D=\frac{\partial}{\partial x}+\sum_{i\geq 0}w_{i+1}\frac{\partial}{\partial w_i}\Big).$$

This comes from the following formula: if $K(t,x,u_0,u_1,\ldots)$ is a function with infinitely many variables $(t,x,u_0,u_1,\ldots)$ and if $u(x)$ is a holomorphic function, then under the relation $u_i=\partial^i u/\partial x^i$ $(i=0,1,2,\ldots)$ we have

$$\frac{\partial}{\partial x}K\Big(t,x,u,\frac{\partial u}{\partial x},\ldots\Big)=\frac{\partial K}{\partial x}+\frac{\partial K}{\partial u_0}u_1+\frac{\partial K}{\partial u_1}u_2+\cdots=D[K].$$

Therefore, for any $m \in \mathbb{N}$ we have

$$D^m[K]\Big(t,x,u,\frac{\partial u}{\partial x},\dots\Big) = \Big(\frac{\partial}{\partial x}\Big)^m \Big[K\Big(t,x,u,\frac{\partial u}{\partial x},\dots\Big)\Big].$$

Definition 3.1. *The coupling of two partial differential equations* (A) *and* (B) *means that we consider the following partial differential equation with infinitely many variables* $(t, x, u_0, u_1, \dots)$

$$(\Phi) \quad \frac{\partial \phi}{\partial t} + \sum_{m\geq 0} D^m[F](t,x,u_0,\dots,u_{m+1})\frac{\partial \phi}{\partial u_m} = G\big(t,x,\phi,D[\phi]\big)$$

(*where* $\phi = \phi(t, x, u_0, u_1, \dots)$ *is the unknown function*), *or the following partial differential equation with infinitely many variables* $(t, x, w_0, w_1, \dots)$

$$(\Psi) \quad \frac{\partial \psi}{\partial t} + \sum_{m\geq 0} D^m[G](t,x,w_0,\dots,w_{m+1})\frac{\partial \psi}{\partial w_m} = F\big(t,x,\psi,D[\psi]\big)$$

(*where* $\psi = \psi(t, x, w_0, w_1, \dots)$ *is the unknown function*).

3.1 The formal meaning of the coupling equation

Let us explain the meaning of the coupling equation (Φ) or (Ψ) in the formal sense. Here, "*in the formal sense*" means that the result is true if the formal calculation makes sense.

The convenience of considering the coupling equation lies in the following proposition.

Proposition 3.2. (1) *If* $\phi(t, x, u_0, u_1, \dots)$ *is a solution of* (Φ) *and if* $u(t, x)$ *is a solution of* (A), *then the function* $w(t, x) = \phi(t, x, u, \partial u/\partial x, \dots)$ *is a solution of* (B).

(2) *If* $\psi(t, x, w_0, w_1, \dots)$ *is a solution of* (Ψ) *and if* $w(t, x)$ *is a solution of* (B), *then the function* $u(t, x) = \phi(t, x, w, \partial w/\partial x, \dots)$ *is a solution of* (A).

Proof. Let us show (1). Set $u_i(t,x) = (\partial/\partial x)^i u(t,x)$ $(i = 0, 1, 2, \dots)$: then we have $w(t,x) = \phi(t,x,u,\partial u/\partial x,\dots) = \phi(t,x,u_0,u_1,\dots)$ and $\partial w/\partial x = D[\phi](t,x,u_0,u_1,\dots)$. Since $u(t,x)$ is a solution of (A) and since $\phi(t,x,u_0, u_1,\dots)$ is a solution of (Φ), we obtain

$$\begin{aligned}\frac{\partial w}{\partial t} &= \frac{\partial \phi}{\partial t} + \sum_{i\geq 0}\frac{\partial \phi}{\partial u_i}\frac{\partial u_i}{\partial t} = \frac{\partial \phi}{\partial t} + \sum_{i\geq 0}\frac{\partial \phi}{\partial u_i}\Big(\frac{\partial}{\partial x}\Big)^i\Big[\frac{\partial u}{\partial t}\Big] \\ &= \frac{\partial \phi}{\partial t} + \sum_{i\geq 0}\frac{\partial \phi}{\partial u_i}\Big(\frac{\partial}{\partial x}\Big)^i\Big[F\Big(t,x,u,\frac{\partial u}{\partial x}\Big)\Big] \\ &= \frac{\partial \phi}{\partial t} + \sum_{i\geq 0}\frac{\partial \phi}{\partial u_i}D^i[F](t,x,u_0,\dots,u_{i+1}) \\ &= G\big(t,x,\phi,D[\phi]\big) = G\Big(t,x,w,\frac{\partial w}{\partial x}\Big). \qquad \square\end{aligned}$$

In order to state the relation between (Φ) and (Ψ), let us introduce the notion of the reversibility of $\phi(t,x,u_0,u_1,\ldots)$.

Definition 3.3. *Let $\phi(t,x,u_0,u_1,\ldots)$ be a function in $(t,x,u_0,u_1,\ldots)$. We say that the relation $w=\phi(t,x,u,\partial u/\partial x,\ldots)$ is reversible with respect to u and w if there is a function $\psi(t,x,w_0,w_1,\ldots)$ in $(t,x,w_0,w_1,\ldots)$ such that the relation*

$$\begin{cases} w_0 = \phi(t,x,u_0,u_1,u_2,\ldots), \\ w_1 = D[\phi](t,x,u_0,u_1,u_2,\ldots), \\ w_2 = D^2[\phi](t,x,u_0,u_1,u_2,\ldots), \\ \quad\cdots\cdots \\ \quad\cdots\cdots \end{cases} \tag{3.1}$$

is equivalent to

$$\begin{cases} u_0 = \psi(t,x,w_0,w_1,w_2,\ldots), \\ u_1 = D[\psi](t,x,w_0,w_1,w_2,\ldots), \\ u_2 = D^2[\psi](t,x,w_0,w_1,w_2,\ldots), \\ \quad\cdots\cdots \\ \quad\cdots\cdots . \end{cases} \tag{3.2}$$

In this case the function $\psi(t,x,w_0,w_1,\ldots)$ is called the reverse function of $\phi(t,x,u_0,u_1,\ldots)$.

Then we have the following result which says that the equation (Ψ) is the reverse of (Φ).

Proposition 3.4. *If $\phi(t,x,u_0,u_1,\ldots)$ is a solution of (Φ) and if the relation $w=\phi(t,x,u,\partial u/\partial x,\ldots)$ is reversible with respect to u and w, then the reverse function $\psi(t,x,w_0,w_1,\ldots)$ is a solution of (Ψ).*

3.2 Equivalence of (A) and (B)

Let $\mathcal{F}$ and $\mathcal{G}$ be function spaces in which we can consider the following two partial differential equations:

$$\text{[A]} \qquad \frac{\partial u}{\partial t} = F\Big(t,x,u,\frac{\partial u}{\partial x}\Big) \quad \text{in } \mathcal{F},$$

$$\text{[B]} \qquad \frac{\partial w}{\partial t} = G\Big(t,x,w,\frac{\partial w}{\partial x}\Big) \quad \text{in } \mathcal{G}.$$

Set

$$\mathcal{S}_A = \text{the set of all solutions of [A] in } \mathcal{F},$$
$$\mathcal{S}_B = \text{the set of all solutions of [B] in } \mathcal{G}.$$

Then, if the coupling equation (Φ) has a solution $\phi(t,x,u_0,u_1\ldots)$ and if $\phi(t,x,u,\partial u/\partial x,\ldots)\in\mathcal{G}$ is well defined for any $u\in\mathcal{S}_A$, then we can define the mapping

$$\Phi : \mathcal{S}_A \ni u(t,x) \longmapsto w(t,x) = \phi(t,x,u,\partial u/\partial x,\dots) \in \mathcal{S}_B. \tag{3.3}$$

If the relation $w = \phi(t,x,u,\partial u/\partial x,\dots)$ is reversible with respect to u and w, and if the reverse function $\psi(t,x,w_0,w_1,\dots)$ satisfies $\psi(t,x,w,\partial w/\partial x,\dots) \in \mathcal{F}$ for any $w \in \mathcal{S}_B$, then we can also define the mapping

$$\Psi : \mathcal{S}_B \ni w(t,x) \longmapsto u(t,x) = \psi(t,x,w,\partial w/\partial x,\dots) \in \mathcal{S}_A. \tag{3.4}$$

Since (3.1) is equivalent to (3.2) we have $\Psi \circ \Phi = \mathit{identity}$ in $\mathcal{S}_A$ and $\Phi \circ \Psi = \mathit{identity}$ in $\mathcal{S}_B$. Thus, we obtain

Theorem 3.5. *Suppose that the coupling equation* (Φ) *has a solution* $\phi(t,x,u_0,u_1\dots)$ *and that the relation* $w = \phi(t,x,u,\partial u/\partial x,\dots)$ *is reversible with respect to* u *and* w. *If both mappings* (3.3) *and* (3.4) *are well defined, we can conclude that the both mappings are bijective and that one is the inverse of the other.*

By this theorem, we may say:

Definition 3.6. (1) *If the coupling equation* (Φ) *(resp.* (Ψ)*) has a solution* $\phi(t,x,u_0,u_1\dots)$ *(resp.* $\psi(t,x,w_0,w_1\dots)$*) and if the relation* $w = \phi(t,x,u,\partial u/\partial x,\dots)$ *(resp.* $u = \psi(t,x,w,\partial w/\partial x,\dots)$*) is reversible with respect to* u *and* w *(or* w *and* u*), then we say that the two equations* (A) *and* (B) *are formally equivalent.*

(2) *In addition, if both mappings* (3.3) *and* (3.4) *are well defined, we say that the two equations* [A] *and* [B] *are equivalent.*

3.3 A sufficient condition for the reversibility

As is seen above, the condition of the reversibility of $\phi(t,x,u_0,u_1,\dots)$ is very important. In this section we will give a sufficient condition for the relation $w = \phi(t,x,u,\partial u/\partial x,\dots)$ to be reversible.

Let us introduce the notations: $D_R = \{x \in \mathbb{C}; |x| \le R\}$, $\mathcal{O}_R$ denotes the ring of holomorphic functions in a neighborhood of D_R, and $\mathcal{O}_R[[u_0,\dots,u_p]]$ denotes the ring of formal power series in $(u_0,\dots,u_p)$ with coefficients in $\mathcal{O}_R$.

Proposition 3.7. *If* $\phi(t,x,u_0,u_1,\dots)$ *is of the form*

$$\phi = u_0 + \sum_{k\ge1} \phi_k(x,u_0,\dots,u_k)\, t^k \in \sum_{k\ge0} \mathcal{O}_R[[u_0,\dots,u_k]]\, t^k, \tag{3.5}$$

the relation $w = \phi(t,x,u,\partial u/\partial x,\dots)$ *is reversible with respect to* u *and* w, *and the reverse function* $\psi(t,x,w_0,w_1,\dots)$ *is also of the form*

$$\psi = w_0 + \sum_{k\ge1} \psi_k(x,w_0,\dots,w_k)\, t^k \in \sum_{k\ge0} \mathcal{O}_R[[w_0,\dots,w_k]]\, t^k. \tag{3.6}$$

4 Equivalence of two partial differential equations

Let $(t,x) \in \mathbb{C}_t \times \mathbb{C}_x$ be the variables, and let $F(t,x,z_1,z_2)$ be a holomorphic function defined in a neighborhood of the origin of $\mathbb{C}_t \times \mathbb{C}_x \times \mathbb{C}_{z_1} \times \mathbb{C}_{z_2}$. In this section we will establish the equivalence of the two equations

$$\frac{\partial u}{\partial t} = F\Big(t,x,u,\frac{\partial u}{\partial x}\Big) \tag{4.1}$$

and

$$\frac{\partial w}{\partial t} = 0. \tag{4.2}$$

4.1 Formal equivalence of (4.1) and (4.2)

The coupling equation of (4.1) and (4.2) is

$$(\Phi) \qquad \frac{\partial \phi}{\partial t} + \sum_{m\ge 0} D^m[F](t,x,u_0,\dots,u_{m+1})\frac{\partial \phi}{\partial u_m} = 0, \quad \text{or}$$

$$(\Psi) \qquad \frac{\partial \psi}{\partial t} = F\big(t,x,\psi,D[\psi]\big).$$

Let us find a formal solution of this equation (Φ) or (Ψ). We have the following result.

Proposition 4.1. (1) *The coupling equation* (Φ) *has a unique formal solution of the form*

$$\phi = u_0 + \sum_{k\ge 1} \phi_k(x,u_0,\dots,u_k)\, t^k \in \sum_{k\ge 0} \mathcal{O}_R[[u_0,\dots,u_k]]\, t^k. \tag{4.3}$$

Moreover we have the following properties: i) $\phi_1(x,u_0,u_1) = -F(0,x,u_0,u_1)$, ii) $\phi_k(x,u_0,\dots,u_k)$ *(for* $k \ge 1$*) is a holomorphic function in a neighborhood of* $\{(x,u_0,\dots,u_k) \in \mathbb{C} \times \mathbb{C}^{k+1}\,;\, |x| \le R, |u_0| \le \rho$ *and* $|u_1| \le \rho\}$ *for some* $R > 0$ *and* $\rho > 0$ *which are independent of* k, *and* iii) $\phi_k(x,u_0,\dots,u_k)$ *(for* $k \ge 2$*) is a polynomial with respect to* $(u_2,\dots,u_k)$.

(2) *The coupling equation* (Ψ) *has a unique formal solution of the form*

$$\psi = w_0 + \sum_{k\ge 1} \psi_k(x,w_0,\dots,w_k)\, t^k \in \sum_{k\ge 0} \mathcal{O}_R[[w_0,\dots,w_k]]\, t^k. \tag{4.4}$$

Moreover we have the following properties: i) $\psi_1(x,w_0,w_1) = F(0,x,w_0,w_1)$, ii) $\psi_k(x,w_0,\dots,w_k)$ *(for* $k \ge 1$*) is a holomorphic function in a neighborhood of* $\{(x,w_0,\dots,w_k) \in \mathbb{C} \times \mathbb{C}^{k+1}\,;\, |x| \le R, |w_0| \le \rho$ *and* $|w_1| \le \rho\}$ *for some* $R > 0$ *and* $\rho > 0$ *which are independent of* k, *and* iii) $\psi_k(x,w_0,\dots,w_k)$ *(for* $k \ge 2$*) is a polynomial with respect to* $(w_2,\dots,w_k)$.

By this result, we have

Corollary 4.2. *The two equations* (4.1) *and* (4.2) *are formally equivalent.*

4.2 Convergence of ϕ and ψ

Let us show the convergence of the formal solution $\phi(t, x, u_0, u_1, \ldots)$ and $\psi(t, x, w_0, w_1, \ldots)$ in Proposition 4.1.

We set

$$U_k(R, \varepsilon, \eta) = \big\{(x, u_0, \ldots, u_k) \in \mathbb{C} \times \mathbb{C}^{k+1};\ |x| \leq R,\ |u_0| \leq 0!\varepsilon/\eta^0, |u_1| \leq 1!\varepsilon/\eta^1, \ldots, |u_k| \leq k!\varepsilon/\eta^k\big\}, \quad \text{and}$$

$$W_k(R, \varepsilon, \eta) = \big\{(x, w_0, \ldots, w_k) \in \mathbb{C} \times \mathbb{C}^{k+1};\ |x| \leq R,\ |w_0| \leq 0!\varepsilon/\eta^0, |w_1| \leq 1!\varepsilon/\eta^1, \ldots, |w_k| \leq k!\varepsilon/\eta^k\big\}$$

($k = 1, 2, \ldots$). By Proposition 4.1 and by taking $R > 0$, $\varepsilon > 0$, $\eta > 0$ so that R, ε and ε/η are sufficiently small we may suppose that each $\phi_k(x, u_0, \ldots, u_k)$ and $\psi_k(x, w_0, \ldots, w_k)$ are holomorphic functions on $U_k = U_k(R, \varepsilon, \eta)$ and $W_k = W_k(R, \varepsilon, \eta)$, respectively; hence

$$\|\phi_k\|_{U_k} = \max_{U_k} \big|\phi_k(x, u_0, \ldots, u_k)\big| < \infty, \quad \text{and}$$

$$\|\psi_k\|_{W_k} = \max_{W_k} \big|\psi_k(x, w_0, \ldots, w_k)\big| < \infty.$$

We have

Proposition 4.3. *Let $R > 0$ be sufficiently small. Then, for any $\eta > 0$ we can find an $\varepsilon > 0$ such that the series*

$$\sum_{k \geq 1} \|\phi_k\|_{U_k} z^k \quad \text{and} \quad \sum_{k \geq 1} \|\psi_k\|_{W_k} z^k$$

(with $U_k = U_k(R, \varepsilon, \eta)$ and $W_k = W_k(R, \varepsilon, \eta)$) are convergent in a neighborhood of $z = 0 \in \mathbb{C}$.

If $u(t, x)$ is a holomorphic function in a neighborhood of D_r, by Cauchy's inequality we have

$$\left\| \frac{\partial^k u}{\partial x^k} \right\|_{r/2} \leq \frac{k! M}{(r/2)^k} \quad (\text{with } M = \|u\|_r), \quad k = 0, 1, 2, \ldots.$$

Therefore, we see:

Corollary 4.4. (1) *If $u(t, x)$ is a holomorphic function in a neighborhood of $(0, 0) \in \mathbb{C}_t \times \mathbb{C}_x$ and if $|u(t, x)|$ is sufficiently small, the function $\phi(t, x, u, \partial u/\partial x, \ldots)$ is well defined and it is also a holomorphic function in a neighborhood of $(0, 0) \in \mathbb{C}_t \times \mathbb{C}_x$.*

(2) *If $w(t, x)$ is a holomorphic function in a neighborhood of $(0, 0) \in \mathbb{C}_t \times \mathbb{C}_x$ and if $|w(t, x)|$ is sufficiently small, the function $\psi(t, x, w, \partial w/\partial x, \ldots)$ is well defined and it is also a holomorphic function in a neighborhood of $(0, 0) \in \mathbb{C}_t \times \mathbb{C}_x$.*

4.3 Equivalence result (1)

Now, let us establish the equivalence of two equations (4.1) and (4.2). Let $r > 0$, $R > 0$ and $\rho > 0$. We suppose:

(C) $F(t,x,z_1,z_2)$ is a holomorphic function on $\{(t,x,z_1,z_2) \in \mathbb{C}^4\,;\ |t| < r, |x| < R, |z_1| < \rho, |z_2| < \rho\}$.

Definition 4.5. (1) *We denote by $\mathscr{X}$ the set of all the functions $w(t,x)$ satisfying the following properties:*

i) *$w(t,x)$ is a holomorphic function on $S_\theta(r_1) \times D_{R_1}$ for some $\theta > 0$, $0 < r_1 < r$ and $0 < R_1 < R$, and*

ii) *$|w(t,x)| \leq \rho_1$ and $|(\partial w/\partial x)(t,x)| \leq \rho_1$ on $S_\theta(r_1) \times D_{R_1}$ for some $0 < \rho_1 < \rho$.*

In the above i) *and* ii) *we used: $S_\theta(r) = \{t \in \mathbb{C}\,;\ 0 < |t| \leq r,\ |\arg t| < \theta\}$.*

(2) *We denote by $\mathscr{H}$ the set of all the functions $w(t,x)$ satisfying the following properties:*

i) *$w(t,x)$ is a holomorphic function on $D_{r_1} \times D_{R_1}$ for some $0 < r_1 < r$, and $0 < R_1 < R$, and*

ii) *$|w(t,x)| \leq \rho_1$ and $|(\partial w/\partial x)(t,x)| \leq \rho_1$ on $D_{r_1} \times D_{R_1}$ for some $0 < \rho_1 < \rho$.*

Then we have

Theorem 4.6. *Suppose the condition* (C). *The following two equations are equivalent:*

$$\frac{\partial u}{\partial t} = F\Big(t,x,u,\frac{\partial u}{\partial x}\Big) \quad \text{in } \mathscr{X} \text{ (resp. in } \mathscr{H}\text{)}, \tag{4.5}$$

$$\frac{\partial w}{\partial t} = 0 \quad \text{in } \mathscr{X} \text{ (resp. in } \mathscr{H}\text{)}. \tag{4.6}$$

This follows from Theorem 3.5 and

Proposition 4.7. (1) *Let $\phi(t,x,u_0,u_1,\ldots)$ be the solution of (Φ) in (4.3). If $u \in \mathscr{X}$ (resp. $u \in \mathscr{H}$), then we have $\phi(t,x,u,\partial u/\partial x,\ldots) \in \mathscr{X}$ (resp. $\phi(t,x,u,\partial u/\partial x,\ldots) \in \mathscr{H}$).*

(2) *Let $\psi(t,x,w_0,w_1,\ldots)$ be the solution of (Ψ) in (4.4). If $w \in \mathscr{X}$ (resp. $w \in \mathscr{H}$), then we have $\psi(t,x,w,\partial w/\partial x,\ldots) \in \mathscr{X}$ (resp. $\psi(t,x,w,\partial w/\partial x,\ldots) \in \mathscr{H}$).*

5 Application

Let ω be an open neighborhood of the origin of $\mathbb{C}_t \times \mathbb{C}_x$. In this section we suppose

(A) $F(t,x,z_1,z_2)$ is a holomorphic function on $\omega \times \mathbb{C}_{z_1} \times \mathbb{C}_{z_2}$

and consider the equation

$$\frac{\partial u}{\partial t} = F\Big(t,x,u,\frac{\partial u}{\partial x}\Big). \tag{5.1}$$

By the Taylor expansion in (z_1,z_2) we can express $F(t,x,z_1,z_2)$ in the form

$$F(t,x,z_1,z_2) = \sum_{(j,\alpha)\in\Delta} a_{j,\alpha}(t,x)\, {z_1}^j\, {z_2}^\alpha,$$

where $(j,\alpha) \in \mathbb{N}\times\mathbb{N}$, Δ is a subset of $\mathbb{N}\times\mathbb{N}$, and $a_{j,\alpha}(t,x)$ are holomorphic functions on ω. Without loss of generality we may suppose that $a_{j,\alpha}(t,x) \not\equiv 0$ for any $(j,\alpha) \in \Delta$; then we can write $a_{j,\alpha}(t,x) = t^{k_{j,\alpha}}\, b_{j,\alpha}(t,x)$, where $k_{j,\alpha}$ is a non-negative integer and $b_{j,\alpha}(0,x) \not\equiv 0$. Using the above, the function $F(t,x,z_1,z_2)$ may now be written as

$$F(t,x,z_1,z_2) = \sum_{(j,\alpha)\in\Delta} t^{k_{j,\alpha}} b_{j,\alpha}(t,x)\, {z_1}^j {z_2}^\alpha.$$

Set $\Delta_2 = \{(j,\alpha) \in \Delta\,;\, j+\alpha \geq 2\}$. We remark that the equation (5.1) is linear if and only if $\Delta_2 = \emptyset$; it is nonlinear otherwise. We assume $\Delta_2 \neq \emptyset$, and we define the index σ by

$$\sigma = \sup_{(j,\alpha)\in\Delta_2} \frac{-k_{j,\alpha}-1}{j+\alpha-1}, \tag{5.2}$$

which was introduced by Kobayashi [3]. Note that this σ is a non-positive real number.

5.1 Equivalence result (2)

Let σ be as above, and define

Definition 5.1. *We denote by $\mathscr{S}_\sigma$ the set of all the functions $w(t,x)$ satisfying the following properties:*

i) *$w(t,x)$ is a holomorphic function on $S_\theta(r) \times D_R$ for some $\theta > 0$, $r > 0$ and $R > 0$, and*

ii) *$\|w(t)\|_R = o(|t|^\sigma)$ (as $t \longrightarrow 0$ in $S_\theta(r)$).*

Then we have

Theorem 5.2. *Suppose the conditions* (A) *and $\Delta_2 \neq \emptyset$; let σ be the one in* (5.2)*. The following two equations are equivalent:*

$$\frac{\partial u}{\partial t} = F\Big(t,x,u,\frac{\partial u}{\partial x}\Big) \quad \text{in } \mathscr{S}_\sigma, \tag{5.3}$$

$$\frac{\partial w}{\partial t} = 0 \quad \text{in } \mathscr{S}_\sigma. \tag{5.4}$$

This follows from Theorem 3.5 and

Proposition 5.3. *Suppose the conditions* (A) *and* $\Delta_2 \neq \emptyset$*; let* σ *be the one in* (5.2). *We have:*

(1) *Let* $\phi(t, x, u_0, u_1, \ldots)$ *be the solution of* (Φ) *in* (4.3). *If* $u \in \mathscr{S}_\sigma$*, then we have* $\phi(t, x, u, \partial u/\partial x, \ldots) \in \mathscr{S}_\sigma$.

(2) *Let* $\psi(t, x, w_0, w_1, \ldots)$ *be the solution of* (Ψ) *in* (4.4). *If* $w \in \mathscr{S}_\sigma$*, then we have* $\psi(t, x, w, \partial w/\partial x, \ldots) \in \mathscr{S}_\sigma$.

5.2 Analytic continuation

Let us give an application. Let $u(t,x) \in \mathscr{S}_\sigma$ be a solution of the equation

$$\frac{\partial u}{\partial t} = F\Big(t, x, u, \frac{\partial u}{\partial x}\Big).$$

Set $w(t,x) = \phi(t, x, u, \partial u/\partial x, \ldots)$. Then, $w(t,x) \in \mathscr{S}_\sigma$ is a solution of

$$\frac{\partial w}{\partial t} = 0$$

and so $w(t,x)$ can be expressed in the form $w(t,x) = h(x)$ for some holomorphic function $h(x)$ in a neighborhood of $x = 0$. Since $w(t,x) = h(x) \in \mathscr{H}$ holds, by the reversibility we have $u(t,x) = \psi(t, x, h, \partial h/\partial x, \ldots) \in \mathscr{H}$. Thus, we have $u(t,x) \in \mathscr{H}$. This proves

Theorem 5.4. *Suppose the conditions* (A) *and* $\Delta_2 \neq \emptyset$*; let* σ *be the one in* (5.2). *If* $u(t,x)$ *is a solution of* (5.1) *belonging in the class* $\mathscr{S}_\sigma$*, then* $u(t,x)$ *can be continued holomorphically up to some neighborhood of the origin of* $\mathbb{C}_t \times \mathbb{C}_x$.

Note that this is just the result of the first order case in Kobayashi [3] and Lope–Tahara [4].

The details and proofs of the results in sections 3, 4 and 5 will be published elsewhere.

References

[1] Gérard, R., Tahara, H.: Singular nonlinear partial differential equations. Aspects of Mathematics, E 28, Vieweg (1996)

[2] Hille, E.: Ordinary differential equations in the complex domain. John Wiley and Sons (1976)

[3] Kobayashi, T.: Singular solutions and prolongation of holomorphic solutions to nonlinear differential equations. Publ. Res. Inst. Math. Sci. Kyoto Univ., **34**, 43–63 (1998)

[4] Lope, J.E.C., Tahara, H.: On the analytic continuation of solutions to nonlinear partial differential equations. J. Math. Pures Appl., **81**, 811–826 (2002)

Instanton-type formal solutions for the first Painlevé hierarchy*

Yoshitsugu Takei

Research Institute for Mathematical Sciences, Kyoto University, Kyoto 606-8502, Japan
takei@kurims.kyoto-u.ac.jp

Summary. Instanton-type formal solutions, which will play an important role in the description of Stokes phenomena, are discussed for the first Painlevé hierarchy. We construct instanton-type solutions by using singular-perturbative reduction of a Hamiltonian system to its Birkhoff normal form. The construction of singular-perturbative reduction to the Birkhoff normal form is also outlined.

Key words: Painlevé hierarchy, Instanton-type solutions, Hamiltonian system, Birkhoff normal form

1 Introduction

Collaborating with Kawai and partly with Aoki, I developed the exact WKB analysis for traditional (i.e., second order) Painlevé equations in the 1990's. (Cf. [5], [1], [6], [16], [17]. See also [9].) To enlarge the scope of its applicability we now try to extend the exact WKB analysis to some hierarchies, particularly the first Painlevé hierarchy $(P_{\mathrm{I}})_m$, of higher order Painlevé equations ("Toulouse Project"). To be more concrete, Toulouse Project is our project to understand the analytic structure of solutions of higher order Painlevé equations, say $(P_{\mathrm{I}})_m$, from the viewpoint of the exact WKB analysis with the following procedure:

Part 1 : Stokes geometry of $(P_{\mathrm{I}})_m$ and its relationship with that of the underlying Lax pair of $(P_{\mathrm{I}})_m$.
Part 2 : Reduction of 0-parameter solutions of $(P_{\mathrm{I}})_m$ to those of the traditional first Painlevé equation $(P_{\mathrm{I}})_1$ near a turning point of the first kind.
Part 3 : Study of the structure of 0-parameter solutions of $(P_{\mathrm{I}})_m$ near a turning point of the second kind.

Received 11 April, 2006. Accepted 24 May, 2006.
* This work is supported in part by JSPS Grant-in-Aid No. 16540148

Part 4 : Construction of instanton-type formal solutions of $(P_{\mathrm{I}})_m$, i.e., $(2m)$-parameter solutions of $(P_{\mathrm{I}})_m$.

Part 5 : Study of the structure of instanton-type solutions of $(P_{\mathrm{I}})_m$ near turning points.

Part 6 : Connection formulas for instanton-type solutions near turning points.

Part 7 : Study of the structure of instanton-type solutions near a crossing point of Stokes curves.

Among the above table of the procedure, Part 1 has already been investigated in detail in [3, 4], Part 2 is established in [7, 8] (which is a generalization of the previous result [5] for traditional Painlevé equations), and Part 3 is also well analyzed (though the results have not yet been published anywhere). Now, the purpose of this paper is to discuss the Toulouse Project Part 4, that is, to discuss the construction of formal solutions of $(P_{\mathrm{I}})_m$ containing sufficiently many (i.e., $2m$) free parameters called "instanton-type solutions".

In the case of traditional Painlevé equations there were two methods for constructing 2-parameter instanton-type formal solutions; the one is to employ the multiple-scale analysis ([1]) and the other is to use reduction of Hamiltonian systems equivalent to Painlevé equations to their Birkhoff normal form ([15]). Here, to construct $(2m)$-parameter instanton-type solutions of the first Painlevé hierarchy $(P_{\mathrm{I}})_m$, we generalize the second method so that it may be applied to $(P_{\mathrm{I}})_m$: After expressing $(P_{\mathrm{I}})_m$ in the form of a Hamiltonian system and localizing it around a 0-parameter solution (where a "0-parameter solution" means a formal solution that is algebraically constructed in a singular-perturbative manner, cf. Section 2 below), we reduce it to its Birkhoff normal form. Instanton-type formal solutions of $(P_{\mathrm{I}})_m$ are then constructed by solving explicitly the Birkhoff normal form thus obtained.

The plan of the paper is as follows: In Section 2 we first recall the explicit form of the first Painlevé hierarchy $(P_{\mathrm{I}})_m$ and state the main result to give the reader a clear image about $(2m)$-parameter instanton-type formal solutions. Next we describe an outline of the proof of the main result, i.e., an outline of the construction of instanton-type solutions of $(P_{\mathrm{I}})_m$ in Section 3. Finally in Section 4 we sketch out the proof of the existence of reduction of a Hamiltonian system in question to its Birkhoff normal form.

In ending this Introduction I would like to express my sincerest gratitude to Prof. T. Kawai for his valuable advice, kind encouragement and really stimulating discussions with him. I am very much pleased to dedicate this paper to him on the occasion of his sixtieth birthday. I also would like to thank many collaborators, especially Prof. T. Aoki and Dr. T. Koike, for stimulating and interesting discussions with them.

2 Main result — The first Painlevé hierarchy $(P_{\mathrm{I}})_m$ and its instanton-type solutions

First of all, let us recall the explicit form of the first Painlevé hierarchy $(P_{\mathrm{I}})_m$ $(m = 1, 2, \ldots)$ with a large parameter η (> 0):

$$\begin{cases} \dfrac{du_j}{dt} = 2\eta v_j \\ \dfrac{dv_j}{dt} = 2\eta(u_{j+1} + u_1 u_j + w_j) \end{cases} \qquad (P_{\mathrm{I}})_m$$

$(j = 1, \ldots, m)$, where u_j and v_j are unknown functions (u_{m+1} is conventionally assumed to be equal to 0) and w_j is a polynomial of u_k and v_l $(1 \leq k, l \leq j)$ determined by the following recursive relation:

$$w_j = \frac{1}{2}\left(\sum_{k=1}^{j} u_k u_{j+1-k}\right) + \sum_{k=1}^{j-1} u_k w_{j-k} - \frac{1}{2}\left(\sum_{k=1}^{j-1} v_k v_{j-k}\right) + c_j + \delta_{jm} t \quad (1)$$

$(j = 1, \ldots, m)$. Here c_j is a constant and δ_{jm} stands for Kronecker's delta.

The above expression of $(P_{\mathrm{I}})_m$ is a slight modification of that of Shimomura, who introduced the hierarchy in his study of the most degenerate Garnier system ([13, 14]). It is essentially the same as the P_{I} hierarchy proposed by Gordoa and Pickering ([2]). See also [11, 12]. Note that the first member of the hierarchy, i.e., $(P_{\mathrm{I}})_1$ is equivalent to (P_{I}), the traditional first Painlevé equation with a large parameter η. This is the reason why this hierarchy is called "the first Painlevé hierarchy" or "the P_{I}-hierarchy".

As is shown in [3], $(P_{\mathrm{I}})_m$ admits the following formal solution $(\hat{u}_j, \hat{v}_j)$ called a "0-parameter solution":

$$\hat{u}_j(t,\eta) = \hat{u}_{j,0}(t) + \eta^{-1}\hat{u}_{j,1}(t) + \cdots, \quad \hat{v}_j(t,\eta) = \hat{v}_{j,0}(t) + \eta^{-1}\hat{v}_{j,1}(t) + \cdots. \quad (2)$$

The 0-parameter solution is algebraically constructed in a singular-perturbative manner; $\hat{u}_{j,0}$ and $\hat{v}_{j,0}$ $(1 \leq j \leq m)$ are first algebraically determined (in particular, $\hat{v}_{j,0} \equiv 0$ holds) and then the other $\hat{u}_{j,k}$'s and $\hat{v}_{j,k}$'s $(k \geq 1)$ are uniquely determined in a recursive manner once (the branch of) $\hat{u}_{j,0}$ is fixed. See [3, Section 2.1] for the details. In [3] the 0-parameter solution is introduced to define the Stokes geometry (i.e., turning points and Stokes curves) of $(P_{\mathrm{I}})_m$.

The construction of 0-parameter solutions is simple and straightforward. In compensation for its simplicity 0-parameter solutions do not contain any free parameters. Thus it is impossible to discuss the Stokes phenomenon, which is observed on a Stokes curve, solely in terms of 0-parameter solutions. As a matter of fact, in the case of the traditional first Painlevé equation (P_{I}), we needed instanton-type formal solutions to describe the connection formula, the concrete expression of the Stokes phenomenon, even for a 0-parameter solution ([16]). The aim of this paper is to construct such instanton-type formal solutions with free parameters also for a higher order Painlevé equation $(P_{\mathrm{I}})_m$.

To state our main theorem, we prepare some notations. Let $(\Delta P_{\mathrm{I}})_m$ denote the linearized equation of $(P_{\mathrm{I}})_m$ at its 0-parameter solution $(\hat{u}_j, \hat{v}_j)$ (sometimes called "Fréchet derivative" for short), that is, the linear part in $(\Delta u_j, \Delta v_j)$ after the substitution $u_j = \hat{u}_j + \Delta u_j$ and $v_j = \hat{v}_j + \Delta v_j$ in $(P_{\mathrm{I}})_m$. Then $(\Delta P_{\mathrm{I}})_m$ becomes a system of linear ordinary differential equations for $(\Delta u_j, \Delta v_j)$ of the following form:

$$\frac{d}{dt}\begin{pmatrix} \Delta u_1 \\ \Delta v_1 \\ \vdots \\ \Delta v_m \end{pmatrix} = \eta C(t,\eta) \begin{pmatrix} \Delta u_1 \\ \Delta v_1 \\ \vdots \\ \Delta v_m \end{pmatrix}, \tag{3}$$

where $C(t,\eta)$ is a formal power series (in η^{-1}) with coefficients of $(2m)\times(2m)$ matrices whose entries are analytic functions of t. Note that, as is verified in [3, Section 2.1], the characteristic equation $\det(\lambda - C_0(t)) = 0$ of the top order part (i.e., the part of order 0 in η) $C_0(t)$ of $C(t,\eta)$ is an m-th degree polynomial of λ^2 (see also Section 3 and Lemma 1 below). In what follows we denote the roots of the characteristic equation $\det(\lambda - C_0(t)) = 0$ by $\pm\lambda_j(t)$ $(j = 1, \ldots, m)$. The turning points of $(P_{\mathrm{I}})_m$ are then defined in terms of $\lambda_j(t)$. There are two kinds of turning points; a turning point of the first kind is a point where λ_j vanishes for some j, and a turning point of the second kind is a point where $\lambda_j - \lambda_k$ or $\lambda_j + \lambda_k$ vanishes for some $j \neq k$.

Now the main result of this paper is the following:

Theorem 1. *Assume that t_0 is not a turning point of $(P_{\mathrm{I}})_m$. Suppose further that*

$$\sum_{j=1}^{m} n_j \lambda_j(t) \ \textit{does not identically vanish for any} \ (n_1, \ldots, n_m) \in \mathbb{Z}^m \setminus \{0\}. \tag{4}$$

Then, in a neighborhood of $t = t_0$, there exists a formal solution of $(P_{\mathrm{I}})_m$ of the following form:

$$\begin{aligned} u_j(t,\eta;\alpha,\beta) &= u_{j,0}(t) + \eta^{-1/2} u_{j,1/2}(t,\Psi,\Phi) + \eta^{-1} u_{j,1}(t,\Psi,\Phi) + \cdots, \\ v_j(t,\eta;\alpha,\beta) &= v_{j,0}(t) + \eta^{-1/2} v_{j,1/2}(t,\Psi,\Phi) + \eta^{-1} v_{j,1}(t,\Psi,\Phi) + \cdots, \end{aligned} \tag{5}$$

$(j = 1, \ldots, m)$. Here $u_{j,l/2}(t,\Psi,\Phi)$ and $v_{j,l/2}(t,\Psi,\Phi)$ $(l = 1, 2, \ldots)$ are polynomials in (Ψ,Φ) of degree at most l with analytic (in t) coefficients (in particular, $u_{j,1/2}$ and $v_{j,1/2}$ are linear combinations of (Ψ,Φ) with analytic coefficients), and $\Psi = (\Psi_1, \ldots, \Psi_m)$ and $\Phi = (\Phi_1, \ldots, \Phi_m)$ are "instantons", that is, formal series of exponential type of the form

$$\begin{aligned} \Psi_j &= \alpha_j \exp\left\{\eta \int^t \left(\sum_{k=0}^{\infty} \eta^{-k} \sum_{|\mu|=k} (\mu_j+1) g_{\mu+e_j}(t,\eta)\gamma^{\mu}\right) dt\right\}, \\ \Phi_j &= \beta_j \exp\left\{-\eta \int^t \left(\sum_{k=0}^{\infty} \eta^{-k} \sum_{|\mu|=k} (\mu_j+1) g_{\mu+e_j}(t,\eta)\gamma^{\mu}\right) dt\right\} \end{aligned} \tag{6}$$

($j = 1,\ldots,m$), where α_j and β_j are free complex constants, γ denotes $\gamma = (\gamma_1,\ldots,\gamma_m) = (\alpha_1\beta_1,\ldots,\alpha_m\beta_m)$, $\mu = (\mu_1,\ldots,\mu_m)$ ($\mu_j \in \mathbb{Z}, \mu_j \geq 0$) and $e_j = (0,\ldots,1,\ldots,0)$ (i.e., only the j-th component is equal to 1 while the others are all 0) are multi-indices, and for each multi-index $\nu = (\nu_1,\ldots,\nu_m)$ $g_\nu(t,\eta)$ is a formal power series of $\eta^{-1/2}$ with analytic coefficients of the following form:

$$g_\nu(t,\eta) = \sum_{l=0}^{\infty} \eta^{-l/2} g_{\nu,l/2}(t). \tag{7}$$

We call the formal solution $(u_j(t,\eta;\alpha,\beta), v_j(t,\eta;\alpha,\beta))$ given in this theorem an "instanton-type solution" of $(P_{\rm I})_m$.

Remark 1. The top order part $(u_{j,0}(t), v_{j,0}(t))$ of $(u_j(t,\eta;\alpha,\beta), v_j(t,\eta;\alpha,\beta))$ is the same as that of the 0-parameter solution $(\hat{u}_j(t,\eta), \hat{v}_j(t,\eta))$. More important is the top order part of the instantons (Ψ_j, Φ_j); it is described by $g_{e_j,0}(t)$, which coincides with the characteristic root $\lambda_j(t)$ of the Fréchet derivative $(\Delta P_{\rm I})_m$. This fact shows the relevance of the instanton-type solutions to the Stokes phenomenon and, at the same time, validates the definition of the Stokes geometry of $(P_{\rm I})_m$ given in [3].

Remark 2. Each coefficient of $u_{j,l/2}$ (resp. $v_{j,l/2}$) may have some singularity in addition to turning points: By the construction of solutions explained below we see that the singular points of $u_{j,l/2}$ (resp. $v_{j,l/2}$) are contained at most in the union of zeros of $\sum_j n_j\lambda_j(t)$ with $|n_1| + \cdots + |n_m| \leq l+1$. Similarly the singular points of a coefficient of $g_\nu(t,\eta)$ are contained in the union of zeros of $\sum_j n_j\lambda_j(t)$ with $|n_1| + \cdots + |n_m| \leq 2|\nu| - 1$.

3 Outline of the construction of instanton-type solutions

As was mentioned in Introduction, we construct instanton-type solutions by using reduction of a Hamiltonian system to its Birkhoff normal form. The concrete procedure of construction consists of the following four steps.

Step 1. First we express $(P_{\rm I})_m$ in the form of a Hamiltonian system.

As is discussed in [13, 14], the first Pianlevé hierarchy $(P_{\rm I})_m$ is obtained by restricting the most degenerate Garnier system onto a one-dimensional complex curve. Since the (degenerate) Garnier system possesses a Hamiltonian structure, the first Pianlevé hierarchy also inherits such a Hamitonian structure. To be more specific, $(P_{\mathrm{I}})_m$ can be expressed in the form of a Hamiltonian system by using the canonical variable (σ_j, τ_j) defined as follows:

$$u_j = (-1)^{j-1} \sum_{k_1 < \cdots < k_j} \sigma_{k_1} \cdots \sigma_{k_j}, \tag{8}$$

$$\tau_j = \frac{1}{2} \left(v_1 \sigma_j^{m-1} + \cdots + v_m \right). \tag{9}$$

(Cf. [13], [10]; u_j is the j-th order fundamental symmetric polynomial of $(\sigma_1, \ldots, \sigma_m)$ (up to the sign) and τ_j is defined as the residue of coefficients of the second order linear differential equation associated with $(P_{\mathrm{I}})_m$ through isomonodromic deformations.) In what follows we use another canonical variable which is more closely attached to the original variable (u_j, v_j): Take q_j as

$$q_j = (-1)^{j-1} u_j \left(= \sum_{k_1 < \cdots < k_j} \sigma_{k_1} \cdots \sigma_{k_j} \right). \tag{10}$$

As the conjugate variable of q_j we choose p_j so that it may satisfy $\sum dq_j \wedge dp_j = \sum d\sigma_j \wedge d\tau_j$ or $\sum p_j dq_j = \sum \tau_j d\sigma_j$. That is, we define p_j in such a way that

$$\tau_j = \sum_k \frac{\partial q_k}{\partial \sigma_j} p_k \tag{11}$$

may be satisfied. More explicitly, p_j is given by the following relation:

$$v_j = 2(-1)^{m-j} (p_{m-j+1} + p_{m-j+2} q_1 + \cdots + p_m q_{j-1}). \tag{12}$$

Remark 3. The explicit relation (12) follows from (9) and (11) in the following way: For $k = 0, 1, \ldots, m-1$ let $s^{(k)}$ and $\tilde{s}_j^{(k)}$ denote the k-th order fundamental symmetric polynomial of $(\sigma_1, \ldots, \sigma_m)$ and that of $(\sigma_1, \ldots, \sigma_{j-1}, \sigma_{j+1}, \ldots, \sigma_m)$, respectively. Then, if we define a k-th degree polynomial $F^{(k)}(z)$ of z by

$$F^{(k)}(z) = s^{(k)} - s^{(k-1)} z + s^{(k-2)} z^2 - \cdots + (-1)^k z^k, \tag{13}$$

the following relation holds for $j = 1, \ldots, m$:

$$F^{(k)}(\sigma_j) = \tilde{s}_j^{(k)}. \tag{14}$$

Taking the relation $\partial q_k / \partial \sigma_j = \partial s^{(k)} / \partial \sigma_j = \tilde{s}_j^{(k-1)}$ into account, we find that (9) and (11) together with (14) entail

$$\frac{1}{2} \left(v_1 z^{m-1} + \cdots + v_{m-1} z + v_m \right) = F^{(m-1)}(z) p_m + \cdots + F^{(1)}(z) p_2 + p_1. \tag{15}$$

Relation (12) immediately follows from comparison of like powers (in z) of (15).

Thus, in the variable (q_j, p_j), $(P_{\mathrm{I}})_m$ can be expressed in the form of the following Hamiltonian system:

$$\frac{dq_j}{dt} = \eta \frac{\partial H}{\partial p_j}, \quad \frac{dp_j}{dt} = -\eta \frac{\partial H}{\partial q_j}. \tag{16}$$

For example, the Hamiltonian is explicitly given by

$$H = -\frac{1}{2}q_1^4 + \frac{3}{2}q_1^2 q_2 - \frac{1}{2}q_2^2 - 2q_1 p_2^2 - 4p_1 p_2 + c_1(-q_1^2 + q_2) - tq_1 \tag{17}$$

for $m = 2$ and by

$$\begin{aligned} H = &-\frac{1}{2}q_1^5 + 2q_1^3 q_2 - \frac{3}{2}q_1^2 q_3 - \frac{3}{2}q_1 q_2^2 + q_2 q_3 + 4q_1 p_2 p_3 + 2q_2 p_3^2 \\ &+ 4p_1 p_3 + 2p_2^2 + c_1(-q_1^3 + 2q_1 q_2 - q_3) + c_2(-q_1^2 + q_2) - tq_1 \end{aligned} \tag{18}$$

for $m = 3$.

Step 2. In the canonical variable (q_j, p_j) there exists the following 0-parameter solution of (16), which corresponds to (2):

$$\hat{q}_j(t,\eta) = \hat{q}_{j,0}(t) + \eta^{-1}\hat{q}_{j,1}(t) + \cdots, \quad \hat{p}_j(t,\eta) = \hat{p}_{j,0}(t) + \eta^{-1}\hat{p}_{j,1}(t) + \cdots. \tag{19}$$

Then we next consider the "localization at the 0-parameter solution" of (16), that is, we introduce a new (formal) variable (ψ_j, φ_j) defined as follows:

$$q_j = \hat{q}_j + \eta^{-1/2}\psi_j, \quad p_j = \hat{p}_j + \eta^{-1/2}\varphi_j. \tag{20}$$

Since (ψ_j, φ_j) is also canonical, in the variable (ψ_j, φ_j) (16) can be expressed again in the Hamiltonian form as

$$\frac{d\psi_j}{dt} = \eta \frac{\partial K}{\partial \varphi_j}, \quad \frac{d\varphi_j}{dt} = -\eta \frac{\partial K}{\partial \psi_j}, \tag{21}$$

where

$$K = \sum_{|\mu+\nu|\geq 2} \frac{1}{\mu!\nu!}\eta^{-(|\mu+\nu|-2)/2} \frac{\partial^{|\mu+\nu|}H}{\partial q^\mu \partial p^\nu}(t, \hat{q}, \hat{p})\psi^\mu \varphi^\nu. \tag{22}$$

Step 3. This is the most important step; we consider the reduction of (21) to its Birkhoff normal form.

As the localization at the 0-parameter solution is done in *Step 2*, the leading part of (21) consequently becomes linear. For example, the coefficient matrix of the top order part (in $\eta^{-1/2}$) of (21) is given by

$$\begin{array}{r} \\ j> \\ \\ j+m> \end{array} \left.\begin{pmatrix} \overset{k}{\vee} & \overset{k+m}{\vee} \\ \dfrac{\partial^2 H}{\partial p_j \partial q_k} & \dfrac{\partial^2 H}{\partial p_j \partial p_k} \\ \hline -\dfrac{\partial^2 H}{\partial q_j \partial q_k} & -\dfrac{\partial^2 H}{\partial q_j \partial p_k} \end{pmatrix}\right|_{\substack{q_l = \hat{q}_{l,0} \\ p_l = \hat{p}_{l,0}}}. \tag{23}$$

Note that the eigenvalues of the matrix (23) exactly coincide with $\pm\lambda_j(t)$, i.e., the characteristic roots of the Fréchet derivative $(\Delta P_{\mathrm{I}})_m$. Therefore they are distinct and non-zero outside the set of turning points.

Making use of this structure peculiar to (21), we can reduce (21) to its Birkhoff normal form, that is, we have

Theorem 2. *We assume that t_0 is not a turning point of $(P_{\mathrm{I}})_m$. We further assume (4). Then, in a neighborhood of $t = t_0$, we can find a canonical transform*

$$\psi_j = \sum_{k=0}^{\infty} \eta^{-k/2} \psi_j^{(k)}(t, \tilde{\psi}, \tilde{\varphi}, \eta^{-1/2}), \quad \varphi_j = \sum_{k=0}^{\infty} \eta^{-k/2} \varphi_j^{(k)}(t, \tilde{\psi}, \tilde{\varphi}, \eta^{-1/2}), \tag{24}$$

where $\psi_j^{(k)}$ and $\varphi_j^{(k)}$ are homogeneous polynomials of degree $(k+1)$ in $(\tilde{\psi}, \tilde{\varphi})$, that transforms (21) into the Birkhoff normal form

$$\frac{d\tilde{\psi}_j}{dt} = \eta \frac{\partial \tilde{K}}{\partial \tilde{\varphi}_j}, \quad \frac{d\tilde{\varphi}_j}{dt} = -\eta \frac{\partial \tilde{K}}{\partial \tilde{\psi}_j}, \tag{25}$$

where

$$\tilde{K} = \tilde{K}(t, \theta_1, \ldots, \theta_m, \eta^{-1/2}) \quad \textit{with} \quad \theta_j = \tilde{\psi}_j \tilde{\varphi}_j. \tag{26}$$

A sketch of the proof of Theorem 2 will be given in Section 4.

Step 4. In view of (26) we find that (25) can be written as

$$\frac{d\tilde{\psi}_j}{dt} = \eta \left.\frac{\partial \tilde{K}}{\partial \theta_j}\right|_{\theta_l = \tilde{\psi}_l \tilde{\varphi}_l} \tilde{\psi}_j, \quad \frac{d\tilde{\varphi}_j}{dt} = -\eta \left.\frac{\partial \tilde{K}}{\partial \theta_j}\right|_{\theta_l = \tilde{\psi}_l \tilde{\varphi}_l} \tilde{\varphi}_j. \tag{27}$$

In particular, this entails that

$$\frac{d}{dt}(\tilde{\psi}_j \tilde{\varphi}_j) = \frac{d\tilde{\psi}_j}{dt} \tilde{\varphi}_j + \tilde{\psi}_j \frac{d\tilde{\varphi}_j}{dt} = 0, \tag{28}$$

that is,

$$\gamma_j := \tilde{\psi}_j \tilde{\varphi}_j \quad \text{does not depend on } t. \tag{29}$$

By substituting (29) into (27) we can explicitly solve (27) to obtain

$$\tilde{\psi}_j = \alpha_j \exp\left(\eta \int^t \left.\frac{\partial \tilde{K}}{\partial \theta_j}\right|_{\theta_l = \gamma_l} dt\right), \quad \tilde{\varphi}_j = \beta_j \exp\left(-\eta \int^t \left.\frac{\partial \tilde{K}}{\partial \theta_j}\right|_{\theta_l = \gamma_l} dt\right), \tag{30}$$

where α_j and β_j are free complex constants of integration. Note that (29) and (30) imply

$$\gamma_j = \alpha_j \beta_j. \tag{31}$$

In this way the Birkhoff normal form (25) has been solved explicitly. If we denote the explicit solution $(\tilde{\psi}_j, \tilde{\varphi}_j)$ of (25) thus obtained by (Ψ_j, Φ_j) and substitute it into the canonical transform (24), we can obtain also a (formal) solution of (21) and consequently an instanton-type solution of $(P_{\mathrm{I}})_m$ with $(2m)$ free parameters $(\alpha_1, \ldots, \alpha_m, \beta_1, \ldots, \beta_m)$. (The solution (Ψ_j, Φ_j) of (25) or (27) gives "instantons".) We have thus finished the construction of instanton-type solutions of $(P_{\mathrm{I}})_m$.

4 A sketch of the proof of Theorem 2

In this section we sketch out the proof of Theorem 2.

Let us denote $\eta^{-1/2}$ by ϵ. We want to construct a canonical transform

$$\psi_j = \sum_{k=0}^{\infty} \epsilon^k \psi_j^{(k)}(t, \tilde{\psi}, \tilde{\varphi}, \epsilon), \quad \varphi_j = \sum_{k=0}^{\infty} \epsilon^k \varphi_j^{(k)}(t, \tilde{\psi}, \tilde{\varphi}, \epsilon) \tag{32}$$

which transforms the Hamiltonian system (21) in question into its Birkhoff normal form. Here $\psi_j^{(k)}$ and $\varphi_j^{(k)}$ are assumed to be of the following form:

$$\psi_j^{(k)} = \sum_{|\mu+\nu|=k+1} \psi_j^{\mu,\nu}(t,\epsilon)\tilde{\psi}^{\mu}\tilde{\varphi}^{\nu}, \quad \varphi_j^{(k)} = \sum_{|\mu+\nu|=k+1} \varphi_j^{\mu,\nu}(t,\epsilon)\tilde{\psi}^{\mu}\tilde{\varphi}^{\nu}. \tag{33}$$

As the construction of $\psi_j^{(0)}$ and $\varphi_j^{(0)}$, i.e., the linear part with respect to $(\tilde{\psi}, \tilde{\varphi})$, is quite different from that of the nonlinear part, we discuss these two parts separately in what follows.

4.1 Construction of the linear part $\psi_j^{(0)}$ and $\varphi_j^{(0)}$

Let us write the quadratic part of the Hamiltonian (22) as

$$K = \frac{1}{2}{}^t\psi M_1 \psi + \frac{1}{2}{}^t\varphi M_2 \varphi + {}^t\varphi M_3 \psi + \cdots, \tag{34}$$

where M_j is a formal power series of ϵ whose coefficients are $m \times m$ matrices of analytic functions of t. Then (the linear part of) the Hamiltonian system (21) can be expressed as

$$\frac{d}{dt}\begin{pmatrix} \psi \\ \varphi \end{pmatrix} = \eta \left(\begin{array}{c|c} M_3 & M_2 \\ \hline -M_1 & -{}^tM_3 \end{array}\right) \begin{pmatrix} \psi \\ \varphi \end{pmatrix} + \cdots. \tag{35}$$

We now want to construct the linear part of a canonical transform

$$\begin{pmatrix} \psi^{(0)} \\ \varphi^{(0)} \end{pmatrix} = A \begin{pmatrix} \tilde{\psi}^{(0)} \\ \tilde{\varphi}^{(0)} \end{pmatrix} \quad \text{with} \quad A = \left(\begin{array}{c|c} a & b \\ \hline c & d \end{array}\right) \tag{36}$$

(where a, b, c and d are also formal power series of ϵ with $m \times m$ matrix coefficients) in such a way that the following two conditions may be satisfied.

(A1) (36) is symplectic,

(A2) (36) diagonalizes the linear part of (35).

First, the top order term (with respect to ϵ) of (36) can be constructed by applying

Lemma 1. *Assume that the top order term of the coefficient of* (35)

$$\left(\begin{array}{c|c} M_{3,0} & M_{2,0} \\ \hline -M_{1,0} & -{}^tM_{3,0} \end{array}\right) = \left.\left(\begin{array}{c|c} M_3 & M_2 \\ \hline -M_1 & -{}^tM_3 \end{array}\right)\right|_{\epsilon=0}, \tag{37}$$

which coincides with (23), *has distinct eigenvalues. Then we can find a symplectic matrix T that satisfies*

$$T^{-1}\left(\begin{array}{c|c} M_{3,0} & M_{2,0} \\ \hline -M_{1,0} & -{}^tM_{3,0} \end{array}\right)T = \left(\begin{array}{ccc|ccc} \lambda_1 & & & & & \\ & \ddots & & & 0 & \\ & & \lambda_m & & & \\ \hline & & & -\lambda_1 & & \\ & 0 & & & \ddots & \\ & & & & & -\lambda_m \end{array}\right), \tag{38}$$

where $\pm\lambda_j(t)$ are eigenvalues of (37), *i.e., of* (23).

As the proof of Lemma 1 is an exercise of the linear algebra, we omit it here. Since the assumption of Lemma 1 is satisfied outside the set of turning points, the existence of the top order term of (36) is guaranteed by this lemma.

Once the top order term is constructed, higher order terms (with respect to ϵ) of (36) are determined in the following manner: If we let X, Y and Z denote bd^{-1}, ca^{-1} and $1-{}^tXY$ $(=1-{}^td^{-1}{}^tbca^{-1})$, respectively, we find that the conditions (A1) and (A2) are equivalent to

$${}^tX = X, \quad M_3X + {}^tX{}^tM_3 + {}^tXM_1X + M_2 - \epsilon^2\frac{\partial X}{\partial t} = 0, \tag{39}$$

$${}^tY = Y, \quad {}^tM_3Y + {}^tYM_3 + M_1 + {}^tYM_2Y + \epsilon^2\frac{\partial Y}{\partial t} = 0, \tag{40}$$

$$d = {}^t(Za)^{-1}, \tag{41}$$

$$a^{-1}Z^{-1}\left[M_3 + {}^tXM_1 + M_2Y + {}^tX{}^tM_3Y + \epsilon^2{}^tX\frac{\partial Y}{\partial t}\right]a - \epsilon^2a^{-1}\frac{\partial a}{\partial t} : \text{diagonal}. \tag{42}$$

Since the top order term has already been constructed, we may assume that

$$a = 1 + O(\epsilon^2), \quad b = O(\epsilon^2), \quad c = O(\epsilon^2), \quad d = 1 + O(\epsilon^2), \tag{43}$$

$$M_1 = O(\epsilon^2), \quad M_2 = O(\epsilon^2), \quad M_3 = \begin{pmatrix} \lambda_1 & & \\ & \ddots & \\ & & \lambda_m \end{pmatrix} + O(\epsilon^2). \tag{44}$$

Equations (39) and (40) then uniquely determine the formal power series X and Y, respectively. Consequently $Z = 1 - {}^tXY$ is also fixed. Furthermore, substituting X, Y and Z thus determined into (42), we may as well determine $a = 1 + \epsilon^2 a_2 + \epsilon^4 a_4 + \cdots$ so that (42) is satisfied. In this way, by using (41) in addition, we can construct higher order terms of a, b, c and d, that is, the higher order terms of (36).

4.2 Construction of the nonlinear part

To construct the nonlinear part of the canonical transform (32), we make use of a generating function of the following form:

$$W(t, \tilde{\psi}, \varphi) = \sum_{|\mu+\nu|\geq 2} \epsilon^{|\mu+\nu|-2} w^{\mu,\nu} \tilde{\psi}^{\mu} \varphi^{\nu}. \tag{45}$$

The canonical transform

$$\psi = \psi(t, \tilde{\psi}, \tilde{\varphi}, \epsilon), \quad \varphi = \varphi(t, \tilde{\psi}, \tilde{\varphi}, \epsilon) \tag{46}$$

induced by the generating function W is determined by

$$\psi_j = -\frac{\partial W}{\partial \varphi_j}, \quad \tilde{\varphi}_j = -\frac{\partial W}{\partial \tilde{\psi}_j}, \tag{47}$$

and the new Hamiltonian $\tilde{K}$ for $(\tilde{\psi}, \tilde{\varphi})$ is described in terms of the original Hamiltonian K and the generating function W as follows:

$$\tilde{K} = K(t, \psi(t, \tilde{\psi}, \tilde{\varphi}, \epsilon), \varphi(t, \tilde{\psi}, \tilde{\varphi}, \epsilon), \epsilon) + \epsilon^2 \frac{\partial W}{\partial t}(t, \tilde{\psi}, \varphi(t, \tilde{\psi}, \tilde{\varphi}, \epsilon), \epsilon). \tag{48}$$

Thus, for the construction of a canonical transform that reduces (21) to its Birkhoff normal form, it suffices to fix each coefficient $w^{\mu,\nu}$ of the generating function W so that

$$\text{any term of the form } \tilde{\psi}^{\mu}\tilde{\varphi}^{\nu} \text{ with } \mu \neq \nu \text{ may not appear in } \tilde{K}. \tag{49}$$

Note that the construction of the linear part of the canonical transform has been already finished in Section 4.1. Hence we may assume that the quadratic part of the original Hamiltonian K has the form (34) where $M_1 = M_2 = 0$ and M_3 is a diagonal matrix whose top order term is given by the right-hand side of (38), and further that

$$w^{\mu,\nu} = -1 \ (\text{for } \mu = \nu), \quad w^{\mu,\nu} = 0 \ (\text{for } \mu \neq \nu) \tag{50}$$

in case $|\mu + \nu| = 2$. Using this "induction hypothesis" and the expression (48) of $\tilde{K}$, we can verify the following Lemma 2 through explicit computations similar to those of [15, Section 2.2].

Lemma 2. *For $|\mu+\nu| \geq 3$ the requirement (49) is equivalent to an equation of the following form:*

$$\left(\sum_{j=1}^{m}(\mu_j-\nu_j)\lambda_j+O(\epsilon^2)\right)w^{\mu,\nu}+\epsilon^2\frac{\partial}{\partial t}w^{\mu,\nu}=R(t,w^{\mu',\nu'},\epsilon^2), \tag{51}$$

where the indices (μ',ν') that appear in $R(t,w^{\mu',\nu'},\epsilon)$ of the right-hand side run in the set $\{(\mu',\nu')\,;\,|\mu'+\nu'|\leq|\mu+\nu|-1\}$.

Thus the terms $w^{\mu,\nu}$ with $\mu\neq\nu$ can be recursively determined.

This completes the proof of Theorem 2.

References

[1] T. Aoki, T. Kawai and Y. Takei, WKB analysis of Painlevé transcendents with a large parameter. II, *Structure of Solutions of Differential Equations*, World Scientific, 1996, pp.1–49.

[2] P.R. Gordoa and A. Pickering, Nonisospectral scattering problems: A key to integrable hierarchies, *J. Math. Phys.*, **40**(1999), 5749–5786.

[3] T. Kawai, T. Koike, Y. Nishikawa and Y. Takei, On the Stokes geometry of higher order Painlevé equations, *Astérisque*, Vol. 297, 2004, pp. 117–166.

[4] T. Kawai, T. Koike, Y. Nishikawa and Y. Takei, On the complete description of the Stokes geometry for the first Painlevé hierarchy, *RIMS Kôkyûroku*, Vol. 1397, 2004, pp. 74–101.

[5] T. Kawai and Y. Takei, WKB analysis of Painlevé transcendents with a large parameter. I, *Adv. Math.*, **118**(1996), 1–33.

[6] T. Kawai and Y. Takei, WKB analysis of Painlevé transcendents with a large parameter. III, *Adv. Math.*, **134**(1998), 178–218.

[7] T. Kawai and Y. Takei, On WKB analysis of higher order Painlevé equations with a large parameter, *Proc. Japan Acad., Ser. A*, **80**(2004), 53–56.

[8] T. Kawai and Y. Takei, WKB analysis of higher order Painlevé equations with a large parameter — Local reduction of 0-parameter solutions for Painlevé hierarchies (P_J) ($J=$ I, II-1 or II-2), *Adv. Math.*, in press.

[9] T. Kawai and Y. Takei, *Algebraic Analysis of Singular Perturbation Theory*, Translations of Mathematical Monographs, Vol. 227, Amer. Math. Soc., 2005. (Originally published in Japanese by Iwanami, Tokyo in 1998.)

[10] H. Kimura, The degeneration of the two dimensional Garnier system and the polynomial Hamiltonian structure, *Ann. Mat. Pura Appl.*, **155**(1989), 25–74.

[11] N. A. Kudryashov, The first and second Painlevé equations of higher order and some relations between them, *Phys. Lett. A*, **224**(1997), 353–360.

[12] N. A. Kudryashov and M. B. Soukharev, Uniformization and transcendence of solutions for the first and second Painlevé hierarchies, *Phys. Lett. A*, **237**(1998), 206–216.

[13] S. Shimomura, Painlevé property of a degenerate Garnier system of (9/2)-type and of a certain fourth order non-linear ordinary differential equation, *Ann. Scuola Norm. Sup. Pisa*, **29**(2000), 1–17.

[14] S. Shimomura, A certain expression of the first Painlevé hierarchy, *Proc. Japan Acad., Ser. A*, **80**(2004), 105–109.

[15] Y. Takei, Singular-perturbative reduction to Birkhoff normal form and instanton-type formal solutions of Hamiltonian systems, *Publ. RIMS, Kyoto Univ.*, **34**(1998), 601–627.

[16] Y. Takei, An explicit description of the connection formula for the first Painlevé equation, *Toward the Exact WKB Analysis of Differential Equations, Linear or Non-Linear*, Kyoto Univ. Press, 2000, pp. 271–296.

[17] Y. Takei, On an exact WKB approach to Ablowitz-Segur's connection problem for the second Painlevé equation, *ANZIAM J.*, **44**(2002), 111–119.

From exact-WKB toward singular quantum perturbation theory II

André Voros**

CEA, Service de Physique Théorique de Saclay (CNRS URA 2306)
F-91191 Gif-sur-Yvette CEDEX, France
voros@spht.saclay.cea.fr

Summary. Following earlier studies, several new features of singular perturbation theory for one-dimensional quantum anharmonic oscillators are computed by exact WKB analysis; former results are thus validated.

Key words: Schrödinger equation, exact WKB analysis, perturbation theory, semi-classical asymptotics, quartic oscillator, complete elliptic integrals, hyperelliptic integrals, Wronskian identity, spectral determinant, spectral zeta function, zeta regularization, divergent integrals.

This note continues our study [1] of singular perturbation theory in one-dimensional (1D) quantum mechanics using exact WKB analysis. Our focus remains the $v \gg 1$ regime for the potentials $V(q) = q^N + vq^M$ on the real line, with $N > M$ positive even integers. Among those, the *quartic oscillator* $q^4 + vq^2$ has been a prime model for the mathematics of quantum perturbation theory [2, 3, 4, 5, 6]. Kawai and Takei [7] pioneered the use of exact WKB analysis in the latter context, followed by [8] (see also [9, Introduction to Part I, and Pham's contribution], [10], and references therein). In spite of those successes, present exact-WKB *quantization conditions* (for $q^4 + vq^2$, say) *fail to tend* toward their harmonic-potential (vq^2) counterparts as $v \to +\infty$, be it analytically or numerically [11]. This worrying observation triggered our present line of work (starting from [12, § 3]): to further probe how consistently exact WKB theory handles the perturbative ($v \gg 1$) regime.

We are happy and honored to dedicate this work to Professor Kawai with gratitude, for his leadership and many essential contributions in exact WKB analysis, but also (earlier, with Professors Sato and Kashiwara) in hyper/micro/function theory; this framework greatly inspired, and its Authors warmly encouraged, our first steps in exact WKB analysis [13].

Received 17 March, 2006. Accepted 20 July, 2006.

** Also at: Institut de Mathématiques de Jussieu–Chevaleret (CNRS UMR 7586), Université Paris 7, F-75251 Paris CEDEX 05, France.

Even though this work is thoroughly tied to [1] (with its bibliography), in § 1 we recall the main background and further strengthen the case for *improper* (divergent) action integrals like $\int_0^\infty \Pi(q)\,\mathrm{d}q$, where $\Pi(q) = (V(q)+\lambda)^{1/2}$ is the classical momentum function. In § 2 we present new cases where $\int_0^\infty \Pi(q)\,\mathrm{d}q$ can be computed exactly for some *trinomial* $\Pi(q)^2$: essentially the *quartic* case $\Pi(q)^2 = q^4 + vq^2 + \lambda$, for which $\int_0^\infty \Pi(q)\,\mathrm{d}q$ reduces to ordinary (i.e., convergent) *complete elliptic* integrals. In § 3 we extend the main outcome of [1], namely the $v \to +\infty$ asymptotic expression of the spectral determinants $D_N^\pm(\lambda, v) \stackrel{\text{def}}{=} \det{}^\pm(-\mathrm{d}^2/\mathrm{d}q^2 + q^N + vq^M + \lambda)$ in terms of $D_M^\pm(\Lambda) \stackrel{\text{def}}{=} \det{}^\pm(-\mathrm{d}^2/\mathrm{d}q^2 + q^M + \Lambda)$, to $v \to \infty$ in a *complex sector.* Thanks to this, finally in § 4 we demonstrate (currently provided $2M+2 > N$) how the fundamental bilinear functional relation satisfied by $D_N^\pm$ *does evolve* into its counterpart for $D_M^\pm$ as $v \to +\infty$, in spite of a discontinuous jump at $v = \infty$ of the main parameter, the *degree* of the potential (from N to M).

1 Background (summarized) [1]

Our model of quantum perturbation theory is the 1D Schrödinger equation

$$\Big[-\frac{\mathrm{d}^2}{\mathrm{d}q^2} + V(q) + \lambda\Big]\Psi(q) = 0, \quad q \in \mathbb{R}, \quad V(q) = q^N + vq^M, \quad v \gg 1, \tag{1}$$

with $N > M$ positive even integers. If we use the unitary equivalence (called *Symanzik scaling*)

$$-\frac{\mathrm{d}^2}{\mathrm{d}q^2} + uq^N + vq^M \approx v^{2/(M+2)}\Big[-\frac{\mathrm{d}^2}{\mathrm{d}x^2} + x^M + uv^{-(N+2)/(M+2)}\,x^N\Big] \tag{2}$$

twice, at $u = 1$ and $u = 0$, the resulting right-hand sides imply that the operator $\hat{H} = -\mathrm{d}^2/\mathrm{d}q^2 + q^N + vq^M$ is a singular perturbation of $\hat{H}_0 = -\mathrm{d}^2/\mathrm{d}q^2 + vq^M$ for $v \gg 1$, and that the degree *drops* from N to M *at* $v = +\infty$.

From the classical dynamics we will use the *momentum* function $(\times\,\mathrm{i})$,

$$\Pi(q) = (V(q)+\lambda)^{1/2} \qquad \text{(real in the classically forbidden region)}, \tag{3}$$

and its *residue* $\mathrm{Res}_{q=\infty}\Pi(q) = \beta_{-1}(0)$, a notation based on the expansion [12]

$$(V(q)+\lambda)^{-s+1/2} \sim \sum_\rho \beta_\rho(s)\,q^{\rho-Ns} \quad (q \to \infty); \quad \rho = \tfrac{1}{2}N,\ \tfrac{1}{2}N-1, \ldots \tag{4}$$

The spectrum of $\hat{H}$ is purely discrete, $0 < \lambda_0 < \lambda_1 < \cdots \uparrow +\infty$, and separates according to parity since V is an even function. Useful spectral functions (labeled by parity) are the *generalized zeta functions*,

$$Z^\pm(s,\lambda) \stackrel{\text{def}}{=} \sum_{n\ \substack{\text{even}\\ \text{odd}}} (\lambda_n + \lambda)^{-s} \qquad (\mathrm{Re}\,s > \tfrac{1}{2} + \tfrac{1}{N}), \tag{5}$$

and the *spectral determinants* $D^{\pm}(\lambda)$, defined through *zeta regularization*,

$$\log D^{\pm}(\lambda) \equiv \log \det{}^{\pm}(\hat{H}+\lambda) \stackrel{\text{def}}{=} [-\partial_s Z^{\pm}(s,\lambda)]_{s \rightsquigarrow 0}, \tag{6}$$

where "$s \rightsquigarrow 0$" implies analytical continuation in s. Scaling laws follow:

$$\det{}^{\pm}[r(\hat{H}+\lambda)] \equiv r^{Z^{\pm}(s=0,\lambda)} \det{}^{\pm}(\hat{H}+\lambda) \qquad (\forall r > 0),$$
$$\text{where} \qquad Z^{\pm}(s=0,\lambda) \equiv -\frac{\beta_{-1}(0)}{N} \pm \frac{1}{4} \tag{7}$$

[11, equations (7), (30)][12, equations (15), (27), (37)].

A more concrete realization of $\log D^{\pm}$ through (6) is, first to formally apply $(\mathrm{d}/\mathrm{d}\lambda)^m$ to (6) with the minimal m such that the result ($\propto Z(m,\lambda)$) converges, i.e., $m > \frac{1}{2} + \frac{1}{N}$, then to integrate back: the separate knowledge that the $\lambda \to +\infty$ expansion of $\log D^{\pm}(\lambda)$ shall only have "canonical" terms [14][12, § 1.1.2] fixes the m integration constants. Here, $N \geq 4$ implies $m = 1$: specifically,

$$\frac{\mathrm{d}}{\mathrm{d}\lambda} \log D^{\pm}(\lambda) \equiv Z^{\pm}(1,\lambda) \tag{8}$$

converges according to (5), and $\log D^{\pm}(\lambda)$ is then *the* unique primitive of $Z^{\pm}(1,\lambda)$ which is devoid of a constant ($\propto \lambda^0$) term in its large-λ expansion.

Classical analogs of those quantum determinants can be defined as well [11, 12, 10], through: $\log D^{\pm}_{\mathrm{cl}}(\lambda) \stackrel{\text{def}}{=}$ {the divergent part of $\log D^{\pm}(\lambda)$ for $\lambda \to +\infty$}, or equivalently [12, § 1.2.1 and equation (46)] through:

$$\log(D^{+}_{\mathrm{cl}}/D^{-}_{\mathrm{cl}})(\lambda) = \log \Pi(0) \equiv \tfrac{1}{2}\log\lambda, \tag{9}$$

$$\log(D^{+}_{\mathrm{cl}}D^{-}_{\mathrm{cl}})(\lambda) = \int_{-\infty}^{+\infty} \Pi(q)\,\mathrm{d}q = 2I, \quad I \stackrel{\text{def}}{=} \int_0^{+\infty} (V(q)+\lambda)^{1/2}\,\mathrm{d}q, \tag{10}$$

where this divergent "improper action integral" gets specified *just like* $\log D^{\pm}$: first,

$$\frac{\mathrm{d}I}{\mathrm{d}\lambda} = \frac{1}{2}\int_0^{+\infty} (V(q)+\lambda)^{-1/2}\,\mathrm{d}q \equiv \frac{1}{4}\int_{-\infty}^{+\infty} (V(q)+\lambda)^{-1/2}\,\mathrm{d}q \tag{11}$$

converges, then $I(\lambda)$ is *defined* as that primitive of (11) which is devoid of a constant ($\propto \lambda^0$) term in its large-λ expansion.

Improper actions as in (10) (i.e., along *infinite* paths) offer *many* benefits for asymptotic and exact WKB analysis. WKB solutions of (1) can now be defined intrinsically: e.g., as $\Psi_{\mathrm{WKB}}(q) = \Pi(q)^{-1/2} \exp \int_{-\infty}^{q} \Pi(q')\,\mathrm{d}q'$, unlike the traditional forms which awkwardly involve extraneous base points. The geometrical analysis no longer requires to set infinite paths apart as it used to [13, 17]. Moreover, the algebra itself is simplified; e.g., consider the full determinant $D(\lambda) \stackrel{\text{def}}{=} (D^{+}D^{-})(\lambda)$: previously, to get the large-λ expansion of $\log D$ in the simplest case $V(q) = q^{2M}$ [13], we had to factor $D = D_{\mathrm{cl}}\,a$ $(a(\lambda)$

is the "Jost function"), then expand $\log a$ using $\log a \equiv \int_{-\infty}^{+\infty} [U - \Pi](q)\,dq$ where $\Psi(q) = U(q)^{-1/2} \exp \int U\,dq$ parametrizes an exact solution of (1), and finally obtain $\log D_{\mathrm{cl}}$ by other means; now that the improper integrals (10) are allowed, all that condenses into a single identity (valid for general V):

$$\log D \equiv \int_{-\infty}^{+\infty} U(q)\,dq. \tag{12}$$

2 Explicit improper actions for trinomial $\Pi(q)^2$

In [1], we computed the improper action integral $I = \int_0^{+\infty} \Pi(q)\,dq$ in closed form for *any binomial* $\Pi(q)^2 = uq^N + vq^M$; then (§ 4.2) we stated that we could no longer do so for a *trinomial* of the general (even) form $\Pi(q)^2 = q^N + vq^M + \lambda$ (with $N > M > 0$), for which we just needed the $v \to +\infty$ behavior of I anyway [1, equation (4.16)], as reproduced in (36) below.

It is nevertheless *wrong* to infer from the above that *strictly no* exact computations can be done in fully trinomial cases, and we now present several examples (still for even $q^N + vq^M + \lambda$ with positive v, λ). After recalling the closed-form results for binomials, we will quote another, trivial and degenerate, instance: perfect-square trinomials. Then, our main new case will be the *quartic* anharmonic oscillator: we can reduce its improper action exactly to standard (i.e., convergent) action integrals, and therefrom to *complete elliptic integrals* [2, 15, 5, 16], as (30)–(32) below; we then verify the abovementioned large-v behavior on this case ($N = 4$). Finally, the same approach must work for higher-degree polynomial $V(q)$, converting $\int_0^{+\infty} \Pi(q)\,dq$ exactly into convergent *hyperelliptic* integrals (as studied in [17, 18]); but since the latter remain not so explicitly understood, we will skip this case ($N > 4$) here.

2.1 Binomial $\Pi(q)^2$: exact evaluation

For $\Pi(q)^2 = uq^N + vq^M$ (with $N > M \geq 0$), $\int_0^{\infty} \Pi(q)\,dq$ was exactly computed in [1, § 4.1]. We recall the main formulae for later convenience:

$$I = \int_0^{+\infty} (uq^N + vq^M)^{1/2}\,dq \overset{\mathrm{def}}{=} \lim_{s \rightsquigarrow 0} I_0(s), \tag{13}$$

$$I_0(s) = \int_0^{+\infty} (uq^N + vq^M)^{1/2 - s}\,dq \qquad (\operatorname{Re} s > \tfrac{1}{2} + \tfrac{1}{N}) \tag{14}$$

$$\equiv \frac{\Gamma(\frac{M(1-2s)+2}{2(N-M)})\Gamma(-\frac{N(1-2s)+2}{2(N-M)})}{(N-M)\,\Gamma(s-1/2)}\, u^{-\frac{M(1-2s)+2}{2(N-M)}}\, v^{\frac{N(1-2s)+2}{2(N-M)}}, \tag{15}$$

where "$s \rightsquigarrow 0$" implies analytical continuation in s, with the result:

$$I = \frac{\Gamma(j - \frac{1}{2})\Gamma(-j)}{(N-M)\Gamma(-\frac{1}{2})}\, u^{-j+1/2}\, v^{j} \quad (j \overset{\mathrm{def}}{=} \tfrac{N+2}{2(N-M)} > \tfrac{1}{2}) \quad \text{when finite;} \tag{16}$$

otherwise, i.e., when $j = 1, 2, \ldots$, a further "canonical" renormalization yields

$$I = -\frac{2j\,\beta_{-1}(0)}{N+2}\left[\log v - \sum_{m=1}^{j}\frac{1}{m} - \frac{2M}{N}\left(\log 2 + \tfrac{1}{2}\log u - \sum_{m=1}^{j-1}\frac{1}{2m-1}\right)\right],$$
$$\beta_{-1}(0) = (-1)^{j-1}\frac{(2j-2)!}{2^{2j-1}(j-1)!\,j!}\,u^{-j+1/2}\,v^{j}. \tag{17}$$

We repeat from [1] the examples we will mostly need:

$$\int_0^{+\infty}(wq^N + \lambda)^{1/2}\,dq = -\frac{\Gamma(1+\frac{1}{N})\Gamma(-\frac{1}{2}-\frac{1}{N})}{2\sqrt{\pi}}\,w^{-\frac{1}{N}}\lambda^{\frac{1}{2}+\frac{1}{N}} \quad (N \neq 2) \tag{18}$$
$$= -\tfrac{1}{4}\,w^{-1/2}\lambda(\log\lambda - 1) \quad (N = 2) \tag{19}$$
$$\int_0^{+\infty}(q^4 + vq^2)^{1/2}\,dq = -\tfrac{1}{3}\,v^{3/2}, \tag{20}$$

all based on (16) except (19), which uses (17) with $j = 1$.

2.2 Perfect-square trinomials: $\Pi(q)^2 = (q^M + \sqrt{\lambda})^2$ (M even)

This degenerate case trivially reduces to a binomial formula like (16), using

$$\int_0^{+\infty}[(q^M + wq^L)^2]^{1/2-s}dq = \frac{\Gamma\left(\frac{L(1-2s)+1}{M-L}\right)\Gamma\left(-\frac{M(1-2s)+1}{M-L}\right)}{(M-L)\Gamma(2s-1)}\,w^{\frac{M(1-2s)+1}{M-L}} \tag{21}$$

for $M > L \geq 0$ and $w > 0$. The singular formula (17) is never needed in our setting ($\Pi(q)^2$ even): for M and L even, no pole can appear in the numerator of (21) at $s = 0$; but one appears in the denominator instead, leading to

$$I = \int_0^{\infty}[(q^{N/2} + \sqrt{\lambda})^2]^{1/2}\,dq \equiv 0 \qquad (\text{for even } N/2 > 0). \tag{22}$$

2.3 The general even quartic case: $\Pi(q)^2 = q^4 + vq^2 + \lambda$

At present, we mean to exploit the large toolbox of results readily available for the *complete elliptic integrals* [19, 20, 21]: specifically here,

$$K(k) \overset{\text{def}}{=} \int_0^1 \left[(1-t^2)(1-k^2t^2)\right]^{-1/2}dt, \quad E(k) \overset{\text{def}}{=} \int_0^1 \left[\frac{(1-k^2t^2)}{(1-t^2)}\right]^{1/2}dt, \tag{23}$$

as functions of the *modulus* k; the *complementary modulus* is $k' \overset{\text{def}}{=} \sqrt{1-k^2}$.

Main needed formulae

- special values: [19, formulae 13.8(5),(6),(15),(16)]

$$K(0) = E(0) = \tfrac{1}{2}\pi \tag{24}$$

$$K(\tfrac{1}{\sqrt{2}}) = \frac{\Gamma(\frac{1}{4})^2}{4\sqrt{\pi}}, \qquad E(\tfrac{1}{\sqrt{2}}) = \frac{1}{2}\left[K(\tfrac{1}{\sqrt{2}}) + \frac{\pi}{2K(\frac{1}{\sqrt{2}})}\right]; \tag{25}$$

- derivatives: [20, formulae 710.00, 710.02][21, formulae 8.123(2),(4)]

$$\frac{\mathrm{d}K}{\mathrm{d}k} = \frac{E(k)}{kk'^2} - \frac{K(k)}{k} \qquad \left(\frac{\mathrm{d}E}{\mathrm{d}k} = \frac{E(k) - K(k)}{k} \text{ is not used here}\right); \tag{26}$$

- expansions for $k \to 1^- \iff k' \to 0^+$ (implying $E(1) = 1$): [22, p. 93–94] [2, footnote 11 p. 1237] [20, formulae 900.05, 900.07]

$$\begin{aligned} K(k) &= \log\frac{4}{k'} + \frac{1}{4}\left[\log\frac{4}{k'} - 1\right] k'^2 + O(k'^4 \log k') \\ E(k) &= 1 + \frac{1}{2}\left[\log\frac{4}{k'} - \frac{1}{2}\right] k'^2 + \frac{3}{16}\left[\log\frac{4}{k'} - \frac{13}{12}\right] k'^4 + O(k'^6 \log k') \end{aligned} \tag{27}$$

- selected transformation formulae: [19, Table 4 p. 319]

$$K(k) = \frac{1+\dot{k}'}{2} K(\dot{k}), \quad E(k) = \frac{E(\dot{k}) + \dot{k}' K(\dot{k})}{1+\dot{k}'} \qquad \text{for } k = \frac{1-\dot{k}'}{1+\dot{k}'} \tag{28}$$

$$K(k) = \tilde{k}' K(\tilde{k}), \qquad E(k) = \frac{E(\tilde{k})}{\tilde{k}'} \qquad \text{for } k = \frac{\mathrm{i}\tilde{k}}{\tilde{k}'}. \tag{29}$$

Our closed-form result

For non-negative v and λ (as in [1], and mainly for simplicity), we find:

$$I \stackrel{\text{def}}{=} \int_0^{+\infty} (q^4 + vq^2 + \lambda)^{1/2}\,\mathrm{d}q \tag{30}$$

$$(v \geq 2\sqrt{\lambda}) : \equiv \tfrac{1}{3}(v + 2\sqrt{\lambda})^{1/2}\left[2\sqrt{\lambda}K(k) - vE(k)\right], \quad k = \left(\frac{v - 2\sqrt{\lambda}}{v + 2\sqrt{\lambda}}\right)^{1/2}; \tag{31}$$

$$(v \leq 2\sqrt{\lambda}) : \equiv \tfrac{1}{3}\lambda^{1/4}\left[(2\sqrt{\lambda} + v)K(\tilde{k}) - 2vE(\tilde{k})\right], \quad \tilde{k} = \frac{(2\sqrt{\lambda} - v)^{1/2}}{2\lambda^{1/4}}. \tag{32}$$

Derivation. We first specify $\mathrm{d}I/\mathrm{d}\lambda$ by means of (11) for $V(q) = q^4 + vq^2$. In contrast to (30), here the integrand $(V(q) + \lambda)^{-1/2}$ *is integrable* at $q = \infty$ in $\mathbb{C}$, allowing to deform the path $(-\infty, +\infty)$ to a *bounded* contour in the complex q-plane:

$$\frac{\mathrm{d}I}{\mathrm{d}\lambda} = \frac{1}{4}\int_{-\infty}^{+\infty} (q^4 + vq^2 + \lambda)^{-1/2}\,\mathrm{d}q = \frac{1}{4}\int_C (q^4 + vq^2 + \lambda)^{-1/2}\,\mathrm{d}q \qquad (33)$$

where C is, e.g., a positive contour encircling the pair of roots $\mathrm{i}q_\pm$ of $\Pi(q)^2$ (*turning points*) that lie in the upper half-plane. We now prefer to pursue explicitly with $v \geq 2\sqrt{\lambda}$ (and analytically continue the result to $v \leq 2\sqrt{\lambda}$ later): then $0 < q_- \leq q_+$, cf. Fig. 1(a). The last integral in (33), being taken over a bounded path, admits a closed-form primitive with respect to λ, as

$$\begin{aligned}\hat{I}(\lambda) &= \frac{1}{2}\int_C (q^4 + vq^2 + \lambda)^{1/2}\,\mathrm{d}q \\ &= -\int_{q_-}^{q_+} (-q^4 + vq^2 - \lambda)^{1/2}\,\mathrm{d}q = -\tfrac{1}{3}\,q_+\big[vE(\dot{k}) - 2q_-^2 K(\dot{k})\big] \quad (34)\end{aligned}$$

where $q_\pm = \big[\tfrac{1}{2}(v \pm \sqrt{v^2 - 4\lambda})\big]^{1/2}$ and $\dot{k} = [1 - q_-^2/q_+^2]^{1/2}$, $\dot{k}' = q_-/q_+$

[19, formula 3.155(1) for $u = b$ and (amplitude) $\lambda = \pi/2$] [2, formula (4.22)[1]].

We cannot rush to conclude that $I = \hat{I}$: the former contour deformation is ill-justified for the divergent integral I itself. On the other hand, we find that it simplifies future steps to use the transformation formula (28) which turns (34) into the expression (31), *but still for* $\hat{I}$.

Next, we continue (31) to the region $\{v \leq 2\sqrt{\lambda}\}$ (k pure-imaginary) by means of the transformation (29), which results in the expression (32) *again for* $\hat{I}$. Only then are we able to probe the $\lambda \to +\infty$ behavior of $\hat{I}$ at fixed v : using $\tilde{k} = \frac{1}{\sqrt{2}} - \frac{1}{4\sqrt{2}}\,v\lambda^{-1/2} + O(\lambda^{-1})$ and (25)–(26), we obtain

$$\begin{aligned}\hat{I}(\lambda) &\sim \tfrac{2}{3}\lambda^{3/4}\big[K(\tfrac{1}{\sqrt{2}}) + \tfrac{\mathrm{d}K}{\mathrm{d}k}(\tfrac{1}{\sqrt{2}})\,\tfrac{-1}{4\sqrt{2}}\,v\lambda^{-1/2} + O(\lambda^{-1})\big] \\ * &\quad +\tfrac{1}{3}\,v\lambda^{1/4}\big[K(\tfrac{1}{\sqrt{2}}) + O(\lambda^{-1/2})\big] - \tfrac{2}{3}\,v\lambda^{1/4}\big[E(\tfrac{1}{\sqrt{2}}) + O(\lambda^{-1/2})\big] \\ &\sim \tfrac{2}{3}\,K(\tfrac{1}{\sqrt{2}})\,\lambda^{3/4} - \frac{\pi}{4K(\frac{1}{\sqrt{2}})}\,v\lambda^{1/4} + O(\lambda^{-1/4}) \qquad (\lambda \to +\infty); \quad (35)\end{aligned}$$

it has no constant ($\propto \lambda^0$) term, hence indeed $\hat{I} \equiv I$, the wanted canonical primitive as defined initially, cf. (10). □

Remark 1. The trivial outcome ($I = \hat{I}$) *seems* to justify the above contour deformation directly for the divergent integral (30), but this is misleading: our $\Pi(q)$ kept a *null residue* $\beta_{-1}(0)$, like all even $\Pi(q)$ with $N \equiv 0 \bmod 4$; but generically, $\beta_{-1}(0) \neq 0$ (e.g., already for trinomial even $\Pi(q)^2$ but with $N = 6, 10, \ldots$), and nontrivial integration constants $I - \hat{I} \neq 0$ ought to follow.

Applications

We can first verify (30)–(32) upon special cases, known earlier:

[1] We think there should be no factor $\rho^{1/2}$ on the left-hand side of this formula.

- $\lambda = 0$: $\quad I = -\frac{1}{3}\, v^{3/2} E(1) = -\frac{1}{3}\, v^{3/2}$ by (27) for $E(1)$, cf. (20);
- $v = 0$: $\quad I = \frac{2}{3}\, K(\frac{1}{\sqrt{2}})\, \lambda^{3/4} = \frac{\Gamma(1/4)^2}{6\sqrt{\pi}}\, \lambda^{3/4}$ by (25), cf. (18) for $N = 4$;
- $v = 2\sqrt{\lambda}$: $I = \frac{\sqrt{2}}{3}\, v^{3/2}[K(0) - E(0)] \equiv 0$ by (24), cf. (22).

But above all, we can use the exact expression (31) to check the $v \to +\infty$ behavior of I directly. Earlier, we predicted the asymptotic form for the general trinomial case to be, for $v \to +\infty$ at fixed λ, [1, equation (4.16)]

$$\int_0^{+\infty} (q^N + vq^M + \lambda)^{1/2}\, \mathrm{d}q \sim \int_0^{+\infty} (q^N + vq^M)^{1/2}\, \mathrm{d}q + \int_0^{+\infty} (vq^M + \lambda)^{1/2}\, \mathrm{d}q$$
$$+ \delta_{M,2} \frac{N}{4(N-2)}\, v^{-1/2} \lambda (\log v + 2 \log 2), \quad (36)$$

where the first line is to be made explicit through (16)–(20), and δ (last line) is the Kronecker delta symbol.

However, our derivation of (36) was quite indirect, and lacked independent tests. Now the present results allow such a test: in the quartic case, we can directly expand I in its exact form (31) for $v \to +\infty$, i.e., $k \to 1^-$, and

$$k' \equiv 2\big(\sqrt{\lambda}/v\big)^{1/2} \big(1 + 2\sqrt{\lambda}/v\big)^{-1/2} \to 0^+. \quad (37)$$

Then, using (27), $I \equiv -\frac{4}{3}\, \lambda^{3/4}\, k'^{-3} \big[(2 - k'^2)\, E(k) - k'^2 K(k)\big]$ expands as

$$I \sim -\tfrac{8}{3}\, \lambda^{3/4}\, k'^{-3} \big[1 - \tfrac{3}{4}\, k'^2 - \tfrac{3}{16} \big(\log \tfrac{4}{k'} - \tfrac{1}{4}\big)\, k'^4 + O(k'^6 \log k')\big]\,; \quad (38)$$

the substitution of k' by (37) yields the desired $v \to +\infty$ expansion in terms of $v^{3/2-n}$ and $v^{-1/2-n} \log v$, $n \in \mathbb{N}$ (no $v^{1/2} \log v$ term!). Remarkably, the next subleading term (of order $v^{1/2}$) also cancels, so that finally

$$I \sim -\tfrac{1}{3}\, v^{3/2} - \tfrac{1}{4} \lambda\, v^{-1/2} \big(\log(\lambda/v^2) - 4 \log 2 - 1\big)\ [+\, O(v^{-3/2} \log v)]. \quad (39)$$

This asymptotic equivalent then *identically reproduces* the prediction made by (36) for $N = 4$ and $M = 2$ with the help of (19)–(20), which confirms our basic earlier result [1, equation (4.16)].

3 The $v \to \infty$ behavior of the determinants

We return to the spectral determinants of the quantum problem (1):

$$D_N^{\pm}(\lambda, v) = \det{}^{\pm}(-\mathrm{d}^2/\mathrm{d}q^2 + q^N + vq^M + \lambda), \quad (40)$$

which are entire functions of $(\lambda, v) \in \mathbb{C}^2$ [24].

3.1 Review of the $v \to +\infty$ results

Our key intermediate result in [1, equations (3.10–12)] was, for $v \to +\infty$:

$$D_N^{\pm}(\lambda, v) \sim \mathrm{e}^{\int_0^{+\infty}(q^N+vq^M+\lambda)^{1/2}\mathrm{d}q}\mathrm{e}^{-\int_0^{+\infty}(vq^M+\lambda)^{1/2}\mathrm{d}q}\det{}^{\pm}(-\mathrm{d}^2/\mathrm{d}q^2+vq^M+\lambda). \tag{41}$$

Now the asymptotic formula (36) reduces this to

$$D_N^{\pm}(\lambda, v) \sim \mathrm{e}^{I(v)}\mathrm{e}^{\delta_{M,2}A_0(\lambda,v)}\det{}^{\pm}(-\mathrm{d}^2/\mathrm{d}q^2 + vq^M + \lambda), \tag{42}$$

$$I(v) = \textstyle\int_0^{+\infty}(q^N + vq^M)^{1/2}\,\mathrm{d}q, \quad A_0(\lambda, v) = \frac{N}{4(N-2)}(\log v + 2\log 2)\, v^{-1/2}\lambda. \tag{43}$$

On the other hand, the exact scaling laws (2) and (7) for $u = 0$, plus

$$\beta_{-1}(0) \equiv \delta_{M,2}\,\Lambda/2 \quad \text{for } \Pi(q) = (q^M + \Lambda)^{1/2}, \tag{44}$$

entail (writing $D_M^{\pm}(\Lambda) \equiv \det^{\pm}(-\mathrm{d}^2/\mathrm{d}q^2 + q^M + \Lambda)$):

$$\det{}^{\pm}(-\mathrm{d}^2/\mathrm{d}q^2 + vq^M + \lambda) \equiv v^{\pm 1/[2(M+2)]}v^{-\delta_{M,2}\,v^{-1/2}\lambda/8}D_M^{\pm}(v^{-2/(M+2)}\lambda); \tag{45}$$

our net asymptotic result was thus [1, equations (5.1–4)]

$$D_N^{\pm}(\lambda, v) \sim \mathrm{e}^{I(v)}\mathrm{e}^{\delta_{M,2}A(\lambda,v)}v^{\pm 1/[2(M+2)]}D_M^{\pm}(\Lambda) \qquad (v \to +\infty), \tag{46}$$

with $I(v) = \int_0^{+\infty}(q^N + vq^M)^{1/2}\,\mathrm{d}q$ given by (16) if $j = \frac{N+2}{2(N-M)} \notin \mathbb{N}$, or by (17) otherwise, and

$$\delta_{M,2}\,A(\lambda, v) = \delta_{M,2}\,\frac{1}{8(N-2)}\,[(N+2)\log v + 4N\log 2]\,\Lambda, \tag{47}$$

$$\Lambda \overset{\text{def}}{=} v^{-2/(M+2)}\lambda \qquad (\equiv v^{-1/2}\lambda \text{ in (47), used when } M = 2). \tag{48}$$

3.2 Extension to a sector in the complex v-plane

The key to our proof of the asymptotic formula (41) for positive $v \to +\infty$ was [1, § 3.2] that a solution $\Psi_\lambda(q, v)$ of (1) with a recessive WKB form for $q \to +\infty$ *connects* all the way down (in that WKB form) to a region $\{1 \ll q \ll v^{1/(N-M)}\}$ – where it then tends to a similarly recessive solution $\Psi_{0,\lambda}(q, v)$ of the *uncoupled* Schrödinger equation $[-(\mathrm{d}^2/\mathrm{d}q^2) + vq^M + \lambda]\Psi_{0,\lambda}(q) = 0$.

In the complex domain, a simple sufficient condition for the WKB form to be preserved is for $q \in \mathbb{C}$ to stay within *one Stokes region* of the momentum function $\Pi(q)$ [13]. In terms of $\theta \overset{\text{def}}{=} \arg v$, the above connection condition then becomes that the Stokes *region* containing $\{1 \ll \mathrm{e}^{\mathrm{i}\theta/(M+2)}q \ll v^{1/(N-M)}\}$ (rotation given by the uncoupled equation) should *link to* $q = +\infty$. E.g., when $v > 0$ the *central* Stokes region does include all of $\mathbb{R}$, cf. Fig. 1(a). We now

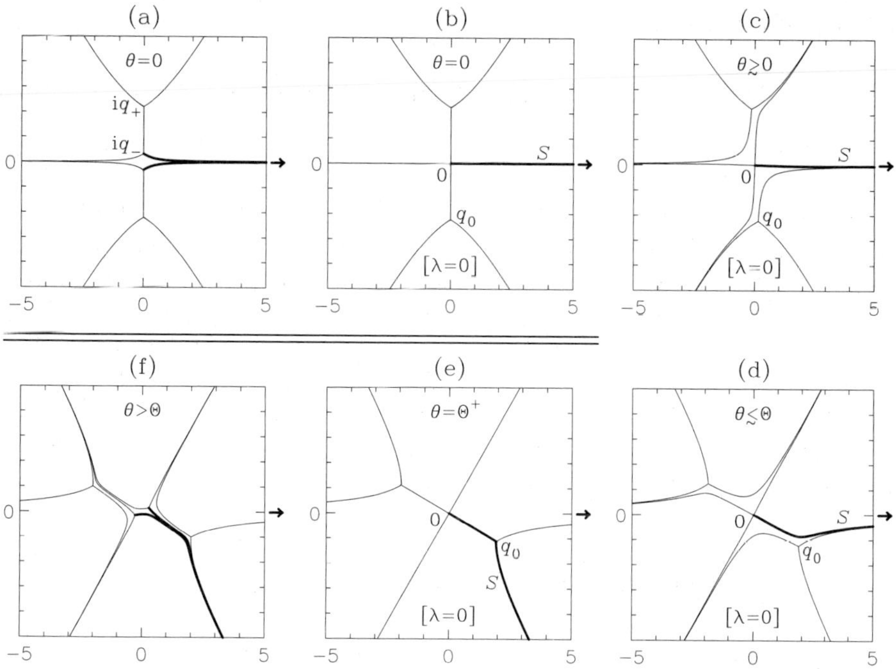

Fig. 1. Plots of the Stokes geometry in the complex q-plane for $\Pi(q)^2 = q^4 + vq^2 + \lambda$ and large complex v, ordered *clockwise* with increasing $\theta = \arg v$ ($|v| = 5$, $\lambda = 0.5$). The intermediate plots (b–e) set $\lambda = 0$ to emulate the $|v| = \infty$ regime at finite q; in that limit the Stokes curve S (*bold line*) stays linked to $q = +\infty$ (*arrow*) for $\theta < \Theta$ (here $\Theta = 2\pi/3$, by (49)).

need to describe the Stokes geometry for $\Pi(q)^2 = q^N + vq^M + \lambda$ with *complex* $v \to \infty$; Fig. 1 illustrates the case $N = 4$, $M = 2$.

When $|v| \to \infty$: the approximate factorization of $\Pi(q)^2$ as $(q^{N-M} + v) \times (q^M + \lambda/v)$ makes M of its complex turning points q_j shrink ($\asymp v^{-1/M}$, "inner" roots) and the other $(N-M)$ grow ($\asymp v^{1/(N-M)}$, "outer" roots); moreover, the central Stokes region contracts to a symmetrical pair of Stokes *curves* from $q = 0$ for the *zero-energy* momentum $\Pi_{\lambda=0}(q)^2 = q^N + vq^M$, and we are to follow the (θ-dependent) Stokes curve S which starts as $S = \mathbb{R}^+$ when $\theta = 0$, cf. Fig. 1(b). In the large-v limit, the connection condition is that θ can be increased above 0 as long as S *remains linked to* $q = +\infty$ (Fig. 1(c–d); the complex-conjugate picture results for $\theta < 0$).

Following [17, § 3], the connection condition breaks (cf. Fig. 1(e)) when the action integral $\mathcal{I} = \int_0^{q_0} (q^N + vq^M)^{1/2}\,\mathrm{d}q$ becomes *real*, where q_0 is the first outer turning point met by S as θ recedes from 0. That action, of *instanton* type [10], is computable in closed form: $q_0 = \mathrm{e}^{-\mathrm{i}\,\pi/(N-M)} v^{1/(N-M)}$, and $\mathcal{I} = \frac{\sqrt{\pi}}{N+2}\,\Gamma(\frac{M+2}{2(N-M)})/\Gamma(\frac{N+2}{2(N-M)})\ [\mathrm{e}^{-\mathrm{i}\,(M+2)\pi} v^{N+2}]^{1/[2(N-M)]}$, which turns real first at $\arg v = \frac{M+2}{N+2}\,\pi$. Consequently, all of § 3.1 extends (at least) to the asymptotic sector

$$\Sigma = \{v \to \infty,\ |\arg v| < \Theta\}, \qquad \Theta = \frac{M+2}{N+2}\,\pi. \tag{49}$$

Remark 2. Some examples (with $\lambda \equiv 0$) make us hope that our end asymptotic formula (46) might actually hold up to $|\arg v| < \pi$. 1) For $\mathrm{Qi}^{\pm}(v) \stackrel{\mathrm{def}}{=} \det^{\pm}(-\mathrm{d}^2/\mathrm{d}q^2 + q^4 + vq^2)$, this was suggested by our numerical observations [12, equation (87) vs Fig. 1] that $\mathrm{Qi}^{\pm}(v)$ behave analogously to the Airy functions $\mathrm{Ai}(v)$, $\mathrm{Ai}'(v)$ for $v \to -\infty$ as well (even though $\Theta = 2\pi/3$ only). 2) The *supersymmetric* determinants $\det^{\pm}(-\mathrm{d}^2/\mathrm{d}q^2 + q^N + vq^{N/2-1})$ are known in closed form, essentially as inverse Γ-functions of v [12, equation (120)]: their large-v asymptotics then amount to the Stirling formula, and the latter definitely holds for $|\arg v| < \pi$ (vs $\Theta = \pi/2$).

4 Asymptotics and the functional relation

An early puzzle of general exact quantization conditions was their *breakdown* (both analytical and numerical) for potentials $q^N + vq^2$ in the regime $v \gg 1$ (as seen for $N = 4$ [11]). Naively, convergence to the elementary harmonic (vq^2) behavior would have been expected. We can now show that singularity to be *unessential*: i.e., the original functional relations (*Wronskian identities*) which produce those quantization conditions behave as well as possible when $v \to +\infty$ for any potential $q^N + vq^M$, *currently under the restriction* $2M + 2 > N$ (which encompasses $q^4 + vq^2$, for instance).

4.1 The basic Wronskian identity

The spectral determinants for a general polynomial potential $V(q)$ of degree N obey the bilinear functional relation: [11, equation (40)]

$$\mathrm{e}^{+\mathrm{i}\varphi_N/4} D^{+[1]} D^{-} - \mathrm{e}^{-\mathrm{i}\varphi_N/4} D^{+} D^{-[1]} \equiv 2\,\mathrm{i}\,\mathrm{e}^{\mathrm{i}\varphi_N \beta_{-1}(0)/2}, \tag{50}$$

where $D^{\pm[1]}$ are the determinants for the *first conjugate problem*: [24, § 7]

$$V(q) \mapsto V^{[1]}(q) \stackrel{\mathrm{def}}{=} \mathrm{e}^{-\mathrm{i}\varphi_N} V(\mathrm{e}^{-\mathrm{i}\varphi_N/2} q) \quad \text{and} \quad \lambda \mapsto \lambda^{[1]} \stackrel{\mathrm{def}}{=} \mathrm{e}^{-\mathrm{i}\varphi_N}\lambda, \tag{51}$$

$$\text{with} \quad \varphi_N \stackrel{\mathrm{def}}{=} \frac{4\pi}{N+2}: \qquad \text{the } \textit{symmetry angle} \text{ in degree } N. \tag{52}$$

Equation (50) is but a *Wronskian identity* for the Schrödinger equation (1), yet it has a key *dynamical* role: while it seems underdetermined, it implies a *complete* set of *exact quantization conditions*, which then solve (1) exactly.

A certain iterate of the transformation (51) is the identity, hence (50) has a cyclic symmetry group, specifically of order $(\frac{1}{2}N+1)$ when V is even.

For the trinomial determinants (40), the first-conjugate parameters are

$$\lambda^{[1]} = \mathrm{e}^{-\mathrm{i}\varphi_N}\lambda, \quad v^{[1]} = \mathrm{e}^{\mathrm{i}\pi/j}v, \quad \text{with } j \equiv \frac{N+2}{2(N-M)} \quad \text{as in (16).} \qquad (53)$$

4.2 The $v \to \infty$ transition

According to (46)–(48) with $\Lambda = v^{-2/(M+2)}\lambda$, $D_N^{\pm}(\lambda, v)$ for finite v is a deformation from $D_M^{\pm}(\Lambda)$ at $v = +\infty$, but the key parameter in the dynamical functional relation (50), namely the degree of V, and often the residue $\beta_{-1}(0)$ as well [1, § 3.1], suffer sharp jumps at $v = \infty$. It is then a non-trivial task to find out whether the basic identity (50) for $D_N^{\pm}$ *continuously evolves* into its counterpart for $D_M^{\pm}$ in the $v \to +\infty$ limit of (46), or not.

Under $\binom{\lambda}{v} \mapsto \binom{\lambda^{[1]}}{v^{[1]}}$ as in (53), the rescaled spectral parameter Λ maps to

$$\Lambda \mapsto \Lambda^{[1]} = \exp\left[-\tfrac{2}{M+2}\tfrac{N-M}{2}\,\mathrm{i}\varphi_N - \mathrm{i}\varphi_N\right]\Lambda = \mathrm{e}^{-\mathrm{i}\varphi_M}\Lambda; \qquad (54)$$

already this is the correct rotation angle for the limiting determinants $D_M^{\pm}$.

To get the asymptotic form of (50) with $D^{\pm} \equiv D_N^{\pm}(\lambda, v)$, we let $v \to \infty$ in its left-hand side with $\arg v = -\pi/2j$, $\arg v^{[1]} = +\pi/2j$, and we invoke (46). *The latter, by (49), requires* $\pi/2j < \Theta \Leftrightarrow j > 1$ or $2M+2 > N$ (otherwise the calculation will still work, but only formally until (49) can grow to a wider sector). The left-hand side of (50) thus displays the asymptotic form

$$\begin{aligned} &\exp\left[I(v) + I(v^{[1]})\right]\ \exp \delta_{M,2}[A(\Lambda, v) + A(\Lambda^{[1]}, v^{[1]})] \times \\ &\qquad [z\, D_M^{+}(\Lambda^{[1]}) D_M^{-}(\Lambda) - z^{-1} D_M^{+}(\Lambda) D_M^{-}(\Lambda^{[1]})], \qquad (55) \\ &z \stackrel{\text{def}}{=} \mathrm{e}^{+\mathrm{i}\varphi_N/4}\left(v^{[1]}/v\right)^{+1/[2(M+2)]} \qquad \text{(a pure phase).} \end{aligned}$$

We now evaluate all the terms in (55): first,

$$\begin{aligned} I(v) \propto v^j \qquad &[\Rightarrow \quad I(v^{[1]}) = -I(v)] \text{ if } j \notin \mathbb{N} \text{ (cf. (16), with } \beta_{-1}(0) = 0) \\ = -\tfrac{2j}{N+2}\,&\beta_{-1}(0)\,[\log v + \text{const.}] \text{ if } j \in \mathbb{N} \text{ (cf. (17), with } \beta_{-1}(0) \propto v^j) \\ &\Rightarrow I(v) + I(v^{[1]}) = \tfrac{2j}{N+2}\,\beta_{-1}(0)\,\mathrm{i}\,\tfrac{\pi}{j} \end{aligned}$$

$$\Rightarrow \qquad I(v) + I(v^{[1]}) \equiv \mathrm{i}\varphi_N \beta_{-1}(0)/2 \quad \text{in all cases, cf. (53).} \qquad (56)$$

Next, just as for (54),

$$z = \mathrm{e}^{\mathrm{i}\varphi_N/4}\mathrm{e}^{\mathrm{i}\pi/[2(M+2)j]} = \mathrm{e}^{\mathrm{i}\varphi_N(N+2)/[4(M+2)]} \equiv \mathrm{e}^{\mathrm{i}\varphi_M/4}. \qquad (57)$$

Finally, and only relevant when $M = 2$, in which case $\Lambda^{[1]} = -\Lambda$,

$$A(\Lambda, v) + A(\Lambda^{[1]}, v^{[1]}) = -\tfrac{N+2}{8(N-2)}\,\mathrm{i}\,\tfrac{\pi}{j}\,\Lambda = -\tfrac{N+2}{16}\,\mathrm{i}\varphi_N\,\Lambda \equiv -\mathrm{i}\pi\Lambda/4. \qquad (58)$$

In the end, substituting (54)–(58) into (50) we indeed get

$$\mathrm{e}^{+\mathrm{i}\varphi_M/4} D_M^+(\mathrm{e}^{-\mathrm{i}\varphi_M}\Lambda) D_M^-(\Lambda) - \mathrm{e}^{-\mathrm{i}\varphi_M/4} D_M^+(\Lambda) D_M^-(\mathrm{e}^{-\mathrm{i}\varphi_M}\Lambda) \equiv 2\,\mathrm{i}\mathrm{e}^{\delta_{M,2}\,\mathrm{i}\pi\Lambda/4}, \tag{59}$$

which *is the correct form* of (50) for $D_M^\pm(\Lambda) = \det^\pm(-\mathrm{d}^2/\mathrm{d}q^2 + q^M + \Lambda)$ (whose $\beta_{-1}(0)$ is given by (44)). For more details: if $M > 2$, see [23, equation (5.32)]; if $M = 2$, then

$$D_2^+(\Lambda) = 2^{1-\Lambda/2}\sqrt{\pi}/\Gamma(\tfrac{1+\Lambda}{4}), \qquad D_2^-(\Lambda) = 2^{-\Lambda/2}\sqrt{\pi}/\Gamma(\tfrac{3+\Lambda}{4}), \tag{60}$$

and (59) with its "anomalous" right-hand side boils down to the reflection formula for the Gamma function; the harmonic-oscillator quantization condition can then also be recovered solely from (59) [12, Appendix A.2.3].

In conclusion, we have verified that the exact functional relation (50), governing both $D_N^\pm(\lambda, v)$ and $D_M^\pm(\Lambda)$ (cf. (59)), is compatible with the general perturbation formula (46), *currently under the restriction* $2M + 2 > N$, or $j > 1$ (which includes the quartic oscillators): this further validates the exact-WKB description of perturbative regimes in [1]. Remaining desirable tasks are: 1) to lift the restriction $j > 1$ (e.g., by extending (46) to $\{|\arg v| < \pi\}$, cf. Remark 2 in § 3.2); and 2) to find exact quantization conditions that themselves behave *continuously* in the zero-coupling limit (here, $v = +\infty$).

References

1. Voros, A.: From exact-WKB towards singular quantum perturbation theory. Publ. RIMS, Kyoto Univ. **40**, 973–990 (2004)
2. Bender, C.M., Wu, T.T.: Anharmonic oscillator. Phys. Rev. **184**, 1231–1260 (1969)
3. Simon, B.: Coupling constant analyticity for the anharmonic oscillator (with an appendix by A. Dicke). Ann. Phys. **58**, 76–136 (1970)
4. Graffi, S., Grecchi, V., Simon, B.: Borel summability: application to the anharmonic oscillator. Phys. Lett. **32B**, 631–634 (1970)
5. Hioe, F.T., Montroll, E.W.: Quantum theory of anharmonic oscillators. I. Energy levels of oscillators with positive quartic anharmonicity. J. Math. Phys. **16**, 1945–1955 (1975)
6. Shanley, P.E.: Spectral properties of the scaled quartic anharmonic oscillator. Ann. Phys. **186**, 292–324 (1988)
7. Kawai, T., Takei, Y.: Secular equations through the exact WKB analysis. In: Boutet de Monvel, L. (ed.): Analyse algébrique des perturbations singulières I. Méthodes résurgentes (Proceedings, CIRM, Marseille–Luminy 1991). Travaux en cours **47**, Hermann, Paris (1994) [pp. 85–102]
8. Delabaere, É., Pham, F.: Unfolding the quartic oscillator. Ann. Phys. **261**, 180–218 (1997)
9. Howls, C.J., Kawai, T., Takei, Y. (eds): Toward the exact WKB analysis of differential equations, linear or non-linear[2] (Proceedings, RIMS, Kyoto 1998). Kyoto University Press (2000)

[2] Our contribution to this volume (pp. 97–108) needs the same corrigendum as [11].

10. Zinn-Justin J., Jentschura U.D.: Multi-instantons and exact results I: conjectures, WKB expansions, and instanton interactions. Ann. Phys. **313**, 197–267 (2004)
11. Voros, A.: Exact resolution method for general 1D polynomial Schrödinger equation. J. Phys. **A32**, 5993–6007 (1999) [corrigendum: **A33**, 5783–5784 (2000)]
12. Voros, A.: Exercises in exact quantization.[3] J. Phys. **A33**, 7423–7450 (2000)
13. Voros, A.: The return of the quartic oscillator. The complex WKB method. Ann. Inst. H. Poincaré **A 39**, 211–338 (1983)
14. Voros, A.: Spectral functions, special functions and the Selberg zeta function. Commun. Math. Phys. **110**, 439–465 (1987)
15. Mathews, P.M., Eswaran, K.: On the energy levels of the anharmonic oscillator. Lett. Nuovo Cimento **5**, 15–18 (1972)
16. Kesarwani, R.N., Varshni, Y.P.: Eigenvalues of an anharmonic oscillator. J. Math. Phys. **22**, 1983–1989 (1981)
17. Delabaere, É., Dillinger, H., Pham, F.: Résurgence de Voros et périodes des courbes hyperelliptiques. Ann. Inst. Fourier (Grenoble) **43**, 163–199 (1993)
18. Elsner, B.: Hyperelliptic action integral. Ann. Inst. Fourier (Grenoble) **49**, 303–331 (1999)
19. A. Erdélyi (ed.): Higher Transcendental Functions (Bateman Manuscript Project). McGraw–Hill, New York (1953) [Vol. II, § 13.8]
20. Byrd, P.F., Friedman, M.D.: Handbook of Elliptic Integrals for Engineers and Scientists. Springer, Berlin Heidelberg New York (1971)
21. Gradshteyn, I.S., Ryzhik, I.M.: Table of integrals, series and products. 5th edition, Jeffrey, A. (ed.), Academic Press, New York (1994)
22. Radon, B.: Sviluppi in serie degli integrali ellittici. Atti Accad. Naz. Lincei, Mem., Cl. Sci. Fis. Mat. Nat., Ser. VIII, **2** Sez. 1, 69–108 (1950)
23. Voros, A.: Spectral zeta functions. In: Kurokawa, N., Sunada, T. (eds): Zeta functions in geometry (Proceedings, Tokyo 1990). Advanced Studies in Pure Mathematics **21**, Math. Soc. Japan, Kinokuniya, Tokyo (1992), pp. 327–358
24. Sibuya, Y.: Global Theory of a Second Order Linear Ordinary Differential Operator with a Polynomial Coefficient. North-Holland, Amsterdam (1975)

[3] In this work we mistakenly used "quasi-exactly solvable" for "supersymmetric" (systems) throughout — this affects none of the results.

WKB analysis and Poincaré theorem for vector fields

Masafumi Yoshino

Department of Mathematics, Graduate School of Science, Hiroshima University, 1-3-1 Kagamiyama, Higashi-Hiroshima, Hiroshima 739-8526, Japan
yoshino@math.sci.hiroshima-u.ac.jp

Summary. A normal form theory of a singular vector field from the viewpoint of an exact WKB analysis is studied. Classical Poincaré's theorem can be proved by a formal WKB solution, a resummation and an analytic continuation.

Key words: normal form of vector fields, Poincaré theorem, WKB analysis
Mathematics Subject Classification (2000): Primary 35C10; Secondary 45E10, 35Q15

1 Introduction

In this note we shall study the normal form theory of a singular vector field from the viewpoint of the exact WKB analysis. Usually, one constructs a change of variables which transforms a given singular vector field to a normal form by solving a system of singular partial differential equations, a homology equation. (See (7)). In this paper, we construct the transformation by a WKB solution, a Borel-Laplace resummation and an analytic continuation. We will see that the classical Poincaré series solution coincides with an analytic continuation of the WKB solution with respect to a certain parameter.

The exact WKB analysis for ordinary differential equations or integrable systems including Painlevé equations has been studied extensively. (cf. [1], [2] and [4]). On the other hand, in the case of semilinear partial differential equations, the theory has not been well developed. We note that the homology equation contains a Painlevé equation as a special case if we restrict independent variables to one variable in an approriate way. (cf. (13)). In view of this, we develop an exact WKB analysis to a homology equation.

For this purpose, we first introduce a parameter by taking the variation of eigenvalues of a transformed vector field as a new parameter. This rather artifical introduction of a parameter agrees with the one introduced to a Painlvé

Received 1 March, 2006. Revised 21 July, 2006. Accepted 8 August 2006.

equation through a monodromy preserving deformation of a Schrödinger equation if a homology equation reduces to a Painlvé equation. (cf. Remark 2.1 in the following). We will construct a WKB solution of a homology equation which is a natural extension of the one introduced by Aoki-Kawai-Takei for the Painlevé equation. Then we will show that a Borel-Laplace transform of a WKB solution gives an exact solution of a homology equation. Moreover, an analytic continuation of a resummed WKB solution with respect to the above parameter coincides with the classical Poincaré series solution.

This paper is organized as follows. In section 2, we introduce a homology equation which appears in the linearization of a singular vector field. In section 3 we study a WKB formal solution of a homology equation. In section 4 the resummation of a formal WKB solution is discussed. In section 5 the main theorem in this paper is proved. In the last section we discuss an analytic continuation of a resummed WKB solution and the relation between a WKB solution and the classical Poincaré solution. (cf. [3] and [5]).

2 Linearization of a vector field and a homology equation

Let $y = (y_1, \ldots, y_n) \in \mathbb{C}^n$, $n \geq 2$ be the variable in $\mathbb{C}^n$. We consider a singular vector field near the origin of $\mathbb{C}^n$

$$\mathcal{X} = \sum_{j=1}^{n} a^j(y) \frac{\partial}{\partial y_j}, \quad a^j(0) = 0, \quad j = 1, \ldots, n, \tag{1}$$

where $a^j(y)\,(j = 1, 2, \ldots, n)$ are holomorphic in some neighborhood of the origin. We set

$$X(y) = (a^1(y), a^2(y), \ldots, a^n(y)), \quad \frac{\partial}{\partial y} = {}^t(\frac{\partial}{\partial y_1}, \ldots, \frac{\partial}{\partial y_n}), \tag{2}$$

and write

$$\mathcal{X} = X(y) \cdot \frac{\partial}{\partial y}, \quad X(y) = y\Lambda + R(y), \tag{3}$$

$$R(y) = (R^1(y), R^2(y), \ldots, R^n(y)), \quad R(y) = O(|y|^2), \tag{4}$$

where Λ is an n-square constant matrix.

We want to linearize $\mathcal{X}$ by a change of variables,

$$y = u(x), \quad u = (u^1, u^2, \ldots, u^n), \tag{5}$$

namely,

$$X(u(x)) \frac{\partial x}{\partial y} \frac{\partial}{\partial x} = X(u(x)) \left(\frac{\partial y}{\partial x} \right)^{-1} \frac{\partial}{\partial x} = x\Lambda \frac{\partial}{\partial x}. \tag{6}$$

It follows that u satisfies the so-called homology equation

$$X(u(x))\left(\frac{\partial u}{\partial x}\right)^{-1} = x\Lambda, \tag{7}$$

that is

$$u\Lambda + R(u) = x\Lambda\frac{\partial u}{\partial x} \equiv \mathcal{L}u. \tag{8}$$

Introduction of a parameter η. Let λ_j $(j = 1, 2, \ldots, n)$ be the eigenvalues of Λ. Then we introduce a parameter η^{-1} in front of the eigenvalues of the linearized vector field by

$$\lambda_j \mapsto \eta^{-1}\lambda_j.$$

More precisely, if Λ is a diagonal matrix, then we introduce a parameter η in the following way

$$X(u(x))\left(\frac{\partial u}{\partial x}\right)^{-1} = \eta^{-1}x\Lambda. \tag{9}$$

Noting that $X(u(x)) = u\Lambda + R(u)$, we can write it in the form

$$\eta^{-1}\mathcal{L}u^j = \lambda_j u^j + R^j(u), \quad j = 1, \ldots, n. \tag{10}$$

In the following, we assume that Λ is semi simple. For the sake of simplicity, we assume that Λ is a diagonal matrix, $\Lambda = \mathrm{diag}\,(\lambda_1, \ldots, \lambda_n)$ for some λ_j. If we set $u = x + v(x)$ in (8), then we have $v\Lambda + R(x+v) = \mathcal{L}v$. By introducing a parameter by the same way as in the above we obtain

$$\eta^{-1}\mathcal{L}v^j = \lambda_j v^j + R^j(x + v(x)), \quad j = 1, \ldots, n. \tag{11}$$

Remark 2.1. Let us assume that λ_j $(j = 1, 2, \ldots, n)$ do not vanish. We introduce a new variable z_j by

$$e^{z_j} = x_j, \qquad j = 1, 2, \ldots, n.$$

We consider a plane wave solution

$$u^j = u^j(t), \quad t = z_1/(n\lambda_1) + \cdots + z_n/(n\lambda_n).$$

Then we have

$$\mathcal{L}u^j(t) = (u^j(t))',$$

and we can write the equation (10) in the form

$$\eta^{-1}(u^j)' = \lambda_j u^j + R^j(u), \quad j = 1, \ldots, n. \tag{12}$$

Let us consider the special case, $n = 3$ and

$$R^1(u) = u^1 \log u^1(\log u^2 - \log u^3),\ R^2(u) = u^2 \log u^2(\log u^3 - \log u^1),$$
$$R^3(u) = u^3 \log u^3(\log u^1 - \log u^2).$$

By a change of unknown functions

$$u^j = \exp(U^j), \quad j = 1, 2, \ldots, n,$$

we obtain a symmetric form of the fourth Painlevé equation

$$\begin{aligned} \eta^{-1}(U^1)' &= \lambda_1 + U^1(U^2 - U^3), \\ \eta^{-1}(U^2)' &= \lambda_2 + U^2(U^3 - U^1), \\ \eta^{-1}(U^3)' &= \lambda_3 + U^3(U^1 - U^2). \end{aligned} \tag{13}$$

On the other hand, by the monodromy preserving deformation, Aoki-Kawai-Takei introduced a parameter η in the Painlevé equation via a parameter in the Schrödinger equation. The above introduction of a large parameter in the homology equation coincides with that of Aoki-Kawai-Takei, if the homology equation is reduced to the fourth Painlevé equation.

3 WKB solution of a homology equation

Let $x = (x_1, x_2, \ldots, x_n)$ $(n \geq 1)$ be the variable in $\mathbb{C}^n$. By the assumption we have

$$\mathcal{L} = \sum_{j=1}^{n} \lambda_j x_j \frac{\partial}{\partial x_j}. \tag{14}$$

For the sake of simplicity we assume that

$$\lambda_j \neq 0, \qquad j = 1, 2, \ldots, n. \tag{15}$$

Hence we have $\det \Lambda \neq 0$. We begin with

Definition 3.1 (WKB solution). A WKB solution $v(x, \eta)$ of (11) is a formal power series solution of η^{-1} in the form

$$v(x, \eta) = \sum_{\nu=0}^{\infty} \eta^{-\nu} v_\nu(x) = v_0(x) + \eta^{-1} v_1(x) + \cdots, \tag{16}$$

where the coefficients $v_\nu(x)$ $(\nu = 0, 1, \ldots)$ are holomorphic vector functions of x in some open set independent of ν.

We set

$$v \equiv v(x, \eta) = (v^1(x, \eta), v^2(x, \eta), \ldots, v^n(x, \eta)), \tag{17}$$

and

$$v_\nu(x) = (v_\nu^1(x), \ldots, v_\nu^n(x)), \ (\nu = 0, 1, \ldots), \ \ v_\nu^j(x) = O(|x|^2), \ j = 1, 2, \ldots, n. \tag{18}$$

By substituting the expansion (16) into (11), we obtain

$$\mathcal{L} v^j = \sum_{\nu=0}^{\infty} \mathcal{L} v_\nu^j(x) \eta^{-\nu}, \tag{19}$$

$$R^j(x+v) = R^j(x + v_0 + v_1\eta^{-1} + v_2\eta^{-2} + \cdots) \\ = R^j(x+v_0) + \eta^{-1}\sum_{k=1}^{n}\left(\frac{\partial R^j}{\partial z_k}\right)(x+v_0)v_1^k + O(\eta^{-2}). \tag{20}$$

By comparing the coefficients of $\eta^0 = 1$ and η^{-1}, we obtain

$$\lambda_j v_0^j(x) + R^j(x_1 + v_0^1, \dots, x_n + v_0^n) = 0, \quad j = 1, 2, \dots, n, \tag{21}$$

$$\mathcal{L}v_0^j = \lambda_j v_1^j + \sum_{k=1}^{n}\left(\frac{\partial R^j}{\partial z_k}\right)(x+v_0)v_1^k, \quad j = 1, 2, \dots, n. \tag{22}$$

These equations give equations for v_0 and v_1. In order to determine $v_\nu(x)$ ($\nu \geq 2$) we compare the coefficients of $\eta^{-\nu}$ of both sides of (11). We differentiate (20) with respect to $\varepsilon = \eta^{-1}$, ν times and we put $\varepsilon = 0$. Then, we obtain, for $j = 1, 2, \dots, n$

$$\mathcal{L}v_{\nu-1}^j = \lambda_j v_\nu^j + \sum_{k=1}^{n}\left(\frac{\partial R^j}{\partial z_k}\right)(x+v_0)v_\nu^k \\ + (\text{terms consisting of } v_i^j,\ i \leq \nu - 1 \text{ and } j = 1, 2, \dots, n). \tag{23}$$

In order to determine v_ν from the above recurrence relations we need a definition. Let Λ be the diagonal matrix with diagonal components given by $\lambda_1, \dots, \lambda_n$ in this order.

Definition 3.2. The point x such that $\det(\Lambda + (\partial R/\partial z)(x + v_0)) = 0$ is called a turning point of the equation (11).

By the assumption (15), the origin $x = 0$ is not a turning point of (11) for every v_0 such that $v_0 = O(|x|^2)$. Then we have

Theorem 1. *Assume that* $\det \Lambda \neq 0$. *Then every coefficient of a WKB solution* $v(x, \eta)$, *(16) is uniquely determined as a holomorphic function in a neighborhood of the origin* $x = 0$ *independent of* ν.

Proof. By the implicit function theorem and (21), $v_0^j(x)$ can be uniquely determined such that $v_0^j(x)$ is holomorhic in some neighborhood of the origin $x = 0$ and satisfies $v_0^j(x) = O(|x|^2)$. Next $v_k^j(x)$, $(k = 1, 2, \dots; j = 1, \dots, n)$ can be uniquely determined as a holomorphic function at the origin by the assumption that the origin $x = 0$ is not a turning point. We note that $v_k^j(x)$ are inductively calculated by differentiation and algebraic manupulations. It follows that all $v_k^j(x)$ are holomorphic in some neighborhood of the origin independent of ν. □

For the later use we need

Definition 3.3 (Resonance condition). We say that η is *resonant*, if

$$\sum_{i=1}^{n} \lambda_i \alpha_i - \eta \lambda_j = 0, \tag{24}$$

for some $\alpha = (\alpha_1, \ldots, \alpha_n) \in \mathbb{Z}_+^n$, $|\alpha| \geq 2$ and $j, 1 \leq j \leq n$. If η is not resonant, we say that η is *nonresonant.*

Definition 3.4 (Poincaré condition) *We say that the set of eigenvalues $\lambda_j, (j = 1, 2, \ldots, n)$ satisfies the Poincaré condition, if the convex hull of $\lambda_j, (j = 1, 2, \ldots, n)$ in the complex plane does not contain the origin.*

4 Resummation of a WKB solution

Let $v(x, \eta) = \sum_{\nu=0}^{\infty} v_\nu(x) \eta^{-\nu}$ be a WKB solution of (11). We define $\tilde{v}(x, \eta) = v(x, \eta) - v_0(x)$. Then the (formal) Borel transform of $\tilde{v}(x, \eta)$ is defined by

$$B(\tilde{v})(x, \zeta) := \sum_{\nu=1}^{\infty} v_\nu(x) \frac{\zeta^{\nu-1}}{(\nu-1)!}. \tag{25}$$

Because $v_\nu(x)$ is holomorphic in some neighborhood of the origin $x = 0$ independent of ν, the expansion $v_\nu(x) = \sum_\alpha v_{\nu,\alpha} x^\alpha$ converges in a common neighborhood of the origin independent of ν. By substituting the expansion into (25) we obtain

$$B(\tilde{v})(x, \zeta) = \sum_{\nu=1}^{\infty} \sum_{\alpha} v_{\nu,\alpha} x^\alpha \frac{\zeta^{\nu-1}}{(\nu-1)!}. \tag{26}$$

Assume that the series (26) absolutely converges in some neighborhood of $(x, \zeta) = (0, 0)$. Then, by changing the order of the summation we obtain

$$B(\tilde{v})(x, \zeta) = \sum_{\alpha} \sum_{\nu=1}^{\infty} v_{\nu,\alpha} \frac{\zeta^{\nu-1}}{(\nu-1)!} x^\alpha. \tag{27}$$

We define the Laplace transform $\tilde{V}(x, \eta)$ of $B(\tilde{v})(x, \zeta)$ by

$$\tilde{V}(x, \eta) := \sum_{\alpha} L \left(\sum_{\nu=1}^{\infty} v_{\nu,\alpha} \frac{\zeta^{\nu-1}}{(\nu-1)!} \right) x^\alpha, \tag{28}$$

where the operator L is given by

$$Lf(\eta) = \int_0^\infty e^{-\zeta\eta} f(\zeta) d\zeta.$$

Here we assume an appropriate growth condition on $f(\zeta)$. We define

$$V(x, \eta) := \tilde{V}(x, \eta) + v_0(x).$$

If we recall that the Borel transform is the inverse of the Laplace transform, $V(x,\eta)$ gives a holomorphic function in η in a sectorial domain with the asymptotic expansion $v(x,\eta)$. We note that the asymptotic solution is not an exact solution in general.

For the direction ξ, $(0 \le \xi < 2\pi)$ and the opening $\theta > 0$ we define the sector $S_{\xi,\theta}$ by

$$S_{\xi,\theta} = \left\{ \eta \in \mathbb{C};\ |\arg \eta - \xi| < \frac{\theta}{2},\ \eta \neq 0 \right\}, \tag{29}$$

where the branch of the argument is the principal value. Then we have

Theorem 2. *(Resummation) Suppose that either the Poincaré condition or the condition*

$$\exists\, \tau_0,\ 0 \le \tau_0 < \pi,\ e^{-i\tau_0}\lambda_j \in \mathbb{R} \setminus \{0\} \quad (j = 1, 2, \ldots, n) \tag{30}$$

is satisfied. Then there exist a direction ξ, an opening $\theta > 0$ and a neighborhood U of the origin $x = 0$ such that $V(x,\eta)$ is holomorphic in $(x,\eta) \in U \times S_{\xi,\theta}$ and satisfies (11).

Remark 4.1. The WKB solution $v(x,\eta)$ is a G^2- asymptotic expansion of $V(x,\eta)$ in $U \times S_{\xi,\theta}$ when $\eta \to \infty$. Namely, for every $N \ge 0$ and $R > 0$, there exist $C > 0$ and $K > 0$ such that

$$\left| V(x,\eta) - \sum_{\nu=0}^{N} \eta^{-\nu} v_\nu(x) \right| \le C K^N N! |\eta|^{-N-1}, \tag{31}$$
$$\forall (x,\eta) \in U \times S_{\xi,\theta},\ |\eta| \ge R.$$

If the Poincaré condition is satisfied, then we can estimate the opening of $S_{\xi,\theta}$. In fact, we have

Theorem 3. *(Summability) Suppose that*

$$|\ arg\ \lambda_j| < \frac{\pi}{4}, \quad j = 1, 2, \ldots, n. \tag{32}$$

Then, there exist a direction ξ, an opening $\theta > \pi$, a neighborhood U of the origin $x = 0$ such that $V(x,\eta)$ is holomorphic in $(x,\eta) \in U \times S_{\xi,\theta}$ and is a unique solution of (11). The function $V(x,\eta)$ is a Borel sum of the WKB solution $v(x,\eta)$ in $U \times S_{\xi,\theta}$.

Remark 4.2. The Poincaré condition follows from the condition (32). The condition $\theta > \pi$ implies that the WKB expansion uniquely determines its Borel sum $V(x,\eta)$.

We give a rough sketch of the proof of Theorem 3. By (32) we can easily see that the complex numbers λ_k/λ_j $(k = 1, 2, \ldots, n)$ are contained in some proper cone Γ_1 in the right half plane $\Re z > 0$. Because $s \equiv \langle \lambda, \alpha \rangle / \lambda_j \in \Gamma_1$ for every $\alpha \in \mathbb{Z}_+^n$, it follows that we can choose $\theta > \pi$ and $c_1 > 0$ such that $|\eta^{-1}s - 1| > c_1$ for all $s \in \Gamma_1$ and $\eta \in S_{\xi,\theta}$. This proves that we have (37) in the following. Then we make a similar argument as the proof of Theorem 2.
□

5 Proof of Theorem 2

For $T > 0$ we denote the polydisk D_T by

$$D_T = \{x_1 \in \mathbb{C}; |x_1| < T\} \times \cdots \times \{x_n \in \mathbb{C}; |x_n| < T\}.$$

We define the set of functions $H(T)$ holomorphic in D_T and continuous up to the boundary by

$$H(T) = \left\{ u(x) = \sum_{\alpha \in \mathbb{Z}_+^n} u_\alpha x^\alpha ; \|u\|_T := \sum_\alpha |u_\alpha| T^{|\alpha|} < \infty \right\}. \tag{33}$$

Note that $\|\cdot\|_T$ is a norm on $H(T)$. The n- product of $H(T)$ is denoted by $(H(T))^n$ with a standard norm of the product space. For the sake of simplicity we denote the norm in $(H(T))^n$ by the same letter $\|\cdot\|_T$. We can easily show that

$$\|uv\|_T \le \|u\|_T \|v\|_T, \qquad \text{for every } u, v \in (H(T))^n.$$

Proof of Theorem 2. We divide the proof into 7 steps.
Step 1. First we shall construct the solution $v \equiv u = (u^1, \ldots, u^n)$ of (11) in the form

$$u(x, \eta) = \sum_{\alpha \in \mathbb{Z}_+^n, |\alpha| \ge 2} u_\alpha(\eta) x^\alpha, \tag{34}$$

where $u_\alpha(\eta)$ is a vector function whose components are rational functions of η. Let E_0 be the set of all resonances. If we substitute the expansion (34) into the homology equation (11), then we obtain the recurrence relation for $u_\alpha(\eta)$

$$(\eta^{-1}\langle \lambda, \alpha \rangle Id - \Lambda) u_\alpha(\eta) = \cdots, \tag{35}$$

where the dots denote the terms consisting of $u_\beta(\eta)$, $(|\beta| < |\alpha|)$. Indeed, the operator $\eta^{-1}\mathcal{L} - \lambda_j$ preserves every power x^α. On the other hand, the nonlinear term $R_j(x + v)$ satisfies that $R_j(x + v) = O(|x + v|^2)$ because of the assumption. Hence we have (35). By the nonresonance condition, every coefficient $u_\alpha(\eta)$ of x^α can be determined successively by dividing both sides of (35) with $\eta^{-1}\langle \lambda, \alpha \rangle - \lambda_j$ for some α and j. In the following, we assume, without loss of generality, that $\Re \lambda_j > 0$ for $j = 1, 2, \ldots, n$ in case the Poincaré condition is satisfied.
Step 2. We set $\xi = \pi$ when the Poincaré condition holds, while, if (30) is satisfied we take ξ such that $0 < \xi < \pi$. First we will show that there exists $\theta > 0$ such that there is no resonance in $S_{\xi,\theta}$, where $S_{\xi,\theta}$ is given by (29). Clearly this is true if (30) is satisfied, because the resonance $\eta = \langle \lambda, \alpha \rangle / \lambda_j$ $(\alpha \in \mathbb{Z}_+^n, 1 \le j \le n)$ is a real number by definition. Hence we consider the case where the Poincaré condition is verified. Let $\alpha \in \mathbb{Z}_+^n, |\alpha| \ge 2$ be arbitrarily given, and let Γ be the convex hull of the points λ_j, $j = 1, 2, \ldots, n$ in the complex plane $\mathbb{C}$. Then, by the assumption there exists $\theta > 0$ such that, if $\eta \in S_{\xi,\theta}$, then we have $\eta^{-1}\Gamma \cap \Gamma = \emptyset$. For $1 \le j \le n$ and $\eta \in S_{\xi,\theta}$, we have

$$|\eta^{-1}\langle\lambda,\alpha\rangle-\lambda_j|=|\eta^{-1}|\alpha|\langle\lambda,\alpha/|\alpha|\rangle-\lambda_j|. \tag{36}$$

Because $\lambda_j\in\Gamma$, $\lambda_j\neq 0$ and $\langle\lambda,\alpha/|\alpha|\rangle\in\Gamma$, it follows that there exists $c_1>0$ such that

$$\left|\eta^{-1}\langle\lambda,\alpha\rangle-\lambda_j\right|\geq c_1,\quad \forall\eta\in S_{\xi,\theta},\ j=1,\dots,n,\ \forall\alpha\in\mathbb{Z}_+^n,|\alpha|\geq 2. \tag{37}$$

This proves that there is no resonance in $S_{\xi,\theta}$.

For the later use we will show that (37) is also valid when (30) is satisfied. We write $\eta^{-1}=\eta'+i\eta''\in S_{-\xi,\theta}$, where η' and η'' are real numbers. We have, for some $c_1>0$

$$\begin{aligned}\left|\eta^{-1}\langle\lambda,\alpha\rangle-\lambda_j\right|&=|\lambda_j|\left|\eta^{-1}\frac{\langle\lambda,\alpha\rangle}{\lambda_j}-1\right| \\ &\geq c_1|\lambda_j|\left(\left|\eta'\frac{\langle\lambda,\alpha\rangle}{\lambda_j}-1\right|+\left|\eta''\frac{\langle\lambda,\alpha\rangle}{\lambda_j}\right|\right).\end{aligned} \tag{38}$$

Because $S_{-\xi,\theta}$ is a proper cone contained in the lower half plane, there exists $c_2>0$ independent of η' and η'' such that, if $\eta'+i\eta''\in S_{-\xi,\theta}$, then $|\eta''|\geq c_2|\eta'|$. It follows that

$$\left|\eta''\frac{\langle\lambda,\alpha\rangle}{\lambda_j}\right|\geq c_2\left|\eta'\frac{\langle\lambda,\alpha\rangle}{\lambda_j}\right|. \tag{39}$$

If $\left|\eta'\langle\lambda,\alpha\rangle\lambda_j^{-1}\right|<2^{-1}$, then we have

$$\left|\eta'\frac{\langle\lambda,\alpha\rangle}{\lambda_j}-1\right|>\frac{1}{2}. \tag{40}$$

On the other hand, if $\left|\eta'\langle\lambda,\alpha\rangle\lambda_j^{-1}\right|\geq\frac{1}{2}$, then we have

$$\left|\eta''\frac{\langle\lambda,\alpha\rangle}{\lambda_j}\right|\geq c_2\left|\eta'\frac{\langle\lambda,\alpha\rangle}{\lambda_j}\right|\geq\frac{c_2}{2}. \tag{41}$$

Therefore, by (38) we have (37).

For $\varepsilon_1>0$ let E_{0,ε_1} be the ε_1- neighborhood of the resonace E_0, and, for $r>0$, let us define the domain S by

$$S:=(\{\eta\in\mathbb{C};|\eta|\leq r\}\setminus E_{0,\varepsilon_1})\cup S_{\xi,\theta}. \tag{42}$$

We can easily show that, there exists $c_2>0$ such that, if $\eta\in S$, then

$$\left|\eta^{-1}\langle\lambda,\alpha\rangle-\lambda_j\right|\geq c_2,\quad\forall\alpha\in\mathbb{Z}_+^n,\ |\alpha|\geq 2,\forall\eta\in S. \tag{43}$$

Step 3. We shall show that there exist $\varepsilon_1>0$ and a neighborhood U of the origin $x=0$ such that if $\eta\in S$ and $x\in U$, then the series (34) converges. Let $\eta\neq 0$. For $\nu=1,\dots,n$, we define

$$P_\nu:=\eta^{-1}\mathcal{L}-\lambda_\nu.$$

Let P be the diagonal matrix with diagonal entries given by $P_1, \ldots, P_n$ in this order. We define the vector function R by (4). We take $T > 0$ so small that $R \in (H(2T))^n$. We define $u_k(x) \equiv u_k(x, \eta) \in (H(T))^n$, $u_k(x,\eta) = (u_k^1(x,\eta), \ldots, u_k^n(x,\eta))$, $(k = 0, 1, 2, \ldots)$ by

$$u_0 = 0, \ Pu_1 = R(x), \tag{44}$$

$$Pu_k = R(x + u_0 + \cdots + u_{k-1}) - R(x + u_0 + \cdots + u_{k-2}), \ k = 2, 3, \ldots \tag{45}$$

Suppose that $u_k(x) \in (H(T))^n$ $(k = 0, 1, 2, \ldots)$, and that the series $\sum_{k=1}^{\infty} u_k =: u$ is convergent in $(H(T))^n$. Then, by (45) we have

$$P \sum_{k=1}^{\ell} u_k = R(x + u_1 + \cdots + u_{\ell-1}).$$

Hence, by letting $\ell \to \infty$, we see that u satisfies $Pu = R(x+u)$. Because the Taylor expansion at $x = 0$ of an analytic solution is uniquely determined by the nonresonance condition, it follows that u coincides with (34) and the series (34) converges.

The convergence of $\sum_{k=1}^{\infty} u_k =: u$ can be proved by a standard argument. Hence we omit the proof. We note that the convergence is uniform with respect to $\eta \in S$. Hence the limit function is holomorphic in $\eta \in S$.

Step 4. We will show that, there exists $\eta_0 > 0$ such that, for every N, $N = 0, 1, 2, \ldots$,

$$u(x,\eta) = \sum_{j=0}^{N} v_j(x)\eta^{-j} + R_N(x,\eta)\eta^{-N-1}, \quad \eta \in S_{\xi,\theta}, \ |\eta| > \eta_0, \tag{46}$$

where $v_j(x)$ $(j = 0, 1, 2, \ldots)$ are given by a WKB solution, and $R_N(x,\eta)$ is holomorphic in x in some neighborhood of the origin.

First we shall show (46) in case $N = 0$. In view of (45) and the above argument, there exists $C > 0$ such that the sum $u \equiv \sum_{k=1}^{\infty} u_k(x,\eta)$ converges uniformly in $\eta \in S_{\xi,\theta}$, $|\eta| \geq C$. It follows that

$$\lim_{\eta \in S_{\xi,\theta}, \eta \to \infty} u(x,\eta) = \lim_{\eta} \sum_{k=1}^{\infty} u_k(x,\eta) = \sum_{k=1}^{\infty} \lim_{\eta} u_k(x,\eta). \tag{47}$$

We will calculate $\lim_\eta u_k(x,\eta)$. By (44) we see that $\lim_\eta u_1(x,\eta) = u_1(x,\infty)$ exists and satisfies the equation

$$\Lambda u_1(x,\infty) + R(x) = 0.$$

Next it follows from (45) that $\lim_\eta u_2(x,\eta) = u_2(x,\infty)$ exists and satisfies

$$\Lambda u_2(x,\infty) + R(x + u_1(x,\infty)) - R(x) = 0.$$

Hence we have

$$\Lambda(u_1(x,\infty)+u_2(x,\infty))+R(x+u_1(x,\infty))=0.$$

Inductively we get, from (44) and (45) that the limit $\lim_\eta u_k(x,\eta)=u_k(x,\infty)$ exists. Moreover we have

$$\begin{aligned}-\Lambda u_k(x,\infty)&=R(x+u_1(x,\infty)+\cdots+u_{k-1}(x,\infty))\\&\quad-R(x+u_1(x,\infty)+\cdots+u_{k-2}(x,\infty)).\end{aligned}$$

Hence we have, for $k\geq 2$

$$\Lambda(u_1(x,\infty)+\cdots+u_k(x,\infty))+R(x+u_1(x,\infty)+\cdots+u_{k-1}(x,\infty))=0.$$

It follows that $\sum_{k=1}^{\infty}u_k(x,\infty)$ satisfies (21). By the uniqueness of a holomorphic solution of (21), we see that $\sum_{k=1}^{\infty}u_k(x,\infty)=v_0(x)$.

Next we set

$$Q(x,\eta)=u(x,\eta)-v_0(x)=\sum_{\alpha}(u_\alpha(\eta)-u_\alpha(\infty))x^\alpha, \tag{48}$$

where $v_0(x)=\sum u_\alpha(\infty)x^\alpha$ is the expansion of $v_0(x)$. We want to show that every component $u_\alpha^j(\eta)$ of $u_\alpha(\eta)$ is a polynomial of $(\lambda_k-\eta^{-1}\langle\lambda,\beta\rangle)^{-1}$. In view of (44) and (45), we see that the degree of u_1 with respect to x is greater than 2. Because P^{-1} preserves every monomial x^α, it follows from (45) that the degree of u_2 is greater than 3. Similarly, we can show that the degree of u_k with respect to x is greater than $k+1$. Moreover, the coefficient of x^α in the expansion of u_k in (45) is obtained by dividing the corresponding coefficient of the right-hand side of (45) with $\eta^{-1}\langle\lambda,\alpha\rangle-\lambda_k$. This implies that if we expand $u=\sum_{k=1}^{\infty}u_k(x,\eta)$ in (34), every component of $u_\alpha(\eta)$ in (34) is a polynomial of $(\eta^{-1}\langle\lambda,\beta\rangle-\lambda_k)^{-1}$ for some k, $1\leq k\leq n$, $\beta\in\mathbb{Z}_+^n$, $|\beta|\leq|\alpha|$.

We recall the following formula

$$(1-\eta^{-1}\lambda_k^{-1}\langle\lambda,\beta\rangle)^{-1}=\sum_{j=0}^{N}\left(\frac{\langle\lambda,\beta\rangle}{\lambda_k\eta}\right)^j+\left(\frac{\langle\lambda,\beta\rangle}{\lambda_k\eta}\right)^{N+1}\left(1-\frac{\eta^{-1}\langle\lambda,\beta\rangle}{\lambda_k}\right)^{-1}, \tag{49}$$

where $N=0,1,2,\ldots$, and $\eta\in S_{\xi,\theta}$. Hence, we get, from (49) that

$$u_\alpha(\eta)-u_\alpha(\infty)=\eta^{-1}\tilde{u}_\alpha(\eta)$$

for some $\tilde{u}_\alpha(\eta)=(\tilde{u}_\alpha^1(\eta),\ldots,\tilde{u}_\alpha^n(\eta))$. It follows that

$$Q(x,\eta)=\eta^{-1}\tilde{Q}(x,\eta),\ \tilde{Q}(x,\eta)=\sum_\alpha\tilde{u}_\alpha(\eta)x^\alpha,\quad \eta\in S_{\xi,\theta},|\eta|>\eta_0, \tag{50}$$

where the equality is understood as a formal power series of x.

In order to obtain the estimate of $\tilde{Q}(x,\eta)$, we substitute (48) and (50) into (11). Then we get, from the definition of v_0 that

$$\begin{aligned}\eta^{-1}\mathcal{L}v_0+\eta^{-2}\mathcal{L}\tilde{Q} &= \Lambda(v_0+\eta^{-1}\tilde{Q})+R(x+v_0+\eta^{-1}\tilde{Q}) \\ &= \eta^{-1}\Lambda\tilde{Q}+R(x+v_0+\eta^{-1}\tilde{Q})-R(x+v_0) \\ &= \eta^{-1}\Lambda\tilde{Q}+\eta^{-1}\int_0^1 \tilde{Q}\cdot\nabla R(x+v_0+\theta\eta^{-1}\tilde{Q})d\theta.\end{aligned} \tag{51}$$

Hence we obtain the following equation for $\tilde{Q}$

$$\eta^{-1}\mathcal{L}\tilde{Q} = -\mathcal{L}v_0+\Lambda\tilde{Q}+\int_0^1 \tilde{Q}\cdot\nabla R(x+v_0+\theta\eta^{-1}\tilde{Q})d\theta. \tag{52}$$

This equation has a similar form as the equation for $u(x,\eta)$. Indeed, the linear part has the same form as (11) The nonlinear term in tho right hand side has the same property as the one in (11). By the same argument as in the proof for the convergence of $u(x,\eta)$ we see that $\tilde{Q}(x,\eta)$ is holomorphic in x in some neighborhood of the origin $x=0$ and $\eta\in S$. We can also see that the convergence of the approximate sequence of $\tilde{Q}(x,\eta)$ is uniform with respect to $\eta\in S$. This proves (46) for $N=0$. Especially, we remark that $\tilde{Q}(x,\eta)$ converges to some holomorphic function of x in some neighborhood of the origin when $\eta\in S_{\xi,\theta}$, $\eta\to\infty$.

Step 5. We will show (46) for $N=1$. We set $R_0(x,\eta)=\tilde{Q}(x,\eta)$ and we define

$$\lim_{\eta\to\infty,\eta\in S_{\xi,\theta}} \tilde{Q}(x,\eta)=U_1(x). \tag{53}$$

The existence of the limit is shown by the same argument as in the case $N=0$. Indeed, the approximate sequence $\tilde{Q}_n$ such that $\lim\tilde{Q}_n=\tilde{Q}$ as in the case $N=0$ converges uniformly with respect to $\eta\in S_{\xi,\theta}$. Therefore we have the following expression

$$R_0(x,\eta)=U_1(x)+\eta^{-1}R_1(x,\eta). \tag{54}$$

We substitute

$$u(x,\eta)=v_0(x)+\eta^{-1}R_0(x,\eta)=v_0(x)+\eta^{-1}U_1(x)+\eta^{-2}R_1(x,\eta) \tag{55}$$

into the equation (11) and we compare the coefficients of η^{-1}. Then we obtain that

$$\mathcal{L}v_0^j=\lambda_j U_1^j+\sum_{k=1}^n\left(\frac{\partial R^j}{\partial z_k}\right)(x+v_0)U_1^k, \tag{56}$$

where $U_1(x)=(U_1^1(x),\ldots,U_1^n(x))$. This is the same equation as (22) with v_1 replaced by U_1. By the uniqueness of $v_1(x)$ in the WKB expansion, we see that $U_1(x)=v_1(x)$.

We look for the equation for $R_1(x,\eta)$. By setting $\tilde{Q}=v_1(x)+\eta^{-1}R_1(x,\eta)$ in (52) we obtain

$$\begin{aligned}&\eta^{-1}\mathcal{L}(v_1+\eta^{-1}R_1)+\mathcal{L}v_0-\Lambda(v_1+\eta^{-1}R_1) \\ &= \int_0^1((v_1+\eta^{-1}R_1))\cdot\nabla R(x+v_0+\theta\eta^{-1}(v_1+\eta^{-1}R_1))d\theta.\end{aligned} \tag{57}$$

Using the relation (22)

$$-\mathcal{L}v_0 + \Lambda v_1 + v_1 \cdot \nabla R(x+v_0) = 0, \tag{58}$$

the right-hand side of (57) is equal to

$$\begin{aligned}\eta^{-1}\Lambda R_1 + \int_0^1 v_1 \cdot \left(\nabla R(x+v_0+\theta\eta^{-1}(v_1+\eta^{-1}R_1)) - \nabla R(x+v_0)\right) d\theta \\ + \eta^{-1}\int_0^1 R_1 \cdot \nabla R(x+v_0+\theta\eta^{-1}(v_1+\eta^{-1}R_1))d\theta.\end{aligned} \tag{59}$$

If we substitute the relation

$$\begin{aligned}&v_1 \cdot \left(\nabla R(x+v_0+\theta\eta^{-1}(v_1+\eta^{-1}R_1)) - \nabla R(x+v_0)\right) \\ &= \int_0^1 v_1 \cdot \nabla^2 R(x+v_0+t\theta\eta^{-1}(v_1+\eta^{-1}R_1)) \cdot \theta\eta^{-1}(v_1+\eta^{-1}R_1)dt\end{aligned}$$

in the right-hand side of (59), then we see that $R_1(x,\eta)$ satisfies the following equation

$$\begin{aligned}&\eta^{-1}\mathcal{L}R_1 = \Lambda R_1 - \mathcal{L}v_1 \\ &+ \int_0^1\int_0^1 v_1 \cdot \nabla^2 R(x+v_0+t\theta\eta^{-1}(v_1+\eta^{-1}R_1)) \cdot \theta(v_1+\eta^{-1}R_1)dtd\theta \\ &+ \int_0^1 R_1 \cdot \nabla R(x+v_0+\theta\eta^{-1}(v_1+\eta^{-1}R_1))d\theta.\end{aligned} \tag{60}$$

This equation has a similar form as the equation (11) for $u(x,\eta)$. The nonlinear term in the right-hand side together with the derivative with respect to η^{-1} are bounded in η when $\eta \to \infty$, $\eta \in S_{\xi,\theta}$. Therefore we see that there exists $\eta_0 > 0$ such that the solution $R_1(x,\eta)$ is holomorphic and bounded in x in some neighborhood of the origin when $\eta \in S_{\xi,\theta}$, $|\eta| \geq \eta_0$. Therefore we have (46) for $N=1$. The general case will be proved in the same way.

Step 6. By definition we have

$$u(x,\eta) - v_0(x) = \sum_\alpha v_\alpha(\eta)x^\alpha, \quad v_\alpha(\eta) = u_\alpha(\eta) - u_\alpha(\infty), \quad \eta \in S_{\xi,\theta}. \tag{61}$$

Now we look for the alternative expression of $u_\alpha(\eta)$. By substituting the expansions (34) and $R^k(z) = \sum_\gamma R^k_\gamma z^\gamma$ into (11) with $v = u$, we obtain the following relation

$$\sum_\alpha (\eta^{-1}\langle\lambda,\alpha\rangle - \lambda_k)u^k_\alpha x^\alpha = \sum_\gamma R^k_\gamma(x + \sum_\alpha u_\alpha x^\alpha)^\gamma. \tag{62}$$

It follows that

$$(\eta^{-1}\langle\lambda,\alpha\rangle-\lambda_k)u^k_\alpha={\sum}' R^k_\gamma\prod_{j=1}^{n}\prod_{i=1}^{\gamma(j)}u^j_{\alpha(j,i)},\tag{63}$$

where we set

$$u^j_\beta=1 \text{ if } \beta=e_j,\quad u^j_\beta=0 \text{ if } \beta\neq e_j, |\beta|=1,\quad j=1,2,\ldots,n,\tag{64}$$

and the summation $\sum'$ is taken over all combinations

$$\sum_{j=1}^{n}\sum_{i=1}^{\gamma(j)}\alpha(j,i)=\alpha,\quad \alpha(j,i)\in\mathbb{Z}^n_+,\ |\alpha(j,i)|\geq 1,\tag{65}$$

$$\gamma=(\gamma(1),\gamma(2),\ldots,\gamma(n))\in\mathbb{Z}^n_+,\quad |\gamma|\geq 2,\tag{66}$$

where $|\gamma|$ corresponds to the number of products in the right-hand side of (62). Because $R(z)=O(|z|^2)$, we see that $|\gamma|\geq 2$. We note that, if $|\alpha|=2$, then, by the condition $|\gamma|\geq 2$ we have $|\alpha(j,i)|=1$. Hence there appear no unknown quantities u^j_α in the right-hand side of (63). Hence we can determine u^k_α ($|\alpha|=2$) if $\eta^{-1}\langle\lambda,\alpha\rangle-\lambda_k\neq 0$. In the case $|\alpha|\geq 3$, we use (63) recurrently, and we obtain

$$u^k_\alpha={\sum}''(\prod_{\nu=1}^{\ell}R^{k(\nu)}_{\gamma^\nu})T_0,\tag{67}$$

where $\gamma^\nu=(\gamma^\nu(1),\gamma^\nu(2),\ldots,\gamma^\nu(n))$, and

$$T_0=(\eta^{-1}\langle\lambda,\alpha\rangle-\lambda_k)^{-1}\prod_{\nu=1}^{\ell}\prod_{j=1}^{n}\prod_{i=1}^{\gamma^\nu(j)}(\eta^{-1}\langle\lambda,\alpha(j,i,\nu)\rangle-\lambda_j)^{-1},\tag{68}$$

and the summation $\sum''$ is taken over all combinations

$$\sum_{\nu=1}^{\ell}\sum_{j=1}^{n}\sum_{i=1}^{\gamma^\nu(j)}\alpha(j,i,\nu)\leq\alpha,\ \ell\geq 1,\ 1\leq k(\nu)\leq n,\ \alpha(j,i,\nu)\in\mathbb{Z}^n_+, |\alpha(j,i,\nu)|\geq 2,\tag{69}$$

where $\beta\leq\alpha$ if $\beta_j\leq\alpha_j$ for $j=1,2,\ldots,n$, where $\alpha=(\alpha_1,\alpha_2,\ldots,\alpha_n)$ and $\beta=(\beta_1,\beta_2,\ldots,\beta_n)$. We note that there exists at least one γ^ν $(1\leq\nu\leq\ell)$ such that $|\gamma^\nu|=2$ in the sum (67), and the integer ℓ is given by the number of times by which the substitutions by (63) are made.

Because $v_\alpha(\eta)=u_\alpha(\eta)-u_\alpha(\infty)$ is holomorphic and vanishes at $\eta=\infty$, we have $v_\alpha(\eta)=\sum_{\nu=1}^{\infty}\tilde{v}_{\nu,\alpha}\eta^{-\nu}$. Hence we get the formal expansion

$$u(x,\eta)-v_0(x)=\sum_\alpha\sum_{\nu=1}^{\infty}\tilde{v}_{\nu,\alpha}\eta^{-\nu}x^\alpha.\tag{70}$$

The series (70) is a formal series, because the radius of convergence of $v_\alpha(\eta)$ is not uniform in α. By (46) and $v_\nu(x)=\sum_\alpha v_{\nu,\alpha}x^\alpha$, we have the following

$$\tilde{v}_{\nu,\alpha} = v_{\nu,\alpha}, \quad \text{for every} \ \ \nu \text{ and } \alpha. \tag{71}$$

Step 7. We will show the convergence of a Borel transform. In order to show the absolute convergence of the right-hand side of (27), we shall show the absolute convergence of the sum

$$\sum_{\alpha}\sum_{\nu=1}^{\infty} v_{\nu,\alpha}\frac{\zeta^{\nu-1}}{(\nu-1)!}x^{\alpha} = \sum_{\alpha}\sum_{\nu=1}^{\infty} \tilde{v}_{\nu,\alpha}\frac{\zeta^{\nu-1}}{(\nu-1)!}x^{\alpha}. \tag{72}$$

Because $v_\alpha(\eta)$ is holomorhic and vanishes at $\eta = \infty$, the formal Borel transform of $v_\alpha(\eta)$ is equal to the Borel transform of the Hankel type for every α, namely

$$\hat{v}_\alpha(\zeta) = \int_{\Gamma} v_\alpha(\eta)e^{\zeta\eta}d\eta, \tag{73}$$

where $\Re\zeta > 0$, and Γ is the path in the domain where $v_\alpha(\eta)$ is holomorphic such that Γ starts from $\eta = \infty$ and passes through the domain $\Im\eta < 0$, $\Re\eta < 0$, encircles the poles of $v_\alpha(\eta)$ in the left-hand side and tends to $\eta = \infty$ through the domain $\Im\eta > 0$, $\Re\eta < 0$.

We take the path Γ depending on α so that for every $\beta \equiv \alpha(j,i,\nu)$ and j appearing in (68) we have

$$|\eta^{-1}\langle\lambda,\beta\rangle| \le (\min_j|\lambda_j|)/2, \ \eta \in \Gamma. \tag{74}$$

Because v_α vanishes at $\eta = \infty$, we can deform the path Γ in the integral (73) to a Jordan curve Γ_0 which encircles all poles of $v_\alpha(\eta)$ and has a distance $C|\alpha|$ from the poles of $v_\alpha(\eta)$ for some $C > 0$ independent of α. Because $\Re\zeta > 0$, the definition of the path Γ implies that $\Re\zeta\eta < 0$ if $\eta \in \Gamma$ and $|\eta|$ is sufficiently large. Because $v_\alpha(\eta)$ is the polynomial of η^{-1} and $(\lambda_k - \eta^{-1}\langle\lambda,\beta\rangle)^{-1}$ for some β, it follows that the deformed Borel transform is an entire function of ζ. By an analytic continuation, we see that the relation (73) is valid for all ζ if the path Γ is replaced by Γ_0. In the following, we put $\Gamma - \Gamma_0$.

We will estimate the growth of the Borel transform (73), assuming that ζ is in some neighborhood of the origin. By definition we may assume that $|\eta| \le C_1|\alpha|$ on $\eta \in \Gamma_0$ for some $C_1 > 0$ independent of $\eta \in \Gamma_0$.

First we shall show that there exists $\rho_1 > 0$ such that

$$|v_\alpha(\eta)| \le \rho_1^{|\alpha|+1}, \ \forall\alpha \in \mathbb{Z}_+^n, \ \eta \in \Gamma_0. \tag{75}$$

Because $v_\alpha(\eta) = u_\alpha(\eta) - u_\alpha(\infty)$ and $u_\alpha(\infty)$ is the Taylor coefficient of the analytic function $v_0(x)$, we may consider $u_\alpha(\eta)$ instead of $v_\alpha(\eta)$. Hence we will estimate (67). In order to estimate T_0 in (68), we get, from (74) that

$$|\eta^{-1}\langle\lambda,\alpha(j,i,\nu)\rangle - \lambda_k| \ge |\lambda_k| - |\eta^{-1}\langle\lambda,\alpha(j,i,\nu)\rangle| \ge |\lambda_k|/2 \ge C_1^{-1},$$

for some $C_1 > 0$. It follows that

$$|T_0| \le C_1 \prod_{\nu=1}^{\ell} \prod_{j=1}^{n} \prod_{i=1}^{\gamma^\nu(j)} C_1 \le C_1 \prod_{\nu=1}^{\ell} \prod_{j=1}^{n} C_1^{\gamma^\nu(j)} = C_1 \prod_{\nu=1}^{\ell} C_1^{|\gamma^\nu|} = C_1^{1+\sum_\nu |\gamma^\nu|}. \tag{76}$$

By (69) we have

$$|\alpha| \ge \sum_{\nu=1}^{\ell} \sum_{j=1}^{n} \sum_{i=1}^{\gamma^\nu(j)} |\alpha(j,i,\nu)| \ge 2 \sum_{\nu=1}^{\ell} \sum_{j=1}^{n} \sum_{i=1}^{\gamma^\nu(j)} = 2 \sum_{\nu=1}^{\ell} |\gamma^\nu|. \tag{77}$$

It follows that there exists $C_2 > 0$ such that

$$|T_0| \le C_2^{1+|\alpha|}. \tag{78}$$

Next we consider $\prod_{\nu=1}^{\ell} R_{\gamma^\nu}^{k(\nu)}$. Because $R(z) = O(|z|^2)$, we may assume that, for every given $\varepsilon_1 > 0$

$$|R_{\gamma^\nu}^{k(\nu)}| \le \varepsilon_1^{|\gamma^\nu|+1}, \quad \text{for all } 1 \le k(\nu) \le n, \gamma^\nu \in \mathbb{Z}_+^n, \tag{79}$$

by making a scale change of variables $x = \varepsilon y$ $(\varepsilon > 0)$, if necessary. Therefore we have

$$\prod_{\nu=1}^{\ell} |R_{\gamma^\nu}^{k(\nu)}| \le \prod_{\nu=1}^{\ell} \varepsilon_1^{|\gamma^\nu|+1} = \varepsilon_1^{\ell + \sum_{\nu=1}^{\ell} |\gamma^\nu|}. \tag{80}$$

We shall show that the number of combinations $\{\alpha(j,i,\nu)\}$ in (69) is bounded by $C_3^{|\alpha|+1}$ for some constant $C_3 > 0$ independent of α. Indeed, because the number of combinations is at most multiplied by n when j runs through 1 to n, we may fix some j, $1 \le j \le n$. On the other hand, noting that

$$\sum{}'' |R_{\gamma^\nu}^{k(\nu)}| \le \sum{}'' \varepsilon_1^{\ell+\sum_{\nu=1}^{\ell}|\gamma^\nu|} \le \sum_{\ell=1}^{\infty} \sum_{n_1,\dots,n_\ell=1}^{\infty} \sum_{|\gamma^\nu|=n_\nu} \varepsilon_1^{\ell+\sum_{\nu=1}^{\ell} n_\nu}, \tag{81}$$

we shall estimate the number of combinations for a given ℓ and $\gamma^\nu(j)$. We define $k = \max_{1\le\nu\le\ell} \gamma^\nu(j)$, and consider the number of multi-indices $\{\alpha(j,i,\nu)\}$ satisfying $\alpha = \sum_{\nu=1}^{\ell} \sum_{j=1}^{n} \sum_{i=1}^{k} \alpha(j,i,\nu)$. Suppressing the index j and rearranging $\{\alpha(j,i,\nu)\}$, we may estimate the number of multi indices $\{\alpha^m\}$ such that $\alpha = \sum_{m=1}^{\mu} \alpha^m$ for some integer μ. Setting

$$\beta^1 = \alpha, \beta^2 = \sum_{m=1}^{\mu-1} \alpha^m, \cdots, \beta^{\mu-1} = \alpha^2 + \alpha^1, \beta^\mu = \alpha^1, \tag{82}$$

the number of the pairs are bounded by the number of paths on $\mathbb{Z}_+^n$ which starts from α and arrives at some point in $\mathbb{Z}_+^n$ with length 2 under the condition that the length $|\gamma|$ on a path is strictly decreasing. Clearly, such number of paths can be bounded by

$$n^{|\alpha|-|\beta^2|} n^{|\beta^2|-|\beta^3|} \times \cdots \times n^{|\beta^{\mu-1}|-|\beta^\mu|} n^{|\beta^\mu|-2} = n^{|\alpha|-2} \le n^{|\alpha|}. \tag{83}$$

Therefore, by (78) we obtain (75).

Hence, by substituting (75) into the right-hand side of (73), we obtain, for some constant $K_2 > 0$, $C_3 > 0$ and $C_4 > 0$ independent of α

$$|\hat{v}_\alpha(\zeta)| \leq K_2 \exp\left(C_3|\alpha| + C_4|\alpha||\zeta|\right). \tag{84}$$

We note that the estimate is valid for ζ in some neighborhood of the origin $\zeta = 0$, because the path of the integration is changed. By the Cauchy estimate we can easily show that

$$|\tilde{v}_{\nu,\alpha}| \leq K_3 \nu! \exp\left(C_3|\alpha| + C_4\nu\right), \quad \nu = 1, 2, \ldots, \tag{85}$$

for, possibly other constants $C_3 > 0$, $C_4 > 0$ and $K_3 > 0$.

By (85), we see that the right-hand side of (72) absolutely converges in some neighborhood of the origin $x = 0$, $\zeta = 0$. Because the right-hand side of (72) is the sum of the Borel transform of $v_\alpha(\eta) = u_\alpha(\eta) - u_\alpha(\infty)$, it follows that the Laplace transform of the right-hand side of (72) is equal to $v_\alpha(\eta) = u_\alpha(\eta) - u_\alpha(\infty)$. It follows that the Borel-Laplace transform $\tilde{V}(x,\eta)$ of the WKB solution $v(x,\eta) - v_0(x)$ is equal to

$$u(x,\eta) - \sum_\alpha u_\alpha(\infty)x^\alpha = u(x,\eta) - v_0(x) = \tilde{V}(x,\eta).$$

Hence we see that $V(x,\eta) \equiv \tilde{V}(x,\eta) + v_0(x) = u(x,\eta)$. This proves the theorem.

6 Analytic continuation of a resummed WKB solution in the Poincaré case

First we remark that there exist an infinte number of resonaces on the right-half plane $\Re\eta > 0$. These resonances accumulate only at infinity in the Poincaré case, and may accumulate on some straight line in the non Poincaré case. Clearly, the solution $V(x,\eta)$ given in Theorem 2 may be singular with respect to η at the resonances. We shall make an analytic continuation (with respect to η) of a resummed WKB solution to the right half plane and reconstruct the classical Poincaré series solution.

Theorem 4. *Suppose that the Poincaré condition is verified. Then $V(x,\eta)$ is analytically continued to the right half plane except for resonances. If $\eta = 1$ is nonresonant, then the analytic continuation of a resummed WKB solution to $\eta = 1$ coincides with the classical Poincaré series solution of (11).*

Remark 6.1. If the Poincaré condition and the nonresonance condition are satisfied, then there is no resonance in some neighborhood of $\eta = 1$ because the set of resonances is discrete. Hence, (11) has a unique solution $u(x,\eta)$ which is holomorphic in (x,η) in some neighborhood of $(x,\eta) = (0,1)$. Let $V(x,\eta)$ be the resummed WKB solution given in Theorem 2. By the construction we see that $V(x,\eta)$ is analytic with respect to $(x,\eta) \in U \times S$ for some neighborhood U of the origin $x = 0$. Indeed, this follows from the representation of $V(x,\eta)$. Hence, $V(x,\eta)$ is a solution of (11) by the analytic continuation along a path with the endpoint $\eta = 1$. By the uniqueness, we have $V(x,\eta) = u(x,\eta)$ when η is in some neighborhood of $\eta = 1$.

We note that $V(x,\eta)(= u(x,\eta))$ may have singularities at the resonances $\eta_{\alpha,j} = \langle \lambda, \alpha \rangle / \lambda_j$ $(\alpha \in \mathbb{Z}_+^n, |\alpha| \geq 2, j = 1, 2, \ldots, n)$. The above argument shows that $V(x,\eta)$ is single-valued at the singularity $\eta_{\alpha,j}$. The recurrence relations among the coefficients of the expansion $u(x,\eta) = \sum_\beta u_\beta(\eta) x^\beta$ show that $u_\beta(\eta)$ has a pole at $\eta = \eta_{\alpha,j}$ for some β, if $V(x,\eta)$ is singular at $\eta = \eta_{\alpha,j}$. The order of the pole $\eta_{\alpha,j}$ of $u_\beta(\eta)$ goes to infinity for some β_ν $(\nu = 1, 2, \ldots)$ such that $|\beta_\nu| \to \infty$ $(\nu \to \infty)$.

References

1. T. Aoki, T. Kawai, T. Koike and Y. Takei: On the exact WKB analysis of operators admitting infinitely many phases, Adv. Math., **181**(2004), 165-189.
2. T. Kawai and Y. Takei: On WKB analysis of higher order Painlevé equations with a large parameter, Proc. Japan Acad., Ser. A, **80**(2004), 53-56.
3. L. Stolovitch: Singular complete integrability. Publ. Math. I.H.E.S., **91**, 134-210 (2000)
4. Y. Takei: Toward the exact WKB analysis for higher-order Painlevé equations — The case of Noumi-Yamada systems, Publ. RIMS, Kyoto Univ., **40**(2004), 709-730.
5. N.T. Zung: Convergence versus integrability in Poincaré-Dulac normal form. Math. Res. Lett. **9**, no. 2-3 (2002), 217–228.